Springer Monographs in Mathematics

From Robert

For my parents, Kurt and Berta Adler, who taught me the most important lesson of all:

> דע מאין באת, ולאן אתה הולך ,ולפני מי אתה עתיד לתן דין וחשבון.
>
> Know from whence you have come, to where you are going,
> and before Whom you must ultimately report.
>
> *Ethics of the Fathers*, Chapter 3 — פרקי אבות, ג

From Jonathan

For my wife, Lee-Ann, and daughter, Isobel, for their tremendous support of my efforts in writing this book and my parents, John and Brenda, for all of their encouragement.

Robert J. Adler
Jonathan E. Taylor

Random Fields and Geometry

With 21 Illustrations

Springer

Robert Adler
Faculty of Industrial Engineering
and Management
Technion – Israel Institute of Technology
Haifa, 32000
Israel

Jonathan Taylor
Department of Statistics
Stanford University
Sequoia Hall
Stanford, CA 94305-4065
U.S.A.

Mathematics Subject Classification (2000): 60-02, 60Gxx, 60G60, 53Cxx, 53C65

Library of Congress Control Number: 2007923465

ISBN-13: 978-0-387-48112-8 e-ISBN-13: 978-0-387-48116-6

Printed on acid-free paper.

9 8 7 6 5 4 3 2 1

springer.com (JLS/EB)

Preface

Since the term "random field" has a variety of different connotations, ranging from agriculture to statistical mechanics, let us start by clarifying that, in this book, a random field is a stochastic process, usually taking values in a Euclidean space, and defined over a parameter space of dimensionality at least 1.

Consequently, random processes defined on countable parameter spaces will not appear here. Indeed, even processes on $\mathbb{R}^1$ will make only rare appearances and, from the point of view of this book, are almost trivial. The parameter spaces we like best are manifolds, although for much of the time we shall require no more than that they be pseudometric spaces.

With this clarification in hand, the next thing that you should know is that this book will have a sequel dealing primarily with applications.

In fact, as we complete this book, we have already started, together with KW (Keith Worsley), on a companion volume [8] tentatively entitled *RFG-A*, or *Random Fields and Geometry: Applications*. The current volume—*RFG*—concentrates on the theory and mathematical background of random fields, while *RFG-A* is intended to do precisely what its title promises. Once the companion volume is published, you will find there not only applications of the theory of this book, but of (smooth) random fields in general.

Making a clear split between theory and practice has both advantages and disadvantages. It certainly eased the pressure on us to attempt the almost impossible goal of writing in a style that would be accessible to all. It also, to a large extent, eases the load on you, the reader, since you can now choose the volume closer to your interests and so avoid either "irrelevant" mathematical detail or the "real world," depending on your outlook and tastes. However, these are small gains when compared to the major loss of creating an apparent dichotomy between two things that should, in principle, go hand-in-hand: theory and application. What is true in principle is particularly true of the topic at hand, and, to explain why, we shall indulge ourselves in a paragraph or two of history.

The precusor to both of the current volumes was the 1981 monograph *The Geometry of Random Fields* (*GRF*) which grew out of RJA's (i.e., Robert Adler's) Ph.D. thesis under Michael Hasofer. The problem that gave birth to the thesis was an applied one, having to do with ground fissures due to water permeating through the earth under a building site. However, both the thesis and *GRF* ended up being directed more to theoreticians than to subject-matter researchers. Nevertheless, the topics there found many applications over the past two decades, in disciplines as widespread as astrophysics and medical imaging.

These applications led to a wide variety of extensions of the material of *GRF*, which, while different in extent to what was there, were not really different in kind. However, in the late 1990s KW found himself facing a brain mapping problem on the cerebral cortex (i.e., the brain surface) that involved looking at random fields on manifolds. Jonathan Taylor (JET) looked at this problem and, in somewhat of a repetition of history, took it to an abstract level and wrote a Ph.D. thesis that completely revolutionized[1] the way one should think about problems involving the geometry generated by smooth random fields. This, and subsequent material, makes up Part III of the current, three-part, book.

In fact, this book is really about Part III, and it is there that most of the new material will be found. Part I is mainly an adaptation of RJA's 1990 IMS lecture notes, *An Introduction to Continuity, Extrema, and Related Topics for General Gaussian Processes*, considerably corrected and somewhat reworked with the intention of providing all that one needs to know about Gaussian random fields in order to read Part III. In addition, Part I includes a chapter on stationarity. En passant, we also included many things that were not really needed for Part III, so that Part I can be (and often has been) used as the basis of a one-quarter course in Gaussian processes. Such a course (and, indeed, this book as a whole) would be aimed at students who have already taken a basic course in measure-theoretic probability and also have some basic familiarity with stochastic processes.

Part II covers material from both integral and differential geometry. However, the material here is considerably less standard than that of Part I, and we expect that few readers other than professional geometers will be familiar with all of it. In addition, some of the proofs are different from what is found in the standard geometry literature in that they use properties of Gaussian distributions.[2]

There are two main aims to Part II. One is to set up an analogue of the critical point theory of Marston Morse in the framework of Whitney stratified manifolds. What makes this nonstandard (at least in terms of what most students of mathematics see as part of their graduate education) is that Morse theory is usually done for smooth manifolds, preferably without boundaries. Whitney stratified manifolds are only piecewise smooth, and are permitted any number of edges, corners, etc. This brings them closer to the objects of integral geometry, to which we devote a chapter. While the results of this specific chapter are actually subsumed by what we shall have to say about Whitney stratified manifolds, they have the advantage that they are easy to state and prove without heavy machinery.

The second aim of Part II is to develop Lipschitz–Killing curvatures in the setting of Whitney stratified manifolds and to describe their role in what are known as "tube

[1] This verb was chosen by RJA and not JET.

[2] After all, since we shall by then have the Gaussian Part I behind us, it seems wasteful not to use it when it can help simplify proofs.

formulas." We shall spend quite some time on this. Some of the material here is "well known" (albeit only to experts) and some, particularly that relating to tube and Crofton formulas in Gauss space, is new. Furthermore, we derive the tube formulas for locally convex Whitney stratified manifolds, which is both somewhat more general than the usual approach for smooth manifolds, and somewhat more practical, since most of the parameter spaces we are interested in have boundaries. In addition, the approach we adopt is often unconventional.

These two aims make for a somewhat unusual combination of material and there is no easily accessible and succinct[3] alternative to our Part II for learning about them. In the same vein, in order to help novice differential geometers, we have included a one-chapter primer on differential geometry that runs quickly, and often unaesthetically, through the basic concepts and notation of this most beautiful part of mathematics.

However, although Parts I and II of this book contain much material of intrinsic interest we would not have written them were it not for Part III, for which they provide necessary background material. What is it in Part III that justifies close to 300 pages of preparation? Part III revolves around the *excursion sets* of smooth, $\mathbb{R}^k$-valued random fields f over piecewise smooth manifolds M. Excursion sets are subsets of M given by

$$A_D \equiv A_D(f, M) \triangleq \{t \in M : f(t) \in D\}$$

for $D \subset \mathbb{R}^k$.

A great deal of the sample function behavior of such fields can be deduced from their excursion sets and a surprising amount from the Euler, or Euler–Poincaré, characteristics of these excursion sets, defined in Part II. In particular, if we denote the Euler characteristic of a set A by $\varphi(A)$, then much of Part III is devoted to finding the following expression for their expectation, when f is Gaussian with zero mean and unit constant variance:

$$\mathbb{E}\{\varphi(A_D)\} = \sum_{j=0}^{\dim M} (2\pi)^{-j/2} \mathcal{L}_j(M) \mathcal{M}_j^k(D). \tag{0.0.1}$$

Here the $\mathcal{L}_j(M)$ are the Lipschitz–Killing curvatures of M with respect to a Riemannian metric induced by the random field f, and the $\mathcal{M}_j^k(D)$ are certain Minkowski-like functionals (closely akin to Lipschitz–Killing curvatures) on $\mathbb{R}^k$ under Gauss measure.

If all of this sounds terribly abstract, the truth is that it both is, and is not. It is abstract, because while (0.0.1) has had many precursors over the last 60 years or so, it has never before been established in the generality described above. It is also abstract in that the tools involved in the derivation of (0.0.1) in this setting require some rather heavy machinery from differential geometry. However, this level of abstraction has

[3] The stress here is on "succinct." With the exception of the material on Gauss space, almost everything that we have to say can be found somewhere in the literatures of integral and differential geometry, for which there are many excellent texts, some of which we shall list later. However, all presume a background knowledge that is beyond what we shall require, and each contains only a subset of the results we shall need.

turned out to pay significant dividends, for not only does it yield insight into earlier results that we did not have before, but it also has practical implications. For example, the approach that we shall employ works just as well for nonstationary processes as it does for stationary ones.[4] However, nonstationarity, even on manifolds as simple as $[0, 1]^2$, was previously considered essentially intractable. Simply put, this is one of those rare but constantly pursued examples in mathematics in which abstraction leads not only to a complete and elegant theory, but also to practical consequences.

An extremely simple and very down-to-earth application of (0.0.1) arises when the manifold is the unit interval $[0, 1]$, f is real-valued, and $D = [u, \infty)$. In that case, $\mathbb{E}\{\varphi(A_D)\}$ is no more than the mean number of upcrossings of the level u by the process f, along with a boundary correction term. Consequently, modulo the boundary term, (0.0.1) collapses to no more than the famous Rice formula, undoubtedly one of the most important results in the applications of smooth stochastic processes. If you are unfamiliar with Rice's formula, then you might want to start reading this book at Section 11.1, where it appears in some detail, together with heuristic, but instructional, proofs and applications.

One of the reasons that Rice's formula is so important is that it has long been used as an approximation, for large u, to the *excursion probability*

$$\mathbb{P}\Big\{ \sup_{t\in[0,1]} f(t) \geq u \Big\},$$

itself an object of major practical importance. The heuristic argument behind this is simple: If f crosses a high level, it is unlikely do so more than once. Thus, in essence, the probability that f crosses the level u is close to the probability that there is an upcrossing of u, along with a boundary correction term. (The correction comes from the fact that one way for $\sup_{t\in[0,1]} f(t)$ to be larger than u is for there to be no upcrossings but $f(0) \geq u$.) Since the number of upcrossings of a high level will always be small, the probability of an upcrossing is well approximated by the mean number of upcrossings. Hence Rice's formula gives an approximation for excursion probabilities.

If (0.0.1) is the main result of Part III, then the second-most-important result is that, at the same level of generality and for a wide choice of D, we can find a bound for the difference

$$|\mathbb{P}\{\exists\, t \in M : f(t) \in D\} - \mathbb{E}\{\varphi(A_D)\}|.$$

A specific case of this occurs when f takes values in $\mathbb{R}^1$, in which case not only can we show that, for large u,

$$\Big|\mathbb{P}\Big\{ \sup_{t\in M} f(t) \geq u \Big\} - \mathbb{E}\{\varphi(A_{[u,\infty)})\}\Big|$$

is small, but we can provide an upper bound to it that is both sharp and explicit.

Given that the second term here is known from (0.0.1), what this inequility gives is an excellent approximation to Gaussian excursion probabilities in a very wide setting, something that has long been a holy grail of Gaussian process theory.

[4] Still assuming marginal stationarity, i.e., zero mean and constant variance.

In the generality in which we shall be working, the bound is determined by geometric properties of the manifold M with the induced Riemannian metric mentioned above. Furthermore, unlike the handwaving argument described above for the simple one-dimensional case, the new tools provide a fully rigorous proof.

At this point we should probably say something about why we chose to take piecewise smooth manifolds as our generic parameter space. This is perhaps best explained via an example. Suppose that we were to take parameter spaces that were smooth, even C^∞, manifolds, with C^∞ boundaries.

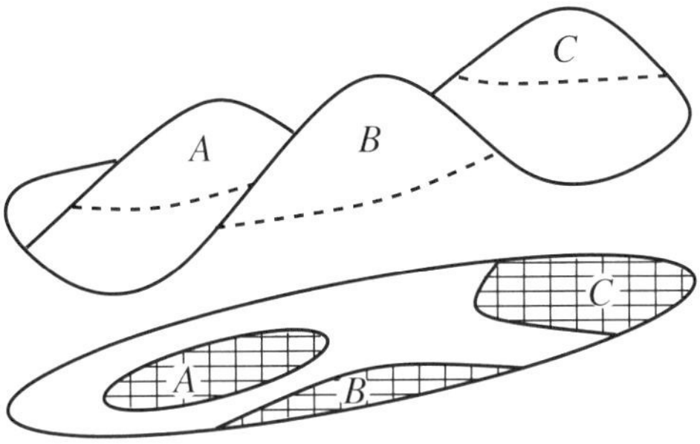

Fig. 0.0.1. A C^∞ function defined over a manifold with a C^∞ boundary gives excursion sets that have sharp, nondifferentiable, corners.

Such an example is shown in Figure 0.0.1, where the parameter space is a disk. The three excursions of a (C^∞) function f above some nominal level are marked on the function surface, and these lie above the three corresponding components of the excursion set $A_{[u,\infty)}$. Note that, despite the smoothness of each component of this example, the excursion set has sharp corners where it intersects with the boundary of the parameter space. In other words, $A_{[u,\infty)}$ is only a piecewise smooth manifold.

It turns out that since we end up with piecewise smooth manifolds for our excursion sets, there is not a lot saved by not starting with them as parameter spaces as well.[5]

So now you know what awaits you at the end of the path through this book. However, traversing the path has value in itself. Wandering, as it does, through the fields of both probability and geometry, it is a path that we imagine not too many of you will have traversed before. We hope that you will enjoy the scenery along the way as much as we have enjoyed describing it. (We also hope, for your sake, that it will be easier and faster in the reading than it was in the writing.)

We are left now with two tasks: Advising how best to read this book, and offering our acknowledgments.

[5] Of course, we could simplify things considerably by working only with parameter spaces that have no boundary, something that would be natural, for example, for a differential geometer. However, this would leave us with a theory that could not handle parameter spaces as simple as the square and the cube, a situation that would be intolerable from the point of view of applications.

The best way to read the book is, of course, to start at the beginning and work through to the end. That was how we wrote it. However, here some other possibilities, depending on what you want to get out of it.

(i) *A course on Gaussian processes*: Chapters 1 through 4 along with Sections 5.1 through 5.4 if you want to learn about stationarity as well. These chapters can be read in more or less any order; see the comments in the introduction to Part I.

(ii) *Random fields on Euclidean spaces, with an accent on geometry*: Sections 1.1–1.4.2 and Chapter 3 for basic Gaussian processes, Sections 4.1, 4.5, and 4.6 for some classical material on extremal distributions, and Chapter 5 on stationarity. Chapter 6 and Section 9.4 give the basic geometry and Chapter 11 the random geometry of Gaussian fields. Section 14.4 gives examples of how the results of Chapter 11 relate to excursion probabilities, and Section 15.10 gives examples of the non-Gaussian theory. (Note that because you have chosen to remain in the Euclidean scenario, and so avoid most of the real challenges of differential geometry, you have been relegated to reading examples instead of the general case!)

(iii) *Probabilisitic problems in, and using, differential geometry*: Sections 1.1, 1.2 and the results (but not proofs) of Chapter 3 to get a bare-bones introduction to Gaussian processes, along with Sections 5.5 and 5.6 for some important notation. As much of Chapter 7 as you need to revise differential-geometric concepts, followed by Chapters 8, 9, and 10. The punch line is then in Chapters 12 and 13 for Gaussian processes and Chapter 15 in general. It is only in this last chapter that you will get to see all the geometric preliminaries of Part II in play at once.

(iv) *Applications without the theory*: Wait for *RFG-A*. We are working on it!

Now for the acknowledgments. Both RJA and JET owe debts of gratitude to KW, and we had better acknowledge them now, since we can hardly do it in the preface of *RFG-A*.

Beyond our personal debts to KW, not least for getting the two us of together, the subject matter of this book also owes him an enormous debt of gratitude. It was during his various extensions and applications of the material of *GRF* that the passage between the old Euclidean theory and its newer manifold version began to take shape. Without his tenacious refusal to leave (applied) problems because the theory (geometry) seemed too hard, the foundations on which our Part III is based would never have been laid.

Back to the personal level, we also owe debts of gratitude to numerous students at the Technion, UC Santa Barbara, Stanford, and the ICE-EM in Brisbane who sat through courses as we put this volume together, as well as the group at McGill that went through the book as a reading course with KW. Their enthusiasm, patience, and refusal to take "it is easy to see that" for an answer when it was not all that easy to see things, not to mention all the typos and errors that they found, has helped iron a lot of wrinkles out of the final product.

In particular, we would like to thank Nicholas Chamandy, Sourav Chatterjee, Steve Huntsman, Farzan Rohani, Alessio Sancetta, Armin Schwartzman, and Sreekar Vadlamani for their questions, comments, and, embarrassingly often, corrections.

The ubiquitous anonymous reviewer also made a number of useful suggestions and we are suitably grateful to him/her.

The generous support of the U.S.–Israel Binational Science Foundation, the Israel Science Foundation, the U.S. National Science Foundation, the Louis and Samuel Seiden Academic Chair, and the Canadian Natural Sciences and Engineering Research Council over the (too long a) period that we worked on this book are all gratefully acknowledged.

Finally, don't forget, after you finish reading this book, to run to your library for a copy of *RFG-A*, to see what all of this theory is *really* good for.

Until such time as *RFG-A* appears in print, preliminary versions will be available on our home pages, which is where we shall also keep a list of typos and/or corrections for this book.

Robert J. Adler
Haifa, Israel
ie.technion.ac.il/Adler.phtml

Jonathan E. Taylor
Stanford, CA, USA
www-stat.stanford.edu/~jtaylo

Contents

Part I

Gaussian Processes

If you have not yet read the preface, then please do so now.

Since you have read the preface, you already know a number of important things about this book, including the fact that Part I is about Gaussian random fields.

The centrality of Gaussian fields to this book is due to two basic factors:

- Gaussian processes have a rich, detailed, and very well-understood general theory, which makes them beloved by theoreticians.
- In applications of random field theory, as in applications of almost any theory, it is important to have specific, explicit formulas that allow one to predict, to compare theory with experiment, etc. As we shall see in Part III, it will be only for Gaussian (and related; cf. Section 1.4.6 and Chapter 15) fields that it is possible to derive such formulas, and then only in the setting of excursion sets.

The main reason behind both these facts is the convenient analytic form of the multivariate Gaussian density, and the related properties of Gaussian fields. This is what Part I is about.

There are five main collections of basic results that will be of interest to us. Rather interestingly, although later in the book we shall be interested in Gaussian fields defined over various types of manifolds, the basic theory of Gaussian fields is actually independent of the specific geometric structure of the parameter space. Indeed, after decades of polishing, even proofs gain little in the way of simplification by restricting to special cases even as simple as $\mathbb{R}$. Thus, at least for a while, we can and shall work in as wide as possible generality, working with fields defined only on topological spaces to which we shall assign a natural (pseudo)metric induced by the covariance structure of the field.

The first set of results that we require, along with related information, form Chapter 1 and are encapsulated in different forms in Theorems 1.3.3 and 1.3.5 and their corollaries. These give a sufficient condition, in terms of metric entropy, ensuring the sample path boundedness and continuity of a Gaussian field along with providing information about moduli of continuity. While this entropy condition is also necessary for *stationary* fields, this is not the case in general, and so for completeness we look briefly at the majorizing measure version of this theory in Section 1.5. However, it will be a rare reader of *this* book who will ever need the more general theory.

To put the seemingly abstract entropy conditions into focus, these results will be followed by a section with a goodly number of extremely varied examples. Despite the fact that these cover only the tip of a very large iceberg, their diversity shows the power of the abstract approach, in that all can be treated via the general theory without further probabilistic arguments. The reader who is not interested in the general Gaussian theory, and cares mainly about the geometry of fields on $\mathbb{R}^N$ or on smooth manifolds, need only read Sections 1.4.1 and 1.4.2 on continuity and differentiability in this scenario, along with the early parts of Section 1.4.3, needed for understanding the spectral representation of Chapter 5.

Chapter 2 contains the Borell–TIS (Borell–Tsirelson–Ibragimov–Sudakov) inequality and Slepian inequalities (along with some of their relatives). The Borell–TIS inequality gives a universal bound for the *excursion probability*

$$\mathbb{P}\left\{ \sup_{t \in T} f(t) \geq u \right\}, \tag{0.0.2}$$

$u > 0$, for *any* centered, continuous Gaussian field. As such, it is a truly basic tool of Gaussian processes, somewhat akin to Chebyshev's inequality in statistics or maximal inequalities in martingale theory. Slepian's inequality and its relatives are just as important and basic, and allow one to use relationships between covariance functions of Gaussian fields to compare excursion probabilities and expectations of suprema.

The main result of Chapter 3 is Theorem 3.1.1, which gives an expansion for a Gaussian field in terms of deterministic eigenfunctions with independent $N(0, 1)$ coefficients. A special case of this expansion is the Karhunen–Loève expansion of Section 3.2, with which we expect many readers will already be familiar. Together with the spectral representations of Chapter 5, they make up what are probably the most important tools in the Gaussian modeler's bag of tricks. However, these expansions are also an extremely important theoretical tool, whose development has far-reaching consequences.

Chapter 4 serves as a basic introduction to what is also one of the central topics of Part III, the computation of the excursion probabilities for a zero-mean Gaussian field and a general parameter space T. In Part III we shall develop highly detailed expansions of the form

$$\mathbb{P}\left\{ \sup_{t \in T} f(t) \geq u \right\} = u^{\alpha} e^{-u^2/2\sigma_T^2} \sum_{j=0}^{n} C_j u^{-j} + \text{error},$$

for large u, appropriate parameters α, σ_T^2, n, and C_j that depend on both f and T. However, to do this, we shall have to place specific assumptions on the parameter space T, in particular assuming that it is a piecewise smooth manifold.[6] Without these assumptions, the best that one can do is to identify α, σ_T^2, and occasionally C_0, and this is what Chapter 4 will do. Furthermore, to keep the treatment down to a reasonable length, we shall generally concentrate only on upper bounds, rather than expansions, for Gaussian excursion probabilities.[7]

Chapter 5, the last of Part I, is somewhat different from the others in that it is not really about Gaussian processes, but about stationarity and isotropy in general. The main reason for the generality is that limiting oneself to the Gaussian scenario gains us so little that it is not worthwhile doing so. The results here, however, will be crucial for many of the detailed calculations of Part III.

---

[6] For those of you who are already comfortable with the theory of stratified manifolds, our "piecewise smooth manifolds" are Whitney stratified manifolds with convex support cones.

[7] There is also a well-developed theory of a Poisson limit nature for probabilities of the form $\mathbb{P}\{\sup_{t \in T_u} f(t) \geq \eta(x, u)\}$, for which $\eta(x, u)$ and the size of T_u grow (to infinity) with u. In this case, one searches for growth rates that give a limit dependent on x, but not u. You can find more about this in Aldous [10], which places the Gaussian theory within a much wider framework of limit results, or in Leadbetter, Lindgren, and Rootzén [97] and Piterbarg [126], which give more detailed and more rigorous accounts for the Gaussian and Gaussian-related situation.

There are a number of ways in which you can read Part I of this book. You should definitely start with Sections 1.1 and 1.2, which have some boring, standard, but important technical material. From then on, it is very much up to you, since the remainder of Chapter 1 and the other four chapters of Part I are more or less independent of one another. A reviewer of the book suggested going from Section 1.2 directly to Chapter 3 to read about how to construct examples of Gaussian processes via orthogonal expansions, a suggestion that certainly makes historical sense and would also be more natural for an analyst rather than a probabilist. With or without Chapter 3, one can go from Section 1.2 to Section 1.3 to learn a little about entropy methods and from there directly to Chapter 4 to get quickly to extremal properties, one of the key topics of this book. Chapter 2, on the Borell and Slepian inequalities, can follow this. We obviously wrote the book in the order that seemed most logical to us, but you do have a fair number of choices in how to read Part I.

Finally, we repeat what was already said in the preface, that there is nothing new in Part I beyond perhaps the way some things are presented, and that as a treament of the basic theory of Gaussian fields it is *not* meant to be exhaustive. There are now many books covering various aspects of this theory, including those by Bogachev [28], Dudley [56], Fernique [67], Hida and Hitsuda [77], Janson [86], Ledoux and Talagrand [99], Lifshits [105], and Piterbarg [126]. In terms of what will be important to us, [56] and [99] stand out from the pack, perhaps augmented with Talagrand's review [154]. Finally, while not as exhaustive[8] as the others, you might find RJA's lecture notes [3], augmented with the corrections in Section 2.1 below, a user-friendly introduction to the subject.

8 Nor, perhaps, as exhausting.

1

Gaussian Fields

1.1 Random Fields

We shall start with some rather dry, but necessary, technical definitions. As mentioned in the penultimate paragraph of the introduction to Part I, once you have read them you have a number of choices as to what to read next.

As you read in the preface, for us a *random field* is simply a stochastic process, usually taking values in a Euclidean space, and defined over a parameter space of dimensionality at least 1. Although we shall be rather loose about exchanging the terms "field" and "process," in general we shall use "field" when the geometric structure of the parameter space is important to us, and shall use "process" otherwise. "Random" and "stochastic" are, of course, completely synonymous.

Here is a formal definition, which states the obvious (and, we hope, the familiar).

Definition 1.1.1. *Let $(\Omega, \mathcal{F}, \mathbb{P})$ be a complete probability space and T a topological space.*[1] *Then a measurable mapping $f : \Omega \to \mathbb{R}^T$ (the space of all real-valued functions on T) is called a real-valued random field.*[2] *Measurable mappings from Ω to $(\mathbb{R}^T)^d$, $d > 1$, are called vector-valued random fields. If $T \subset \mathbb{R}^N$, we call f an (N, d) random field, and if $d = 1$, simply an N-dimensional random field.*

Thus, $f(\omega)$ is a function, and $(f(\omega))(t)$ its value at time t. In general, however, we shall not distinguish among

---

[1] The use of T comes from the prehistory of Gaussian processes, and probably stands for "time." While the whole point of this book is to get away from the totally ordered structure of $\mathbb{R}$, the notation is too deeply entombed in the collective psyche of probabilists to change it now. Later on, however, when we move to manifolds as parameter spaces, we shall emphasize this by replacing T by M. Nevertheless, points in M will still be denoted by t. We hereby make the appropriate apologies to geometers.

[2] More on notation: While we shall follow the standard convention of denoting random variables by uppercase Latin characters, we shall use lowercase to denote random functions. The reason for this will be become clear in Parts II and III, where we shall need the former for tangent vectors.

$$f_t \equiv f(t) \equiv f(t,\omega) \equiv (f(\omega))(t),$$

etc., unless there is some special need to do so.

Before going further, we shall make one technical blanket assumption on all random fields that will appear in this book, although we shall rarely mention it again. Throughout, we shall demand that all random fields be *separable*.[3] If you are familiar with this assumption, you will know that it solves many measurability problems. For example, without separability, it is not necessarily the case that the supremum of a random field is a well-defined random variable. If you are not familiar with the concept, it probably would not have occurred to you that such problems might exist. Either way, you need not worry about it anymore.

1.2 Gaussian Variables and Fields

A real-valued random variable X is said to be *Gaussian* (or *normally distributed*) if it has the density function

$$\varphi(x) \triangleq \frac{1}{\sqrt{2\pi}\,\sigma} e^{-(x-m)^2/2\sigma^2}, \qquad x \in \mathbb{R},$$

for some $m \in \mathbb{R}$ and $\sigma > 0$. It is elementary calculus that the mean of X is m and the variance σ^2, and that the characteristic function is given by

$$\phi(\theta) = \mathbb{E}\{e^{i\theta X}\} = e^{i\theta m - \sigma^2\theta^2/2}.$$

We abbreviate this by writing $X \sim N(m, \sigma^2)$. The case $m = 0$, $\sigma^2 = 1$ is rather special, and in this situation we say that X has a *standard* normal distribution. In general, if a random variable or process has zero mean, we call it *centered*.

Since the indefinite integral of φ is not a simple function, we also need notation (Φ) for the distribution function and (Ψ)for the tail probability function of a standard normal variable:

$$\Phi(x) \triangleq 1 - \Psi(x) \triangleq \frac{1}{\sqrt{2\pi}} \int_{-\infty}^{x} e^{-u^2/2}\, du. \tag{1.2.1}$$

---

[3] This property, due originally to Doob [50], implies conditions on both T and f. In particular, An $\mathbb{R}^d$-valued random field f, on a topological space T, is called separable if there exists a countable dense subset $D \subset T$ and a fixed event N with $\mathbb{P}\{N\} = 0$ such that, for any closed $B \subset \mathbb{R}^d$ and open $I \subset T$,

$$\{\omega : f(t,\omega) \in B\ \forall t \in I\} \Delta \{\omega : f(t,\omega) \in B\ \forall t \in I \cap D\} \subset N.$$

Here Δ denotes the usual symmetric difference operator, so that

$$A \Delta B = (A \cap B^c) \cup (A^c \cap B), \tag{1.1.1}$$

where A^c is the complement of A.

While Φ and Ψ may not be explicit, there are simple, and rather important, bounds that hold for every $x > 0$ and become sharp very quickly as x grows. In particular, in terms of Ψ we have[4]

$$\left(\frac{1}{x} - \frac{1}{x^3}\right)\varphi(x) < \Psi(x) < \frac{1}{x}\varphi(x). \tag{1.2.2}$$

An $\mathbb{R}^d$-valued random variable X is said to be *multivariate Gaussian* if, for every $\alpha = (\alpha_1, \dots, \alpha_d) \in \mathbb{R}^d$, the real-valued variable $\langle \alpha, X' \rangle = \sum_{i=1}^d \alpha_i X_i$ is Gaussian.[5] In this case there exists a mean vector $m \in \mathbb{R}^d$ with $m_j = \mathbb{E}\{X_j\}$ and a nonnegative definite[6] $d \times d$ covariance matrix C, with elements $c_{ij} = \mathbb{E}\{(X_i - m_i)(X_j - m_j)\}$, such that the probability density of X is given by

$$\varphi(x) = \frac{1}{(2\pi)^{d/2}|C|^{1/2}} e^{-\frac{1}{2}(x-m)C^{-1}(x-m)'}, \tag{1.2.3}$$

where $|C| = \det C$ is the determinant[7] of C. Consistently with the one-dimensional case, we write this as $X \sim N(m, C)$, or $X \sim N_d(m, C)$ if we need to emphasize the dimension.

---

[4] The upper bound in (1.2.2) follows from the observation that

$$\int_x^{-\infty} e^{-u^2/2}\, du \le \int_x^{-\infty} \frac{u}{x} e^{-u^2/2}\, du = x^{-1} e^{-x^2/2}$$

after integration by parts. For the lower bound, make the substitution $u \mapsto x + v/x$ to note that

$$\begin{aligned} \int_x^{-\infty} e^{-u^2/2}\, du &= \int_0^{\infty} e^{-(x^2+2v+v^2/x^2)/2} x^{-1}\, dv \\ &= x^{-1} e^{-x^2/2} \int_0^{\infty} e^{-(2v+v^2/x^2)/2}\, dv \\ &\ge x^{-1} e^{-x^2/2} \int_0^{\infty} e^{-v}(1 - v^2/(2x^2))\, dv \end{aligned}$$

on using the fact that $e^{-z} > 1 - z$ for all $z \ge 0$. It is now trivial to check that the remaining integral is at least $1 - x^{-2}$.

[5] Throughout the book, vectors are taken to be row vectors, and a prime indicates transposition. The inner product between x and y in $\mathbb{R}^d$ is usually denoted by $\langle x, y \rangle$ or, occasionally, by $x \cdot y$ and even (x, y) when this makes more sense.

[6] A $d \times d$ matrix C is called nonnegative definite (or positive semidefinite) if $xCx' \ge 0$ for all $x \in \mathbb{R}^d$. A function $C : T \times T \to \mathbb{R}$ is called nonnegative definite if the matrices $(C(t_i, t_j))_{i,j=1}^n$ are nonnegative definite for all $1 \le n < \infty$ and all $(t_1, \dots, t_n) \in T^n$.

[7] At various places we shall use the notation $|\ |$ to denote any of "absolute value," "Euclidean norm," "determinant," or "Lebesgue measure," depending on the argument, in a natural fashion. The notation $\|\ \|$ is used only for either the norm of complex numbers or for special norms, when it usually appears with a subscript. This is unless it is used, as in Chapter 2 in particular, as $\|f\|$ to denote the supremum of a function f, which is not a norm at all. Despite this multitude of uses of a simple symbol, its meaning should always be clear from the context.

In view of (1.2.3) we have that Gaussian distributions are completely determined by their first- and second-order moments and that uncorrelated Gaussian variables are independent. Both of these facts will be of crucial importance later on.

While the definitions are fresh, note for later use that it is relatively straightforward to check from (1.2.3) that the characteristic function of a multivariate Gaussian X is given by

$$\phi(\theta) = \mathbb{E}\{e^{i\langle\theta, X\rangle}\} = e^{i\langle\theta, m\rangle - \frac{1}{2}\theta C \theta'}, \tag{1.2.4}$$

where $\theta \in \mathbb{R}^d$.

One consequence of the simple structure of ϕ is the fact that if $\{X_n\}_{n\geq 1}$ is an L^2 convergent[8] sequence of Gaussian vectors, then the limit X must also be Gaussian. Furthermore, if $X_n \sim N(m_n, C_n)$, then

$$|m_n - m|^2 \to 0 \quad \text{and} \quad \|C_n - C\|^2 \to 0 \tag{1.2.5}$$

as $n \to \infty$, where m and C are the mean and covariance matrix of the limiting Gaussian. The norm on vectors is Euclidean and that on matrices any of the usual. The proofs involve only (1.2.4) and the continuity theorem for convergence of random variables.

One immediate consequence of either (1.2.3) or (1.2.4) is that if A is any $d \times d$ matrix and $X \sim N_d(m, C)$, then

$$AX \sim N(mA, A'CA). \tag{1.2.6}$$

A judicious choice of A then allows us to compute conditional distributions as well. If $n < d$ make the partitions

$$\begin{aligned}
X &= (X^1, X^2) = ((X_1, \dots, X_n), (X_{n+1}, \dots X_d)),\\
m &= (m^1, m^2) = ((m_1, \dots, m_n), (m_{n+1}, \dots m_d)),\\
C &= \begin{pmatrix} C_{11} & C_{12} \\ C_{21} & C_{22} \end{pmatrix},
\end{aligned}$$

where C_{11} is an $n \times n$ matrix. Then each X^i is $N(m^i, C_{ii})$ and the conditional distribution[9] of X^i given X^j is also Gaussian, with mean vector

$$m_{i|j} = m^i + C_{ij}C_{jj}^{-1}(X^j - m^j)' \tag{1.2.7}$$

and covariance matrix

[8] That is, there exists a random vector X such that $\mathbb{E}\{|X_n - X|^2\} \to 0$ as $n \to \infty$.

[9] To prove this, take

$$A = \begin{pmatrix} I_n & -C_{12}C_{22} \\ 0 & I_{d-n} \end{pmatrix}$$

and define $Y = (Y^1, Y^2) = AX$. Check using (1.2.6) that Y^1 and $Y^2 \equiv X^2$ are independent and use this to obtain (1.2.7) and (1.2.8) for $i = 1,\ j = 2$.

$$C_{i|j} = C_{ii} - C_{ij} C_{jj}^{-1} C_{ji}. \tag{1.2.8}$$

We can now define a real-valued *Gaussian (random) field* or *Gaussian (random) process* to be a random field f on a parameter set T for which the (finite-dimensional) distributions of $(f_{t_1}, \ldots, f_{t_n})$ are multivariate Gaussian for each $1 \le n < \infty$ and each $(t_1, \ldots, t_n) \in T^n$. The functions $m(t) = \mathbb{E}\{f(t)\}$ and

$$C(s,t) = \mathbb{E}\{(f_s - m_s)(f_t - m_t)\}$$

are called the *mean* and *covariance functions* of f. Multivariate[10] Gaussian fields taking values in $\mathbb{R}^d$ are fields for which $\langle \alpha, f_t' \rangle$ is a real-valued Gaussian field for every $\alpha \in \mathbb{R}^d$.

In fact, one can also go in the other direction as well. Given *any* set T, a function $m: T \to \mathbb{R}$, and a nonnegative definite function $C: T \times T \to \mathbb{R}$ there exists[11] a Gaussian process on T with mean function m and covariance function C.

Putting all this together, we have the important principle that *for a Gaussian process, everything about it is determined by the mean and covariance functions.* The fact that no real structure is required of the parameter space T is what makes Gaussian fields such a useful model for random processes on general spaces. To build an appreciation for this, you may want to jump ahead to Section 1.4 to look at some examples. However, you will get more out of that section if you first bear with us to answer one of the most fundamental questions in the theory of Gaussian processes: When are Gaussian processes (almost surely) bounded and/or continuous?

1.3 Boundedness and Continuity

The aim of this section is to develop a useful *sufficient* condition for a centered Gaussian field on a parameter space T to be almost surely bounded and/or continuous, i.e., to determine conditions for which

$$\mathbb{P}\Big\{\sup_{t \in T} |f(t)| < \infty\Big\} = 1 \quad \text{or} \quad \mathbb{P}\Big\{\lim_{s \to t} |f(t) - f(s)| = 0,\ \forall t \in T\Big\} = 1.$$

Of course, in order to talk about continuity—i.e., for the notation $s \to t$ above to have some meaning—it is necessary that T have some topology, so we assume that (T, τ) is a metric space, and that continuity is in terms of the τ-topology. Our first step is to show that τ is irrelevant to the question of continuity.[12] This is rather useful, since

[10] Similary, Gaussian fields taking values in a Banach space B are fields for which $\alpha(f_t)$ is a real-valued Gaussian field for every α in the topological dual B^* of B. The covariance function is then replaced by a family of operators $C_{st} : B^* \to B$, for which $\mathrm{Cov}(\alpha(f_t), \beta(f_s)) = \beta(C_{st}\alpha)$, for $\alpha, \beta \in B^*$.

[11] This is a consequence of the Kolmogorov existence theorem, which, at this level of generality, can be found in Dudley [55]. Such a process is a random variable in $\mathbb{R}^T$ and may have terrible properties, including lack of measurability in t. However, it will *always* exist.

[12] However, τ will come back into the picture when we talk about moduli of continuity later in this section.

we shall also soon show that boundedness and continuity are essentially the same problem for Gaussian fields, and formulating the boundedness question requires no topological demands on T.

To start, define a new metric d on T by

$$d(s,t) \stackrel{\Delta}{=} \{\mathbb{E}[(f(s)-f(t))^2]\}^{1/2}, \tag{1.3.1}$$

in a notation that will henceforth remain fixed. Actually, d is only a pseudometric, since although it satisfies all the other demands of a metric, $d(s,t) = 0$ does not necessarily[13] imply that $s = t$. Nevertheless, we shall abuse terminology by calling d the *canonical metric* for T and/or f.

It will be convenient for us always to assume that d is continuous in the τ-topology, and we shall indeed do so, since in the environment of f continuity that will interest us, d continuity costs us nothing. To see this, suppose that $\sup_T \mathbb{E}\{f_t^2\} < \infty$, and that f is a.s. continuous. Then

$$\begin{aligned}\lim_{s\to t} d^2(s,t) &= \lim_{s\to t} \mathbb{E}\{(f(s)-f(t))^2\} \\ &= \mathbb{E}\Big\{\lim_{s\to t}(f(s)-f(t))^2\Big\} \\ &= 0,\end{aligned}$$

the exchange of limit and expectation justified by the uniform integrability provided by the fact that since f is Gaussian, boundedness in L^2 implies boundness in all L_p. In other words, a.s. continuity of f implies the continuity of d.

Here is the lemma establishing the irrelevance of τ to the continuity question.

Lemma 1.3.1. *Let f be a centered Gaussian process on a compact metric space (T,τ). Then f is a.s. continuous with respect to the τ-topology if and only if it is a.s. continuous with respect to the d (pseudo)topology. More precisely, with probability one, for all $t \in T$,*

$$\lim_{s:\tau(s,t)\to 0} |f(s)-f(t)| = 0 \iff \lim_{s:d(s,t)\to 0} |f(s)-f(t)| = 0.$$

Proof. Since d is (always) assumed continuous in the τ topology, it is immediate that if f is d-continuous then it is τ-continuous.

Suppose, therefore, that f is τ-continuous. For $\eta \geq 0$, let

$$A_\eta = \{(s,t) \in T \times T : d(s,t) \leq \eta\}.$$

Since d is continuous, this is a τ-closed subset of $T \times T$. Furthermore, $\bigcap_{\eta>0} A_\eta = A_0$. Fix $\varepsilon > 0$. Then, by the τ-compactness of T, there is a finite set $B \subset A_0$ (the number of whose elements will in general depend on ε) such that

[13] For a counterexample, think of a periodic process on $\mathbb{R}$, with period p. Then $d(s,t) = 0$ implies no more than $s - t = kp$ for some $k \in \mathbb{Z}$.

$$\bigcup_{(s',t')\in B} \left\{(s,t) \in T \times T : \max(\tau(s,s'), \tau(t,t')) \le \varepsilon\right\}$$

covers A_η for some $\eta = \eta(\varepsilon) > 0$, with $\eta(\varepsilon) \to 0$ as $\varepsilon \to 0$. That is, whenever $(s,t) \in A_\eta$ there is an $(s',t') \in B$ with $\tau(s,s'), \tau(t,t') \le \varepsilon$. Note that

$$|f_t - f_s| \le |f_s - f_{s'}| + |f_{s'} - f_{t'}| + |f_{t'} - f_t|.$$

Since $(s',t') \in B \subset A_0$, we have $f(s') = f(t')$ a.s. Thus

$$\sup_{d(s,t)\le\eta(\varepsilon)} |f_t - f_s| \le 2 \sup_{\tau(s,t)\le\varepsilon} |f_t - f_s|,$$

and the τ-continuity of f implies its d-continuity. □

The astute reader will have noted that in the statement of Lemma 1.3.1 the parameter space T was quietly assumed to be compact, and that this additional assumption was needed in the proof. Indeed, from now on we shall assume that this is always the case, and shall rely on it heavily. Fortunately, however, it is not a serious problem. As far as continuity is concerned, if T is σ-compact[14] then a.s. continuity on its compact subsets immediately implies a.s. continuity over T itself. We shall not go beyond σ-compact spaces in this book. The same is not true for boundedness, nor should it be.[15] However, we shall see that, at least on compact T, boundedness and continuity are equivalent problems. Furthermore, both depend on how "large" the parameter set T is when size is measured in a metric that comes from the process itself.[16] One way to measure the size of T is via the notion of metric entropy.

Definition 1.3.2. *Let f be a centered Gaussian field on T, and d the canonical metric* (1.3.1). *Assume that T is d-compact, and write*

$$B_d(t,\varepsilon) \triangleq \{s \in T : d(s,t) \le \varepsilon\} \tag{1.3.2}$$

for the d ball centered on $t \in T$ and of radius ε. Let $N(T,d,\varepsilon) \equiv N(\varepsilon)$ denote the smallest number of such balls that cover T, and set

$$H(T,d,\varepsilon) \equiv H(\varepsilon) = \ln(N(\varepsilon)). \tag{1.3.3}$$

Then N and H are called the (*metric*) *entropy and log-entropy functions for T* (*or f*). *We shall refer to any condition or result based on N or H as an entropy condition/result.*

---

[14] T is σ-compact if it can be represented as the countable union of compact sets.

[15] Think of simple Brownian motion on $\mathbb{R}_+$. While bounded on every finite interval, it is a consequence of the law of the iterated logarithm that it is unbounded on $\mathbb{R}_+$.

[16] A particularly illuminating example of this comes from Brownian noise processes of Section 1.4.3. There we shall see that the *same* process can be continuous, or discontinuous, depending on how we specify its parameter set.

Note that since we are assuming that T is d-compact, it follows that $H(\varepsilon) < \infty$ for all $\varepsilon > 0$. The same need not be (nor generally is) true for $\lim_{\varepsilon \to 0} H(\varepsilon)$. Furthermore, note for later use that if we define

$$\operatorname{diam}(T) = \sup_{s,t \in T} d(s,t), \tag{1.3.4}$$

then $N(\varepsilon) = 1$ and so $H(\varepsilon) = 0$ for $\varepsilon \geq \operatorname{diam}(T)$.

Here then are the main results of this section, all of which will be proven soon.

Theorem 1.3.3. *Let f be a centered Gaussian field on a d-compact T, d the canonical metric, and H the corresponding entropy. Then there exists a universal constant K such that*

$$\mathbb{E}\left\{ \sup_{t \in T} f_t \right\} \leq K \int_0^{\operatorname{diam}(T)/2} H^{1/2}(\varepsilon)\, d\varepsilon. \tag{1.3.5}$$

This result has immediate consequences for continuity. Define the *modulus of continuity* ω_F of a real-valued function F on a metric space (T, τ) as

$$\omega_F(\delta) \equiv \omega_{F,\tau}(\delta) \stackrel{\Delta}{=} \sup_{\tau(s,t) \leq \delta} |F(t) - F(s)|, \quad \delta > 0. \tag{1.3.6}$$

The modulus of continuity of f can be thought of as the supremum of the random field $f_{s,t} = f_t - f_s$ over a certain neighborhood of $T \times T$, in that

$$\omega_{f,\tau}(\delta) = \sup_{\substack{(s,t) \in T \times T \\ \tau(s,t) \leq \delta}} f(s,t). \tag{1.3.7}$$

(Note we can drop the absolute value sign since the supremum here is always non-negative.)

More precisely, write $d^{(2)}$ for the canonical metric of $f_{s,t}$ on $T \times T$. Then

$$\begin{aligned} d^{(2)}((s,t),(s',t')) &= \left[\mathbb{E}\left\{ ((f_t - f_s) - (f_{t'} - f_{s'}))^2 \right\} \right]^{1/2} \\ &\leq 2 \max(d(s,t), d(s',t')), \end{aligned}$$

and so

$$N(\{(s,t) \in T \times T : d(s,t) \leq \delta\}, d^{(2)}, \delta) \leq N(T, d, \delta/2).$$

From these observations, it is immediate that Theorem 1.3.3 implies the following corollary.

Corollary 1.3.4. *Under the conditions of Theorem* 1.3.3 *there exists a universal constant K such that*

$$\mathbb{E}\left\{ \omega_{f,d}(\delta) \right\} \leq K \int_0^{\delta} H^{1/2}(\varepsilon)\, d\varepsilon. \tag{1.3.8}$$

Note that this is not quite enough to establish the a.s. continuity of f. Continuity is, however, not far away, since the same construction used to prove Theorem 1.3.3 will also give us the following, which, with the elementary tools we have at hand at the moment,[17] neither follows from, nor directly implies, (1.3.8).

[17] See, however, Theorem 2.1.3 below to see what one can do with better tools.

Theorem 1.3.5. *Under the conditions of Theorem* 1.3.3 *there exists a random* $\eta \in (0, \infty)$ *and a universal constant* K *such that*

$$\omega_{f,d}(\delta) \leq K \int_0^\delta H^{1/2}(\varepsilon)\, d\varepsilon, \tag{1.3.9}$$

for all $\delta < \eta$.

Note that (1.3.9) is expressed in terms of the d modulus of continuity. Translating this to a result for the τ modulus is trivial.

We shall see later that if f is stationary,[18] then the convergence of the entropy integral is also necessary for continuity and that continuity and boundedness always occur together (Theorem 1.5.4). Now, however, we shall prove Theorems 1.3.3 and 1.3.5 following the approach of Talagrand [154]. The original proof of Theorem 1.3.5 is due to Dudley [51], and in fact, things have not really changed very much since then. Immediately following the proofs, in Section 1.4, we shall look at examples, to see how entropy arguments work in practice. You may want to skip to the examples before going through the proofs first time around.

We start with the following almost trivial, but important, observations.

Observation 1. If f is a separable process on T then $\sup_{t\in T} f_t$ is a well-defined (i.e., measurable) random variable.

Measurability follows directly from Definition 1.1.1 of separability, which gave us a countable dense subset $D \subset T$ for which

$$\sup_{t\in T} f_t = \sup_{t\in D} f_t.$$

The supremum of a countable set of measurable random variables is always measurable.

One can actually manage without separability for the rest of this section, in which case

$$\sup\left\{\mathbb{E}\Big[\sup_{t\in F} f_t\Big] : F \subset T,\ F \text{ finite}\right\}$$

should be taken as the definition of $\mathbb{E}\{\sup_T f_t\}$.

Observation 2. If f is a separable process on T and X a centered random variable (not necessarily independent of f), then

$$\mathbb{E}\Big\{\sup_{t\in T}(f_t + X)\Big\} = \mathbb{E}\Big\{\sup_{t\in T} f_t\Big\}.$$

---

[18] We shall treat stationarity in detail in Chapter 5. For the moment all you need to know is that under stationarity the expectation $\mathbb{E}\{f(t)\}$ is constant, while the covariance $\mathbb{E}\{f(t)f(s)\}$ is a function of $s - t$ only.

As trite as this observation is, it ceases to be valid if, for example, we investigate $\sup_t |f_t + X|$ rather than $\sup_t (f_t + X)$.

Proof of Theorem 1.3.3. Fix a point $t_0 \in T$ and consider $f_t - f_{t_0}$. In view of Observation 2, we can work with $\mathbb{E}\{\sup_T (f_t - f_{t_0})\}$ rather than $\mathbb{E}\{\sup_T f_t\}$. Furthermore, in view of separability, it will suffice to take these suprema over the countable separating set $D \subset T$. To save on notation, we might therefore just as well assume that T is countable, which we now do.

We shall represent the difference $f_t - f_{t_0}$ via a telescoping sum, in what is called a *chaining* argument, and is in essence an approximation technique. We shall keep track of the accuracy of the approximations via entropy and simple union bounds.

To build the approximations, first fix some $r \geq 2$ and choose the largest $i \in \mathbb{Z}$ such that $\mathrm{diam}(T) \leq r^{-i}$, where the $\mathrm{diam}(T)$ is measured in terms of the canonical metric of f. For $j > i$, take a finite subset Π_j of T such that

$$\sup_{t \in T} \inf_{s \in \Pi_j} d(s,t) \leq r^{-j}$$

(possible by d-compactness), and define a mapping $\pi_j : T \to \Pi_j$ satisfying

$$\sup_{t \in T} d(t, \pi_j(t)) \leq r^{-j}, \tag{1.3.10}$$

so that

$$\sup_{t \in T} d(\pi_j(t), \pi_{j-1}(t)) \leq 2r^{-j+1}. \tag{1.3.11}$$

For consistency, set $\Pi_i = \{t_0\}$ and $\pi_i(t) = t_0$ for all t. Consistent with the notation of Definition 1.3.2, we can choose Π_j to have no more than $N_j \triangleq N(r^{-j})$ points, and so entropy has now entered into the argument.

The idea of this construction is that the points $\pi_j(t)$ are successive approximations to t, and that as we move along the "chain" $\pi_j(t)$ we have the decomposition[19]

$$f_t - f_{t_0} = \sum_{j>i} f_{\pi_j(t)} - f_{\pi_{j-1}(t)}. \tag{1.3.12}$$

[19] We actually need to check two things: that the potentially infinite sum in (1.3.12) is well defined, with probability one, and that it indeed converges to $f_t - f_{t_0}$. The first is straightforward, the second requires a little measure theoretic legerdemain.

By (1.3.11) and (1.3.13) it follows that

$$\begin{aligned} \mathbb{P}\{|f_{\pi_j(t)} - f_{\pi_{j-1}(t)}| \geq (r/\sqrt{2})^{-j}\} &\leq 2\exp\left(\frac{-(r/\sqrt{2})^{-2j}}{2(2r^{-j+1})^2}\right) \\ &= 2\exp\left(-2^j/8r^2\right), \end{aligned}$$

which is emminently summable. By Borel–Cantelli, and recalling that $r \geq 2$, we have that the sum in (1.3.12) converges absolutely, with probability one.

For the second step, fix a $t \in T$ and consider the sum

To start the proof we look for a bound on the tail of the distribution of $\sup_T(f_t - f_{t_0})$, for which we need a little notation (carefully chosen to make the last lines of the proof come out neatly). Define

$$M_j = N_j N_{j-1}, \qquad a_j = 2^{3/2} r^{-j+1} \sqrt{\ln(2^{j-i} M_j)}, \qquad S = \sum_{j>i} a_j.$$

Then M_j is the maximum number of possible pairs $(\pi_j(t), \pi_{j-1}(t))$ as t varies through T and a_j was chosen so as to make later formulas simplify. Recall (1.2.2), which implies

$$\mathbb{P}\{X \geq u\} \leq e^{-u^2/2\sigma^2}, \tag{1.3.13}$$

for $X \sim N(0, \sigma^2)$ and $u > 0$. Applying (1.3.13) we have, for all $u > 0$,

$$\mathbb{P}\left\{\exists\, t \in T : f_{\pi_j(t)} - f_{\pi_{j-1}(t)} > u a_j\right\} \leq M_j \exp\left(\frac{-u^2 a_j^2}{2(2r^{-j+1})^2}\right), \tag{1.3.14}$$

and so

$$\mathbb{P}\left\{\sup_{t \in T} f_t - f_{t_0} \geq uS\right\} \leq \sum_{j>i} M_j \exp\left(\frac{-u^2 a_j^2}{2(2r^{-j+1})^2}\right)$$
$$= \sum_{j>i} M_j \left(2^{j-i} M_j\right)^{-u^2}.$$

For $u > 1$ this is at most

$$\sum_{j>i} \left(2^{j-i}\right)^{-u^2} \leq 2^{-u^2} \sum_{j>i} 2^{-(j-i)+1} = 2 \cdot 2^{-u^2}.$$

The basic relationship that, for nonnegative random variables X,

$$\mathbb{E}\{X\} = \int_0^\infty \mathbb{P}\{X \geq u\}\, du,$$

together with the observation that $\sup_{t \in T}(f_t - f_{t_0}) \geq 0$ since $t_0 \in T$, immediately yields

---

$$f_t^{(n)} \stackrel{\Delta}{=} f_{t_0} + \sum_{j=i+1}^{n} f_{\pi_j(t)} - f_{\pi_{j-1}(t)} \equiv f_{\pi_n(t)}.$$

From the construction of the π_j it is immediate that $\mathbb{E}\{|f(t) - f_t^{(n)}|^2\} \to 0$ as $n \to \infty$, and so the fact that $f_t^{(n)}$ converges almost surely implies that it must also converge to f_t. Since this must be true for any countable subset of points in T, separability gives that it is true throughout T (cf. footnote 3).

$$\mathbb{E}\Big\{\sup_{t\in T} f_t\Big\} \le KS, \tag{1.3.15}$$

with $K = 1 + 2\int_1^\infty 2^{-u^2}\,du$. Thus, all that remains to complete the proof is to compute S.

Using the definition of S, along with the elementary inequality that $\sqrt{ab} \le \sqrt{a} + \sqrt{b}$, gives

$$\begin{aligned} S &\le 2^{3/2} \sum_{j>i} r^{-j+1}\left(\sqrt{j-i}\sqrt{\ln 2} + \sqrt{\ln N_j} + \sqrt{\ln N_{j-1}}\right) \\ &\le K\left(r^{-i} + \sum_{j\ge i} r^{-j}\sqrt{\ln N_j}\right) \\ &\le K \sum_{j\ge i} r^{-j}\sqrt{\ln N_j}, \end{aligned}$$

where K is a constant that may change from line to line, but depends only on r. The last inequality follows from absorbing the lone term of r^{-i} into the second term of the sum, which is possible since the very definition of i implies that $N_{i+1} \ge 2$, and changing the multiplicative constant K accordingly.

Recalling now the definition of N_j as $N(r^{-j})$, we have that

$$\varepsilon \le r^{-j} \Rightarrow N(\varepsilon) \ge N_j.$$

Thus

$$\int_0^{r^{-i}} \sqrt{\ln N(\varepsilon)}\,d\varepsilon \ge \sum_{j\ge i}\left(r^{-j} - r^{-j-1}\right)\sqrt{\ln N_j} = K\sum_{j\ge i} r^{-j}\sqrt{\ln N_j}.$$

Putting this together with the bound on S and substituting into (1.3.15) gives

$$\mathbb{E}\Big\{\sup_{t\in T} f_t\Big\} \le K\int_0^{r^{-i}} H^{1/2}(\varepsilon)\,d\varepsilon.$$

Finally, note that $N(\varepsilon) = 1$ (and so $H(\varepsilon) = 0$) for $\varepsilon \ge 2r^{-i} \ge \operatorname{diam}(T)$, to establish (1.3.5) and so complete the proof. □

Proof of Theorem 1.3.5. The proof starts with the same construction as in the proof of Theorem 1.3.3. Note that from the same principles behind the telescoping sum (1.3.12) defining $f_t - f_{t_0}$ we have that for all $s, t \in T$ and $J > i$,

$$\begin{aligned} f_t - f_s = f_{\pi_J(t)} - f_{\pi_J(s)} &+ \sum_{j>J}[f_{\pi_j(t)} - f_{\pi_{j-1}(t)}] \\ &- \sum_{j>J}[f_{\pi_j(s)} - f_{\pi_{j-1}(s)}]. \end{aligned} \tag{1.3.16}$$

From (1.3.13) we have that for all $u > 0$,

$$\mathbb{P}\{|f_{\pi_j(t)} - f_{\pi_j(s)}| \geq ud(\pi_j(t), \pi_j(s))\} \leq e^{-u^2/2}.$$

Arguing as we did to obtain (1.3.14), and the line or two following, we now see that

$$\mathbb{P}\left\{\exists\, s, t \in T : |f_{\pi_j(t)} - f_{\pi_j(s)}| \geq \sqrt{2}d(\pi_j(t), \pi_j(s))\sqrt{\ln(2^{j-i}N_j^2)}\right\} \leq 2^{i-j}.$$

Since this is a summable series, Borel–Cantelli gives the existence of a random $j_0 > i$ for which, with probability one,

$$j > j_0 \Rightarrow |f_{\pi_j(t)} - f_{\pi_j(s)}| \leq \sqrt{2}d(\pi_j(t), \pi_j(s))\sqrt{\ln(2^{j-i}N_j^2)}$$

for all $s, t \in T$.

Essentially the same argument also gives that

$$j > j_0 \Rightarrow |f_{\pi_j(t)} - f_{\pi_{j-1}(t)}| \leq \sqrt{2}d(\pi_j(t), \pi_{j-1}(t))\sqrt{\ln(2^{j-i}M_j)}$$

for all $t \in T$.

Putting these into (1.3.16) gives that

$$\begin{aligned} |f_t - f_s| &\leq Kd(\pi_{j_0}(t), \pi_{j_0}(s))\sqrt{\ln(2^{j_0-i}N_{j_0}^2)} \\ &\quad + K\sum_{j>j_0} d(\pi_j(t), \pi_{j-1}(t))\sqrt{\ln(2^{j-i}M_j)}. \end{aligned}$$

Note that $d(\pi_j(t), \pi_{j-1}(t)) \leq 2r^{-j+1}$ and

$$\pi_{j_0}(s)) \leq d(s,t) + 2r^{-j_0} \leq 3r^{-j_0}$$

if we take $d(s,t) \leq \eta \stackrel{\Delta}{=} r^{-j_0}$. The above sums can be turned into integrals just as we did at the end of the previous proof, which leads to (1.3.9) and so completes the argument. □

Before leaving to look at some examples, you should note one rather crucial fact: The *only* Gaussian ingredient in the preceding two proofs was the basic inequality (1.3.13) giving $\exp(-u^2/2)$ as a tail bound for a single $N(0,1)$ random variable. The remainder of the proof used little more than the union bound on probabilities and some clever juggling. Furthermore, it does not take a lot of effort to see that the square root in the entropy integrals such as (1.3.5) is related to "inverting" the square in $\exp(-u^2/2)$, while the logarithm comes from "inverting" the exponential. If this makes you feel that there is a far more general non-Gaussian theory behind all this, and that it is not going to be very different from the Gaussian one, then you are right. A brief explanation of how it works is in Section 1.4.6.

1.4 Examples

Our choice of examples was determined primarily by the type of random fields that we need later in the book. Thus we shall start by looking at random fields on $\mathbb{R}^N$, followed by looking at how to turn questions of differentiability into questions of continuity. Following these, we shall look at the Brownian family of processes, needed for setting up the spectral representation theory of Chapter 5. As mentioned above, this family is also very instructive in understanding *why* the finiteness of entropy integrals is a natural condition for continuity. The section on generalized fields is useful as a recipe for providing examples, and the section on set-indexed processes is there for fun. Neither of these will appear elsewhere in the book. The non-Gaussian examples of Section 1.4.6 are worth knowing about already at this stage.

1.4.1 Fields on $\mathbb{R}^N$

Returning to Euclidean space after the abstraction of entropy on general metric spaces, it is natural to expect that conditions for continuity and boundedness will become so simple to both state and prove that there was really no need to introduce such abstruse general concepts.

This expectation is both true and false. It turns out that avoiding the notion of entropy does not make it any easier to establish continuity theorems, and indeed, reliance on the specific geometry of the parameter space often confounds the basic issues. On the other hand, the following important result is easy to state without specifically referring to any abstract notions. To formulate it, let f_t be a centered Gaussian process on a compact $T \subset \mathbb{R}^N$ and define

$$p^2(u) = \sup_{|s-t| \le u} \mathbb{E}\left\{|f_s - f_t|^2\right\}, \tag{1.4.1}$$

where $|\cdot|$ is the usual Euclidean metric. If f is stationary, then

$$p^2(u) = 2 \sup_{|t| \le u} [C(0) - C(t)]. \tag{1.4.2}$$

Theorem 1.4.1. *If, for some $\delta > 0$, either*

$$\int_0^\delta (-\ln u)^{\frac{1}{2}}\, dp(u) < \infty \quad or \quad \int_\delta^\infty p\left(e^{-u^2}\right) du < \infty, \tag{1.4.3}$$

then f is continuous and bounded on T with probability one. A sufficient condition for either integral in (1.4.3) *to be finite is that for some $0 < C < \infty$ and $\alpha, \eta > 0$,*

$$\mathbb{E}\{|f_s - f_t|^2\} \le \frac{C}{|\log|s-t||^{1+\alpha}}, \tag{1.4.4}$$

for all s, t with $|s - t| < \eta$. Furthermore, there exists a constant K, dependent only on the dimension N, and a random $\delta_0 > 0$ such that for all $\delta < \delta_0$,

$$\omega_f(\delta) \le K \int_0^{p(\delta)} (-\ln u)^{\frac{1}{2}}\, dp(u), \tag{1.4.5}$$

where the modulus of continuity ω_f is taken with respect to the Euclidean metric. A similar bound, in the spirit of (1.3.8), *holds for* $\mathbb{E}\{\omega_f(\delta)\}$.

Proof. Note first that since $p(u)$ is obviously nondecreasing in u, the Riemann–Stieltjes integral (1.4.3) is well defined. The proof that both integrals in (1.4.3) converge and diverge together and that the convergence of both is assured by (1.4.4) is simple calculus and left to the reader. Of more significance is relating these integrals to the entropy integrals of Theorems 1.3.3 and 1.3.5 and Corollary 1.3.4. Indeed, all the claims of the theorem will follow from these results if we show that

$$\int_0^{\delta} H^{1/2}(\varepsilon)\, d\varepsilon \le K \int_0^{p(\delta)} (-\ln u)^{\frac{1}{2}}\, dp(u) \tag{1.4.6}$$

for small enough δ.

Since T is compact, we can enclose it in an N-cube C_L of side length $L = \max_{i=1,\dots,N} \sup_{s,t\in T} |t_i - s_i|$. Since p is nondecreasing, there is no problem in defining

$$p^{-1}(\varepsilon) \triangleq \sup\{u : p(u) \le \varepsilon\}.$$

Now note that for each $\varepsilon > 0$, the cube C_L, and so T, can be covered by $[1 + L\sqrt{N}/(2p^{-1}(\varepsilon))]^N$ (Euclidean) N-balls, each of which has radius no more than ε in the canonical metric d. Thus,

$$\begin{aligned}
\int_0^{\delta} H^{1/2}(\varepsilon)\, d\varepsilon &\le \sqrt{N} \int_0^{\delta} (\ln(1 + L\sqrt{N}/(2p^{-1}(\varepsilon))))^{\frac{1}{2}}\, d\varepsilon \\
&= \sqrt{N} \int_0^{p(\delta)} (\ln(1 + L\sqrt{N}/2u))^{\frac{1}{2}}\, dp(u) \\
&\le 2\sqrt{N} \int_0^{p(\delta)} (-\ln u)^{\frac{1}{2}}\, dp(u)
\end{aligned}$$

for small enough δ. This completes the proof. □

The various sufficient conditions for continuity of Theorem 1.4.1 are quite sharp, but not necessary. There are two stages at which necessity is lost. One is simply that entropy conditions, in general, need not be necessary in the nonstationary case. The second is that something is lost in the passage from entropy to the conditions on p.

As an example of the latter, take standard Brownian motion W on $[0, 1]$, take the homeomorphism $h(t) = \ln 4/(\ln(4/t))$ from $[0, 1]$ to $[0, 1]$, and define a new process on $[0, 1]$ by $f(t) = f(h^{1/2}(t))$. It is trivial that f is a.s. continuous, since it is obtained from W via a continuous transformation. It is also easy to verify that f and W have identical entropy functions. However, while the p function for W satisfies (1.4.3), this is not true for that of f. The reason for this is the high level of nonstationarity of the increments of f as opposed to those of W.

Despite these drawbacks, the results of Theorem 1.4.1 are, from a practical point of view, reasonably definitive. For example, we shall see below (Corollary 1.5.5) that if f is stationary and

$$\frac{K_1}{(-\ln|t|)^{1+\alpha_1}} \le C(0) - C(t) \le \frac{K_2}{(-\ln|t|)^{1+\alpha_2}}, \tag{1.4.7}$$

for $|t|$ small enough, then f will be sample path continuous if $\alpha_2 > 0$ and discontinuous if $\alpha_1 < 0$.

Before leaving the Euclidean case, it is also instructive to see how the above conditions on covariance functions translate to conditions on the spectral measure ν of (5.4.1) when f is stationary. Although this involves using concepts yet to be defined, this is a natural place to describe the result. If you are familiar with spectral theory, the following will be easy to follow. If not, then you can return here after reading Chapter 5.

The translation is via standard Tauberian theory, which translates the behavior of C at the origin to that of ν at infinity (cf., for example, [25]). A typical result is the following, again in the centered, Gaussian case: If the integral

$$\int_0^\infty \big(\log(1+\lambda)\big)^{1+\alpha}\,\nu(d\lambda)$$

converges for some $\alpha > 0$, then f is a.s. continuous, while if it diverges for some $\alpha < 0$, then f is a.s. discontinuous.

In other words, it is the "high-frequency oscillations" in the spectral representation that are controlling the continuity/discontinuity dichotomy. This is hardly surprising. What is perhaps somewhat more surprising, since we have seen that for Gaussian processes continuity and boundedness come together, is that it is these same oscillations that are controlling boundedness as well.

1.4.2 Differentiability on $\mathbb{R}^N$

We shall stay with Euclidean $T \subset \mathbb{R}^N$ for the moment, and look at the question of a.s. differentiability of centered, Gaussian f.

Firstly, however, we need to define the L^2, or *mean square, (partial) derivatives* of a random field.

Choose a point $t \in \mathbb{R}^N$ and a sequence of k "directions" $t'_1, \ldots, t'_k$ in $\mathbb{R}^N$, so that $t' = (t'_1, \ldots, t'_k) \in \otimes^k \mathbb{R}^N$, the k-fold tensor product $\mathbb{R}^N$ with itself. We say that f has a kth-order L^2 partial derivative at t, in the direction t', which we denote by $D^k_{L^2} f(t,t')$, if the limit

$$D^k_{L^2} f(t,t') \triangleq \lim_{h_1,\ldots,h_k \to 0} \frac{1}{\prod_{i=1}^k h_i} F\left(t, \sum_{i=1}^k h_i t'_i\right) \tag{1.4.8}$$

exists in L^2, where $F(t,t')$ is the symmetrized difference

$$F(t,t') = \sum_{s\in\{0,1\}^k} (-1)^{k-\sum_{i=1}^k s_i} f\left(t + \sum_{i=1}^k s_i t'_i\right)$$

and the limit above is interpreted sequentially, i.e., first send h_1 to 0, then h_2, etc. A simple sufficient condition for L^2 partial differentiability of order k in all directions and throughout a region $T \in \mathbb{R}^N$ is that

$$\lim_{h_1,\dots,h_k,\widehat{h}_i,\dots,\widehat{h}_k\to 0} \frac{1}{\prod_{i=1}^k h_i\widehat{h}_i} \mathbb{E}\left\{F\left(t, \sum_{i=1}^k h_i t'_i\right) F\left(s, \sum_{i=1}^k \widehat{h}_i s'_i\right)\right\} \tag{1.4.9}$$

exists[20] for all $s,t \in T$, all directions $s', t' \in \otimes^k\mathbb{R}^N$, and all sequences $h_1,\dots,h_k$ and $\widehat{h}_i,\dots,\widehat{h}_k$, where the limits are again to be interpreted sequentially. Note that if f is Gaussian then so are its L^2 derivatives, when they exist.

With L^2 partial derivatives defined, we can now turn to sample path, or almost sure, derivatives and ask, for example, whether they are continuous and bounded. It turns out that the general structure that we have so far built actually makes this a very simple question to answer.

To see why, endow the space $\mathbb{R}^N \times \otimes^k\mathbb{R}^N$ with the norm

$$\|(s,s')\|_{N,k} \stackrel{\Delta}{=} |s| + \|s'\|_{\otimes^k\mathbb{R}^N} = |s| + \left(\sum_{i=1}^k |s'_i|^2\right)^{1/2},$$

and write $B_{N,k}(y,h)$ for the h-ball centered at $y=(t,t')$ in the metric induced by $\|\cdot\|_{N,k}$. Furthermore, write

$$T_{k,\rho} \stackrel{\Delta}{=} T \times \left\{t' : \|t'\|_{\otimes^k\mathbb{R}^N} \in (1-\rho, 1+\rho)\right\}$$

for the product of T with the ρ-tube around the unit sphere in $\otimes^k\mathbb{R}^N$. This is enough to allow us to formulate the following theorem.

Theorem 1.4.2. *Suppose f is a centered Gaussian random field on an open $T \in \mathbb{R}^N$, possessing kth-order partial derivatives in the L^2 sense in all directions everywhere inside T. Suppose, furthermore, that there exist $0 < K < \infty$ and $\rho, \delta, h_0 > 0$ such that for $0 < \eta_1, \eta_2, h < h_0$,*

$$\begin{aligned}&\mathbb{E}\{[F(t,\eta_1 t') - F(s,\eta_2 s')]^2\} \\ &\quad < K(-\ln(\|(t,t')-(s,s')\|_{N,k} + |\eta_1-\eta_2|))^{-(1+\delta)},\end{aligned} \tag{1.4.10}$$

for all

$$((t,t'),(s,s')) \in T_{k,\rho} \times T_{k,\rho} : (s,s') \in B_{N,k}((t,t'),h).$$

Then, with probability one, f is k-times continuously differentiable; that is, $f \in C^k(T)$.

[20] This is an immediate consequence of the fact that a sequence X_n of random variables converges in L^2 if and only if $\mathbb{E}\{X_n X_m\}$ converges to a constant as $n,m\to\infty$.

Proof. Recalling that we have assumed the existence of L^2 derivatives, we can define the Gaussian field

$$\widehat{F}(t, t', \eta) = \begin{cases} F(t, \eta t'), & \eta \neq 0, \\ D^k_{L^2} f(t, t'), & \eta = 0, \end{cases}$$

on $\widehat{T} \stackrel{\Delta}{=} T_{k,\rho} \times (-h, h)$, an open subset of the finite-dimensional vector space $\mathbb{R}^N \times \otimes^k \mathbb{R}^N \times \mathbb{R}$ with norm

$$\|(t, t', \eta)\|_{N,k,1} = \|(t, t')\|_{N,k} + |\eta|.$$

Whether $f \in C^k(T)$ is clearly the same issue as whether $\widehat{F} \in C(\widehat{T})$, with the issue of the continuity of f really only being on the hyperplane where $\eta = 0$. But this puts us back into the setting of Theorem 1.4.1, and it is easy to check that condition (1.4.4) there translates to (1.4.10) in the current scenario. □

1.4.3 The Brownian Family of Processes

Perhaps the most basic of all random fields is a collection of independent Gaussian random variables. While it is simple to construct such random fields for finite and even countable parameter sets, deep technical difficulties obstruct the construction for uncountable parameter sets. The path that we shall take around these difficulties involves the introduction of random measures, which, at least in the Gaussian case, are straightforward to formulate.

Let $(T, \mathcal{T}, \nu)$ be a σ-finite measure space and denote by $\mathcal{T}_\nu$ the collection of sets of $\mathcal{T}$ of finite ν measure. A *Gaussian noise* based on ν, or "Gaussian ν-noise," is a random field $W : \mathcal{T}_\nu \to \mathbb{R}$ such that for all $A, B \in \mathcal{T}_\nu$,

$$W(A) \sim N(0, \nu(A)), \tag{1.4.11}$$

$$A \cap B = \emptyset \Rightarrow W(A \cup B) = W(A) + W(B) \text{ a.s.}, \tag{1.4.12}$$

$$A \cap B = \emptyset \Rightarrow W(A) \text{ and } W(B) \text{ are independent.} \tag{1.4.13}$$

Property (1.4.12) encourages one to think of W as a random (signed) measure, although it is not generally σ-finite.[21] We describe (1.4.13) by saying that W has *independent increments*.

Theorem 1.4.3. *If $(T, \mathcal{T}, \nu)$ is a measure space, then there exists a real-valued Gaussian noise, defined for all $A \in \mathcal{T}_\nu$, satisfying* (1.4.11)–(1.4.13).

Proof. In view of the closing remarks of the preceding section, all we need do is provide an appropriate covariance function on $\mathcal{T}_\nu \times \mathcal{T}_\nu$. Try

[21] While the notation "W" is inconsistent with our determination to use lowercase Latin characters for random functions, we retain it as a tribute to Norbert Wiener, who is the mathematical father of these processes.

$$C_\nu(A,B) \triangleq \nu(A\cap B). \tag{1.4.14}$$

This is nonnegative definite, since for any $A_i \in \mathcal{T}_\nu$ and $\alpha_i \in \mathbb{R}$,

$$\begin{aligned}\sum_{i,j}\alpha_i C_\nu(A_i,A_j)\alpha_j &= \sum_{i,j}\alpha_i\alpha_j\int_T I_{A_i}(x)I_{A_j}(x)\nu(dx)\\ &= \int_T\left(\sum_i \alpha_i I_{A_i}(x)\right)^2\nu(dx)\\ &\geq 0.\end{aligned}$$

Consequently, there exists a centered Gaussian random field on $\mathcal{T}_\nu$ with covariance C_ν. It is immediate that this field satisfies (1.4.11)–(1.4.13), and so we are done. □

A particularly simple example of a Gaussian noise arises when $T=\mathbb{Z}$. Take ν a discrete measure of the form $\nu(A)=\sum_k a_k\delta_k(A)$, where the a_k are nonnegative constants and δ_k is the Dirac measure on $\{k\}$. For $\mathcal{T}$ take all subsets of $\mathbb{Z}$. In this case, the Gaussian noise can actually be defined on points $t\in T$ and extended to a signed measure on sets in $\mathcal{T}_\nu$ by additivity. What we get is a collection $\{W_k\}_{k\in\mathbb{Z}}$ of independent, centered, Gaussian variables, with $\mathbb{E}\{W_k^2\}=a_k$. If the a_k are all equal, this is classical Gaussian "white" noise on the integers.

A more interesting case is $T=\mathbb{R}^N$, $\mathcal{T}=\mathcal{B}^N$, the Borel σ-algebra on $\mathbb{R}^N$, and $\nu(\cdot)=|\cdot|$, Lebesgue measure. This gives us a Gaussian white noise defined on the Borel subsets of $\mathbb{R}^N$ of finite Lebesgue measure, which is also a field with orthogonal increments, in the sense of (1.4.13). It is generally called the *set-indexed Brownian sheet*. It is not possible, in this case, to assign nontrivial values to given points $t\in\mathbb{R}^N$, as was the case in the previous example.

It also turns out that working with the Brownian sheet on all of $\mathcal{B}^N$ is not really the right thing to do, since, as will follow from Theorem 1.4.5, this process is rather badly behaved. Restricting the parameter space to various classes of subsets of $\mathcal{B}^N$ is the right approach. Doing so gives us a number of interesting examples, with which the remainder of this section is concerned.

As a first step, restrict W to rectangles of the form $[0,t]\subset\mathbb{R}_+^N$, where $t\in\mathbb{R}_+^N=\{(t_1,\dots,t_N):t_i\geq 0\}$. It then makes sense to define a random field on $\mathbb{R}_+^N$ itself via the equivalence

$$W(t)=W([0,t]). \tag{1.4.15}$$

W_t is called the Brownian sheet on $\mathbb{R}_+^N$, or *multiparameter Brownian motion*. It is easy to check that this field is the centered Gaussian process with covariance

$$\mathbb{E}\{W_sW_t\}=(s_1\wedge t_1)\times\cdots\times(s_N\wedge t_N), \tag{1.4.16}$$

where $s\wedge t\triangleq\min(s,t)$.

When $N=1$, W is the standard Brownian motion on $[0,\infty)$. When $N>1$, if we fix $N-k$ of the indices, it is a scaled k-parameter Brownian sheet in the remaining variables. (This is easily checked via the covariance function.) Also, when $N > 1$, it follows immediately from (1.4.16) that $W_t = 0$ when $\min_k t_k = 0$, i.e., when t is on one of the axes. It is this image, with $N = 2$, of a sheet tucked in at two sides and given a good shake, that led Ron Pyke [128] to introduce the name.

A simple simulation of a Brownian sheet, along with its contour lines, is shown in Figure 1.4.1.

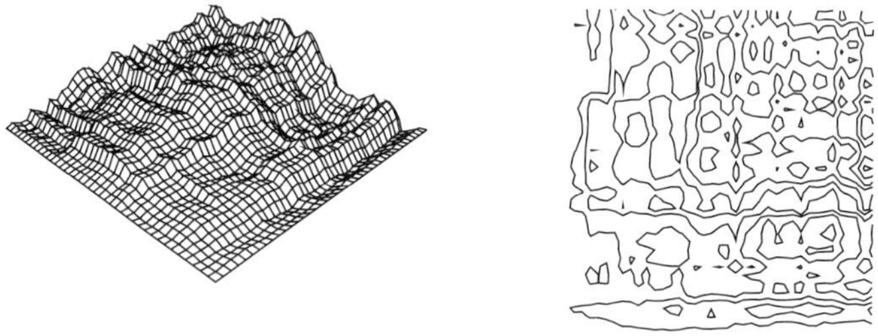

Fig. 1.4.1. A simulated Brownian sheet on $[0, 1]^2$, along with its contour lines at the zero level.

One of the rather interesting aspects of the contour lines of Figure 1.4.1 is that they are predominantly parallel to the axes. There is a rather deep reason for this, and it has generated a rather massive literature. Many fascinating geometrical properties of the Brownian sheet have been discovered over the years (e.g., [36, 44, 45] and references therein), and a description of the potential theoretical aspects of the Brownian sheet is well covered in [91], where you will also find more references. Nevertheless, the geometrical properties of fields of this kind fall beyond the scope of our interests, since we shall be concerned with the geometrical properties of smooth (i.e., at least differentiable) processes only. Since the Brownian motion on $\mathbb{R}^1_+$ is well known to be nondifferentiable at all points, it follows from the above comments relating the sheet to the one-dimensional case that Brownian sheets too are far from smooth.

Nevertheless, we shall still have need of these processes, primarily since they hold roughly the same place in the theory of multiparameter stochastic processes that the standard Brownian motion does in one dimension. The Brownian sheet is a multiparameter martingale (e.g., [36, 84, 170, 171]) and forms the basis of the multiparameter stochastic calculus. There is a nice review of its basic properties in [165], which also develops its central role in the theory of stochastic partial differential equations, and describes in what sense it is valid to describe the derivative

$$\frac{\partial^N W(t_1, \ldots, t_N)}{\partial t_1 \cdots \partial t_N}$$

as Gaussian white noise.

The most basic of the sample path properties of the Brownian sheets are the continuity results of the following theorem. Introduce a partial order on $\mathbb{R}^k$ by writing $s \prec (\preceq) t$ if $s_i < (\leq) t_i$ for all $i = 1, \ldots, k$, and for $s \preceq t$ let $\Delta(s, t) = \prod_1^N [s_i, t_i]$. Although $W(\Delta(s, t))$ is already well defined via the original set-indexed process, it is also helpful to think of it as the "increment" of the point-indexed W_t over $\Delta(s,t)$, that is,

$$W(\Delta(s,t)) = \sum_{\alpha \in \{0,1\}^N} (-1)^{N-\sum_{i=1}^N \alpha_i} W\left(s + \sum_{i=1}^N \alpha_i (t_i - s_i)\right). \qquad (1.4.17)$$

We call $W(\Delta(s,t))$ the *rectangle-indexed* version of W.

Theorem 1.4.4. *The point and rectangle-indexed Brownian sheets are continuous*[22] *over compact $T \subset \mathbb{R}^N$.*

Proof. We need only consider the point-indexed sheet, since by (1.4.17) its continuity immediately implies that of the rectangle-indexed version. Furthermore, we lose nothing by taking $T = [0,1]^N$. Thus, consider

$$w(s,t) \triangleq \mathbb{E}\left\{|W(s) - W(t)|^2\right\}$$

for $s, t \in [0,1]^N$. We shall show that

$$w(s,t) \le 2N|t-s|, \qquad (1.4.18)$$

from which it follows, in the notation of (1.4.1), that $p^2(u) \le 2Nu$. Since $\int_1^\infty e^{-u^2/2}\,du < \infty$, Theorem 1.4.1 (cf. (1.4.3)) immediately yields the continuity and boundedness of W.

To establish (1.4.18) write $u \vee v$ for $\max(u,v)$ and note that

$$w(s,t) \le 2\prod_{i=1}^N (s_i \vee t_i) - 2\prod_{i=1}^N (s_i \wedge t_i).$$

Set $a = 2\prod_{i=1}^{N-1}(s_i \vee t_i)$ and $b = 2\prod_{i=1}^{N-1}(s_i \wedge t_i)$. Then $2 \ge a > b$ and

$$w(s,t) \le a(s_N \vee t_N) - b(s_N \wedge t_N).$$

If $s_N > t_N$, the right-hand side is equal to

$$as_N - bt_N = a(s_N - t_N) + t_N(a-b) \le 2|s_N - t_N| + |a-b|.$$

Similarly, if $s_N < t_N$ the right-hand side equals

$$at_N - bs_N = a(t_N - s_N) + s_N(a-b) \le 2|t_N - s_N| + |a-b|,$$

so that

[22] For the rectangle-indexed case, we obviously need a metric on the sets forming the parameter space. The symmetric difference metric of (1.1.1) is natural, and so we use it.

$$w(s,t) \le 2|t_N - s_N| + \prod_{i=1}^{N-1}(s_i \vee t_i) - \prod_{i=1}^{N-1}(s_i \wedge t_i).$$

Continuing this process yields

$$w(s,t) \le 2\sum_{i=1}^{N} |t_i - s_i| \le 2N|t-s|,$$

which establishes (1.4.18) and so the theorem. □

In the framework of the general set-indexed sheet, Theorem 1.4.4 states that the Brownian sheet is continuous over $\mathcal{A} = \{\text{all rectangles in } T\}$ for compact T, and so bounded. This is far from a trivial result, for enlarging the parameter set, for the *same* process, can lead to unboundedness. The easiest way to see this is with an example.

An interesting, but quite simple example is given by the class of lower layers in $[0,1]^2$. A set A in $\mathbb{R}^N$ is called a *lower layer* if $s \prec t$ and $t \in A$ implies $s \in A$. In essence, restricted to $[0,1]^2$ these are sets bounded by the two axes and a nonincreasing line. A specific example in given in Figure 1.4.2, which is part of the proof of the next theorem.

Theorem 1.4.5. *The Brownian sheet on lower layers in* $[0,1]^2$ *is discontinuous and unbounded with probability one.*

Proof. We start by constructing some examples of lower layers. Write a generic point in $[0,1]^2$ as (s,t) and let T_{01} be the upper right triangle of $[0,1]^2$, i.e., those points for which which $s \le 1$ and $t \le 1 \le s+t$. Let C_{01} be the largest square in T_{01}, i.e., those points for which which $\frac{1}{2} < s \le 1$ and $\frac{1}{2} \le t \le 1$.

Continuing this process, for $n = 1, 2, \dots,$ and $j = 1, \dots, 2^n$, let T_{nj} be the right triangle defined by $s + t \ge 1$, $(j-1)2^{-n} \le s < j2^{-n}$, and $1 - j2^{-n} < t \le 1 - (j-1)2^{-n}$. Let C_{nj} be the square filling the upper right corner of T_{nj}, in which $(2j-1)2^{-(n+1)} \le s < j2^{-n}$ and $1 - (2j-1)2^{-(n+1)} \le t < 1 - (j-1)2^{-n}$.

The class of lower layers in $[0,1]^2$ certainly includes all sets made up by taking those points that lie between the axes and one of the step-like structures of Figure 1.4.2, where each step comes from the horizontal and vertical sides of some T_{nj} with, perhaps, different n.

Note that since the squares C_{nj} are disjoint for all n and j, the random variables $W(C_{nj})$ are independent. Also $|C_{nj}| = 4^{-(n+1)}$ for all n, j.

Let D be the negative diagonal $\{(s,t) \in [0,1]^2 : s+t = 1\}$ and $L_{nj} = D \cap T_{nj}$. For each $n \ge 1$, each point $p = (s,t) \in D$ belongs to exactly one such interval $L_{n,j(n,p)}$ for some unique $j(n,p)$.

For each $p \in D$ and $M < \infty$ the events

$$E_{np} \triangleq \{W(C_{n,j(n,p)}) > M2^{-(n+1)}\}$$

are independent for $n = 0, 1, 2, \dots,$ and since $W(C_{nj})/2^{-(n+1)}$ is standard normal for all n and j, they also have the same positive probability. Thus, for each p we

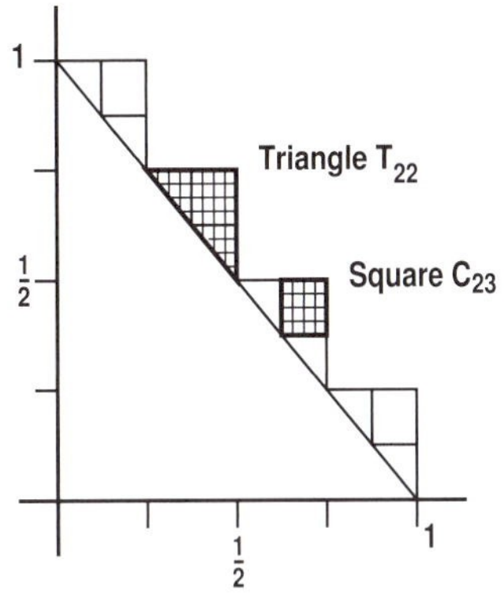

Fig. 1.4.2. Construction of some lower layers.

have that for almost all ω, the events E_{np} occur infinitely often. Let $n(p) = n(p, \omega)$ be the least such n.

Since the events $E_{np}(\omega)$ are measurable jointly in p and ω, Fubini's theorem implies that with probability one, for almost all $p \in D$ (with respect to Lebesgue measure on D) some E_{np} occurs, and $n(p) < \infty$. Let

$$V_\omega = \bigcup_{p \in D} T_{n(p), j(n(p), p)},$$

$$A_\omega \triangleq \{(s, t) : s + t \leq 1\} \cup V_\omega,$$

$$B_\omega \triangleq A_\omega \setminus \bigcup_{p \in D} C_{n(p), j(n(p), p)}.$$

Then A_ω and B_ω are lower layers. Furthermore, almost all $p \in D$ belong to an interval of length $2^{\frac{1}{2} - n(p)}$, which is the hypotenuse of a triangle with the square $C_p = C_{n(p), j(n(p), p)}$ in its upper right corner, for which $2W(C_p) > M2^{-n(p)}$. Consequently,

$$W(A_\omega) - W(B_\omega) \geq \sum_p M2^{-n(p)}/2,$$

where the sum is over those $p \in D$ corresponding to distinct intervals $L_{n(p), j(n(p), p)}$. Since the union of the countably many such intervals is almost all of the diagonal, the sum of $\sum 2^{-n(p)}$ is precisely 1.

Hence $W(A_\omega) - W(B_\omega) \geq M/2$, implying that $\max\{|W(A_\omega)|, |W(B_\omega)|\} \geq M/4$. Sending $M \to \infty$ we see that W is unbounded and so, a fortiori, discontinuous with probability one over lower layers in $[0, 1]^2$. □

The above argument is due to Dudley [54], and a similar argument shows that W is unbounded over the convex subsets of $[0, 1]^3$. A fortiori, W is also unbounded over the convex subsets of $[0, 1]^k$ for all $k \geq 4$ and (just to make sure that you don't confuse sample path properties with topological properties of the parameter space) W *is* continuous over convex subsets of the unit square. For details see [56].

These examples should be enough to convince you that the relationship between a Gaussian process and its parameter space is, as far as continuity and boundedness are concerned, an important and delicate subject.

1.4.4 Generalized Fields

We start with an example: Take a centered, Gaussian random field f on $\mathbb{R}^N$, with covariance function C. Let $\mathcal{F}$ be a family of functions on $\mathbb{R}^N$, and for $\varphi \in \mathcal{F}$ define

$$f(\varphi) = \int_{\mathbb{R}^N} \varphi(t) f(t)\, dt. \tag{1.4.19}$$

We thus obtain a centered Gaussian process indexed by functions in $\mathcal{F}$, whose covariance functional is given by

$$C(\varphi, \psi) = \mathbb{E}\{f(\varphi) f(\psi)\} = \int_{\mathbb{R}^N} \int_{\mathbb{R}^N} \varphi(s) C(s,t) \psi(t)\, ds\, dt. \tag{1.4.20}$$

The above construction makes sense only when $C(t,t) < \infty$, for otherwise f has infinite variance and the integrand in (1.4.19) is not defined. Nevertheless, there are occasions when (1.4.20) makes sense, even though $C(t,t) = \infty$. In this case we shall refer to C as a *covariance kernel*, rather than as a covariance function.

We shall look at a very important example of such processes in Section 5.3 when we consider moving averages of Gaussian ν-noise, in which case $\mathcal{F}$ will be made up of translations of the form $\varphi(s) = F(t-s)$, for $F \in L^2(\nu)$.

Another important example arises in the spectral representation of stationary processes. In Section 5.4 we shall take $\mathcal{F}$ to the family of complex exponentials $\exp(it \cdot \lambda)$ for fields on $\mathbb{R}^N$, or as the family of characters for fields on a general group.[23]

While moving averages and stationary processes afford two classes of examples, there are many more, some of which we shall describe at the end of this section. In particular, given any positive definite function C on $\mathbb{R}^N \times \mathbb{R}^N$, not necessarily finite, one can define a *function-indexed* process on

[23] On the assumption that you might already be familiar with basic spectral theory, consider the spectral distribution theorem, Theorem 5.4.1. This is written in the setting of complex-valued fields. For simplicity, assume that the spectral measure ν has a spectral density g. Then (5.4.1) gives us that stationary covariances over $\mathbb{R}^N$ can be formally written as

$$\begin{aligned} C(s,t) &= \int_{\mathbb{R}^N} e^{i(t-s)\cdot\lambda} g(\lambda)\, d\lambda \\ &= \int_{\mathbb{R}^N} \int_{\mathbb{R}^N} e^{it\cdot\lambda_1} \left[g^{1/2}(\lambda_1) \delta(\lambda_1, \lambda_2) g^{1/2}(\lambda_2) \right] e^{-is\cdot\lambda_2}\, d\lambda_1\, d\lambda_2, \end{aligned}$$

which is in the form of (1.4.20), with $\varphi = e^{it\cdot}$, $\psi = e^{-is\cdot}$. The covariance kernel in the integrand now involves the Dirac delta function, which is certainly not finite on the diagonal. It is this issue that will lead to our having to be careful in defining the stochastic integral in the spectral representation of Theorem 5.4.2.

$$\mathcal{F}_C \triangleq \left\{ \varphi : \int_{\mathbb{R}^N} \int_{\mathbb{R}^N} \varphi(s) C(s,t) \varphi(t) \, ds \, dt < \infty \right\}.$$

The proof requires no more than checking that given such a C on $\mathbb{R}^N \times \mathbb{R}^N$, the corresponding C defined by (1.4.20) determines a *finite* positive definite, and so covariance, function on $\mathcal{F}_C \times \mathcal{F}_C$.

In general, function-indexed processes of this kind, for which the covariance kernel $C(s,t)$ is infinite on the diagonal, are known as *generalized random fields*.[24]

The question that we shall now look at is when such processes are continuous and bounded. The answer involves a considerable amount of work, but it is worthwhile at some stage to go through the argument carefully. It is really the only non-Euclidean case for which we shall give an involved entropy calculation in reasonably complete detail.

We start by describing potential function spaces to serve as parameter spaces for continuous generalized Gaussian fields with such covariance kernels.

Take $T \subset \mathbb{R}^N$ compact, $q > 0$, and $p = \lfloor q \rfloor$. Let $C_0, \ldots, C_p$ and C_q be finite, positive constants, and let $\mathcal{F}^{(q)} = \mathcal{F}^{(q)}(T, C_0, \ldots, C_p, C_q)$ be the class of functions on T whose partial derivatives of orders $1, \ldots, p$ are bounded by $C_0, \ldots, C_p$, and for which the partial derivatives of order p satisfy Hölder conditions of order $q - p$ with constant C_q. Thus for each $\varphi \in \mathcal{F}^{(q)}$ and $t, t+\tau \in T$,

$$\varphi(t+\tau) = \sum_{n=0}^{p} \frac{\Phi_n(t,\tau)}{n!} + \Delta(t,\tau), \tag{1.4.21}$$

where each $\Phi_n(t,\tau)$ is a homogeneous polynomial of degree n in $\tau = (\tau_1, \ldots, \tau_N)$ of the form

---

[24] The terminology comes from deterministic analogues. There are many partial differential equations for which no pointwise solution exists, but "smoothed" or "weak" versions of the equations do have solutions. This is analogous to the nonexistence of a pointwise-defined $f(t)$ in (1.4.19), while a function-indexed version of f does make sense. The function φ plays the role of a local mollifier, or smoothing function.

A particularly important class of examples, arising in physics, is given by covariance kernels whose behavior near the diagonal is bounded, for $\alpha < N$, by

$$C(s,t) \le \frac{C}{|s-t|^\alpha}.$$

When $\alpha = N-2$, $N > 2$, these kernels include the so-called free field of Euclidean quantum field theory. (When $N = 2$ the free field has a covariance kernel with a logarithmic singularity at 0, and when $N = 1$ the free field is no longer generalized, but is the real-valued, stationary Markov Gaussian process with covariance function $C(t) = e^{-\beta|t|}$, for some $\beta > 0$.) These processes, along with a large number of related generalized fields whose covariance kernels satisfy similar conditions, possess a type of multidimensional Markov property. For details on this see, for example, Dynkin [57, 58, 59, 60] and Adler and Epstein [6] and references therein. For structural and renormalization properties of generalized fields of this kind, presented among a much wider class of examples, see Dobrushin [49], who also treats a large variety of non-Gaussian fields.

$$\Phi_n(t,\tau) = \sum_{j_1=1}^{N} \cdots \sum_{j_n=1}^{N} \frac{\partial^n \varphi(t)}{\partial t_{j_1} \cdots \partial t_{j_n}} \tau_{j_1} \cdots \tau_{j_n}, \tag{1.4.22}$$

and where

$$\sup_{t\in T} \left| \frac{\partial^n \varphi(t)}{\partial t_{j_1} \cdots \partial t_{j_n}} \right| \le C_n \quad \text{and} \quad |\Delta(t,\tau)| \le C_q |\tau|^q. \tag{1.4.23}$$

For these parameter spaces, we have the following result.[25]

Theorem 1.4.6. *Let $T \subset \mathbb{R}^N$ be compact, and suppose that C is a covariance kernel satisfying*

$$\int_{T\times T} C(s,t)\, ds\, dt < \infty.$$

A centered Gaussian process with covariance function satisfying (1.4.20) *will be continuous on $\mathcal{F}^{(q)}(T, C_0, \ldots, C_p, C_q)$ if $q > N$.*

A couple of comments are in order before we start the proof. Firstly, note that since we have not specified any other metric on $\mathcal{F}^{(q)}$, the continuity claim of the theorem is in relation to the topology induced by the canonical metric d. There are, of course, more natural metrics on $\mathcal{F}^{(q)}$, but recall from Lemma 1.3.1 that mere continuity is independent of the metric, as long as $C(\varphi, \psi)$ is continuous in φ and ψ. More detailed information on moduli of continuity will follow immediately from Theorem 1.3.5, the relationship of the chosen metric to d, and the entropy bound (1.4.32) below.

Secondly, the condition $q > N$, while sufficient, is far from necessary. To see this one need only take the case of $f(t) \equiv X$, where X is centered Gaussian. In this case, defining $f(\varphi)$ by (1.4.19) gives a generalized process that is continuous over $\mathcal{F}^{(q)}$ for all $q > 0$, regardless of the dimension N. To obtain sharper results one needs to assume more on the specific form of the covariance kernel, which we shall not do here.

Proof of Theorem 1.4.6. The proof relies on showing that the usual entropy integral converges, where the entropy of $\mathcal{F}^{(q)}$ is measured in the canonical metric d given by

$$d^2(\varphi,\psi) = \int_T \int_T (\varphi(s) - \psi(s))\, C(s,t)\, (\varphi(t) - \psi(t))\, ds\, dt. \tag{1.4.24}$$

We shall obtain a bound on the entropy by explicitly constructing, for each $\varepsilon > 0$, a finite family $\mathcal{F}_\varepsilon^{(q)}$ of functions that serve as an ε-net for $\mathcal{F}^{(q)}$ in the d metric. To make life notationally easier, we shall assume throughout that $T = [0,1]^N$.

Fix $\varepsilon > 0$ and set

$$\delta = \delta(\varepsilon) = \varepsilon^{1/q}. \tag{1.4.25}$$

Let $\mathbb{Z}_\delta$ denote the grid of the $(1 + \lfloor \delta^{-1} \rfloor)^N$ points in $[0,1]^N$ of the form

[25] There is a similar-looking but incorrect result ("Theorem" 1.7) in the lecture notes [3]. We thank Leonid Mytnik for pointing out that it had to be wrong.

$$t_\eta = (\eta_1\delta, \ldots, \eta_N\delta), \quad \eta_i = 0, 1, \ldots, \lfloor \delta^{-1} \rfloor, \ i = 1, \ldots, N. \tag{1.4.26}$$

Set

$$\delta_n = \delta_n(\varepsilon) = \delta^{q-n}, \qquad n = 0, \ldots, p = \lfloor q \rfloor,$$

and for each $\varphi \in \mathcal{F}^{(q)}$, $n = 0, \ldots, p$, and t_η of the form (1.4.26) let $A_\eta^{(n)}(\varphi)$ denote the vector formed by taking the integer part of δ_n^{-1} times the partial derivatives of φ of order n evaluated at the point t_η. (The index η here is, of course, N-dimensional.) Thus, a typical element of $A_\eta^{(n)}$ is of the form

$$\left\{A_\eta^{(n)}(\varphi)\right\}_i = \left\lfloor \frac{\varphi^{(n_1,\ldots,n_N)}(t_\eta)}{\delta_n} \right\rfloor, \quad n_1 + \cdots + n_N = n, \tag{1.4.27}$$

where we have written $\varphi^{(n_1,\ldots,n_N)}$ for the derivative $\partial^n\varphi/\partial^{n_1}x_1 \cdots \partial^{n_N}x_N$, and the index i runs from 1 to $\binom{n+N-1}{N-1}$, the number of partitions of n into N parts.

Finally, for $\varphi \in \mathcal{F}^{(q)}$, let $A^{(n)} = A^{(n)}(\varphi)$ denote the vector-valued function on $\mathbb{Z}_\delta$ defined by $A^{(n)}(t_\eta) = A_\eta^{(n)}(\varphi)$. For each $\varphi \in \mathcal{F}^{(q)}$, let $F_{A^{(n)}(\varphi)}$ denote the set of $\psi \in \mathcal{F}^{(q)}$ with fixed matrix $A^{(n)}(\varphi)$. Our first task will be to show that the d-radius of $F_{A^{(n)}(\varphi)}$ is not greater than $C\varepsilon$, where C is a constant dependent only on q and N. All that will then remain will be to calculate how many different collections $F_{A^{(n)}(\varphi)}$ are required to cover $\mathcal{F}^{(q)}$. In other words, we need to find how many φ's are needed to approximate, in terms of the metric (1.4.24), all functions in $\mathcal{F}^{(q)}$.

Thus, take $\varphi_1, \varphi_2 \in F_{A^{(n)}(\varphi)}$, and set

$$\varphi = \varphi_1 - \varphi_2. \tag{1.4.28}$$

Let $\|\ \|_d$ be the norm induced on $\mathcal{F}^{(q)}$ by the metric d, and $\|\ \|_\infty$ the usual sup norm. Then

$$\|\varphi\|_d^2 = \int_{[0,1]^N}\int_{[0,1]^N} \varphi(s)C(s,t)\varphi(t)\,ds\,dt, \qquad \|\varphi\|_\infty = \sup_{[0,1]^k} |\varphi(t)|.$$

We want to show that the φ of (1.4.28) has d-norm less than $C\varepsilon$, for some finite constant C.

Note first, however, that in view of the definition (1.4.27) of the matrix A_δ we have for each $t_\eta \in \mathbb{Z}_\delta$ and each partial derivative $\varphi^{(n_1,\ldots,n_N)}$ of such a φ of order $n_1 + \cdots + n_N = n \le p$ that

$$\varphi^{(n_1,\ldots,n_N)}(t_\eta) \le \delta_n.$$

Putting this inequality together with the Taylor expansion (1.4.21)–(1.4.22), we find that for all $t \in [0, 1]^N$,

$$|\varphi(t)| \le \sum_{n=0}^{p} \frac{N^n}{n!}\delta_n\delta^n + C\delta^q = C(N, p)\delta^q,$$

the equality following from the definition of the δ_n and the fact that each polynomial Φ_n of (1.4.22) has less than N^n distinct terms.

Thus, for φ of the form (1.4.28),

$$\|\varphi\|_\infty \le C\delta^q. \tag{1.4.29}$$

We now turn to $\|\varphi\|_d$. Allowing the constant C to change from line to line, we have

$$\begin{aligned}\|\varphi\|_d^2 &= \int_{[0,1]^N\times[0,1]^N} \varphi(s)C(s,t)\varphi(t)\,ds\,dt \\ &\le C\delta^{2q}\int_{[0,1]^N\times[0,1]^N} C(s,t)\,ds\,dt \\ &\le C\delta^{2q}.\end{aligned}$$

Thus, by (1.4.25),

$$\|\varphi\|_d < C\varepsilon, \tag{1.4.30}$$

as required.

It remains to determine how many collections $F_{A^{(n)}(\varphi)}$ are required to cover $\mathcal{F}^{(q)}$. Since this is a calculation that is now independent of both Gaussian processes in general and the above covariance function in particular, we shall only outline how it is done. The details, which require somewhat cumbersome notation, can be found in Kolmogorov and Tihomirov [94], which is a basic reference for general entropy computations.

Consider, for fixed δ, the matrix A_δ, parameterized, as in (1.4.26), by $\eta_i = 0, 1, \dots, [\delta^{-1}]$, $i = 1, \dots, N$, and $n = 0, 1, \dots, p$. Fix, for the moment, $\eta_2 = \cdots = \eta_N = 0$. It is clear from the restrictions (1.4.23), (1.4.27), the definition of δ_n, and the fact that each vector $A_\eta^{(n)}$ has no more than $\binom{n+N-1}{N-1}$ distinct elements that there are no more than

$$O\left(\frac{1}{\delta_0}\left(\frac{1}{\delta_1}\right)^{\binom{N}{N-1}}\cdots\left(\frac{1}{\delta_p}\right)^{\binom{p+N}{N-1}}\right) = O\left(\delta^{-\xi}\right)$$

(for an appropriate and eventually unimportant ξ) ways to fill in the row of A_δ corresponding to $(n_1, \dots, n_N) = (0, \dots, 0)$.

What remains to show is that because of the rigid continuity conditions on the functions in $\mathcal{F}^{(q)}$, there exists an absolute constant $M = M(q, C_0, \dots, C_p, C_q)$ such that once this first row is determined, there are no more than M ways to complete the row corresponding to $(n_1, \dots, n_N) = (1, \dots, 0)$, and similarly no more than M^2 ways to complete the row corresponding to $(n_1, \dots, n_N) = (2, \dots, 0)$, etc. Thus, all told, there are no more than

$$O(\delta^{-\xi} M^{(1+\delta^{-1})^N}) \tag{1.4.31}$$

ways to fill the matrix A_δ, and so we have a bound for the number of different collections $F_{A^{(n)}(\varphi)}$.

Modulo a constant, it now follows from (1.4.25), (1.4.30), and (1.4.31) that the log entropy function for our process is bounded above by

$$C_1 + \frac{C_2\xi}{(q + (N - \alpha)/2)} \ln\left(\frac{1}{\varepsilon}\right) + C_3\left(\frac{1}{\varepsilon}\right)^{1/q}. \tag{1.4.32}$$

Since this is integrable if $q > N$, we are done. □

Before leaving function-indexed processes, there are a number of comments that are worth making that relate them to other problems both within and outside of the theory of Gaussian processes.

In most of the literature pertaining to generalized Gaussian fields the parameter space used is the Schwarz space $\mathcal{S}$ of infinitely differentiable functions decaying faster than any polynomial at infinity. Since this is a very small class of functions (at least in comparison to the classes $\mathcal{F}^{(q)}$ that Theorem 1.4.6 deals with), continuity over $\mathcal{S}$ is automatically assured and therefore not often explicitly treated. However, considerations of continuity and smaller parameter spaces are of relevence in the treatment of infinite-dimensional diffusions arising as the solutions of stochastic partial differential equations, in which solutions over very specific parameter spaces are often sought. For more on this see, for example, [82, 166, 165].

A class of examples of particular importance to the theory of empirical processes is that in which the covariance kernel is the product of a Dirac delta δ and a bounded "density," g, in the sense that

$$\begin{aligned}\mathbb{E}\{f(\varphi)f(\psi)\} &= \int \varphi(t)\psi(t)g(t)\,dt \\ \text{“}{=}\text{”} &\iint \varphi(s)[g^{1/2}(s)\delta(s,t)g^{1/2}(t)]\psi(t)\,dt.\end{aligned}$$

Such processes will arise as the stochastic integrals $W(\varphi)$ in Section 5.2, where W is a Gaussian ν-noise where ν is (now) a probability measure with density g. For more on this setting, in which the computations are similar in spirit to those made above, see Dudley [56].

Secondly, it is worth noting that much of what has been said above regarding generalized fields—i.e., function-indexed processes—can be easily extended to Gaussian processes indexed by families of measures. For example, if we consider the function φ in (1.4.19) to be the (positive) density of a measure μ on $\mathbb{R}^N$, then by analogy it makes sense to write

$$f(\mu) = \int_{\mathbb{R}^N} f(t)\mu(dt),$$

with the corresponding covariance functional

$$C(\mu,\nu) = \mathbb{E}\{f(\mu)f(\nu)\} = \int_{\mathbb{R}^N}\int_{\mathbb{R}^N} \mu(ds)C(s,t)\nu(dt).$$

Again, as was the case for generalized Gaussian fields, the process $X(\mu)$ may be well defined even if the covariance kernel C diverges on the diagonal. In fact, $f(\mu)$ will be well defined for all $\mu \in M_C$, where

$$M_C \triangleq \left\{ \mu : \int_{\mathbb{R}^N} \int_{\mathbb{R}^N} \mu(ds) C(s,t) \mu(dt) < \infty \right\}.$$

Similar arguments to those used above to characterize the continuity of a family of Gaussian fields on $\mathcal{F}^{(q)}$ can be used to ascertain continuity of measure-indexed processes on suitably smooth classes of measures. We leave both the details and an attempt to formulate the appropriate results to the interested reader. The general idea of how to proceed can be gleaned from the treatment of set-indexed processes in the following section.

1.4.5 Set-Indexed Processes

We have already met some set-indexed processes in dealing with the Brownian family of processes in Section 1.4.3, in which we concentrated on Gaussian ν-noise (cf. (1.4.11)–(1.4.13)) indexed by sets. We saw, for example, that while the Brownian sheet indexed by rectangles was continuous (Theorem 1.4.4), when indexed by lower layers in $[0, 1]^2$ it was discontinuous and unbounded (Theorem 1.4.5).

In this section we shall remain in the setting of ν-noise, and look at two classes of set-indexed processes. The first will be Euclidean, and the parameter spaces will be classes of sets in compact $T \subset \mathbb{R}^N$ with smooth boundaries. For this example we also add the assumption that ν has a density bounded away from zero and infinity on T. For the second eaxmple, we look at Vapnick–Červonenkis classes of sets, of singular importance in statistical learning theory and image processing, and characterized by certain combinatorial properties. Here the ambient space (in which the sets lie) can be any measure space. We shall skimp on proofs when they have nothing qualitatively new to offer. In any case, all that we have to say is done in full detail in Dudley [53, 56], where you can also find a far more comprehensive treatment and many more examples.

Our first family is actually closely related to the family $\mathcal{F}^{(q)}$ of functions we have just studied in detail. While we shall need some of the language of manifolds and homotopy to describe this example, which will be developed only later, in Chapter 6, it will be basic enough that the average reader should have no trouble following the argument.

With S^{N-1} denoting the unit sphere in $\mathbb{R}^N$, recall the basic fact that we can cover it by two patches V_1 and V_2, each of which maps via a C^∞ diffeomorphism $F_j : V_j \to B^{N-1}$ to the open ball $B^{N-1} = \{t \in \mathbb{R}^{N-1} : |t|^2 < 1\}$.

Adapting slightly the notation of the previous example, let $\mathcal{F}^{(q)}(V_j, M)$ be the set of all real-valued functions φ on V_j such that

$$\varphi \circ F_j^{-1} \in \mathcal{F}^{(q)}(B^{N-1}, M, \dots, M)$$

(cf. (1.4.21)–(1.4.23)). Furthermore, let $\mathcal{F}^{(q)}(S^{N-1}, M)$ denote the set of all real-valued functions φ on S^{N-1} such that the restriction of φ to V_j is in $\mathcal{F}^{(q)}(V_j, M)$. Now taking the N-fold Cartesian product of copies of $\mathcal{F}^{(q)}(S^{N-1}, M)$, we obtain a family of functions from S^{N-1} to $\mathbb{R}^N$, which we denote by $D(N, q, M)$, where the D stands for Dudley, who introduced this family in [52].

Each $\varphi \in D(N, q, M)$ defines an $(N-1)$-dimensional surface in $\mathbb{R}^N$, and a simple algebraic geometric construction[26] enables one to "fill in" the interior of this surface to obtain a set I_φ. We shall denote the family of sets obtained in this fashion by $I(N, q, M)$, and call them the *Dudley sets with q-times differentiable boundaries*.

Theorem 1.4.7. *The Brownian sheet is continuous on a bounded collection of Dudley sets in $\mathbb{R}^N$ with q-times differentiable boundaries if $q > N-1 \geq 1$. If $N-1 > 1 \geq q > 0$ or if $N-1 > q \geq 1$, then the Brownian sheet is unbounded with probability* 1.

Outline of proof. The proof of the unboundedness part of the result is beyond us, and so you are referred to [56]. As far as the proof of continuity is concerned, what we need are the following inequalities for the log-entropy, on the basis of which continuity follows from Theorem 1.3.5:

$$H(I(N, q, M), \varepsilon) \leq \begin{cases} C\varepsilon^{-2(N-1)/(Nq-N+1)}, & (N-1)/N < q \leq 1, \\ C\varepsilon^{-2(N-1)/q}, & 1 \leq q. \end{cases}$$

These inequalities rely on the "simple algebraic geometric construction" noted above, and so we shall not consider them in detail. The details are in [56]. The basic idea, however, requires little more than noting that there are basically as many sets in $I(N, q, M)$ as there are functions in $D(N, q, M)$, and we have already seen, in the previous example, how to count the number of functions in $D(N, q, M)$.

There are also equivalent lower bounds for the log-entropy for certain values of N and q, but these are not important to us. □

We now turn to the so-called *Vapnick–Červonenkis*, or VC, sets, due, not surprisingly, to Vapnick and Červonenkis [163, 164]. These sets arise in a very natural way in many areas including statistical learning theory and image analysis. The recent book [162] by Vapnick is a good place to see why.

The arguments involved in entropy calculations for VC classes of sets are of an essentially combinatoric nature, and so somewhat different from those we have met so far. We shall therefore look at them somewhat more closely than we did for Dudley sets. For more details, however, including a discussion of the importance of VC classes to the problem of finding universal Donsker classes in the theory of empirical processes, see [56].

Let $(E, \mathcal{E})$ be a measure space. Given a class $\mathcal{C}$ of sets in $\mathcal{E}$ and a finite set $F \subset E$, let $\Delta^{\mathcal{C}}(F)$ be the number of different sets $A \cap F$ for $A \in \mathcal{C}$. If the number of such sets is $2^{|F|}$, then $\mathcal{C}$ is said to *shatter* F. For $n = 1, 2, \ldots,$ set

$$m^{\mathcal{C}}(n) \stackrel{\Delta}{=} \max\{\Delta^{\mathcal{C}}(F) : F \text{ has } n \text{ elements}\}.$$

---

[26] The construction works as follows: For given $\varphi : S^{N-1} \to \mathbb{R}^N$ in $D(N, q, M)$, let R_φ be its range and A_φ the set of all $t \in \mathbb{R}^N$, $t \notin R_\varphi$, such that among mappings of S^{N-1} into $\mathbb{R}^N \setminus \{t\}$, φ is not homotopic (cf. Definition 6.1.3) to any constant map $\psi(s) \equiv r \neq t$. Then define $I_\varphi = R_\varphi \cup A_\varphi$. For an example, try untangling this description for φ the identity map from S^{N-1} to itself to see that what results is $I_\varphi = B^N$.

Clearly, $m^{\mathcal{C}}(n) \le 2^n$ for all n. Also, set

$$V(\mathcal{C}) \triangleq \begin{cases} \inf\{n : m^{\mathcal{C}}(n) < 2^n\}, & m^{\mathcal{C}}(n) < 2^n \text{ for some } n, \\ \infty, & m^{\mathcal{C}}(n) = 2^n \text{ for all } n. \end{cases} \tag{1.4.33}$$

The class $\mathcal{C}$ is called a Vapnick–Červonenkis class if $m^{\mathcal{C}}(n) < 2^n$ for *some* n, i.e., if $V(\mathcal{C}) < \infty$. The number $V(\mathcal{C})$ is called the VC index of $\mathcal{C}$, and $V(\mathcal{C}) - 1$ is the largest cardinality of a set shattered by $\mathcal{C}$.

Two extreme but easy examples that you can check for yourself are $E = \mathbb{R}$ and $\mathcal{C}$ all half-lines of the form $(-\infty, t]$, for which $m^{\mathcal{C}}(n) = n + 1$ and $V(\mathcal{C}) = 2$, and $E = [0, 1]$ with $\mathcal{C}$ all the open sets in $[0, 1]$. Here $m^{\mathcal{C}}(n) = 2^n$ for all n, and so $V(\mathcal{C}) = \infty$ and $\mathcal{C}$ is not a VC class.

A more instructive example, which also leads into the general theory we are after, is $E = \mathbb{R}^N$ and $\mathcal{C}$ is the collection of half-spaces of $\mathbb{R}^N$. Let $\Phi(N, n)$ be the maximal number of components into which it is possible to partition $\mathbb{R}^N$ via n hyperplanes. Then, by definition, $m^{\mathcal{C}}(n) = \Phi(N, n)$. It is not hard to see that Φ must satisfy the recurrence relation

$$\Phi(N, n) = \Phi(N, n-1) + \Phi(N-1, n-1), \tag{1.4.34}$$

with the boundary conditions $\Phi(0, n) = \Phi(N, 0) = 1$.

To see this, note that if $\mathbb{R}^N$ has already been partitioned into $\Phi(N, n-1)$ subsets via $n-1$ ($(N-1)$-dimensional) hyperplanes, $H_1, \dots, H_{n-1}$, then adding one more hyperplane H_n will cut in half as many of these subsets as intersect H_n. There can be no more such subsets, however, than the maximal number of subsets formed on H_n by partitioning with the $n-1$ $(N-2)$-dimensional hyperplanes $H_1 \cap H_n, \dots, H_{n-1} \cap H_n$; i.e., $\Phi(N-1, n-1)$. Hence (1.4.34).

Induction then shows that

$$\Phi(N, n) = \begin{cases} \sum_{j=0}^{N} \binom{n}{j}, & n > N, \\ 2^n, & n \le N, \end{cases}$$

where we adopt the usual convention that $\binom{n}{j} = 0$ if $n < j$.

From either the above or (1.4.34) you can now check that

$$\Phi(N, n) \le n^N + 1 \quad \text{for all } N, n > 0. \tag{1.4.35}$$

It thus follows, from (1.4.33), that the half-spaces of $\mathbb{R}^N$ form a VC class for all N.

What is somewhat more surprising, however, is that an inequality akin to (1.4.35), which we developed only for this special example, holds in wide (even non-Euclidean) generality.

Lemma 1.4.8. *Let $\mathcal{C}$ be a collection of sets in $\mathcal{E}$ such that $V(\mathcal{C}) \le v$. Then*

$$m^{\mathcal{C}}(n) < \Phi(v, n) \le n^v + 1 \quad \textit{for all } n \ge v. \tag{1.4.36}$$

Since the proof of this result is combinatoric rather than probabilistic, and will be of no further interest to us, you are referred to either of [163, 56] for a proof.

The importance of Lemma 1.4.8 is that it enables us to obtain bounds on the entropy function for Gaussian ν-noise over VC classes that are independent of ν.

Theorem 1.4.9. *Let W be the Gaussian ν-noise on a probability space $(E, \mathcal{E}, \nu)$. Let $\mathcal{C}$ be a Vapnick–Červonenkis class of sets in $\mathcal{E}$ with $V(\mathcal{C}) = v$. Then there exists a constant $K = K(v)$ (not depending on ν) such that for $0 < \varepsilon \le \frac{1}{2}$, the entropy function for W satisfies*

$$N(\mathcal{C}, \varepsilon) \le K\varepsilon^{-2v} |\ln \varepsilon|^{v}.$$

Proof. We start with a little counting, and then turn to the entropy calculation proper. The counting argument is designed to tell us something about the maximum number of $\mathcal{C}$ sets that are a certain minimum distance from one another and can be packed into E.

Fix $\varepsilon > 0$ and suppose $A_1, \dots, A_m \in \mathcal{C}$, $m \ge 2$, and $\nu(A_i \Delta A_j) \ge \varepsilon$ for $i \ne j$. We need an upper bound on m. Sampling with replacement, select n points at random from E. The ν-probability that at least one of the sets $A_i \Delta A_j$ contains none of these n points is at most

$$\binom{m}{2}(1-\varepsilon)^n. \tag{1.4.37}$$

Choose $n = n(m, \varepsilon)$ large enough so that this bound is less than 1. Then

$$P\{\text{all symmetric differences } A_i \Delta A_j \text{ are nonempty}\} > 0,$$

and so for at least one configuration of the n sample points the class $\mathcal{C}$ picks out at least m distinct subsets (since, with positive probability, given any two of the A_i there is at least one point not in both of them). Thus, by (1.4.36),

$$m \le m^{\mathcal{C}}(n) \le n^v = \big(n(m,\varepsilon)\big)^v. \tag{1.4.38}$$

Take now the smallest n for which (1.4.37) is less than 1. For this n we have $m^2(1-\varepsilon)^{n-1} \ge 2$, so that

$$n - 1 \le \frac{2\ln m - \ln 2}{|\ln(1-\varepsilon)|},$$

and $n \le (2\ln m)/\varepsilon$. Furthermore, by (1.4.38), $m \le (2\ln m)^v \varepsilon^{-v}$.

For some $m_0 = m_0(v) < \infty$, $(2\ln m)^v \le m^{1/(v+1)}$ for $m \ge m_0$, and then $m \le \varepsilon^{-v-1}$, so $\ln m \le (v+1)|\ln \varepsilon|$. Hence

$$m \le K(v)\varepsilon^{-v}|\ln \varepsilon|^v \quad \text{for } 0 < \varepsilon \le \frac{1}{2},$$

if $K(v) = \max(m_0, 2^{v+1}(v+1)^v)$.

This concludes the counting part of the proof. We can now do the entropy calculation. Recall that the canonical distance between sets in $\mathcal{E}$ is given by $d_\nu(A, B) = [\nu(A \Delta B)]^{\frac{1}{2}}$.

Fix $\varepsilon > 0$. In view of the above, there can be no more than $m = K(v)\varepsilon^{2v}|2 \ln \varepsilon|^v$ sets $A_1, \dots, A_m$ in $\mathcal{C}$ for which $d_\nu(A_i, A_j) \geq \varepsilon$ for all i, j. Take an ε-neighborhood of each of the A_i in the d_ν metric. (Each such neighborhood is a collection of sets in $\mathcal{E}$.) The union of these neighborhoods covers $\mathcal{C}$, and so we have constructed an ε-net of the required size, and are done. □

An immediate consequence of the entropy bound of Theorem 1.4.9 is the following.

Corollary 1.4.10. *Let W be Gaussian ν-noise based over a probability space $(E, \mathcal{E}, \nu)$. Then W is continuous over any Vapnick–Červonenkis class of sets in $\mathcal{E}$.*

For more about set-indexed processes, particularly from the martingale viewpoint, see [84].

1.4.6 Non-Gaussian Processes

A natural question to ask is whether the results and methods that we have seen in this section extend naturally to non-Gaussian fields. We already noted, immediately after the proof of the central Theorem 1.3.5, that the proof there used normality only once, and so the general techniques of entropy should be extendable to a far wider setting.

For most of the processes that will concern us, this will not be terribly relevant. In Chapter 15 we shall concentrate on random fields that can be written in the form

$$f(t) = F(g^1(t), \dots, g^d(t)),$$

where the g^i are i.i.d. Gaussian and $F : \mathbb{R}^d \to \mathbb{R}$ is smooth. In this setting, continuity and boundedness of the non-Gaussian f follow deterministically from similar properties on F and the g^i, and so no additional theory is needed.

Nevertheless, there are many processes that are not attainable in this way, for which one might a priori expect that the random geometry of Part III of this book might apply. In particular, we are thinking of the smooth stable fields of Samorodnitsky and Taqqu [138]. With this in mind, and for completeness, we state Theorem 1.4.11 below. However, other than the "function of Gaussian" non-Gaussian scenario, we know of no cases for which this random geometry has even the beginnings of a parallel theory.

To set up the basic result, let f_t be a random process defined on a metric space (T, τ) and taking values in a Banach space $(B, \|\ \|_B)$. Since we are no longer in the Gaussian case, there is no reason to assume that there is any longer a canonical metric of T to replace τ. Also, recall that a function $\varphi : \mathbb{R} \to \mathbb{R}$ is called a *Young function* if it is even, continuous, convex, and satisfies

$$\lim_{x \to 0} \frac{\varphi(x)}{x} = 0, \qquad \lim_{x \to \infty} \frac{\varphi(x)}{x} = \infty.$$

Theorem 1.4.11. *Take f as above, and assume that the real-valued process $\|f_t - f_s\|_B$ is separable. Let N_τ be the metric entropy function for T with respect to the metric τ. If there exists an $\alpha \in (0, 1]$ and a Young function φ such that the following two conditions are satisfied, then f is continuous with probability one:*

$$\mathbb{E}\left\{\varphi\left(\frac{\|f(t) - f(s)\|_B^\alpha}{\tau(s,t)}\right)\right\} \leq 1,$$

$$\int_{N_\tau(u)>1} \varphi^{-1}\left(N_\tau(u)\right)\, du < \infty.$$

The best place to read about this is in Ledoux and Talagrand [99].

1.5 Majorizing Measures

Back in Section 1.3 we used the notion of entropy integrals to determine sufficient conditions for the boundedness and continuity of Gaussian fields. We claimed there that these arguments were sharp for stationary processes, but need not be sharp in general. That is, there are processes that *are* a.s. continuous, but whose entropy integrals diverge. We also noted that the path to solving these issues lay via *majorizing measures*, and so we shall now explain what these are, and how they work.

Our plan here is to give only the briefest of introductions to majorizing measures. You can find the full theory in the book by Ledoux and Talagrand [99] and the more recent papers and book [152, 154, 155, 156] by Talagrand. In particular, [154] gives a very user friendly introduction to the subject, and the proof of Theorem 1.5.1 below is taken from there.

We include this section for mathematical completeness[27] and will not use its results anywhere else in the book. It is here mainly to give you an idea of how to improve on the weaknesses of entropy arguments, and to whet your appetite to turn to the sources for more. Here is the main result.

Theorem 1.5.1. *If f is a centered Gaussian process on (d-compact) T, then there exists a universal constant K such that*

$$\mathbb{E}\Big\{\sup_{t\in T} f_t\Big\} \leq K \inf_\mu \sup_{t\in T} \int_0^\infty \sqrt{\ln \frac{1}{\mu(B_d(t,\varepsilon))}}\, d\varepsilon, \tag{1.5.1}$$

where B_d is the d-ball of (1.3.2) *and the infimum on μ is taken over all probability measures μ on T.*

Furthermore, if f is a.s. bounded, then there exists a probability measure μ and a universal constant K' such that

$$K' \sup_{t\in T} \int_0^\infty \sqrt{\ln \frac{1}{\mu(B_d(t,\varepsilon))}}\, d\varepsilon \leq \mathbb{E}\Big\{\sup_{t\in T} f_t\Big\}. \tag{1.5.2}$$

[27] "Mathematical completeness" should be understood in a relative sense, since our proofs here will most definitely be incomplete!

A measure μ for which the integrals above are finite for all t is called a majorizing measure.

Note that the upper limit to the integrals in the theorem is really $\text{diam}(T)$, since

$$\varepsilon > \text{diam}(T) \quad \Rightarrow \quad T \subset B_d(t,\varepsilon) \quad \Rightarrow \quad \mu(B_d(t,\varepsilon)) = 1,$$

and so the integrand is zero beyond this limit.

Theorem 1.5.1 is the majorizing measure version of Theorem 1.3.3,[28] which gave an upper bound for $\mathbb{E}\{\sup_t f_t\}$ based on an entropy integral, but which did not have a corresponding lower bound. Theorem 1.5.1, however, takes much more work to prove than its entropic relative. Nevertheless, by building on what we have already done, it is not all that hard to see where the upper bound (1.5.1) comes from.

Outline of proof. Start by rereading the proof of Theorem 1.3.3 as far as (1.3.14). The argument leading up to (1.3.14) was that, eventually, increments $f_{\pi_j(t)} - f_{\pi_{j-1}}(t)$ would be smaller than ua_j. However, on the way to this we could have been less wasteful in a number of our arguments.

For example, we could have easily arranged things so that

$$\forall\, s,t \in T, \quad \pi_j(t) = \pi_j(s) \Rightarrow \pi_{j-1}(t) = \pi_{j-1}(s), \tag{1.5.3}$$

which would have given us fewer increments to control. We could have also assumed that

$$\forall\, t \in \Pi_j, \quad \pi_{j(t)} = t, \tag{1.5.4}$$

so that, by (1.5.3),

$$\pi_{j-1}(t) = \pi_{j-1}(\pi_j(t)). \tag{1.5.5}$$

So let us assume both (1.5.3) and (1.5.4). Then controlling the increments $f_{\pi_j(t)} - f_{\pi_{j-1}}(t)$ actually means controlling the increments $f_t - f_{\pi_{j-1}}(t)$ for $t \in \Pi_j$. There are only N_j such increments, which improves on our previous estimate of $N_j N_{j-1}$. This does not make much of a difference, but what does, and this is the core of the majorizing measure argument, is replacing the original a_j by families of nonnegative numbers $\{a_j(t)\}_{t\in\Pi_j}$, and then to ask when

$$\forall\, t \in \Pi_j, \quad f_t - f_{\pi_{j-1}(t)} \le ua_j(t) \tag{1.5.6}$$

for large enough j. Note that under (1.5.6),

$$\forall\, t \in T, \quad f_t - f_{t_0} \le uS,$$

where

[28] There is a similiar extension of Theorem 1.3.5, but we shall not bother with it.

$$S \triangleq \sup_{t \in T} \sum_{j>i} a_j(\pi_j(t)).$$

Thus

$$\begin{aligned}\mathbb{P}\Big\{\sup_{t\in T}(f_t - f_{t_0}) \geq uS\Big\} &\leq \sum_{j>i}\sum_{t\in\Pi'_j} \mathbb{P}\{f_t - f_{\pi_{j-1}(t)} \geq u a_j(t)\} \\ &\leq \sum_{j>i}\sum_{t\in\Pi'_j} 2\exp\left(\frac{-u^2 a_j^2(t)}{(2r^{-j+1})^2}\right), \qquad (1.5.7)\end{aligned}$$

where $\Pi'_j \triangleq \Pi_j \setminus \bigcup_{i<k\leq j-1} \Pi_k$. (The move from the Π_j to the disjoint Π'_j is crucial, and made possible by (1.5.5).) This bound is informative only if the right-hand side is less than or equal to one. Let us see how to ensure this when $u = 1$. Setting

$$w_j(t) = 2\exp\left(\frac{-a_j^2(t)}{(2r^{-j+1})^2}\right), \qquad (1.5.8)$$

we want to have $\sum_j \sum_{t\in\Pi'_j} w_j(t) \leq 1$. We are now getting close to our majorizing measure.

Recall that T has long ago been assumed countable. Suppose we have a probability measure μ supported on T and for all $j > i$ and all $t \in \Pi_j$ set $w_j(t) = \mu(\{t\})$. Undo (1.5.8) to see that this means that we need to take

$$a_j(t) = 2r^{-j+1}\sqrt{\ln\frac{2}{\mu(\{t\})}}.$$

With this choice, the last sum in (1.5.7) is no more than 2^{1-u^2}, and S is given by

$$S = 2\sup_{t\in T}\sum_{j>i} r^{-j+1}\sqrt{\ln\frac{2}{\mu(\{\pi_j(t)\})}},$$

all of which ensures that for the *arbitrary* measure μ we now have

$$\mathbb{E}\Big\{\sup_{t\in T} f_t\Big\} \leq K \sup_{t\in T}\sum_{j>i} r^{-j+1}\sqrt{\ln\frac{2}{\mu(\{\pi_j(t)\})}}. \qquad (1.5.9)$$

This is, in essence, the majorizing measure upper bound. To make it look more like (1.5.1), note that each map π_j defines a partition $\mathcal{A}_j$ of T comprising the sets

$$A_t \triangleq \{s \in T : \pi_j(s) = t\}, \quad t \in \Pi_j.$$

With $A_j(t)$ denoting the unique element of $\mathcal{A}_j$ that contains $t \in T$, it is not too hard (but also not trivial) to reformulate (1.5.9) to obtain

$$\mathbb{E}\left\{\sup_{t\in T} f_t\right\} \le K \sup_{t\in T}\sum_{j>i} r^{-j+1}\sqrt{\ln\frac{2}{\mu(\{A_j(t)\})}}, \tag{1.5.10}$$

which is now starting to look a lot more like (1.5.1). To see why this reformulation works, you should go to [154], which you should now be able to read without even bothering about notational changes. You will also find there how to turn (1.5.10) into (1.5.1), and also how to get the lower bound (1.5.2). All of this takes quite a bit of work, but at least now you should have some idea of how majorizing measures arose. □

Despite the elegance of Theorem 1.5.1, it is not always easy, given a specific Gaussian process, to find the "right" majorizing measure for it. To circumvent this problem, Talagrand recently [155] gave a recipe for how to wield the technique without the need to explicitly compute a majorizing measure. However, we are already familiar with one situation in which there is a simple recipe for building majorizing measures. This is when entropy integrals are finite.

Thus, let be $H(\varepsilon)$ be our old friend (1.3.3), and set

$$g(t) \triangleq \sqrt{\ln\frac{1}{t}}, \qquad 0 < t \le 1. \tag{1.5.11}$$

Then here is a useful result linking entropy and majorizing measures.

Lemma 1.5.2. *If $\int_0^\infty H^{1/2}(\varepsilon)\,d\varepsilon < \infty$, then there exists a majorizing measure μ and a universal constant K such that*

$$\sup_{t\in T}\int_0^\eta g(\mu(B(t,\varepsilon)))\,d\varepsilon < K\left(\eta|\log\eta| + \int_0^\eta H^{1/2}(\varepsilon)\,d\epsilon\right), \tag{1.5.12}$$

for all $\eta > 0$.

This lemma, together with Theorem 1.5.1, provides an alternative proof of Theorems 1.3.3 and 1.3.5, since the additional term of $\eta|\log\eta|$ in (1.5.12) can be absorbed into the integral (for small enough η) by changing K. In other words, the lemma shows how entropy results follow from those for majorizing measures. What is particularly interesting, however, is the proof of the lemma, which actually shows how to construct a majorizing measure when the entropy integral is a priori known to be finite.

Proof. For convenience we assume that $\mathrm{diam}(T) = 1$. Let $\{A_{n,1},\ldots,A_{n,N(2^{-n})}\}$, for $n \ge 0$, be a minimal family of d-balls of radius 2^{-n} that cover T. Set

$$B_{n,k} = A_{n,k} \setminus \bigcup_{j<k} A_{n,j}, \tag{1.5.13}$$

so that $\mathcal{B}_n = \{B_{n,1},\ldots,B_{n,N(2^{-n})}\}$ is a partition of T and each B_i is contained in a d-ball of radius 2^{-n}. For each pair n, k choose a point $t_{n,k} \in B_{n,k}$ and then define a probability measure μ on T by

$$\mu(A) = \sum_{n=0}^{\infty} 2^{-(n+1)}(N(2^{-n}))^{-1} \sum_{k=1}^{N(2^{-n})} \delta_{t_{n,k}}(A),$$

where δ_{t_k} is the measure giving unit mass to t_k. This will be our majorizing measure. To check that it satisfies (1.5.12), note first that if $\varepsilon \in (2^{-(n+1)}, 2^{-n}]$, then

$$\mu(B(t,\varepsilon)) \geq (2^{n+1}N(2^{-(n+1)}))^{-1},$$

for all $t \in T$. Consequently,

$$\begin{aligned}
\int_0^{2^{-n}} \sqrt{\ln \frac{1}{\mu(B(t,\varepsilon))}}\, d\varepsilon &\leq \sum_{k=n+1}^{\infty} 2^{-k}(\ln(2^k N(2^{-k})))^{\frac{1}{2}} \\
&\leq \sum_{k=n+1}^{\infty} 2^{-k}(k \ln 2)^{\frac{1}{2}} + 2\int_0^{2^{-n}} (\ln(N(\varepsilon))^{\frac{1}{2}}\, d\varepsilon \\
&\leq (n+2)2^{-n}\sqrt{\ln 2} + 2\int_0^{2^{-n}} H^{1/2}(\varepsilon)\, d\varepsilon,
\end{aligned}$$

the last line following from a little elementary algebra.

This establishes (1.5.12) for dyadic η. The passage to general η follows via a monotonicity argument. □

Another class of examples for which it is easy to find a majorizing measure is given by that of stationary fields over compact Abelian groups. Once again, we shall assume that you are either familiar with the notion of stationarity or will return after having read Chapter 5. Under stationarity it is perhaps not suprising that Haar measure is a majorizing measure.

Theorem 1.5.3. *If f is stationary over a compact Abelian group T, then* (1.5.1) *and* (1.5.2) *hold with μ taken to be normalized Haar measure on T.*

A very similar result holds if T is only locally compact. You can find the details in [99].

Proof. Since (1.5.1) is true for any probability measure on T, it also holds for Haar measure. Thus we need only prove the lower bound (1.5.2).

Thus, assume that f is bounded, so that by Theorem 1.5.1 there exists a majorizing measure μ satisfying (1.5.2). We need to show that μ can be replaced by Haar measure on T, which we denote by ν.

Set

$$D_\mu \stackrel{\Delta}{=} \sup\{\eta : \mu(B(t,\eta)) < 1/2, \text{ for all } t \in T\},$$

with D_ν defined analogously. With g as in (1.5.11), (1.5.2) can be rewritten as

$$\int_0^{D_m} g(\mu(B(t,\varepsilon))\, d\varepsilon \leq K\mathbb{E}\Big\{\sup_{t\in T} f_t\Big\},$$

for all $t \in T$. Let τ be a random variable with distribution ν; i.e., τ is uniform on T. For each $\varepsilon > 0$, set $Z(\varepsilon) = \mu(B(\tau, \varepsilon))$. Then, for any $t_0 \in T$,

$$\begin{aligned}\mathbb{E}\{Z(\varepsilon)\} &= \int_T \mu(B(t,\varepsilon))\nu(dt) \\ &= \int_T \mu(t + B(t_0,\varepsilon))\nu(dt) \\ &= \int_T \nu(t + B(t_0,\varepsilon))\mu(dt) \\ &= \nu(B(t_0,\varepsilon)),\end{aligned}$$

where the second equality comes from the stationarity of f and the third and fourth from the properties of Haar measures.

Now note that $g(x)$ is convex over $x \in (0, \frac{1}{2})$, so that it is possible to define a function $\widehat{g}$ that agrees with it on $(0, \frac{1}{2})$, is bounded on $(\frac{1}{2}, \infty)$, and convex on all of $\mathbb{R}_+$. By Jensen's inequality,

$$\widehat{g}(\mathbb{E}\{Z(\varepsilon)\}) \le \mathbb{E}\{\widehat{g}(Z(\varepsilon))\}.$$

That is,

$$\widehat{g}(\nu(B(t_0,\varepsilon))) \le \int_T \widehat{g}(\mu(B(t,\varepsilon))\nu(dt).$$

Finally, set $\Lambda = \min(D_\mu, D_\nu)$. Then

$$\begin{aligned}\int_0^\Lambda \widehat{g}(\nu(B(t_0,\varepsilon))\, d\varepsilon &\le \int_0^\Lambda d\varepsilon \int_T \widehat{g}(\mu(B(t,\varepsilon))\nu(dt) \\ &= \int_T \int_0^\Lambda \nu(dt)\widehat{g}(\mu(B(t,\varepsilon))\, d\varepsilon \\ &\le K\mathbb{E}\Big\{\sup_{t\in T} f_t\Big\}.\end{aligned}$$

This is the crux of (1.5.2). The rest is left to you. □

This more or less completes our discussion of majorizing measures. However, we still have two promises to fulfill before completing this chapter. The first of these is to show why entropy conditions are necessary, as well as sufficient, for boundedness and continuity if f is stationary. This is Theorem 1.5.4 below. Its proof is really a little dishonest, since it relies on the main majorizing measure result Theorem 1.5.1, which we have not proven. It could in fact be proven without resort to majorizing measures, and was done so originally. However, for a chapter that was supposed to be only an "introduction" to the general theory of Gaussian processes, we have already used up more than enough space, and so shall make do with what we have at hand already.

Theorem 1.5.4. *Let f be a centered stationary Gaussian process on a compact group T. Then the following three conditions are equivalent:*

$$f \text{ is a.s. continuous on } T, \tag{1.5.14}$$

$$f \text{ is a.s. bounded on } T, \tag{1.5.15}$$

$$\int_0^\infty H^{1/2}(\varepsilon)\, d\varepsilon < \infty. \tag{1.5.16}$$

Proof. That (1.5.14) implies (1.5.15) is obvious. That (1.5.16) implies (1.5.14) is Lemma 1.5.2 together with Theorem 1.5.1. Thus it suffices to show that (1.5.15) implies (1.5.16), which we shall now do.

Note firstly that by Theorem 1.5.3 we know that

$$\sup_{t\in T} \int_0^\infty g\big(\mu\big(B(t,\varepsilon)\big)\big)\, d\varepsilon < \infty \tag{1.5.17}$$

for μ normalized Haar measure on T. Furthermore, by stationarity, the value of the integral must be independent of t.

For $\varepsilon \in (0,1)$ let $M(\varepsilon)$ be the maximal number of points $\{t_k\}_{k=1}^{M(\varepsilon)}$ in T for which

$$\min_{1\le j,k\le M(\varepsilon)} d(t_j, t_k) > \varepsilon.$$

It is easy to check that

$$N(\varepsilon) \le M(\varepsilon) \le N(\varepsilon/2).$$

Thus, since μ is a probability measure, we must have

$$\mu(B(t,\varepsilon)) \le (N(2\varepsilon))^{-1}.$$

Consequently, by (1.5.17) and, in particular, its independence on t,

$$\infty > \int_0^\infty g(\mu(B(t,\varepsilon)))\, d\varepsilon \ge \int_0^\infty (\ln N(2\varepsilon))^{\frac{1}{2}}\, d\varepsilon = 2\int_0^\infty H^{1/2}(\varepsilon)\, d\varepsilon,$$

which proves the theorem. □

Our second and final responsibility is to prove that the claims we made back at (1.4.7) are correct. That they are is the statement of the following corollary to Theorem 1.5.4.

Corollary 1.5.5. *If f is centered and stationary on open $T \subset \mathbb{R}^N$ with covariance function C, and*

$$\frac{K_1}{(-\ln|t|)^{1+\alpha_1}} \le C(0) - C(t) \le \frac{K_2}{(-\ln|t|)^{1+\alpha_2}}, \tag{1.5.18}$$

for $|t|$ small enough, then f will be sample path continuous if $\alpha_2 > 0$ and discontinuous if $\alpha_1 < 0$.

Proof. Recall from Theorem 1.4.1 that the basic energy integral in (1.5.16) converges or diverges together with

$$\int_{\delta}^{\infty} p(e^{-u^2})\,du, \tag{1.5.19}$$

where

$$p^2(u) = 2 \sup_{|t|\le u} [C(0) - C(t)]$$

(cf. (1.4.2)). Applying the bounds in (1.5.18) to evaluate (1.5.19) and applying the extension of Theorem 1.5.3 to noncompact groups easily proves the corollary. □

2

Gaussian Inequalities

Basic statistics has its Chebyshev inequality, martingale theory has its maximal inequalities, Markov processes have large deviations, but all pale in comparison to the power and simplicity of the coresponding basic inequality of Gaussian processes. This inequality was discovered independently, and established with very different proofs, by Borell [30] and Tsirelson, Ibragimov, and Sudakov (TIS) [160]. For brevity, we shall call it the Borell–TIS inequality. In the following section we shall treat it in some detail.

A much older inequality, which allows comparison between the suprema of different Gaussian processes, is due to Slepian, and we shall describe it and some of its newer relatives in Section 2.2.

2.1 Borell–TIS Inequality

One of the first facts we learned about Gaussian processes was that if $X \sim N(0, \sigma^2)$, then for all $u > 0$,

$$\left(\frac{\sigma}{\sqrt{2\pi}u} - \frac{\sigma^3}{\sqrt{2\pi}u^3}\right) e^{-\frac{1}{2}u^2/\sigma^2} \leq \mathbb{P}\{X > u\} \leq \frac{\sigma}{\sqrt{2\pi}u} e^{-\frac{1}{2}u^2/\sigma^2} \tag{2.1.1}$$

(cf. (1.2.2)). One immediate consequence of this is that

$$\lim_{u\to\infty} u^{-2} \ln \mathbb{P}\{X > u\} = -(2\sigma^2)^{-1}. \tag{2.1.2}$$

There is a classical result of Landau and Shepp [96] and Marcus and Shepp [110] that gives a result closely related to (2.1.2), but for the supremum of a general centered Gaussian process. If we assume that f_t is a.s. bounded, then they showed that

$$\lim_{u\to\infty} u^{-2} \ln \mathbb{P}\left\{\sup_{t\in T} f_t > u\right\} = -(2\sigma_T^2)^{-1}, \tag{2.1.3}$$

where

$$\sigma_T^2 \stackrel{\Delta}{=} \sup_{t\in T} \mathbb{E}\{f_t^2\}$$

is a notation that will remain with us throughout this section. An immediate consequence of (2.1.3) is that for all $\varepsilon > 0$ and large enough u,

$$\mathbb{P}\Big\{\sup_{t\in T} f_t > u\Big\} \le e^{\varepsilon u^2 - u^2/2\sigma_T^2}. \tag{2.1.4}$$

Since $\varepsilon > 0$ is arbitrary, comparing (2.1.4) and (2.1.1), we reach the rather surprising conclusion that the supremum of a centered, bounded Gaussian process behaves much like a single Gaussian variable with a suitably chosen variance.

In Chapter 4 we shall see that it requires little more than the basic techniques of the current chapter, along with the notion of entropy from Chapter 1, to considerably improve on (2.1.4), in that the exponential term $e^{\varepsilon u^2}$ can be replaced by a power term of the form u^α. Later, in Part III, we shall how to do even better, although then we shall have to assume more on the random fields.

Now, however, we want to see what can be done with minimal assumptions, and to see from where (2.1.4) comes. In fact, (2.1.4) and its consequences are all special cases of a *nonasymptotic* result due, as mentioned above, independently, and with very different proofs, to Borell [30] and Tsirelson, Ibragimov, and Sudakov (TIS) [160].

Theorem 2.1.1 (Borell–TIS inequality). *Let f_t be a centered Gaussian process, a.s. bounded on T. Write*

$$\|f\| = \|f\|_T = \sup_{t\in T} f_t.$$

Then

$$\mathbb{E}\{\|f\|\} < \infty,$$

and for all $u > 0$,

$$\mathbb{P}\{\|f\| - \mathbb{E}\{\|f\|\} > u\} \le e^{-u^2/2\sigma_T^2}. \tag{2.1.5}$$

Before we look at the proof of (2.1.5), which we refer to as the Borell–TIS inequality,[1] we take a moment to look at some of its conseqences, which are many and major. It is no exaggeration to say that this inequality is today the single most important tool in the general theory of Gaussian processes.

An immediate and trivial consequence of (2.1.5) is that for all $u > \mathbb{E}\{\|f\|\}$,

$$\mathbb{P}\{\|f\| > u\} \le e^{-(u-\mathbb{E}\|f\|)^2/2\sigma_T^2},$$

so that both (2.1.3) and (2.1.4) follow from the Borell–TIS inequality.

[1] Actually, Theorem 2.1.1 is not in the same form as Borell's original inequality, in which $\mathbb{E}\{\|f\|\}$ was replaced by the median of $\|f\|$. However, the two forms are equivalent. For this and other variations of (2.1.5), including extensions to Banach space–valued processes for which $\|\ \|$ is the norm, see the more detailed treatments of [28, 56, 67, 99, 105]. To see how the Borell–TIS inequality fits into the wider theory of concentration inequalities, see the recent book [98] by Ledoux.

Indeed, a far stronger result is true, for (2.1.4) can now be replaced by

$$\mathbb{P}\{\|f\| > u\} \leq e^{Cu-u^2/2\sigma_T^2},$$

where C is a constant depending only on $\mathbb{E}\{\|X\|\}$, and we know at least how to bound this quantity from Theorem 1.3.5.

Note that, despite the misleading notation, $\| \; \| \equiv \sup$ is not a norm, and that very often one needs bounds on the tail of $\sup_t |f_t|$, which does give a norm. However, symmetry immediately gives

$$\mathbb{P}\Big\{\sup_t |f_t| > u\Big\} \leq 2\mathbb{P}\Big\{\sup_t f_t > u\Big\}, \tag{2.1.6}$$

so that the Borell–TIS inequality helps out here as well.

Here is a somewhat more significant consequence of the Borell–TIS inequality.

Theorem 2.1.2. *For f centered, Gaussian,*

$$\begin{aligned}\mathbb{P}\{\|f\| < \infty\} = 1 &\iff \mathbb{E}\{\|f\|\} < \infty \\ &\iff \mathbb{E}\Big\{e^{\alpha\|f\|^2}\Big\} < \infty\end{aligned} \tag{2.1.7}$$

for sufficiently small α.

Proof. The existence of the exponential moments of $\|f\|$ implies the existence of $\mathbb{E}\{\|f\|\}$, and this in turn implies the a.s. finiteness of $\|f\|$. Furthermore, since by Theorem 2.1.1 we already know that the a.s. finiteness of $\|f\|$ entails that of $\mathbb{E}\{\|f\|\}$, all that remains to prove is that the a.s. finiteness of $\|f\|$ also implies the existence of exponential moments.

But this is an easy consequence of the Borell–TIS inequality, since, with both $\|f\|$ and $\mathbb{E}\{\|f\|\}$ now finite,

$$\begin{aligned}\mathbb{E}\{e^{\alpha\|f\|^2}\} &= \int_0^\infty \mathbb{P}\Big\{e^{\alpha\|f\|^2} > u\Big\}\,du \\ &\leq e^\alpha + \mathbb{E}\{\|f\|\} + 2\int_{e^\alpha \vee \mathbb{E}\{\|f\|\}}^\infty \mathbb{P}\Big\{\|f\| > \sqrt{\ln u^{1/\alpha}}\Big\}\,du \\ &\leq e^\alpha + \mathbb{E}\{\|f\|\} \\ &\quad + 2\int_{e^\alpha \vee \mathbb{E}\{\|f\|\}}^\infty \exp\left(\frac{-\big(\sqrt{\ln u^{1/\alpha}} - \mathbb{E}\{\|f\|\}\big)^2}{2\sigma_T^2}\right) du \\ &\leq e^\alpha + \mathbb{E}\{\|f\|\} \\ &\quad + 4\alpha\int_0^\infty u\exp\left(\frac{-(u-\mathbb{E}\{\|f\|\})^2}{2\sigma_T^2}\right)\exp\{\alpha u^2\}\,du,\end{aligned}$$

which is clearly finite for small enough α. □

Recall Theorems 1.3.3 and 1.3.5, which established, respectively, the a.s. boundedness of $\|f\|$ and a bound on the modulus of continuity $\omega_{f,d}(\delta)$ under essentially identical entropy conditions. It was rather irritating back there that we had to establish each result independently, since it is "obvious" that one should imply the other. A simple application of the Borell–TIS inequality almost does this.

Theorem 2.1.3. *Suppose that f is a.s. bounded on T. Then f is also a.s uniformly continuous (with respect to the canonical metric d) if and only if*

$$\lim_{\eta \to 0} \phi(\eta) = 0, \tag{2.1.8}$$

where

$$\phi(\eta) \triangleq \mathbb{E}\Big\{ \sup_{d(s,t)<\eta} (f_s - f_t) \Big\}. \tag{2.1.9}$$

Furthermore, under (2.1.8), *for all $\varepsilon > 0$ there exists an a.s. finite random variable $\eta > 0$ such that*

$$\omega_{f,d}(\delta) \le \phi(\delta)\, |\ln \phi(\delta)|^{\varepsilon}, \tag{2.1.10}$$

for all $\delta \le \eta$.

Proof. We start with necessity. For almost every ω we have

$$\lim_{\eta \to 0} \sup_{d(s,t)<\eta} |f_s(\omega) - f_t(\omega)| = 0.$$

But (2.1.8) now follows from dominated convergence and Theorem 2.1.2.

For sufficiency, note that from (2.1.8) we can find a sequence $\{\delta_n\}$ with $\delta_n \to 0$ such that $\phi(\delta_n) \le 2^{-n}$. Set $\delta_n' = \min(\delta_n, 2^{-n})$, and consider the event

$$A_n = \Big\{ \sup_{d(s,t)<\delta_n'} |f_s - f_t| > 2^{-n/2} \Big\}.$$

The Borell–TIS inequality (cf. (2.1.6)) gives that for $n \ge 3$,

$$\mathbb{P}\{A_n\} \le 2 \exp\left(-\frac{1}{2}(2^{-n/2} - 2^{-n})^2 / 2^{-2n} \right) \le K \exp(-2^{n-1}).$$

Since $P\{A_n\}$ is an admirably summable series, Borel–Cantelli gives us that f is a.s. uniformly d-continuous, as required.

To complete the proof we need to establish the bound (2.1.10) on $\omega_{f,d}$. Note first that

$$\begin{aligned} \mathrm{diam}(S) &= \sup_{s,t \in S} (\mathbb{E}\{|f_t - f_s|^2\})^{1/2} \\ &= \sqrt{2\pi} \sup_{s,t \in S} \mathbb{E}\{|f_t - f_s|\} \\ &\le \sqrt{2\pi}\, \mathbb{E}\Big\{ \sup_{s,t \in S} f_t - f_s \Big\} \\ &= \sqrt{2\pi}\, \phi(\mathrm{diam}(S)), \end{aligned}$$

where the second line is an elementary Gaussian computation and the third uses the fact that $\sup_{s,t\in S}(f_t - f_s)$ is nonnegative. Consequently, we have that

$$\delta \le \sqrt{2\pi}\,\phi(\delta), \tag{2.1.11}$$

for all $\delta > 0$.

Now define the numbers

$$\delta_n = \sup\{\delta : \phi(\delta) \le e^{-n}\}$$

and, for $\varepsilon > 0$, the events

$$B_n = \Big\{ \sup_{d(s,t)<\delta_n} |f_s - f_t| > \phi(\delta_n) |\ln \phi(\delta_n)|^{\varepsilon/2} \Big\}.$$

Then the Borell–TIS inequality gives

$$\mathbb{P}\{B_n\} \le 4 \exp\left\{ -\frac{1}{2}(|\ln \phi(\delta_n)|^{\varepsilon/2} - 1)^2 \frac{\phi^2(\delta_n)}{\delta_n^2} \right\} \le K_1 \exp\{-K_2 n^{\epsilon}\},$$

the second inequality following from (2.1.11) and the definition of δ_n.

Since $\sum_n \mathbb{P}\{B_n\} < \infty$, we have that for $n \ge N(\omega)$,

$$\omega_{f,d}(\delta_n) \le \phi(\delta_n) |\ln \phi(\delta_n)|^{\varepsilon/2}.$$

Monotonicity of $\omega_{f,d}$, along with separability, complete the proof. □

We now turn to the proof of the Borell–TIS inequality. There are essentially three quite different ways to tackle this proof. Borell's original proof relied on isoperimetric inequalities.[2] While isoperimetric inequalities may be natural for a book with the word "geometry" in its title, we shall avoid them, since they involve setting up a number of concepts for which we shall have no other need. The proof of Tsirelson, Ibragimov, and Sudakov used Itô's formula from stochastic calculus. This is one of our[3] favorite proofs, since as one of the too few links between the Markovian and Gaussian worlds of stochastic processes, it is to be prized.

We shall, however, take a more direct route, which we learned about from the excellent collection of exercises [38], although its roots are much older. The first step in this route involves the following two lemmas, which are of independent interest.

Lemma 2.1.4. *Let X and Y be independent k-dimensional vectors of centered, unit-variance, independent, Gaussian variables. If $f, g : \mathbb{R}^k \to \mathbb{R}$ are bounded C^2 functions then*

[2] Borell has a newer, very recent version of this approach, to be found in [31, 32].

[3] Or at least one of RJA's favorite proofs. Indeed, this approach was used in RJA's lecture notes [3]. There, however, there is a problem, for in the third line from the bottom of page 46 appear the words "To complete the proof note simply that. . . ." The word "simply" is simply out of place, and in fact, Lemma 2.2 there is false as stated. A correction (with the help of Amir Dembo) appears, en passant, in the "Proof of Theorem 2.1.1" below.

$$\operatorname{Cov}(f(X), g(X)) = \int_0^1 \mathbb{E}\{\langle \nabla f(X), \nabla g(\alpha X + \sqrt{1-\alpha^2}Y)\rangle\}\, d\alpha, \tag{2.1.12}$$

where $\nabla f(x) = (\frac{\partial f}{\partial x_i} f(x))_{i=1,\dots,k}$.

Proof. It suffices to prove the lemma with $f(x) = e^{i\langle t,x\rangle}$ and $g(x) = e^{i\langle s,x\rangle}$, with $s, t, x \in \mathbb{R}^k$. Standard approximation arguments (which is where the requirement that f is C^2 appears) will do the rest. Write

$$\varphi(t) \stackrel{\Delta}{=} \mathbb{E}\{e^{i\langle t,X\rangle}\} = e^{|t|^2/2}$$

(cf. (1.2.4)). It is then trivial that

$$\operatorname{Cov}(f(X), g(X)) = \varphi(s+t) - \varphi(s)\varphi(t).$$

On the other hand, computing the integral in (2.1.12) gives

$$\begin{aligned}
&\int_0^1 \mathbb{E}\{\langle \nabla f(X), \nabla g(\alpha X + \sqrt{1-\alpha^2}Y)\rangle\}\, d\alpha \\
&\quad = \int_0^1 d\alpha \mathbb{E}\left\{\left\langle \left(it_j e^{i\langle t,X\rangle}\right)_j, \left(is_j e^{i\langle s,\alpha X + \sqrt{1-\alpha^2}Y\rangle}\right)_j\right\rangle\right\} \\
&\quad = -\int_0^1 d\alpha \sum_j s_j t_j \mathbb{E}\{e^{i\langle t+\alpha s, X\rangle}\}\mathbb{E}\{e^{i\langle s, \sqrt{1-\alpha^2}Y\rangle}\} \\
&\quad = -\int_0^1 d\alpha\, \langle s,t\rangle\, e^{(|t|^2 + 2\alpha\langle s,t\rangle + |s|^2)/2} \\
&\quad = -\varphi(s)\varphi(t)\left(1 - e^{\langle s,t\rangle}\right) \\
&\quad = \varphi(s+t) - \varphi(s)\varphi(t),
\end{aligned}$$

which is all that we need. □

Lemma 2.1.5. *Let X be as in Lemma* 2.1.4. *If $h : \mathbb{R}^k \to \mathbb{R}$ is C^2 with Lipschitz constant*[4] 1 *and if* $\mathbb{E}\{h(X)\} = 0$, *then for all* $t > 0$,

$$\mathbb{E}\{e^{th(X)}\} \le e^{t^2/2}. \tag{2.1.13}$$

Proof. Let Y be an independent copy of X and α a uniform random variable on $[0, 1]$. Define the pair (X, Z_α) via

$$(X, Z_\alpha) \stackrel{\Delta}{=} \left(X, \alpha X + \sqrt{1-\alpha^2}Y\right).$$

---

[4] Recall that the Lipschitz constant of a function on $\mathbb{R}^k$ is given by

$$\|f\|_{\mathrm{Lip}} \stackrel{\Delta}{=} \sup_x |\nabla f(x)| = \sup_x \left\{\frac{|f(x) - f(y)|}{|x-y|}\right\}.$$

Take h as in the statement of the lemma, $t \geq 0$ fixed, and define $g = e^{th}$. Applying (2.1.12) gives

$$\begin{aligned}
\mathbb{E}\{h(X)g(X)\} &= \int_0^1 \mathbb{E}\{\langle \nabla g(X), \nabla h(Z_\alpha)\rangle\}\, d\alpha \\
&= t \int_0^1 \mathbb{E}\big\{\big\langle \nabla h(X), \nabla h(Z_\alpha)\big\rangle e^{th(X)}\big\}\, d\alpha \\
&\leq t\mathbb{E}\big\{e^{th(X)}\big\},
\end{aligned}$$

using the Lipschitz property of h. Let u be the function defined by

$$e^{u(t)} = \mathbb{E}\big\{e^{th(X)}\big\}.$$

Then

$$\mathbb{E}\big\{h(X)e^{th(X)}\big\} = u'(t)e^{u(t)},$$

so that from the preceding inequality, $u'(t) \leq t$. Since $u(0) = 0$ it follows that $u(t) \leq t^2/2$, and we are done. □

The following lemma gives the crucial step toward proving the Borell–TIS inequality.

Lemma 2.1.6. *Let X be a k-dimensional vector of centered, unit-variance, independent Gaussian variables. If $h : \mathbb{R}^k \to \mathbb{R}$ has Lipschitz constant σ, then for all $u > 0$,*

$$\mathbb{P}\{h(X) - \mathbb{E}\{h(X)\} > u\} \leq e^{-\frac{1}{2}u^2/\sigma^2}. \tag{2.1.14}$$

Proof. By scaling it suffices to prove the result for $\sigma = 1$. Assume for the moment that $f \in C^2$. Then, for every $t, u > 0$,

$$\begin{aligned}
\mathbb{P}\{h(X) - \mathbb{E}\{h(X)\} > u\} &\leq \int_{h(x)-\mathbb{E}\{h(X)\}>u} e^{t(h(x)-\mathbb{E}\{h(X)\}-u)}\, dP(x) \\
&\leq e^{-tu}\mathbb{E}\big\{e^{t(h(x)-\mathbb{E}\{h(X)\})}\big\} \\
&\leq e^{\frac{1}{2}t^2 - tu},
\end{aligned}$$

the last inequality following from (2.1.13). Taking the optimal choice of $t = u$ gives (2.1.14) for $f \in C^2$.

To remove the C^2 assumption, take a sequence of C^2 approximations to f each one of which has Lipschitz coefficient no greater than σ and apply Fatou's inequality. This completes the proof. □

We now have all we need to prove Theorem 2.1.1.

Proof of Theorem 2.1.1. There will be two stages to the proof. Firstly, we shall establish Theorem 2.1.1 for finite T. We then lift the result from finite to general T.

Thus, let T be finite, so that we can write it as $\{1, \dots, k\}$. In this case we can replace sup by max, which has Lipshitz constant 1. Theorem 2.1.1 then follows immediately from 2.1.6. The general case is a little more delicate.

Let C be the $k \times k$ covariance matrix of f on T, with components $c_{ij} = \mathbb{E}\{f_i f_j\}$, so that

$$\sigma_T^2 = \max_{1 \le i \le k} c_{ii} = \max_{1 \le i \le k} \mathbb{E}\{f_i^2\}.$$

Let W be a vector of independent, standard Gaussian variables, and A such that $A'A = C$. Thus $f \stackrel{\mathcal{L}}{=} AW$, and $\max_i f_i \stackrel{\mathcal{L}}{=} \max_i (AW)_i$, where $\stackrel{\mathcal{L}}{=}$ indicates equivalence in distribution (law).

Consider the function $h(x) = \max_i (Ax)_i$, which is trivially C^2. Then

$$\begin{aligned}
\left| \max_i (Ax)_i - \max_i (Ay)_i \right| &= \left| \max_i (e_i Ax) - \max_i (e_i Ay) \right| \\
&\le \max_i \left| e_i A(x-y) \right| \\
&\le \max_i \left| e_i A \right| \cdot |x-y|,
\end{aligned}$$

where as usual e_i is the vector with 1 in position i and zeros elsewhere. The first inequality above is elementary, and the second is Cauchy–Schwarz. But

$$|e_i A|^2 = e_i' A' A e_i = e_i' C e_i = c_{ii},$$

so that

$$\left| \max_i (Ax)_i - \max_i (Ay)_i \right| \le \sigma_T |x-y|.$$

In view of the equivalence in law of $\max_i f_i$ and $\max_i (AW)_i$ and Lemma 2.1.6, this establishes the theorem for finite T.

We now turn to lifting the result from finite to general T. This is, almost, an easy exercise in approximation. For each $n > 0$ let T_n be a finite subset of T such that $T_n \subset T_{n+1}$ and T_n increases to a dense subset of T. By separability,

$$\sup_{t \in T_n} f_t \stackrel{\text{a.s.}}{\to} \sup_{t \in T} f_t,$$

and since the convergence is monotone, we also have that

$$\mathbb{P}\Big\{ \sup_{t \in T_n} f_t \ge u \Big\} \to \mathbb{P}\Big\{ \sup_{t \in T} f_t \ge u \Big\} \quad \text{and} \quad \mathbb{E}\Big\{ \sup_{t \in T_n} f_t \Big\} \to \mathbb{E}\Big\{ \sup_{t \in T} f_t \Big\}.$$

Since $\sigma_{T_n}^2 \to \sigma_T^2 < \infty$ (again monotonically), this would be enough to prove the general version of the Borell–TIS inequality from the finite-T version if only we knew that the one worrisome term, $\mathbb{E}\{\sup_T f_t\}$, were definitely finite, as claimed in the statement of the theorem. Thus if we show that the assumed a.s. finiteness of $\|f\|$ implies also the finiteness of its mean, we shall have a complete proof to both parts of the theorem.

We proceed by contradiction. Thus, assume $E\{\|f\|\} = \infty$, and choose $u_0 > 0$ such that

$$e^{-u_0^2/\sigma_T^2} \leq \frac{1}{4} \quad \text{and} \quad \mathbb{P}\left\{\sup_{t\in T} f_t < u_0\right\} \geq \frac{3}{4}.$$

Now choose $n \geq 1$ such that $\mathbb{E}\{\|f\|_{T_n}\} > 2u_0$, which is possible since $\mathbb{E}\{\|f\|_{T_n}\} \to \mathbb{E}\{\|f\|_T\} = \infty$. The Borell–TIS inequality on the finite space T_n then gives

$$\begin{aligned}\frac{1}{2} \geq 2e^{-u_0^2/\sigma_T^2} &\geq 2e^{-u_0^2/\sigma_{T_n}^2} \geq \mathbb{P}\{|\|f\|_{T_n} - \mathbb{E}\{\|f\|_{T_n}\}| > u_0\} \\ &\geq \mathbb{P}\{\mathbb{E}\{\|f\|_{T_n}\} - \|f\|_T > u_0\} \geq \mathbb{P}\{\|f\|_T < u_0\} \geq \frac{3}{4}.\end{aligned}$$

This provides the required contradiction, and so we are done. □

2.2 Comparison Inequalities

The theory of Gaussian processes is rich in comparison inequalities, where by this term we mean results of the form "if f is a "rougher" process than g, and both are defined over the same parameter space, then $\|f\|$ will be "larger" than $\|g\|$." The most basic of these is Slepian's inequality.

Theorem 2.2.1 (Slepian's inequality). *If f and g are a.s. bounded, centered Gaussian processes on T such that $\mathbb{E}\{f_t^2\} = \mathbb{E}\{g_t^2\}$ for all $t \in T$ and*

$$\mathbb{E}\{(f_t - f_s)^2\} \leq \mathbb{E}\{(g_t - g_s)^2\} \tag{2.2.1}$$

for all $s, t \in T$, then for all real u,

$$\mathbb{P}\{\|f\| > u\} \leq \mathbb{P}\{\|g\| > u\}. \tag{2.2.2}$$

Furthermore,

$$\mathbb{E}\{\|f\|\} \leq \mathbb{E}\{\|g\|\}. \tag{2.2.3}$$

Slepian's inequality is so natural that it hardly seems to require a proof, and hardly the rather analytic, nonprobabilistic one that will follow. To see that there is more to the story than meets the eye, one need only note that (2.2.2) does *not* follow from (2.2.1) if we replace[5] $\sup_T f_t$ by $\sup_T |f_t|$ and $\sup_T g_t$ by $\sup_T |g_t|$.

Slepian's inequality is based on the following technical lemma, the proof of which, in all its important details, goes back to Slepian's original paper [144].

[5] For a counterexample to "Slepian's inequality for absolute values" take $T = \{1, 2\}$, with f_1 and f_2 standard normal with correlation ρ. Writing $P_\rho(u)$ for the probability under correlation ρ that $\max(|f_1|, |f_2|) > u$, it is easy to check that for all $u > 0$, $P_{-1}(u) < P_0(u)$, while $P_0(u) > P_1(u)$, which negates the monotonicity required by Slepian's inequality.

Lemma 2.2.2. *Let $f_1, \dots, f_k$ be centered Gaussian variables with covariance matrix $C = (c_{ij})_{i,j=1}^k$, $c_{ij} = \mathbb{E}\{f_i f_j\}$. Let $h : \mathbb{R}^k \to \mathbb{R}$ be C^2, and assume that, together with its derivatives, it satisfies a $o(|x|^d)$ growth condition at infinity for some finite d.*[6] *Let*

$$H(C) = \mathbb{E}\{h(f_1, \dots, f_k)\}, \tag{2.2.4}$$

and assume that for a pair (i, j), $1 \le i < j \le k$,

$$\frac{\partial^2 h(x)}{\partial x_i \partial x_j} \ge 0 \tag{2.2.5}$$

for all $x \in \mathbb{R}^k$. Then $H(C)$ is an increasing function of c_{ij}.

Proof. We have to show that

$$\frac{\partial H(C)}{\partial c_{ij}} \ge 0$$

whenever $\partial^2 h/\partial x_i \partial x_j \ge 0$.

To make our lives a little easier we assume that C is nonsingular, so that it makes sense to write $\varphi(x) = \varphi_C(x)$ for the centered Gaussian density on $\mathbb{R}^k$ with covariance matrix C. Straightforward algebra shows that[7]

$$\frac{\partial \varphi}{\partial c_{ii}} = \frac{1}{2}\frac{\partial^2 \varphi}{\partial x_i^2}, \qquad \frac{\partial \varphi}{\partial c_{ij}} = \frac{\partial^2 \varphi}{\partial x_i \partial x_j}, \qquad i \ne j. \tag{2.2.6}$$

Applying this and our assumptions on h to justify two integrations by parts, we obtain, for $i \ne j$,

$$\frac{\partial H(C)}{\partial c_{ij}} = \int_{\mathbb{R}^k} h(x) \frac{\partial \varphi(x)}{\partial c_{ij}}\, dx = \int_{\mathbb{R}^k} \frac{\partial^2 h(x)}{\partial x_i \partial x_j} \varphi(x)\, dx \ge 0.$$

This completes the proof for the case of nonsingular C. The general case can be handled by approximating a singular C via a sequence of nonsingular covariance matrices. □

Proof of Theorem 2.2.1. By separability, and the final argument in the proof of the Borell–TIS inequality, it suffices to prove (2.2.2) for T finite. Note that since $\mathbb{E}\{f_t^2\} = \mathbb{E}\{g_t^2\}$ for all $t \in T$, (2.2.1) implies that $\mathbb{E}\{f_s f_t\} \ge \mathbb{E}\{g_s g_t\}$ for all $s, t \in T$. Let $h(x) = \prod_{i=1}^k h_i(x_i)$, where each h_i is a positive nonincreasing, C^2 function satisfying the growth conditions placed on h in the statement of Lemma 2.2.2, and k is the number of points in T. Note that for $i \ne j$,

[6] We could actually manage with h twice differentiable only in the sense of distributions. This would save the approximation argument following (2.2.7) below, and would give a neater, albeit slightly more demanding, proof of Slepian's inequality, as in [99].

[7] This is, of course, little more than the heat equation of PDE theory.

$$\frac{\partial^2 h(x)}{\partial x_i \partial x_j} = h_i'(x_i) h_j'(x_j) \prod_{\substack{n \neq i \\ n \neq j}} h_n(x_n) \geq 0,$$

since both h_i' and h_j' are nonpositive. It therefore follows from Lemma 2.2.2 that

$$\mathbb{E}\left\{\prod_{i=1}^{k} h_i(f_i)\right\} \geq \mathbb{E}\left\{\prod_{i=1}^{k} h_i(g_i)\right\}. \tag{2.2.7}$$

Now take $\{h_i^{(n)}\}_{n=1}^{\infty}$ to be a sequence of positive, nonincreasing C^2 approximations to the indicator function of the interval $(-\infty, \lambda]$, to derive that

$$\mathbb{P}\{\|f\| < u\} \geq \mathbb{P}\{\|g\| < u\},$$

which implies (2.2.2).

To complete the proof, all that remains is to show that (2.2.2) implies (2.2.3). But this is a simple consequence of integration by parts, since

$$\begin{aligned}
\mathbb{E}\{\|f\|\} &= \int_0^{\infty} \mathbb{P}\{\|f\| > u\}\, du - \int_{-\infty}^{0} \mathbb{P}\{\|f\| < u\}\, du \\
&\leq \int_0^{\infty} \mathbb{P}\{\|g\| > u\}\, du - \int_{-\infty}^{0} \mathbb{P}\{\|g\| < u\}\, du \\
&= \mathbb{E}\{\|g\|\}.
\end{aligned}$$

This completes the proof. □

As mentioned above, there are many extensions of Slepian's inequality, the most important of which is probably the following.

Theorem 2.2.3 (Sudakov–Fernique inequality). *Let f and g be a.s. bounded Gaussian processes on T. If*

$$\mathbb{E}\{f_t\} = \mathbb{E}\{g_t\}$$

and

$$\mathbb{E}\{(f_t - f_s)^2\} \leq \mathbb{E}\{(g_t - g_s)^2\}$$

for all $s, t \in T$, then

$$\mathbb{E}\{\|f\|\} \leq \mathbb{E}\{\|g\|\}. \tag{2.2.8}$$

In other words, a Slepian-like inequality holds without a need to assume either zero mean or identical variance for the compared processes. However, in this case we have only the weaker ordering of expectations of (2.2.3) and not the stochastic domination of (2.2.2).

The original version of this inequality assumed zero means, and its proof involved considerable calculus, in spirit not unlike that in the proof of Slepian's inequality (cf. [66] or [85]). The proof that we give is due to Sourav Chatterjee, [37] who, at the time of writing, is a gifted young graduate student at Stanford. It starts with a simple and well-known lemma.

Lemma 2.2.4. *Let $X = (X_1, \dots, X_k)$ be a vector of centered Gaussian variables with arbitrary covariance matrix, and let $h : \mathbb{R}^k \to \mathbb{R}$ be C^1, with h and its first-order derivatives satisfying a $o(|x|^d)$ growth condition at infinity for some finite d. Then, for $1 \le i \le k$,*

$$\mathbb{E}\{X_i h(X)\} = \sum_{j=1}^{k} \mathbb{E}\{X_i X_j\}\, \mathbb{E}\{h_j(X)\}. \tag{2.2.9}$$

where $h_j \triangleq \partial h/\partial x_j$.

Proof. Assume firstly that the X_j are independent and have unit variance. Then,

$$\begin{aligned}\mathbb{E}\{X_i h(X)\} &= \int_{\mathbb{R}^k} x_i h(x) e^{-|x|^2/2}\, dx \\ &= \int_{\mathbb{R}^k} h_i(x) e^{-|x|^2/2}\, dx \\ &= \mathbb{E}\{h_i(X)\},\end{aligned} \tag{2.2.10}$$

where the second equality is via a simple integration by parts. This is (2.2.9) for this case.

For the general case, write $C = A^T A$ for the covariance matrix of X, and define $h'(x) = h(Ax)$. Let $X' = (X_1,'' \dots, X'_k)$ be a vector of independent, standard normal variables. Then, $X \sim AX'$, and writing a_{mn} and a^T_{mn} for the elements of A and A^T and applying (2.2.10),

$$\begin{aligned}\mathbb{E}\{X_i h(X)\} &= \mathbb{E}\left\{\sum_{m=1}^{k} a_{im} X''_m h'(X')\right\} = \sum_{m=1}^{k} a_{im} \mathbb{E}\{h'_m(X')\} \\ &= \sum_{m=1}^{k} a_{im} \mathbb{E}\left\{\sum_{j=1}^{k} a_{jm} h_j(X)\right\} = \sum_{j=1}^{k} \mathbb{E}\{h_j(X)\} \sum_{m=1}^{k} a_{im} a^T_{mj} \\ &= \sum_{j=1}^{k} \mathbb{E}\{X_i X_j\} \mathbb{E}\{h_j(X)\},\end{aligned}$$

as required. □

The Sudakov–Fernique inequality on finite spaces is a special case of the following result, whose proof, en passant, actually provides more. For general (nonfinite) spaces the arguments that we have previously used to go from finite to general parameter work here as well, giving the classic Sudakov–Fernique inequality with no extra work.

Theorem 2.2.5. *Let* $X = (X_1, \ldots, X_k)$ *and* $Y = (Y_1, \ldots, Y_k)$ *be Gaussian vectors with* $\mathbb{E}\{X_i\} = \mathbb{E}\{Y_i\}$ *for all* i. *Set*

$$\gamma_{ij}^X = \mathbb{E}\{(X_i - X_j)^2\}, \qquad \gamma_{ij}^Y = \mathbb{E}\{(Y_i - Y_j)^2\},$$

and

$$\gamma \triangleq \sup_{1 \le i, j \le k} |\gamma_{ij}^X - \gamma_{ij}^Y|.$$

Then

$$\left|\mathbb{E}\big\{\max_i X_i\big\} - \mathbb{E}\big\{\max_i Y_i\big\}\right| \le \sqrt{2\gamma \log k}. \tag{2.2.11}$$

If, in addition, $\gamma_{ij}^X \le \gamma_{ij}^Y$ *for* $1 \le i, j \le k$, *then*

$$\mathbb{E}\big\{\max_i X_i\big\} \le \mathbb{E}\big\{\max_i Y_i\big\}. \tag{2.2.12}$$

Proof. Without loss of generality, we may assume that X and Y are defined on the same probability space and are independent. Fix $\beta > 0$ and define $h_\beta : \mathbb{R}^k \to \mathbb{R}$ by

$$h_\beta(x) \triangleq \beta^{-1} \log \left(\sum_{i=1}^k e^{\beta x_i} \right).$$

Note that

$$\begin{aligned}
\max_i x_i &= \beta^{-1} \log(e^{\beta \max_i x_i}) \le \beta^{-1} \log \left(\sum_{i=1}^k e^{\beta x_i} \right) \\
&\le \beta^{-1} \log(k e^{\beta \max_i x_i}) = \beta^{-1} \log k + \max_i x_i.
\end{aligned}$$

Thus

$$\sup_{x \in \mathbb{R}^k} \left| h_\beta(x) - \max_i x_i \right| \le \beta^{-1} \log k. \tag{2.2.13}$$

Now let $\mu_i = \mathbb{E}\{X_i\} = \mathbb{E}\{Y_i\}$ and set

$$\widetilde{X}_i = X_i - \mu_i, \qquad \widetilde{Y}_i = Y_i - \mu_i.$$

Furthermore, write $\sigma_{ij}^X = \mathbb{E}\{\widetilde{X}_i \widetilde{X}_j\}$, $\sigma_{ij}^Y = \mathbb{E}\{\widetilde{X}_i \widetilde{X}_j\}$, and for $t \in [0, 1]$ define the random vector Z_t by

$$Z_{t,i} \triangleq \sqrt{1-t}\widetilde{X}_i + \sqrt{t}\widetilde{Y}_i + \mu_i.$$

Now fix $\beta > 0$ and, for all $t \in [0, 1]$, set

$$\varphi(t) \triangleq \mathbb{E}\left\{h_\beta(Z_t)\right\}.$$

Then φ is differentiable, and

$$\varphi'(t) = \mathbb{E}\left\{\sum_{i=1}^{k} \frac{\partial h_\beta}{\partial x_i}(Z_t)\left(\frac{\widetilde{Y}_i}{2\sqrt{t}} - \frac{\widetilde{X}_i}{2\sqrt{1-t}}\right)\right\}.$$

Again, for any i, Lemma 2.2.4 gives us that

$$\mathbb{E}\left\{\frac{\partial h_\beta}{\partial x_i}(Z_t)\widetilde{X}_i\right\} = \sqrt{1-t}\sum_{j=1}^{k} \sigma_{ij}^X \mathbb{E}\left\{\frac{\partial^2 h_\beta}{\partial x_i \partial x_j}(Z_t)\right\}$$

and

$$\mathbb{E}\left\{\frac{\partial h_\beta}{\partial x_i}(Z_t)\widetilde{Y}_i\right\} = \sqrt{t}\sum_{j=1}^{k} \sigma_{ij}^Y \mathbb{E}\left\{\frac{\partial^2 h_\beta}{\partial x_i \partial x_j}(Z_t)\right\},$$

so that

$$\varphi'(t) = \frac{1}{2}\sum_{i,j=1}^{k} \mathbb{E}\left\{\frac{\partial^2 h_\beta}{\partial x_i \partial x_j}(Z_t)\right\}(\sigma_{ij}^Y - \sigma_{ij}^X). \tag{2.2.14}$$

In a moment we shall show that if $\gamma_{ij}^X \le \gamma_{ij}^Y$ then $\varphi' \ge 0$ throughout its range. This will prove (2.2.12), since it implies that for all $\beta > 0$,

$$\mathbb{E}\{h_\beta(Y)\} = \varphi(1) \ge \varphi(0) = \mathbb{E}\{h_\beta(X)\}.$$

Taking $\beta \to \infty$ and applying (2.2.13) establishes (2.2.12). The more delicate inequality (2.2.11) requires more information on φ, which comes from a little elementary calculus.

Note that

$$\frac{\partial h_\beta}{\partial x_i}(x) = \frac{e^{\beta x_i}}{\sum_{j=1}^k e^{\beta x_j}} \triangleq p_i(x),$$

where for each $x \in \mathbb{R}^k$, the $p_i(x)$ define a probability measure on $\{1, \dots, k\}$. It is then straightforward to check that

$$\frac{\partial^2 h_\beta}{\partial x_i \partial x_j}(x) = \begin{cases} \beta(p_i(x) - p_i^2(x)), & i = j, \\ -\beta p_i(x) p_j(x), & i \neq j. \end{cases}$$

Thus,

$$
\begin{aligned}
&\sum_{i,j=1}^{k} \frac{\partial^2 h_\beta}{\partial x_i \partial x_j}(x)(\sigma_{ij}^Y - \sigma_{ij}^X) \qquad (2.2.15)\\
&= \beta \sum_{i=1}^{k} p_i(x)(\sigma_{ij}^Y - \sigma_{ij}^X) - \beta \sum_{i,j=1}^{k} p_i(x)p_j(x)(\sigma_{ij}^Y - \sigma_{ij}^X)\\
&= \frac{\beta}{2} \sum_{i,j=1}^{k} p_i(x)p_j(x)[(\sigma_{ii}^Y + \sigma_{jj}^Y - 2\sigma_{ij}^Y) - (\sigma_{ii}^X + \sigma_{jj}^X - 2\sigma_{ij}^X)]\\
&= \frac{\beta}{2} \sum_{i,j=1}^{k} p_i(x)p_j(x)(\gamma_{ij}^Y - \gamma_{ij}^X).
\end{aligned}
$$

The second equality here uses the fact that $\sum_i p_i(x) = 1$, while the third one relies on the fact that

$$
\begin{aligned}
\sigma_{ii}^X + \sigma_{jj}^X - 2\sigma_{ij}^X &= \mathbb{E}\{(\widetilde{X}_i - \widetilde{X}_j)^2\}\\
&= \mathbb{E}\{(X_i - X_j)^2\} - (\mu_i - \mu_j)^2 = \gamma_{ij}^X - (\mu_i - \mu_j)^2,
\end{aligned}
$$

along with a similar equality for the σ^Y.

Substituting (2.2.15) into (2.2.14) gives us that $\varphi'(t) \geq 0$ whenever $\gamma_{ij}^X \leq \gamma_{ij}^Y$, as we claimed above.

All that remains, therefore, is to establish (2.2.11), which we do by getting a bound on $\varphi'(t)$.

Combining (2.2.15) with the definition of γ, it is immediate that

$$
\left| \sum_{i,j=1}^{k} \frac{\partial^2 h_\beta}{\partial x_i \partial x_j}(x) \left(\sigma_{ij}^Y - \sigma_{ij}^X\right) \right| \leq \frac{\beta\gamma}{2} \sum_{i,j=1}^{k} p_i(x)p_j(x) = \frac{\beta\gamma}{2}.
$$

Substituting this into (2.2.14) gives

$$
|\mathbb{E}\{h_\beta(Y)\} - \mathbb{E}\{h_\beta(X)\}| = |\varphi(1) - \varphi(0)| \leq \frac{\beta\gamma}{4}.
$$

Combining this with (2.2.13) gives

$$
\left|\mathbb{E}\left\{\max_i X_i\right\} - \mathbb{E}\left\{\max_i Y_i\right\}\right| \leq \frac{\beta\gamma}{4} + \frac{2\log k}{\beta},
$$

and choosing $\beta = \sqrt{8 \log k / \gamma}$ gives (2.2.11), as required. □

There are many extensions to the Sudakov–Fernique inequality that we shall not need in this book, but you can find them in the references in the description of Part I. From those sources you can also find out how to extend the above arguments to obtain conditions on covariance functions that allow statements of the form

$$\mathbb{P}\{\min_{1\le i\le n}\max_{1\le j\le m} X_{ij} \ge u\} \ge \mathbb{P}\{\min_{1\le i\le n}\max_{1\le j\le m} Y_{ij} > u\},$$

along with even more extensive variations due, originally, to Gordon [69]. Gordon [70] also shows how to extend the essentially Gaussian computations above to elliptically contoured distributions.

As we mentioned earlier, Chapter 4, which does not require the material of Chapter 3, continues with the theme of suprema distributions.

3

Orthogonal Expansions

While most of what we shall have to say in this brief chapter is rather theoretical, it actually covers one of the most important practical aspects of Gaussian modeling. The basic result is Theorem 3.1.1, which states that *every* centered Gaussian process with a continuous covariance function has an expansion of the form

$$f(t) = \sum_{n=1}^{\infty} \xi_n \varphi_n(t), \tag{3.0.1}$$

where the ξ_n are i.i.d. $N(0, 1)$, and the φ_n are certain functions on T determined by the covariance function C of f. In general, the convergence in (3.0.1) is in $L^2(\mathbb{P})$ for each $t \in T$, but (Theorem 3.1.2) if f is a.s. continuous then the convergence is uniform over T, with probability one.

There are many theoretical conclusions that follow from this representation. For one example, note that since continuity of C will imply that of the φ_n (cf. Lemma 3.1.4), it follows from (3.0.1) that sample path continuity of f is a "tail event" on the σ-algebra determined by the ξ_n, from which one can show that centered Gaussian processes are either continuous with probability one, or discontinuous with probability one. There is no middle ground. A wide variety of additional zero–one laws also follow from this representation. Perhaps the most notable is the following,

$$\begin{aligned} &\mathbb{P}\left\{\lim_{s\to t} f_s = f_t \text{ for all } t \in T\right\} = 1 \\ &\qquad \Longleftrightarrow \mathbb{P}\left\{\lim_{s\to t} f_s = f_t\right\} = 1 \text{ for each } t \in T. \end{aligned}$$

That is, under our ubiquitous assumption that the covariance function C is continuous, pointwise and global a.s. continuity are equivalent for Gaussian processes. A proof, in the spirit of what follows, can be found in [3].

The practical implications of (3.0.1) are mainly in the area of simulation. By truncating the sum (3.0.1) at a point appropriate to the problem at hand, one needs "only" to determine the φ_n. However, these arise as the orthonormal basis of a particular Hilbert space, and can generally be found by solving an eigenfunction problem involving C. In particular, if T is a nice subset of $\mathbb{R}^N$, this leads to the famous Karhunen–Loève expansion, which we shall treat briefly in Section 3.2. While even in the Euclidean situation there are only a handful of situations for which this eigenfunction problem can be solved analytically, from the point of view of computing it is a standard problem, and the approach is practical.

3.1 The General Theory

As usual, we shall take T to be compact in the canonical metric d generated by f.

The first step toward establishing the expansion (3.0.1) lies in setting up the so-called *reproducing kernel Hilbert space* (RKHS) of a centered Gaussian process with covariance function C.

In essence, the RKHS is made up of functions that have about the same smoothness properties that $C(s,t)$ has, as a function in t for fixed s, or vice versa. Start with

$$S = \left\{ u : T \to \mathbb{R} : u(\cdot) = \sum_{i=1}^{n} a_i C(s_i, \cdot),\ a_i \text{ real}, s_i \in T,\ n \geq 1 \right\}.$$

Define an inner product on S by

$$(u, v)_H = \left(\sum_{i=1}^{n} a_i C(s_i, \cdot), \sum_{j=1}^{m} b_j C(t_j, \cdot) \right)_H = \sum_{i=1}^{n} \sum_{j=1}^{m} a_i b_j C(s_i, t_j). \qquad (3.1.1)$$

The fact that C is nonnegative definite implies $(u,u)_H \geq 0$ for all $u \in S$. Furthermore, note that the inner product (3.1.1) has the following unusual property:

$$(u, C(t, \cdot))_H = \left(\sum_{i=1}^{n} a_i C(s_i, \cdot), C(t, \cdot) \right)_H = \sum_{i=1}^{n} a_i C(s_i, t) = u(t). \qquad (3.1.2)$$

This is the *reproducing kernel* property.

For the sake of exposition, assume that the covariance function, C, is positive definite (rather than merely nonnegative definite), so that $(u,u)_H = 0$ if and only if $u(t) \equiv 0$. In this case (3.1.1) defines a norm $\|u\|_H = (u,u)_H^{1/2}$.

For $\{u_n\}_{n\geq 1}$ a sequence in S we have

$$\begin{aligned} |u_n(t) - u_m(t)| &= |(u_n - u_m, C(t, \cdot))_H| \leq \|u_n - u_m\|_H \|C(t, \cdot)\|_H \\ &\leq \|u_n - u_m\|_H C(t,t), \end{aligned}$$

the last line following directly from (3.1.1). Thus it follows that if $\{u_n\}$ is Cauchy in $\|\cdot\|_H$ then it is pointwise Cauchy. The closure of S under this norm is a space of real-valued functions, denoted by $H(C)$, and called the RKHS of f or of C, since every $u \in H(C)$ satisfies (3.1.2) by the separability of $H(C)$. (The separability of $H(C)$ follows from the separability of T and the assumption that C is continuous.)

Since all this seems at first rather abstract, consider two concrete examples. Take $T = \{1, \ldots, n\}$ finite, and f centered Gaussian with covariance matrix $C = (c_{ij})$, $c_{ij} = \mathbb{E}\{f_i f_j\}$. Let $C^{-1} = (c^{ij})$ denote the inverse of C, which exists by positive definiteness. Then the RKHS of f is made up of all n-dimensional vectors $u = (u_1, \ldots, u_n)$ with inner product

$$(u, v)_H = \sum_{i=1}^{n} \sum_{j=1}^{n} u_i c^{ij} v_j.$$

To prove this, we need only check that the reproducing kernel property (3.1.2) holds.[1] However, with $\delta(i, j)$ the Kronecker delta function, and C_k denoting the kth row of C,

$$(u, C_k)_H = \sum_{i=1}^{n} \sum_{j=1}^{n} u_i c^{ij} c_{kj} = \sum_{i=1}^{n} u_i \delta(i, k) = u_k,$$

as required.

For a slightly more interesting example, take $f = W$ to be standard Brownian motion on $T = [0, 1]$, so that $C(s, t) = \min(s, t)$. Note that $C(s, \cdot)$ is differentiable everywhere except at s, so that following the heuristics developed above we expect that $H(C)$ should be made up of a subset of functions that are differentiable almost everywhere.

To both make this statement more precise and prove it, we start by looking at the space S. Thus, let

$$u(t) = \sum_{i=1}^{n} a_i C(s_i, t), \qquad v(t) = \sum_{i=1}^{n} b_i C(t_i, t),$$

be two elements of S, with inner product

$$(u, v)_H = \sum_{i=1}^{n} \sum_{j=1}^{n} a_i b_j \min(s_i, t_j).$$

Since the derivative of $C(s, t)$ with respect to t is $\mathbb{1}_{[0,s]}(t)$, the derivative of u is $\sum_{i=1}^{n} a_i \mathbb{1}_{[0,s_i]}(t)$. Therefore,

$$\begin{aligned}(u, v)_H &= \sum \sum a_i b_j \int_0^1 \mathbb{1}_{[0,s_i]}(t) \mathbb{1}_{[0,t_j]}(t)\, dt \\ &= \int_0^1 \Big(\sum a_i \mathbb{1}_{[0,s_i]}(t) \sum b_j \mathbb{1}_{[0,t_j]}(t) \Big)\, dt \\ &= \int_0^1 \dot{u}(t) \dot{v}(t)\, dt.\end{aligned}$$

---

[1] A simple proof by contradiction shows that there can never be two different inner products on S with the reproducing kernel property.

With S under control, we can now look for an appropriate candidate for the RKHS. Define

$$H = \left\{ u : u(t) = \int_0^t \dot{u}(s)\,ds,\ \int_0^1 (\dot{u}(s))^2\,ds < \infty \right\}, \tag{3.1.3}$$

equipped with the inner product

$$(u, v)_H = \int_0^1 \dot{u}(s)\dot{v}(s)\,ds. \tag{3.1.4}$$

Since it is immediate that $C(s, \cdot) \in H$ for $t \in [0, 1]$, and

$$(u, C(t, \cdot))_H = \int_0^1 \dot{u}(s)\mathbb{1}_{[0,t]}(s)\,ds = u(t),$$

it follows that the H defined by (3.1.3) is indeed our RKHS. This H is also known, in the setting of diffusion processes, as a *Cameron–Martin* space.

With a couple of examples under our belt, we can now return to our main task: setting up the expansion (3.0.1). The first step is finding a countable orthonormal basis for the separable Hilbert space $H(C)$. We start with $\mathcal{H} = \text{span}\{f_t,\ t \in T\}$, with the (covariance) inner product that $\mathcal{H}$ inherits as a subspace of $L^2(\mathbb{P})$. Next we define a linear mapping $\Theta : S \to \mathcal{H}$ by

$$\Theta(u) = \Theta\left(\sum_{i=1}^n a_i C(t_i, \cdot)\right) = \sum_{i=1}^n a_i f(t_i).$$

Clearly, $\Theta(u)$ is Gaussian for each $u \in S$, and Θ is norm-preserving. That is,

$$\left\|\Theta\left(\sum a_i C(t_i, \cdot)\right)\right\|_H^2 = \sum_{i,j} a_i C(t_i, t_j) a_j = \left\|\sum a_i f(t_i)\right\|_2^2.$$

Consequently, Θ extends to all of $H(C)$ with range equal to all of $\mathcal{H}$, with all limits remaining Gaussian. This extension is called the *canonical isomorphism* between these spaces.

Since $H(C)$ is separable, we now also know that $\mathcal{H}$ is, and proceed to build an orthonormal basis for it. If $\{\varphi_n\}_{n\geq 1}$ is an orthonormal basis for $H(C)$, then setting $\xi_n = \Theta(\varphi_n)$ gives $\{\xi_n\}_{n\geq 1}$ as an orthonormal basis for $\mathcal{H}$. In particular, we must have that the ξ_n are $N(0, 1)$ and

$$f_t = \sum_{n=1}^\infty \xi_n \mathbb{E}\{f_t \xi_n\}, \tag{3.1.5}$$

where the series converges in $L^2(\mathbb{P})$. Since Θ was an isometry, it follows from (3.1.5) that

$$\mathbb{E}\{f_t \xi_n\} = (C(t, \cdot), \varphi_n)_H = \varphi_n(t), \tag{3.1.6}$$

the last equality coming from the reproducing kernel property of $H(C)$. Putting (3.1.6) together with (3.1.5) now establishes the following central result.

Theorem 3.1.1. *If $\{\varphi_n\}_{n\geq 1}$ is an orthonormal basis for $H(C)$, then f has the L^2 representation*

$$f_t = \sum_{n=1}^{\infty} \xi_n \varphi_n(t), \tag{3.1.7}$$

where $\{\xi_n\}_{n\geq 1}$ is the orthonormal sequence of centered Gaussian variables given by $\xi_n = \Theta(\varphi_n)$.

The equivalence in (3.1.7) is only in L^2; i.e., the sum is, in general, convergent, for each t, only in mean square. The following result shows that much more is true if we know that f is a.s. continuous.

Theorem 3.1.2. *If f is a.s. continuous, then the sum in* (3.1.7) *converges uniformly on T with probability* 1.[2]

We need two preliminary results before we can prove Theorem 3.1.2. The first is a convergence result due to Itô and Nisio [83], which, since it is not really part of a basic probability course,[3] we state in full, and the second an easy lemma.

Lemma 3.1.3. *Let $\{Z_n\}_{n\geq 1}$ be a sequence of symmetric independent random variables, taking values in a separable, real Banach space B, equipped with the norm topology. Let $X_n = \sum_{i=1}^n Z_i$. Then X_n converges with probability one if and only if there exists a B-valued random variable X such that $\langle X_n, x^*\rangle \to \langle X, x^*\rangle$ in probability for every $x^* \in B^*$, the topological dual of B.*

Lemma 3.1.4. *Let $\{\varphi_n\}_{n\geq 1}$ be an orthonormal basis for $H(C)$. Then, under our global assumption of continuity of the covariance function, each φ_n is continuous and*

$$\sum_{n=1}^{\infty} \varphi_n^2(t) \tag{3.1.8}$$

converges uniformly in $t \in T$ to $C(t,t)$.

Proof. Note that

$$\begin{aligned}
|\varphi_n(s) - \varphi_n(t)| &= \left|(C(s,\cdot), \varphi_n(\cdot))_H - (C(t,\cdot), \varphi_n(\cdot))_H\right| \\
&= \left|([C(s,\cdot) - C(t,\cdot)], \varphi_n(\cdot))_H\right| \\
&\leq \|\varphi_n\|_H \|C(s,\cdot) - C(t,\cdot)\|_H \\
&= \|C(s,\cdot) - C(t,\cdot)\|_H \\
&= C(s,s) - 2C(s,t) + C(t,t) \\
&= d^2(s,t),
\end{aligned}$$

---

[2] There is also a converse to Theorem 3.1.2, that the a.s. uniform convergence of a sum like (3.1.7) implies the continuity of f, and some of the earliest derivations (e.g., [68]) of sufficient conditions for continuity actually used this approach. Entropy-based arguments, however, turn out to give much easier proofs.

[3] When the Banach space of the Itô–Nisio theorem is the real line, this *is* a textbook result. In that case, however, it goes under the name of Lévy's theorem.

where the first and last-but-one equalities use the reproducing kernel property and the one inequality is Cauchy–Schwarz. The continuity of φ_n now follows from that of C.

To establish the uniform convergence of (3.1.8), note that the orthonormal expansion and the reproducing kernel property imply

$$C(t,\cdot)=\sum_{n=1}^{\infty}\varphi_n(\cdot)(C(t,\cdot),\varphi_n)_H=\sum_{n=1}^{\infty}\varphi_n(\cdot)\varphi_n(t), \tag{3.1.9}$$

convergence of the sum being in the $\| \; \|_H$ norm. Hence, $\sum_{n=1}^{\infty}\varphi_n^2(t)$ converges to $C(t,t)$ for every $t\in T$. Furthermore, the convergence is monotone, and so it follows that it is also uniform ($\equiv$ Dini's theorem). □

Proof of Theorem 3.1.2. We know that for each $t\in T$, $\sum_{n=1}^{\infty}\xi_n\varphi_n(t)$ is a sum of independent variables converging in $L^2(\mathbb{P})$. Thus, by Lemma 3.1.3, applied to real-valued random variables, it converges with probability one to a limit, which we denote by f_t. The limit process is, by assumption, almost surely continuous.

Now consider both f and each function $\xi_n\varphi_n$ as random variables in the Banach space $C(T)$, with sup-norm topology. Elements of the dual $C^*(T)$ are therefore finite, signed Borel measures μ on T, and $\langle f,\mu\rangle=\int f\,d\mu$. Define

$$f_n(\cdot)=\sum_{i=1}^{n}\xi_i\varphi_i(\cdot)=\sum_{i=1}^{n}\Theta(\varphi_i)\varphi_i(\cdot).$$

By Lemma 3.1.3, it suffices to show that for every $\mu\in C^*(T)$ the random variables $\langle f_n,\mu\rangle$ converge in probability to $\langle f,\mu\rangle$. However,

$$\begin{aligned}
\mathbb{E}\{|\langle f_n,\mu\rangle-\langle f,\mu\rangle|\}&=\mathbb{E}\left\{\left|\int_T(f_n(t)-f(t))\mu(dt)\right|\right\}\\
&\le\int_T\mathbb{E}\{|f_n(t)-f(t)|\}\,|\mu|(dt)\\
&\le\int_T[\mathbb{E}(f_n(t)-f(t))^2]^{\frac12}\,|\mu|(dt)\\
&=\int_T\left(\sum_{j=n+1}^{\infty}\varphi_j^2(t)\right)^{\frac12}|\mu|(dt),
\end{aligned}$$

where $|\mu|(A)$ is the total variation measure for μ.

Since $\sum_{j=n+1}^{\infty}\varphi_j^2(t)\to 0$ uniformly in $t\in T$ by Lemma 3.1.4, the last expression above tends to zero as $n\to\infty$. Since this implies the convergence in probability of $\langle f_n,\mu\rangle$ to $\langle f,\mu\rangle$, we are done. □

3.2 The Karhunen–Loève Expansion

As we have already noted, applying the orthogonal expansion (3.1.7) in practice relies on being able to find the orthonormal functions φ_n. When T is a compact subset of $\mathbb{R}^N$ there is a special way in which to do this, leading to what is known as the Karhunen–Loève expansion.

For simplicity, take $T = [0, 1]^N$. Let $\lambda_1 \geq \lambda_2 \geq \cdots$, and $\psi_1, \psi_2, \ldots$, be, respectively, the eigenvalues and normalized eigenfunctions of the operator $\mathcal{C} : L^2(T) \to L^2(T)$ defined by $(\mathcal{C}\psi)(t) = \int_T C(s,t)\psi(s)\,ds$. That is, the λ_n and ψ_n solve the integral equation

$$\int_T C(s,t)\psi(s)\,ds = \lambda\psi(t), \tag{3.2.1}$$

with the normalization

$$\int_T \psi_n(t)\psi_m(t)\,dt = \begin{cases} 1, & n = m, \\ 0, & n \neq m. \end{cases}$$

These eigenfunctions lead to a natural expansion of C, known as Mercer's theorem (see [132, 181] for a proof).

Theorem 3.2.1 (Mercer). *Let C, $\{\lambda_n\}_{n\geq 1}$, and $\{\psi_n\}_{n\geq 1}$ be as above. Then*

$$C(s,t) = \sum_{n=1}^{\infty} \lambda_n \psi_n(s)\psi_n(t), \tag{3.2.2}$$

where the series converges absolutely and uniformly on $[0,1]^N \times [0,1]^N$.

The key to the Karhunen–Loève expansion is the following result.

Lemma 3.2.2. *For f on $[0,1]^N$ as above, $\{\sqrt{\lambda_n}\psi_n\}$ is a complete orthonormal system in $H(C)$.*

Proof. Set $\varphi_n = \sqrt{\lambda_n}\psi_n$ and define

$$H = \left\{ h : h(t) = \sum_{n=1}^{\infty} a_n \varphi_n(t),\ t \in [0,1]^N,\ \sum_{n=1}^{\infty} a_n^2 < \infty \right\}.$$

Give H the inner product

$$(h, g)_H = \sum_{n=1}^{\infty} a_n b_n,$$

where $h = \sum a_n \varphi_n$ and $g = \sum b_n \varphi_n$.

To check that H has the reproducing kernel property, note that

$$\begin{aligned} (h(\cdot), C(t,\cdot))_H &= \left(\sum_{n=1}^{\infty} a_n \varphi_n(\cdot), \sum_{n=1}^{\infty} \sqrt{\lambda_n}\psi_n(t)\varphi_n(\cdot) \right)_H \\ &= \sum_{n=1}^{\infty} \sqrt{\lambda_n} a_n \psi_n(t) = h(t). \end{aligned}$$

It remains to be checked that H is in fact a Hilbert space, and that $\{\sqrt{\lambda_n}\psi_n\}$ is both complete and orthonormal. But all this is standard, given Mercer's theorem. □

We can now start rewriting things to get the expansion we want. Remaining with the basic notation of Mercer's theorem, we have that the RKHS, $H(C)$, consists of all square-integrable functions h on $[0,1]^N$ for which

$$\sum_{n=1}^{\infty} \lambda_n \left| \int_T h(t)\psi_n(t)\,dt \right|^2 < \infty,$$

with inner product

$$(h,g)_H = \sum_{n=1}^{\infty} \lambda_n \int_T h(t)\psi_n(t)\,dt \int_T g(t)\psi_n(t)\,dt.$$

The Karhunen–Loève expansion of f is obtained by setting $\varphi_n = \lambda_n^{\frac{1}{2}}\psi_n$ in the orthonormal expansion (3.1.7), so that

$$f_t = \sum_{n=1}^{\infty} \lambda_n^{\frac{1}{2}} \xi_n \psi_n(t), \tag{3.2.3}$$

where the ξ_n are i.i.d. $N(0,1)$.

As simple as this approach might seem, it is generally limited by the fact that it is usually not easy to solve the integral equation (3.2.1) analytically. There is, however, at least one classic, example for which this can be done, the Brownian sheet of Section 1.4.3, which we recall has covariance function

$$\mathbb{E}\{W_s W_t\} = (s_1 \wedge t_1) \times \cdots \times (s_N \wedge t_N) \tag{3.2.4}$$

on $I^N = [0,1]^N$.

For the moment, we set $N = 1$ and so have the standard Brownian motion on $[0,1]$, for which we have already characterized the corresponding RKHS as the Cameron–Martin space. For Brownian motion (3.2.1) becomes

$$\lambda\psi(t) \int_0^1 \min(s,t)\psi(s)\,ds = \int_0^t s\psi(s)\,ds + t\int_t^1 \psi(s)\,ds.$$

Differentiating both sides with respect to t gives

$$\lambda\psi'(t) = \int_t^1 \psi(s)\,ds,$$
$$\lambda\psi''(t) = -\psi(t),$$

together with the boundary conditions $\psi(0) = 0$ and $\psi'(1) = 0$.

The solutions of this pair of differential equations are given by

$$\psi_n(t) = \sqrt{2}\sin\left(\frac{1}{2}(2n+1)\pi t\right), \qquad \lambda_n = \left(\frac{2}{(2n+1)\pi}\right)^2, \tag{3.2.5}$$

as is easily verified by substitution. Thus, the Karhunen–Loève expansion of Brownian motion on [0, 1] is given by

$$W_t = \frac{\sqrt{2}}{\pi} \sum_{n=0}^{\infty} \xi_n \left(\frac{2}{2n+1} \right) \sin \left(\frac{1}{2}(2n+1)\pi t \right).$$

Returning now to the general Brownian sheet, it is now clear from the product form (3.2.4) of the covariance function that rather than taking a single sum in the Karhunen–Loève expansion, it is natural to work with an N-dimensional sum and a multi-index $n = (n_1, \dots, n_N)$. With this in mind, it follows then from (3.2.5) that the eigenfunctions and eigenvalues are given by

$$\psi_n(t) = 2^{N/2} \prod_{i=1}^{N} \sin \left(\frac{1}{2} (2n_i + 1) \pi t_i \right),$$

$$\lambda_n = \prod_{i=1}^{N} \left(\frac{2}{(2n_i+1)\pi} \right)^2 .$$

As an aside, there is also an elegant expansion of the Brownian sheet using the Haar functions, which also works in the set-indexed setting, due to Ron Pyke [129].

Another class of examples that can almost be handled via the Karhunen–Loève approach is that of stationary fields defined on all of (noncompact) $\mathbb{R}^N$. Since what is about to come is rather imprecise, we shall allow ourselves the standard notational luxury of the stationary theory of taking our field to be complex-valued, despite the fact that all that we have done in this chapter was for real-valued processes. In this case the covariance function is $C(s,t) = \mathbb{E}\{f_s \overline{f_t}\} = C(t-s)$ and is a function of $t-s$ only. It is then easy to find eigenfunctions for (3.2.1) via complex exponentials. Note that for any $\lambda \in \mathbb{R}^N$, the function $e^{i\langle t,\lambda\rangle}$ (a function of $t \in \mathbb{R}^N$) satisfies

$$\begin{aligned} \int_{\mathbb{R}^N} C(s,t) e^{i\langle s,\lambda\rangle}\, ds &= \int_{\mathbb{R}^N} C(t-s) e^{i\langle s,\lambda\rangle}\, ds \\ &= e^{i\langle t,\lambda\rangle} \int_{\mathbb{R}^N} C(u) e^{-i\langle u,\lambda\rangle}\, du = K_\lambda e^{i\langle t,\lambda\rangle}, \end{aligned}$$

for some, possibly zero, K_λ.

Suppose that $K_\lambda \neq 0$ for only a countable number of $\lambda \in \mathbb{R}^N$. Then the Mercer expansion (3.2.2) can be written as

$$C(t) = \sum_{\lambda} K_\lambda e^{i\langle t,\lambda\rangle}, \tag{3.2.6}$$

while the Karhunen–Loève expansion becomes

$$f(t) = \sum_{\lambda} K_\lambda^{1/2} \xi_\lambda e^{i\langle t,\lambda\rangle}. \tag{3.2.7}$$

These are, respectively, special cases of the spectral distribution theorem (cf. (5.4.1)) and the spectral representation theorem (cf. (5.4.6)) of Chapter 5 when the spectral measure is discrete.

Despite the minor irregularity of assuming that f is complex-valued, the above argument is completely rigorous. On the other hand, what follows is not.

If $K_\lambda \neq 0$ on an uncountable set, then the situation becomes more delicate, but is nevertheless worth looking at. In this case, one could imagine replacing the summations in (3.2.6) and (3.2.7) by integrals, to obtain

$$C(t) = \int_{\mathbb{R}^N} K_\lambda e^{i\langle t,\lambda\rangle}\, d\lambda$$

and

$$f(t) = \int_{\mathbb{R}^N} K_\lambda^{1/2} \xi_\lambda e^{i\langle t,\lambda\rangle}\, d\lambda. \tag{3.2.8}$$

Everything is well defined in the first of these integrals, but in the second we have the problem that the ξ_λ should be independent for each λ, and it is well known that there is no measurable way to construct such a process.

Nevertheless, we shall see in Chapter 5 that there are ways to get around these problems, and that when properly formulated, (3.2.8) actually makes sense.

4

Excursion Probabilities

As we have already mentioned more than once before, one of the oldest and most important problems in the study of stochastic processes of any kind is to evaluate the *excursion probabilities*

$$\mathbb{P}\Big\{ \sup_{t\in T} f(t) \geq u \Big\},$$

where f is a random process over some parameter set T. As usual, we shall restrict ourselves to the case in which f is centered and Gaussian and T is compact for the canonical metric of (1.3.1).

While the Borell–TIS inequality, Theorem 2.1.1, gives an easy and universal bound to Gaussian excursion probabilities, it is far from optimal. In fact, computing excursion probabilities for general Gaussian processes is a surprisingly difficult problem. Even the simple case of $T = [0, 1]$ is hard. In this one-dimensional case, there is hope for obtaining an explicit expression for the excursion probability if f is Markovian. For example, if f is Brownian motion, then, as any elementary textbook on stochastic processes will tell you, the so-called reflection principle easily yields that $\mathbb{P}\{\sup_{0\leq t\leq 1} f(t) \geq u\} = 2\Psi(u)$, where Ψ is the Gaussian tail probability (1.2.1).

However, if we turn to stationary processes on $\mathbb{R}$, then there are only five nontrivial[1] Gaussian processes for which the excursion probability has a known analytic form, and the resulting formulas, while amenable to numerical computation, are not generally very pretty.[2]

In each of these five examples the process is either Markovian or close to Markovian, and this is what makes the computation feasible. Similarly, none of the processes has even a mean square derivative, and so none are particularly smooth. In the case of smooth processes, or processes over parameter spaces richer than the unit interval, the very specific tools that work for these very special cases fail completely. What one can do in the general case is the subject of this chapter and of Chapter 14.

[1] There is also one close-to-trivial but extremely enlightening case, with covariance function $\cos(\omega t)$, for which elementary calculations give the excursion probability. We shall discuss this in some detail in Section 14.4.4.

[2] The five covariance functions are

Our ultimate aim will be to develop, when possible and appropriate, expansions of the form

$$\mathbb{P}\Big\{\sup_{t\in T} f(t) \geq u\Big\} = u^{\alpha} e^{-u^2/2\sigma_T^2}\left[\sum_{j=0}^{n} C_j u^{-j} + \text{error}\right] \tag{4.0.1}$$

for large u and appropriate parameters α, σ_T^2, n, and C_j that depend on both f and T. Furthermore, we would like to be able to identify the constants C_j and also to be able to say something about the error term.

In Chapter 14 we shall indeed establish such a result, with a full identification of all constants. However, we shall have to make two major assumptions. One is that the parameter space T is a piecewise smooth manifold, and the second is that f is almost surely C^2. The current chapter will make fewer assumptions, and will concentrate on general approaches relying heavily on the concept of entropy from Chapter 1. As a consequence, we shall generally be able to identify only the first term in (4.0.1). Even then, while we shall be able to determine α and σ_T^2, only under rare cases shall we also be able to identify C_0.

Furthermore, so as to keep the treatment down to a reasonable length, we shall generally concentrate on upper bounds, rather than expansions, for Gaussian excursion probabilities.

This chapter contains six sections. The first four have just been described. The remaining two are both brief, and are meant as quick introductions to techniques that can lead to sharper results under additional assumptions. They lie between the general results of the first four sections and the very sharp results, for smooth Gaussian fields on structured domains, of Chapter 14.

4.1 Entropy Bounds

As we saw in Section 2.1, an immediate consequence of the Borell–TIS inequality is the fact that for all $\varepsilon > 0$ and large enough u,

---

$$C(t) = e^{-|t|},$$

$$C(t) = \begin{cases} 1-|t|, & |t|\leq 1,\\ 0, & |t|\geq 1,\end{cases}$$

$$C(t) = \tfrac{3}{2}\exp[-\tfrac{1}{3}|t|]\{1-\tfrac{1}{3}\exp[-2|t|/\sqrt{3}]\},$$

$$C(t) = \begin{cases} 1-\alpha|t|, & |t|\leq 1,\\ C(t-1), & -\infty < t < \infty,\end{cases}$$

$$C(t) = \begin{cases} 1-|t|/(1-\beta), & |t|\leq 1,\\ -\beta/(1-\beta), & 1<|t|<(1-\beta)/\beta,\end{cases}$$

where in the last case, $\beta \in (0,1)$, and the computation of the excursion probability is known only for $f(0)$ conditioned to be 0. For details, see [145, 43] and the review [26].

$$\mathbb{P}\left\{\sup_{t\in T} f_t \ge u\right\} \le e^{\varepsilon u^2 - u^2/2\sigma_T^2}, \tag{4.1.1}$$

where $\sigma_T^2 = \sup_T \mathbb{E}\{f_t^2\}$ (cf. (2.1.4)). This is in the classic form of a large-deviation result, and holds with no assumptions beyond the almost sure boundedness of f. Our aim is to improve on (4.1.1) by placing side conditions on the entropy (cf. Definition 1.3.2) of f over T. Here is a very easy, and definitely suboptimal, result.

Theorem 4.1.1. *Let f be a centered, a.s. continuous Gaussian field over T with entropy function N. If $N(\varepsilon) \le K\varepsilon^{-\alpha}$, then for all sufficiently large u,*

$$\mathbb{P}\left\{\sup_{t\in T} f(t) \ge u\right\} \le C_\alpha u^{\alpha+\eta} e^{-u^2/2\sigma_T^2}, \tag{4.1.2}$$

for every $\eta > 0$, where $C_\alpha = C(K, \alpha, \sigma_T^2)$ is a finite constant.

Proof. Take $\varepsilon > 0$ and define

$$\mu(t,\varepsilon) = \mathbb{E}\left\{\sup_{s\in B_d(t,\varepsilon)} f_s\right\}$$

and

$$\mu(\varepsilon) = \sup_{t\in T} \mu(t,\varepsilon), \tag{4.1.3}$$

where $B_d(t,\varepsilon)$ is a ball of radius ε around t in the canonical metric d of (1.3.1). Since $N(\varepsilon)$ balls of radius ε cover T, it is an immediate consequence of the Borell–TIS inequality that for $u > \mu(\varepsilon)$,

$$\mathbb{P}\left\{\sup_{t\in T} f(t) \ge u\right\} \le N(\varepsilon) e^{-\frac{1}{2}(u-\mu(\varepsilon))^2/\sigma_T^2}. \tag{4.1.4}$$

Allowing $C = C(\alpha)$ to denote a constant, dependent only on α and that may change from line to line, we have from Theorem 1.3.3 that

$$\mu(t,\varepsilon) \le C\int_0^\varepsilon (\log N(\varepsilon))^{\frac{1}{2}}\, d\varepsilon \le C\int_0^\varepsilon (\log(1/\varepsilon))^{\frac{1}{2}}\, d\varepsilon \le C\varepsilon\sqrt{\log(1/\varepsilon)}. \tag{4.1.5}$$

Set $\varepsilon = \varepsilon(u) = u^{-1}$, choose u large enough so that $u > Cu^{-1}\sqrt{\log u}$, and substitute into (4.1.4) to obtain

$$\begin{aligned}\mathbb{P}\left\{\sup_{t\in T} f(t) \ge u\right\} &\le C_1 u^\alpha e^{-\frac{1}{2}\left(u - C_2 u^{-1}\sqrt{\log u}\right)^2/\sigma_T^2} \\ &\le C_3 u^\alpha e^{C_4\sqrt{\log u}} e^{-u^2/2\sigma_T^2}.\end{aligned} \tag{4.1.6}$$

Since for $\eta > 0$ and u large enough $e^{C\sqrt{\log u}} < u^\eta$, this gives us (4.1.2) and so completes the proof. □

The bound in (4.1.2) can be improved slightly, to $C(u\log u)^\alpha e^{-u^2/2\sigma_T^2}$, by choosing $\varepsilon=\varepsilon(u)$ in the above proof to satisfy $u^{-1}=\varepsilon(\log(1/\varepsilon))^{1/2}$. The gain, however, is rather small, in view of the fact that far stronger results hold. In particular, in a series of papers [135, 136, 150, 137, 153] by Samorodnitsky and Talagrand, with a leapfrogging of ideas and techniques,[3] the results of the following three theorems, among others, were obtained. In each, f is a priori assumed to have continuous sample paths with probability one, $\sigma_T^2 \triangleq \sup_T \mathbb{E}\{f_t^2\}$, and at each appearance, K represents a universal constant that may differ between appearances.

Theorem 4.1.2. *If for some $A>\sigma_T$, some $\alpha>0$, and some $\varepsilon_0\in[0,\sigma_T]$ we have*

$$N(T,d,\varepsilon)\le (A/\varepsilon)^\alpha \tag{4.1.7}$$

for all $\varepsilon<\varepsilon_0$, then for $u\ge \sigma_T^2[(1+\sqrt{\alpha})/\varepsilon_0]$ we also have

$$\mathbb{P}\left\{\sup_{t\in T} f_t\ge u\right\}\le \left(\frac{KAu}{\sqrt{\alpha}\sigma_T^2}\right)^\alpha \Psi\left(\frac{u}{\sigma_T}\right), \tag{4.1.8}$$

where K is a universal constant.[4]

Theorem 4.1.3. *Set*

$$T_\delta=\left\{t\in T:\mathbb{E}\{f_t^2\}\ge \sigma_T^2-\delta^2\right\}. \tag{4.1.9}$$

Suppose there exist $\alpha>\beta>1$ such that for all $\delta>0$, $\varepsilon\in(0,\delta(1+\sqrt{\alpha})/\sqrt{\beta})$, we have

$$N(T_\delta,d,\varepsilon)\le A\delta^\beta\varepsilon^{-\alpha}. \tag{4.1.10}$$

Then for $u\ge 2\sigma_T\sqrt{\beta}$,

$$\mathbb{P}\left\{\sup_{t\in T} f_t\ge u\right\}\le \frac{A\beta^{\beta/2}}{\alpha^{\alpha/2}}K^{\alpha+\beta}\left(\frac{u}{\sigma_T^2}\right)^{\alpha-\beta}\Psi\left(\frac{u}{\sigma_T}\right). \tag{4.1.11}$$

Theorem 4.1.4. *Suppose there exist $A,B>0$ and $\alpha\in(0,2)$ such that*

$$N(T,\varepsilon)<A\exp(B\varepsilon^{-\alpha}).$$

---

[3] While the Samorodnitsky and Talagrand papers are what led to the final form presented here, the key arguments are much older, and two of the key players were Simeon Berman (e.g., [20, 21, 22, 23]) and Michel Weber (e.g., [167]).

[4] Note that at first sight, the right-hand side of (4.1.8) seems to scale incorrectly, since in the power term we would expect to see u/σ_T rather than u/σ_T^2. This is an artifact of the condition $A>\sigma_T$, which implies that increasing σ_T requires increasing A as well, thus ensuring correct scaling. The same phenomenon appears in (4.1.11), albeit a little more subtly.

Then, for all $u > 0$,

$$\mathbb{P}\left\{\sup_{t\in T} f_t > u\right\} \le K_1 \exp\left(K_2 u^{2\alpha/(2+\alpha)}\right) \Psi\left(\frac{u}{\sigma_T}\right),$$

where K_1 and K_2 are universal.

In a moment we shall turn to proofs of the first two of these results,[5] adopting (and occasionally correcting typos and oversights in) the proofs in [153]. Note that (4.1.10) raises an interesting question whether we (formally) take $\alpha = \beta$. In that case the upper bound is of the form $C(\alpha, \beta)\Psi(u/\sigma_T)$, and this also serves as a trivial lower bound when $C = 1$. Thus it is natural to ask if there are scenarios in which one can in fact also take $C = 1$ in the upper bound. It turns out that there are, and we shall look at these in Section 4.2.

Theorems 4.1.2 and 4.1.3 treat entropy functions for which the growth of N in ε is of a power form, which, at least for processes indexed by points in Euclidean spaces or smooth manifolds, are the most common. Theorem 4.1.4, which we shall not prove, handles the scenario of exponential entropy and is a qualitatively different result. Note that you cannot set $\alpha = 0$ in this result to recover either Theorem 4.1.2 or 4.1.3. The upper bound given here is, under mild side conditions, also a lower bound.

Before reading further, you might want to turn now to Section 4.3 to see some concrete examples for which the above results apply. Here, however, we shall give the proofs of Theorems 4.1.2 and 4.1.3. The central idea is old, and goes back at least to the works of Berman [18, 19] and Pickands [123, 124, 125] in the mid to late 1960s, which themselves have roots in Cramér's classic paper [41].

The basic approach lies in looking for a subset $T_{\max} \subset T$ (very often a unique point in T) where the maximal variance is achieved, and then studying two things: the "size" of $T_{\max}$ (e.g., as measured in terms of metric entropy) and how rapidly $E\{f_t^2\}$ decays as one moves out of $T_{\max}$. The underlying idea is that the supremum of the process is most likely to occur in or near the region of maximal variance, and the rate of decay of $E\{f_t^2\}$ outside that region determines how best to account for the impact of nearby regions.

A key lemma, on which all the proofs we shall give rely, is the following result of [153]. Recall throughout the assumption of Chapter 1 that the parameter space T is compact for the pseudometric d, which we shall assume is still in place.

Lemma 4.1.5. *Let f be a centered, a.s. continuous Gaussian process on T and X a centered Gaussian variable such that the family $\{X, f_t, t \in T\}$ is jointly Gaussian. Assume that*

$$\mathbb{E}\{(f_t - X)X\} \le 0 \tag{4.1.12}$$

[5] A proof of a lower bound for (4.1.11) can be found in [136]. You can also find quite general arguments, for both upper and lower bounds and for non-Gaussian as well as Gaussian *stationary* processes, in a number of papers by Albin (e.g., [9]). We shall see more in the way of lower bounds for specific cases soon.

for all $t \in T$. Set

$$\mu = \mathbb{E}\Big\{\sup_{t\in T} f_t\Big\}, \qquad \sigma^2 = \mathbb{E}\{X^2\}, \qquad a^2 = \sup_{t\in T} \mathbb{E}\{(f_t - X)^2\}.$$

Then, if $a < \sigma$ and $u > \mu$,

$$\mathbb{P}\Big\{\sup_{t\in T} f_t \geq u\Big\} \leq \frac{1}{2}\exp\left(-\frac{(u-\mu)^2}{2a^2}\right) + \frac{a}{\sqrt{a^2+\sigma^2}}\exp\left(-\frac{(u-\mu)^2}{2(a^2+\sigma^2)}\right) + \Psi\left(\frac{u-\mu}{\sigma}\right). \tag{4.1.13}$$

If, in addition, $u > \max(2\mu, \mu+\sigma)$, then

$$\mathbb{P}\Big\{\sup_{t\in T} f_t \geq u\Big\} \leq \Psi\left(\frac{u}{\sigma}\right)\left[1 + K\frac{au}{\sigma^2}\exp\left(\frac{a^2u^2}{2\sigma^4}\right)\right]\exp\left(\frac{2u\mu}{\sigma^2}\right), \tag{4.1.14}$$

where K is a universal constant.

In a typical application of Lemma 4.1.5, X is usually the value of f at a point $t_0 \in T$ of maximal variance. In that case (4.1.12) trivially holds, and $a < \sigma$ will also hold if T is sufficiently small.

Proof. While the proof is rather technical, the idea behind it is actually quite simple. It starts by defining the function a_t by

$$a_t = \frac{\mathbb{E}\{(f_t - X)X\}}{\sigma^2} \leq 0,$$

and the random process g_t by

$$f_t = g_t + (1 + a_t)X.$$

Note that $\mathbb{E}\{g_t X\} = 0$. That is, X and the full process g are independent.

The argument then works as follows. There are essentially three ways for $\sup_T f_t$ to be larger than u. One is for $\sup_T g_t$ to be large. One is for each of X and $\sup_T g_t$ to be large enough for the sum to be large, and the last is for X to be large. The three terms in (4.1.13) correspond, in order, to these three cases, and each is obtained by conditioning on the distribution of X and then using the Borell–TIS inequality to handle $\sup_T g_t$. That there is only one term in (4.1.14) comes from the fact that here u is even larger, and then the last term actually dominates all others. The reason for this is that from the assumption $a_t \leq 0$ and $a < \sigma$ it follows that $E\{X^2\} > \sup_T E\{g_t^2\}$, and so it is "easier" for X to reach larger values than it is for g. Now read on.

It is trivial to check that for all $s, t \in T$,

$$\mathbb{E}\left\{(g_t - g_s)^2\right\} \leq \mathbb{E}\left\{(f_t - f_s)^2\right\},$$

so that by the Sudakov–Fernique inequality (Theorem 2.2.3),

$$\mathbb{E}\left\{\sup_{t\in T} g_t\right\} \le \mathbb{E}\left\{\sup_{t\in T} f_t\right\} = \mu. \tag{4.1.15}$$

Preparing for the conditioning argument, write

$$\begin{aligned}\mathbb{P}\Big\{\sup_{t\in T} f_t \ge u \Big| X = x\Big\} &= \mathbb{P}\Big\{\sup_{t\in T}(f_t - X) \ge u - x \Big| X = x\Big\} \\ &= \mathbb{P}\Big\{\sup_{t\in T}(g_t + a_t X) \ge u - x \Big| X = x\Big\}.\end{aligned}$$

For $x > 0$, using the fact that $a_t \le 0$, we have

$$\sup_{t\in T}\,(g_t + a_t x) \le \sup_{t\in T} g_t,$$

from which it trivially follows that for $x \ge 0$,

$$\mathbb{P}\Big\{\sup_{t\in T} f_t \ge u \Big| X = x\Big\} \le \mathbb{P}\Big\{\sup_{t\in T} g_t \ge u - x \Big| X = x\Big\}.$$

However, since X and the process g are independent, we can drop the last conditioning, and since

$$\mathbb{E}\left\{g_t^2\right\} \le \mathbb{E}\left\{(g_t + a_t X)^2\right\} = \mathbb{E}\left\{(f_t - X)^2\right\} \le a^2,$$

it follows from (4.1.15) and the Borell–TIS inequality (Theorem 2.1.1) that for $0 \le x \le u - \mu$,

$$\mathbb{P}\left\{\sup_{t\in T} f_t \ge u \Big| X = x\right\} \le \exp\left(-\frac{(u - x - \mu)^2}{2a^2}\right). \tag{4.1.16}$$

We get a slightly different bound for $x \le 0$. Note that since

$$|a_t| = \frac{|\mathbb{E}\{(f_t - X)X\}|}{\sigma^2} \le \frac{a}{\sigma} \le 1,$$

we have

$$\sup_{t\in T}(g_t + a_t x) \le |x| + \sup_{t\in T} g_t = -x + \sup_{t\in T} g_t.$$

Thus, again applying the Borell–TIS inequality, we have

$$\begin{aligned}\mathbb{P}\Big\{\sup_{t\in T}(g_t + a_t X) \ge u - x \Big| X = x\Big\} &\le \mathbb{P}\Big\{\sup_{t\in T} g_t \ge u\Big\} \\ &\le \exp\left(-\frac{(u-\mu)^2}{2a^2}\right),\end{aligned} \tag{4.1.17}$$

which is independent of x. We now have all we need to establish (4.1.13). Write, for $u \ge \mu$,

$$\begin{aligned}
\mathbb{P}\Big\{\sup_{t\in T} f_t \geq u\Big\} &= \int_{-\infty}^{\infty} \mathbb{P}\Big\{\sup_{t\in T} f_t \geq u \Big| X = x\Big\} \frac{e^{-x^2/2\sigma^2}}{\sqrt{2\pi}\sigma}\, dx \\
&= \int_{-\infty}^{0} + \int_{0}^{u-\mu} + \int_{u-\mu}^{\infty} \cdots \\
&\stackrel{\Delta}{=} I_1 + I_2 + I_3.
\end{aligned}$$

The term I_3 is trivially bounded by the last term in (4.1.13). For I_1, bound the integrand by (4.1.17) and integrate out x to obtain the first term on the right-hand side of (4.1.13). For the intermediate term, apply (4.1.16) and note, with $s = u - \mu$, that

$$\begin{aligned}
I_2 &\leq \frac{1}{\sqrt{2\pi}\sigma} \int_{-\infty}^{\infty} \exp\left(-\frac{(s-x)^2}{2a^2} - \frac{x^2}{2\sigma^2}\right) dx \\
&= \frac{a}{\sqrt{a^2+\sigma^2}} \exp\left(-\frac{s^2}{2(\sigma^2+a^2)}\right),
\end{aligned}$$

evaluating the integral by completing the square. Thus the first part of the lemma, (4.1.13), is established.

It remains to prove (4.1.14). We start by observing that

$$\exp\left(-\frac{(u-\mu)^2}{2a^2}\right) = \exp\left(-\frac{(u-\mu)^2}{2(a^2+\sigma^2)}\right) \exp\left(-\frac{(u-\mu)^2\sigma^2}{2a^2(a^2+\sigma^2)}\right).$$

Since $\sigma > a$ and $u > \mu + \sigma$,

$$\exp\left(-\frac{(u-\mu)^2\sigma^2}{2a^2(a^2+\sigma^2)}\right) \leq \exp\left(-\frac{\sigma^2}{4a^2}\right) \leq \frac{a}{\sigma}.$$

Thus the sum of the first two terms on the right-hand side of (4.1.13) is bounded above by

$$\frac{2a}{\sigma} \exp\left(-\frac{(u-\mu)^2}{2(a^2+\sigma^2)}\right).$$

Since

$$\frac{1}{\alpha^2+\sigma^2} \geq \frac{1}{\sigma^2} - \frac{a^2}{\sigma^4},$$

the previous expression is at most

$$\begin{aligned}
&\frac{2a}{\sigma} \exp\left(-\frac{(u-\mu)^2}{2\sigma^2} + \frac{u^2a^2}{2\sigma^4}\right) \\
&\quad \leq \frac{2au(u-\mu)}{\sigma^3} \exp\left(\frac{u^2a^2}{2\sigma^4}\right) \exp\left(-\frac{(u-\mu)^2}{2\sigma^2}\right),
\end{aligned}$$

on applying the various inequalities between a, σ, u and μ. Apply the basic Gaussian tail bound (1.2.2) to bound the above by

$$\frac{Kua}{\sigma^2}\Psi\left(\frac{u-\mu}{\sigma}\right)\exp\left(\frac{u^2a^2}{2\sigma^4}\right),$$

for a universal K. To complete the argument note that for $x \geq 1$, $y \geq 0$,

$$\Psi(x-y) \leq e^{2xy}\Psi(x).$$

(This is easily proven from the fact that $f(y) \triangleq e^{2xy}\Psi(x) - \Psi(x-y)$ satisfies $f(0) = 0$ and has positive derivative, since

$$f'(y) = 2xe^{2xy}\Psi(x) - \frac{1}{\sqrt{2\pi}}e^{-(x-y)^2/2} \geq 0,$$

again applying the Gaussian tail bound (1.2.2).) □

While Lemma 4.1.5 will be the main tool for the proofs of Theorems 4.1.2 and 4.1.3, we first need a counting result about partitioning our (pseudo)metric space (T,d).

Lemma 4.1.6. *Let (T,d) be a metric space and take p, q integers, $p < q$. Let $\mathcal{P}_q$ be a partition of T of sets of diameter no more than 4^{-q+1}. Take a set of integers k_l, $p \leq l \leq q$. Then there exists an increasing sequence $\{\mathcal{P}_l\}_{p \leq l \leq q}$ of partitions of T with the following properties:*

$$\mathcal{P}_{l+1} \text{ is a refinement of } \mathcal{P}_l. \text{ i.e., For each } A \in \mathcal{P}_{l+1} \text{ there is an } A' \in \mathcal{P}_l \text{ such that } A \subseteq A'. \tag{4.1.18}$$

$$\text{For each set } A \in \mathcal{P}_l,\ \mathrm{diam}(A) \leq 4^{-l+1}. \tag{4.1.19}$$

$$\text{Each set of } \mathcal{P}_l \text{ contains at most } k_l \text{ sets of } \mathcal{P}_{l+1}. \tag{4.1.20}$$

$$|\mathcal{P}_l| \leq N(4^{-l}) + \frac{|\mathcal{P}_{l+1}|}{k_l}, \quad \text{for all } l < q, \tag{4.1.21}$$

where $N(\varepsilon)$ is the metric entropy function for (T,d) and $|\mathcal{P}_l|$ is the number of sets in $\mathcal{P}_l$.

Proof. We shall construct the partitions $\mathcal{P}_{\updownarrow}$ by decreasing induction over l; i.e., we show how to construct $\mathcal{P}_{\updownarrow}$ once $\mathcal{P}_{\updownarrow+\infty}$ is given.

Set $N = N(4^{-l})$ and consider points $\{t_i\}_{1 \leq i \leq N}$ of T such that

$$\sup_{t \in T}\inf_i d(t,t_i) \leq 4^{-l}.$$

For $i \leq N$ let A_i be the union of all the sets in $\mathcal{P}_{\updownarrow+\infty}$ that intersect $B_d(t_i, 4^{-l})$. Thus, since the sets in $\mathcal{P}_{\updownarrow+\infty}$ have diameter no larger than 4^{-l}, A_i has diameter at most

$$2 \cdot 4^{-l} + 2 \cdot 4^{-l} = 4^{-l+1}.$$

Now define $C_i = A_i \setminus \bigcup_{j<i} A_j$. These form a partition of T that is coarser than $\mathcal{P}_{\updownarrow+\infty}$. Some of these sets might contain more that k_l sets of $\mathcal{P}_{\updownarrow+\infty}$. In such a case, these sets can be repartitioned into smaller sets, all but at most one of which is the union of exactly k_l of the sets in $\mathcal{P}_{\updownarrow+\infty}$. The exception will contain fewer than k_l of the sets in $\mathcal{P}_{\updownarrow+\infty}$. This constructs $\mathcal{P}_{\updownarrow}$ and (4.1.21) is immediate. □

Corollary 4.1.7. *Suppose that in Lemma* 4.1.6 *we have, for some* $A, \alpha > 0$, $N(4^{-l}) \le (A4^l)^\alpha$ *if* $l \ge p$ *and* $|\mathcal{P}_q| \le (A4^q)^\alpha$. *Then, if* $k_l \ge 2 \cdot 4^\alpha$ *for all* l, *we have* $|\mathcal{P}_l| \le 2(A4^l)^\alpha$.

Proof. Using the construction given in the proof of the lemma, and by decreasing induction over l,

$$|\mathcal{P}_{\updownarrow}| \le (A4^l)^\alpha + \frac{2(A4^{l+1})^\alpha}{k_l} \le 2(A4^l)^\alpha. \qquad \square$$

Proof of Theorem 4.1.2. The argument is by partitioning, much as for the proof of Theorem 4.1.1. The improvement in the result relies on being more careful as to how the partitioning is carried out, and from a judicious application of Lemma 4.1.5, which we did not previously have.

Start by choosing $a < \sigma_T$ and $\mu > 0$. Then choose a partition $\{T_i\}_{i \le N}$ of T into N compact sets, each of diameter no more than a and for each of which $\mathbb{E}\{\sup_{T_i} f_t\} \le \mu$. Then, for $u > \max(2\mu, \mu + \sigma_T)$, it follows from (4.1.14) (by taking X to be f taken at a point of maximal variance in each T_i) that

$$\mathbb{P}\left\{\sup_{t \in T} f_t \ge u\right\} \le N \left(1 + K\frac{au}{\sigma_T^2} \exp\left(\frac{a^2u^2}{2\sigma_T^4}\right)\right) \exp\left(\frac{2u\mu}{\sigma_T^2}\right) \Psi\left(\frac{u}{\sigma_T}\right), \tag{4.1.22}$$

where K is a universal constant. The free variables here are μ and a, and the problem is that $N \to \infty$ as $a, \mu \to 0$. The trick therefore is to find a sequence of partitions for which N does not grow too fast.

As usual, our first step is to assume that T is finite and once again apply the argument at the end of the proof of the Borell–TIS inequality for moving from finite to infinite T.[6] Thus we can, and do, assume that T is finite, and can now start the serious part of the proof.

Since T is finite, we can consider it as a partition of itself, and find a q large enough so that $d(s,t) \ge 4^{-q+1}$ for all $s, t \in T$. Applying Corollary 4.1.7 with $k_l = \lfloor 2 \cdot 4^\alpha \rfloor + 1 \le 3 \cdot 4^\alpha$, and for each $l \le q$, with $4^{-l} \le \varepsilon_0$, we find a further partition of T into $N \le 2(A4^l)^\alpha$ sets $\{T_i\}_{i \le N}$ such that for $m \ge l$,

$$N(T_i, d, 4^{-m+1}) \le (3 \cdot 4^\alpha)^{m-l}$$

for all $i \le N$. The last inequality follows from repeated applications of (4.1.20). Using this bound in Theorem 1.3.3, it is a simple calculation to see that

$$\mathbb{E}\left\{\sup_{t \in T_i} f_t\right\} \le K\sqrt{\alpha}4^{-l}, \tag{4.1.23}$$

for some universal K.

[6] The one issue that is slightly more involved this time is whether a bound on the entropy function for the original parameter space T translates to an equivalent bound for the process defined on a finite subset of T. This, however, is easy to check.

Now take for l the smallest integer such that $4^{-l} \le \sqrt{\alpha}\sigma_T^2/4Ku$, with the K of (4.1.23). Assuming, for no loss of generality, that $K \ge 1$, we have $4^{-l} \le \varepsilon_0$, provided that $u \ge \sqrt{\alpha}\sigma_T^2/\varepsilon_0$. Since $4^{-l} \ge \sqrt{\alpha}\sigma_T^2/16Ku$ we have $N \le (16KAu/\sqrt{\alpha}\sigma_T^2)^\alpha$. We also have, in the notation of Theorem 4.1.2, that

$$\mu \le \frac{\alpha\sigma_T^2}{4u}, \qquad a \le 4^{-l+1} \le \frac{\sqrt{\alpha}\sigma_T^2}{u}.$$

Now observe that since $u > \sqrt{\alpha}\sigma_T$, we have $\mu \le \sqrt{\alpha}\sigma_T \le u/4$ and $a < \sigma_T$, so that the result now follows from (4.1.22), as promised. □

Proof of Theorem 4.1.3. As before, the proof uses a partitioning argument. This time, however, we start by partioning the space into regions depending on the size of $\mathbb{E}\{f_t^2\}$. In each of these regions we apply Theorem 4.1.2 to bound the supremum, so that the hard part of the bounding has already been done. The rest of the proof is really just accounting, to make sure that the various pieces add up appropriately.

To start, fix $u > 2\sigma_T\sqrt{\beta}$. Set $\delta_0 = 0$, $\delta_1 = \sqrt{\beta}\sigma_T^2/u$ and, for $k \ge 1$, set $\delta_k = 2^{k-1}\delta_1$. For $k \ge 1$ set $U_k = T_{\delta_k} \setminus T_{\delta_{k-1}}$. Set $\varepsilon_0 = \delta_1(1+\sqrt{\alpha})/\sqrt{\beta}$, and note that

$$\frac{\sigma_T^2\left(1+\sqrt{\alpha}\right)}{\varepsilon_0} \le \frac{\sigma_T^2\sqrt{\beta}}{\delta_1} = u.$$

Thus, setting $\sigma_k^2 = \sigma_T^2 - \delta_{k-1}^2$, and applying Theorem 4.1.2 to U_k, we obtain

$$\mathbb{P}\left\{\sup_{t\in U_k} \ge u\right\} \le A\delta_k^\beta \left(\frac{Ku}{\sqrt{\alpha}\sigma_k^2}\right)^\alpha \Psi\left(\frac{u}{\sigma_k}\right).$$

Let k_0 be the largest integer such that $\delta_{k_0-1} \le \sigma_T/2$. Since $u \ge 2\sqrt{\beta}\sigma_T$, we have that $k_0 \ge 2$ and $\delta_{k_0-1} \ge \sigma_T/4$. For $k \le k_0$, we have $\sigma_k^2 \ge 3\sigma_T^2/4$, so that

$$\mathbb{P}\left\{\sup_{t\in U_k} f_t \ge u\right\} \le A\delta_k^\beta \left(\frac{Ku}{\sqrt{\alpha}\sigma_T^2}\right)^\alpha \left(\frac{\sigma_T}{u}\right) \exp\left(-\frac{u^2}{2(\sigma_T^2-\delta_{k-1}^2)}\right).$$

Since we also have that

$$\frac{1}{\sigma_T^2-\delta_{k-1}^2} \ge \frac{1}{\sigma_T^2} + \frac{\delta_{k-1}^2}{\sigma_T^4},$$

we see that

$$\mathbb{P}\left\{\sup_{t\in U_k} f_t \ge u\right\} \le A\delta_k^\beta \left(\frac{Ku}{\sqrt{\alpha}\sigma_T^2}\right)^\alpha \exp\left(-\frac{u^2\delta_{k-1}^2}{2\sigma_T^4}\right) \Psi\left(\frac{u}{\sigma_T}\right).$$

Note that

$$
\begin{aligned}
\sum_{k\geq 1} \delta_k^\beta \exp\left(-\frac{u^2\delta_{k-1}^2}{2\sigma_T^4}\right) &= \delta_1^\beta + \sum_{k\geq 2} \delta_k^\beta \exp\left(-\frac{u^2\delta_{k-1}^2}{2\sigma_T^4}\right) \\
&\leq \delta_1^\beta \left(1 + \sum_{k\geq 2} 2^{\beta k} \exp(-\beta 2^{2k-3})\right) \\
&\leq (K\delta_1)^\beta,
\end{aligned}
$$

the last line requiring $\beta > 1$. Consequently, if we set $T' = \bigcup_{k\leq k_0} U_k$, we have, recalling the value of δ_1, that

$$
\mathbb{P}\Big\{\sup_{t\in T'} f_t \geq u\Big\} \leq A\frac{\beta^{\beta/2}}{\alpha^{\alpha/2}} K^{\beta+\alpha} \left(\frac{u}{\sigma_T^2}\right)^{\alpha-\beta} \Psi\left(\frac{u}{\sigma_T}\right). \tag{4.1.24}
$$

For $t \notin T'$ we have

$$
\mathbb{E}\left\{f_t^2\right\} \leq \sigma_T^2 - \delta_{k_0-1}^2 \leq \sigma_T^2 - \sigma_T^2/16 = 15\sigma_T^2/16.
$$

Thus, if we now apply the entropy bound (4.1.10), we see that for $u > \sigma_T\sqrt{\beta}$ we have

$$
\mathbb{P}\Big\{\sup_{t\in T\setminus T'} f_t \geq u\Big\} \leq A\sigma_T^\beta \left(\frac{Ku}{\sqrt{\alpha}\sigma_T^2}\right)^\alpha \Psi\left(\frac{4u}{\sigma_T\sqrt{15}}\right). \tag{4.1.25}
$$

To conclude, it suffices to check that for $x \geq \sqrt{\beta}$ we have $\Psi(4x/\sqrt{15}) \leq (K\beta)^{\beta/2}x^{-\beta}\Psi(x)$, so that the left-hand side of (4.1.25) is dominated by the left-hand side of (4.1.11). □

4.2 Processes with a Unique Point of Maximal Variance

We shall now look at a rather interesting special case. We know already from the discussion following Theorem 4.1.3 that there would seem to be cases in which the trivial lower bound

$$
\mathbb{P}\left\{\sup_{t\in T} f(t) \geq u\right\} \geq \sup_{t\in T} \mathbb{P}\left\{f(t) \geq u\right\} = \Psi\left(\frac{u}{\sigma_T}\right)
$$

comes close (up to a constant) to serving as an upper bound as well.

A class of examples in which this holds with the constant 1 arises when there is a *unique* point $t_0 \in T$ at which the maximal variance is achieved and some additional side conditions are assumed. In that case, we have the following elegant result of Talagrand [151].

Theorem 4.2.1. *Suppose f is a centered, bounded Gaussian process on T, and that there is a unique point $t_0 \in T$ such that*

$$\mathbb{E}\left\{f_{t_0}^2\right\} = \sigma_T^2 = \sup_T \mathbb{E}\left\{f_t^2\right\}. \tag{4.2.1}$$

For $\delta > 0$, set

$$T_\delta = \left\{t \in T : \mathbb{E}\left\{f_t f_{t_0}\right\} \geq \sigma_T^2 - \delta^2\right\}.$$

Suppose that

$$\lim_{\delta \to 0} \frac{\mathbb{E}\left\{\sup_{t \in T_\delta} f_t\right\}}{\delta} = 0. \tag{4.2.2}$$

Then

$$\lim_{u \to \infty} \frac{\mathbb{P}\left\{\sup_{t \in T} f(t) \geq u\right\}}{\Psi(u/\sigma_T)} = 1. \tag{4.2.3}$$

There is actually a converse to this result, which states that (4.2.3) implies the existence of a $t_0 \in T$ for which (4.2.1) and (4.2.2) hold. You can find a proof of the converse in [151].

Proof. Since it is a triviality that

$$\mathbb{P}\left\{\sup_{t \in T} f_t \geq u\right\} \geq \mathbb{P}\left\{f_{t_0} \geq u\right\} = \Psi\left(\frac{u}{\sigma_T}\right),$$

the theorem will be proven once we establish (4.2.3) with a limsup rather than lim.

To start, take $\eta \in (0, 1)$ and, by (4.2.2), δ_0 small enough so that

$$\delta \leq 2\delta_0 \Rightarrow \mathbb{E}\left\{\sup_{t \in T_\delta} f_t\right\} \leq \eta^2 \delta. \tag{4.2.4}$$

We can and do assume that $\delta_0 \leq \sigma_T \eta^2$. Note also that

$$\sup_{t \notin T_{\delta_0}} \mathbb{E}\left\{f_t^2\right\} < \sigma_T^2,$$

since otherwise there would be a $t \notin T_{\delta_0}$ (so that $t \neq t_0$) for which $\mathbb{E}\left\{f_t^2\right\} = \sigma_T^2$, which contradicts the assumption of there being a unique point of maximal variance. It therefore follows immediately from the Borell–TIS inequality that

$$\lim_{u \to \infty} \frac{\mathbb{P}\left\{\sup_{t \in T \setminus T_{\delta_0}} f_t \geq u\right\}}{\Psi(u/\sigma_T)} = 0, \tag{4.2.5}$$

and so to complete the proof it will suffice to show that for some universal K, and for u large enough,

$$\mathbb{P}\Big\{\sup_{t\in T_{\delta_0}} f_t \geq u\Big\} \leq \Psi\left(\frac{u}{\sigma_T}\right)(1+K\eta). \tag{4.2.6}$$

Fix u and set $\alpha = \sigma_T^2/\eta u$. Adopting and adapting the main idea in the proof of Theorem 4.1.3, define a nondecreasing sequence of subsets of T, and a sequence of "annuli," by setting $V_{-1} = \emptyset$ and, for $k \geq 0$,

$$V_k = T_{2^k\alpha}, \qquad U_k = V_k \setminus V_{k-1}.$$

If p is the smallest integer for which $2^p\alpha \geq \delta_0$, then

$$T_{\delta_0} \subset \bigcup_{0\leq k\leq p} U_k,$$

and we shall now try to obtain the bound in (4.2.6) by obtaining a similar bound for each U_k and then combining them via a union bound.

Setting $\mu_k = \mathbb{E}\left\{\sup_{t\in V_k} f_t\right\}$, (4.2.4) implies that $\mu_k \leq \alpha\eta^2 2^k$ for $k \leq p$. Setting

$$\omega_k = \sup_{t\in V_k}\left(\mathbb{E}\left\{\left(f_t - f_{t_0}\right)^2\right\}\right)^{1/2},$$

we have

$$\omega_k = \sup_{t\in V_k} K\mathbb{E}\left\{\left|f_t - f_{t_0}\right|\right\} \leq K\mathbb{E}\left\{\sup_{t\in V_k}\left|f_t - f_{t_0}\right|\right\} \leq K\mu_k \leq K\alpha\eta^2 2^k, \tag{4.2.7}$$

the first line being a standard Gaussian identity and the second inequality coming from the fact that $t_0 \in V_k$.

By (4.1.14) with $X = f_{t_0}$ we therefore have

$$\mathbb{P}\left\{\sup_{t\in U_0} f_t \geq u\right\} \leq \Psi\left(\frac{u}{\sigma_T}\right)(1+K\eta e^{K\eta^2/2})e^{2\eta} \leq \Psi\left(\frac{u}{\sigma_T}\right)(1+K\eta).$$

We use a similar argument for the remaining U_k. Take the random variable X of Lemma 4.1.5 to be $X_k = (1-(\alpha 2^{k-1})^2/\sigma_T^2)f_{t_0}$. It is then easy to check that the definition of T_δ implies that (4.1.12) holds. Furthermore, since $k \leq p$ and $\delta_0 \leq \sigma_T\eta^2$,

$$\left(\mathbb{E}\left\{\left(X_k - f_{t_0}\right)^2\right\}\right)^{1/2} = \frac{\left(\alpha 2^{k-1}\right)^2}{\sigma_T} \leq \frac{\delta_0}{\sigma_T}\alpha 2^{k-1} \leq \eta^2\alpha 2^k.$$

It follows from (4.2.7) and the triangle inequality that if we set

$$a_k^2 = \sup_{t\in V_k}\mathbb{E}\left\{(f_t - X_k)^2\right\},$$

then $a_k \leq K\eta^2\alpha 2^k$. Thus, by (4.1.14) and the fact that $\eta < 1$ we have

$$\mathbb{P}\Big\{\sup_{t\in U_k} f_t \geq u\Big\} \leq \Psi\left(\frac{u}{\sigma_T(1-(\alpha 2^{k-1})^2/\sigma_T^2)}\right)\left(1+K\eta 2^k e^{K\eta^2 2^{2k}}\right)e^{K\eta 2^k}$$
$$\leq \Psi\left(\frac{u}{\sigma_T(1-(\alpha 2^{k-1})^2/\sigma_T^2)}\right)e^{K\eta 2^{2k}}.$$

Using the fact that for $x \leq 1$ we know that $(1-x)^{-1} \geq 1+x$, we have

$$\frac{u}{\sigma_T(1-(\alpha 2^{k-1})^2/\sigma_T^2)} \geq \frac{u}{\sigma_T} + \frac{(\alpha 2^{k-1})^2 u}{2\sigma_T^3}.$$

Also, since a simple change of variables shows that

$$\Psi(x+y) \leq e^{-xy}\Psi(x),$$

for $y \geq 0$, it follows, on recalling the definition of α, that

$$\Psi\left(\frac{u}{\sigma_T(1-(\alpha 2^{k-1})^2/\sigma_T^2)}\right) \leq \Psi\left(\frac{u}{\sigma_T}\right)\exp\left(-\frac{(\alpha 2^{k-1})^2 u^2}{\sigma_T^4}\right)$$
$$\leq \Psi\left(\frac{u}{\sigma_T}\right)\exp\left(-\frac{2^{2k}}{4\eta}\right).$$

Putting all this together, we thus have

$$\sum_{1\leq k\leq p} \mathbb{P}\Big\{\sup_{t\in U_k} f_t \geq u\Big\} \leq \Psi\left(\frac{u}{\sigma_T}\right)\sum_{1\leq k\leq p}\exp\left(-2^{2k}\left(\frac{1}{4\eta}-K\eta\right)\right).$$

Since the last sum can be made arbitrarily small by taking η small enough, we have established (4.2.6) and so the theorem. □

4.3 Examples

Theorem 4.2.1 provided us with conditions under which Gaussian processes have supremum distribution, at least in the tail, which behaves the same as that of a single Gaussian variable. Here is a class of examples for which this happens.

Example 4.3.1. *Let f_t be a centered Gaussian random field on $\mathbb{R}^N$, with $f(0)=0$ and with stationary and isotropic increments, in the sense that the distribution of $f(t+s)-f(s)$ depends only on $|t|$. Suppose that*

$$p^2(t) \stackrel{\Delta}{=} \mathbb{E}\left\{(f_{t+s}-f_s)^2\right\} = \mathbb{E}\left\{f_t^2\right\}$$

is convex, and that

$$\lim_{t\to 0}\frac{p^2(t)}{|t|}=0. \tag{4.3.1}$$

With some abuse of notation, let $T=[0,T]$ *be an* N*-dimensional rectangle. Then*

$$\lim_{u\to\infty}\frac{\mathbb{P}\left\{\sup_{t\in T} f_t\geq u\right\}}{\Psi(u/\sigma_T)}=1, \tag{4.3.2}$$

where $\sigma_T^2=\sup_{t\in T}\mathbb{E}\left\{f^2(t)\right\}=\mathbb{E}\left\{f^2(T)\right\}$.

Without doubt the best known class of random fields statisfying the conditions of the example are the so-called *fractional Brownian motions* or *fields*, which have covariance function

$$C(s,t)=\frac{1}{2}\left(|t|^{2\alpha}+|s|^{2\alpha}-|t-s|^{2\alpha}\right).$$

While these processes have isotropic increments for all $\alpha\in(0,1)$, it is only when $\alpha>\frac{1}{2}$ that p^2 is convex. More information on these processes can be found, for example, in [138].

Proof. The proof requires no more than checking that conditions (4.2.1) and (4.2.2) of Theorem 4.2.1 hold. Since f has stationary increments, (4.2.1) is clearly satisfied by taking $t_0=T$. Thus we need only show that (4.2.2) holds, namely,

$$\lim_{\delta\to 0}\frac{\mathbb{E}\left\{\sup_{t\in T_\delta} f_t\right\}}{\delta}=0.$$

However, since $p^2(t)$ is convex, it has a left derivative at each point, and so it is easy to check (draw a picture) that for $t\in T_\delta$ we have $|T-t|\leq K\delta^2$ for some finite K. Thus, since f has isotropic increments, it suffices to show that

$$\lim_{\delta\to 0}\frac{\mathbb{E}\left\{\sup_{0\leq t\leq\delta^2}|f_t|\right\}}{\delta}=0.$$

But this follows immediately from an entropy bound such as Theorem 1.3.3 along with the inequality (1.4.6) and condition (4.3.1). □

We now look at two applications of Theorem 4.1.3. The first, which treats a nonstationary process on $I^N=[0,1]^N$, is designed to show how sample roughness and nonstationarity interact to determine excursion probabilties. Setting $\beta=\infty$ gives a result for stationary processes, and setting $\alpha=2$ gives a result for (relatively) smooth processes of a form that we shall study in far more detail in Chapter 14.

The second example shows how to use the general theory to handle a Brownian sheet problem.

Example 4.3.2. *Let* f *be a centered, unit-variance, stationary Gaussian process on* I^N *with covariance function* C *satisfying*

$$1 - C(t) \le a|t|^{\alpha}, \tag{4.3.3}$$

when $|t| \le \gamma$, *for some* $a, \gamma > 0$ *and* $\alpha \in [0, 2]$. *Let* $\sigma(t)$ *be positive, continuous, and nondecreasing (under the usual partial order) on* I^N *such that*

$$|\sigma(t) - \sigma(s)| \le b|t - s|^{\beta} \tag{4.3.4}$$

for all $s, t \in I^N$ *and some* $b, \beta \ge 0$, *and define a new, nonstationary, process* h *by*

$$h(t) = \sigma(t) f(t), \qquad t \in I^N. \tag{4.3.5}$$

Finally, let σ, *without a parameter, denote* $\sigma(1, \dots, 1)$. *Then there exists a (computable) finite* $C = C(N, a, \alpha, b, \beta, \sigma) > 0$ *such that for sufficiently large* u,

$$\mathbb{P}\Big\{ \sup_{t \in I^N} h(t) > u \Big\} \le \begin{cases} C u^{-\frac{2}{\beta} + \frac{2N}{\alpha}} \Psi\left(\frac{u}{\sigma}\right), & 0 < \alpha < 2\beta, \\ C \Psi\left(\frac{u}{\sigma}\right), & 0 < 2\beta \le \alpha. \end{cases} \tag{4.3.6}$$

Proof. Note firstly that if σ is strictly increasing, then h has a unique point of maximum variance, at $t_0 = (1, \dots, 1)$. Nevertheless, even in this case, the assumptions we have made are not strong enough to imply that (4.2.2) holds, so that Theorem 4.2.1 (which is designed for examples in which there is a unique point of maximal variance) need not apply. Consequently, we apply Theorem 4.1.3, so that the proof relies on finding a good bound for the entropy $N(T_\delta, d, \varepsilon)$ of the set

$$T_\delta = \Big\{ t \in I^N : \mathbb{E}\{h_t^2\} = \sigma^2(t) \ge \sigma^2 - \delta^2 \Big\}.$$

It is an easy calculation from (4.3.4) that T_δ can be covered by a cube of side length no more than $C\delta^{2/\beta} b^{-1/\beta}$, where $C = C(N)$ is a dimension-dependent constant.

Furthermore, in view of the definition (4.3.5) of h,

$$\begin{aligned} \mathbb{E}\left\{ (h_t - h_s)^2 \right\} &= (\sigma_t - \sigma_s)^2 + 2\sigma_t\sigma_s\,(1 - C(t - s)) \\ &\le b^2|t - s|^{2\beta} + 2\sigma^2 a|t - s|^{\alpha}, \end{aligned} \tag{4.3.7}$$

the inequality holding for all $|t - s| < \gamma$, by (4.3.5).

Moving to the canonical metric on I^N in order to calculate entropies, it follows that if $0 < \alpha < 2\beta$, then (4.3.7) implies

$$d(s, t) \le C|t - s|^{\alpha/2}.$$

Combining this with the comments above on the size of T_δ, we have that

$$N(T_\delta, d, \varepsilon) \le C\varepsilon^{-2N/\alpha}\delta^{2/\beta}.$$

The first case in (4.3.6) now follows from Theorem 4.1.3. The second is similar and left to you. □

Our last example deals with the pinned Brownian sheet over N-dimensional rectangles. Recall from Section 1.4.3 that the Brownian sheet is a Gaussian noise defined on the Borel subsets of $\mathbb{R}^N$. The *set-indexed, pinned Brownian sheet*, based on a probability measure ν on the Borel sets of I^N, is the set-indexed, zero-mean Gaussian process B with covariance function

$$\mathbb{E}\left\{B(A)B(A')\right\} = \nu(A \cap A') - \nu(A)\nu(A'). \tag{4.3.8}$$

The *point-indexed, pinned Brownian sheet* is obtained by looking only at rectangles of the form $[0, t]$ for some $t \in I^N$, in which case we can index it simply by t. Both the set and point-indexed pinned sheets play a crucial role in the theory of empirical processes.

Example 4.3.3. *Let B be the set-indexed, pinned Brownian sheet based on a probability measure ν and defined over the collection $\mathcal{A}_N$ of N-dimensional rectangles in I^N. Assume that ν have a bounded density that is everywhere positive.*[7] *Then there exists a finite $C = C(\nu) > 0$ such that for large enough u,*

$$\mathbb{P}\Big\{\sup_{A\in\mathcal{A}_N} B(A) > u\Big\} \le Cu^{2(2N-1)}e^{-2u^2}. \tag{4.3.9}$$

For the point-indexed pinned sheet,[8]

[7] The condition on ν can be relaxed to demanding that ν has a strictly positive density on some interval $[s-\varepsilon, t+\varepsilon]$ $(s, t, \varepsilon \in I^N)$ for which $\nu([s,t]) \ge \frac{1}{2}$. Indeed, the conditions can be relaxed even further to those used by [5] in obtaining corresponding bounds for the point-indexed sheet. See Samorodnitsky [136, 137] for details.

[8] Using the "double-sum" techniques of Section 4.5 below, one can do considerably better than (4.3.10). In a series of works [78, 62, 63, 161] various versions of this technique were used to show that

$$\mathbb{P}\Big\{\sup_{t\in I^N} B(t) > u\Big\} \asymp \frac{(4\ln 2)^{N-1}}{(N-1)!}u^{2(N-1)}e^{-2u^2},$$

where $a(u) \asymp b(u) \iff \lim_{u\to\infty} a(u)/b(u)$ is well defined. (Note: the factor $(N-1)!$ is missing in [62, 63].)

In the one-dimensional case there is no loss of generality in taking ν to be uniform measure on $[0, 1]$, in which case B is the Brownian bridge. We can then do better than (4.3.10). In particular, any standard graduate probability textbook will prove—via an iterative use of the classic reflection principle—that

$$\mathbb{P}\Big\{\sup_{t\in[0,1]} |B(t)| > u\Big\} = 2\sum_{k=1}^{\infty}(-1)^{k-1}e^{-2k^2u^2},$$

the first term of which, of course, we have in (4.3.10), albeit without a precise constant. Note that the factor of 2 here comes from the fact that we have looked at the supremum of the absolute value of B. The fact that it is precisely 2 comes from the "well-known" fact that high maxima and low minima of Gaussian processes are asymptotically indpendent (e.g., [97]).

$$\mathbb{P}\left\{ \sup_{t \in I^N} B(t) > u \right\} \leq C u^{2(N-1)} e^{-2u^2}. \tag{4.3.10}$$

Proof. The first point to note is that since ν is a probability measure, it follows from (4.3.8) that

$$\sup_{A \in \mathcal{A}_N} \mathbb{E}\left\{ B(A)^2 \right\} = \sup_{t \in I^N} \mathbb{E}\left\{ B(t)^2 \right\} = \frac{1}{4}.$$

Assuming all its conditions hold, Theorem 4.1.3 therefore immediately gives the exponent of the exponentials in (4.3.9) and (4.3.10). The remainder of the proof involves an unentertaining amount of straightforward algebra, so we shall be content with only an outline. In particular, we shall discuss only (4.3.9). The point-indexed result (4.3.10) is easier.

As in the proof of Example 4.3.2 we need to find a bound for the entropy of $N(T_\delta, d, \varepsilon)$, where T_δ is now given by

$$\begin{aligned} T_\delta &= \left\{ A \in \mathcal{A}_N : \mathbb{E}\left\{ B^2(A) \right\} \geq \frac{1}{4} - \delta^2 \right\} \\ &= \left\{ A \in \mathcal{A}_N : \nu(A) - \nu^2(A) \geq \frac{1}{4} - \delta^2 \right\}. \end{aligned}$$

If ν is Lebesgue measure, which we now assume to keep the notation manageable, this simplifies to

$$T_\delta = \left\{ A \in \mathcal{A}_N : \frac{1}{2}(1-\delta) \leq \nu(A) \leq \frac{1}{2}(1+\delta) \right\}. \tag{4.3.11}$$

To approximate the sets in T_δ in terms of the canonical metric, take N-dimensional rectangles whose endpoints sit on the points of the lattice

$$L_\varepsilon^N = \left\{ t \in [0,1]^N : t_i = n_i \varepsilon^2,\ n_i \in \left(0, 1, \ldots, [\varepsilon^{-2}]\right),\ i = 1, \ldots, N \right\}.$$

We can choose each of the first $N-1$ coordinates n_i in at most ε^{-2} ways, and the last (in view of (4.3.11)) in at most $C_N \delta \varepsilon^{-2}$ ways, so that

$$N\left(T_\delta, d, \varepsilon\right) \leq C'_N \varepsilon^{-4N} \delta.$$

Substituting this into (4.1.10) and (4.1.11) is all we need in order to establish (4.3.9) and so complete the argument.

If ν is not Lebesgue measure, but has an everywhere positive bounded density, the same argument works with a few more constants. □

4.4 Extensions

There is almost no limit to the number of extensions and variations that exist of the results of the preceding two sections. Obviously, the more one is prepared to assume, the more one can obtain.

For example, in the notation of Theorem 4.2.1, set

$$L(h) = \mathbb{E}\Big\{ \sup_{t\in T_h} (f_t - \mathbb{E}\{f_t f_{t_0}\} f_{t_0}) \Big\}.$$

It is not hard to see that condition (4.2.2) of Theorem 4.2.1 is equivalent to

$$\lim_{h\to 0} \frac{L(h)}{h} = 0.$$

Dobric, Marcus, and Weber [48] have shown that if we assume a little more about $L(h)$ we can improve Theorem 4.2.1 as follows.

Theorem 4.4.1. *Suppose there exists a unique $t_0 \in T$ such that $\mathbb{E}\left\{f_{t_0}^2\right\} = 1 = \sup_T \mathbb{E}\left\{f_t^2\right\}$, and two functions ω_1, ω_2, concave for $h \in [0, \bar{h}]$ for some $\bar{h} > 0$ with $\omega_i(0) = 0$. Define*

$$h_i(u) = \sup\left\{ h : \frac{\omega_i(h)}{h^2} = u \right\}, \quad i = 1, 2.$$

If

$$\omega_1(h) \le L(h) \le \omega_2(h)$$

and

$$\sup_{t\in T_h} \mathbb{E}\left\{ \left[f_t - f_{t_0}\mathbb{E}(f_t f_{t_0}) \right]^2 \right\} < (2-\varepsilon)h^2$$

for $h \in [0, \bar{h}]$ for some $\varepsilon > 0$, then there exist constants C_1 and C_2 such that for all u large enough,

$$e^{C_1 u \omega_1(h_1(u))} \le \frac{\mathbb{P}\{\sup_T f_t \ge u\}}{\Psi(u)} \le e^{C_2 u \omega_2(h_2(u))}.$$

If, for example,

$$\limsup_{h\to 0} \frac{L(h)}{\omega_1(h)} \ge 1,$$

then for all $\varepsilon > 0$,

$$\limsup_{u\to\infty} \frac{\mathbb{P}\{\sup_{t\in T} f(t) \ge u\}}{\Psi(u) \exp(k_1(1-\varepsilon)u\omega_1(h_1(u)))} \ge 1.$$

You can find a proof of this result in [48]. It differs in detail, but not in kind, from what we have seen here.

There are also asymptotic bounds of a somewhat different nature due to Lifshits [104], who, in a quite general setting, has shown that

$$\mathbb{P}\left\{\sup_{t\in T} f_t > u\right\} \asymp (2-p)^{1/2}\Psi_p\left(\frac{u^{2-p}-du^{1-p}}{p\sigma_T^2}\right)\Psi\left(\frac{u-d}{\sigma_T}\right)$$
$$\times \exp\left\{-\frac{(2-p)u^2}{2p\sigma_T^2}-\frac{d(p-1)u}{p\sigma_T^2}+\frac{d^2}{2\sigma_T^2}\right\},$$

for all $1 \le p < 2$, where $\Psi_p(x) \equiv \mathbb{E}\{\exp(x \sup_{t\in T} f_t{}^p)$ and d a constant (not easily) determined by f. Of course, this result is somewhat circular, since one needs the Laplace transform of $\sup_{t\in T} f_t{}^p$ before being able to compute anything. Nevertheless, the above inequality leads to some nice theoretical results about ℓ^p-valued Gaussian processes.

We close this section with a much less general case than we have hitherto considered. Suppose that f is a stationary, zero-mean Gaussian field on $\mathbb{R}^N$ and has a covariance function that near the origin can be written as

$$C(t) = 1 - t\Lambda t' + o(|t|^2), \tag{4.4.1}$$

where $\Lambda = \{\lambda_{ij}\}_{1\le i,j\le N}$ is the matrix of second-order spectral moments (cf. (5.5.2)). In this case, one can show that

$$\lim_{u\to\infty} \frac{\mathbb{P}\left\{\sup_{t\in[0,T]^N} f_t \ge u\right\}}{u^N\Psi(u)} = \frac{(\det\Lambda)^{1/2}}{(2\pi)^{N/2}}. \tag{4.4.2}$$

Why this result should be true is something we shall soon investigate in detail and over parameter spaces far more general than cubes in Part III of the book. Indeed, we shall obtain far more precise results, of the ilk of (4.0.1).

4.5 The Double-Sum Method

Since we have mentioned it already a number of times, it is only reasonable that we at least also briefly describe the so-called *double-sum method* for computing asymptotic exceedance probabilities.

The description will be very brief, since the topic is treated in detail in both [97] and [126]. In particular, Piterbarg's monograph [126] has an excellent and remarkably readable treatment of this rather technical material in the setting of random fields.

In essence, the technique is not qualitatively different from what we have been doing throughout this chapter. The differences lie only in the details and in the concentration on lower, as well as upper, bounds for exceedance probabilities. The basic idea is to break up the parameter space T into a finite union of small sets T_k, where the size of the T_k generally depends on the exceedance level u. The T_k need not be disjoint, although any overlap should be small in relation to their sizes. While initially there is no need to assume any particular structure for T, this usually is needed for the most precise results.

It is then elementary that

$$\sum_k \mathbb{P}\Big\{\sup_{t\in T_k} f(t) \geq u\Big\} \geq \mathbb{P}\Big\{\sup_{t\in T} f(t) \geq u\Big\}$$
$$\geq \sum_k \mathbb{P}\Big\{\sup_{t\in T_k} f(t) \geq u\Big\} \tag{4.5.1}$$
$$-\sum\sum_{j\neq k} \mathbb{P}\Big\{\sup_{t\in T_j} f(t) \geq u,\ \sup_{t\in T_k} f(t) \geq u\Big\}.$$

The single summations are treated much as we did before, by choosing a point $t_k \in T_k$, writing

$$\mathbb{P}\Big\{\sup_{t\in T_k} f(t) \geq u\Big\} = \int_{-\infty}^{\infty} \mathbb{P}\Big\{\sup_{t\in T_k} f(t) \geq u \big| f(t_k) = u\Big\} p_{t_k}(x)\,dx,$$

and arguing much as we did in the proof of Lemma 4.1.5.

The crux of the double-sum method lies in showing that the remaining double-sum is of lower order than the single sum and, where possible, estimating its size.

If one could write the joint probabilities in (4.5.1) as the product

$$\mathbb{P}\Big\{\sup_{t\in T_j} f(t) \geq u,\ \sup_{t\in T_k} f(t) \geq u\Big\} = \mathbb{P}\Big\{\sup_{t\in T_j} f(t) \geq u\Big\}\mathbb{P}\Big\{\sup_{t\in T_k} f(t) \geq u\Big\},$$

then we would be basically done, since then the double-sum term would easily be seen to be of lower order than the single sum. Such independence obviously does not hold, but if we choose the sizes of the T_k in such a fashion that a "typical component" of an excursion set is considerably smaller than this size, and manage to show that high extrema are independent of one another, then we are well on the way to a proof.

The details, which are heavy, are all in Piterbarg [126]. Piterbarg's monograph is also an excellent source of worked examples, and includes a number of rigorous computations of excursion probabilities for many interesting examples of processes and fields.

4.6 Local Maxima and Excursion Probabilities

We close this chapter with somewhat of a red herring, in that we shall describe an approach that makes a lot of sense, that has been used and developed with considerable success by Piterbarg [126] and others, but which we shall nevertheless not explore further. Despite this, it does give a good heuristic feeling for results we shall develop in Part III, particularly those of Chapter 14.

We leave the general setting of the previous sections, and concentrate on smooth random fields over structured parameter spaces. In particular, we shall adopt the regularity conditions of Chapter 12, in which f will be a smooth Gaussian process defined over a Whitney stratified manifold M. Since we shall get around to defining these manifolds only in Section 8.1, you can either return to the current discussion after reading that section, or simply assume that "an N-dimensional Whitney stratified manifold" is no more than another way of saying "a unit cube $I^N \subset \mathbb{R}^N$." You will not lose much in the way of intuition in doing so.

Furthermore, since the current section deals with a general idea rather than specific results, we shall not be more precise about exactly what conditions we require and shall simply assume that everything we write is well defined. The appropriate rigor will come in Part III of the book.

Our aim is to connect the excursion probability of f with the mean number of its local maxima of a certain type.

We start by noting that an N-dimensional Whitney stratified manifold M can be written as the disjoint union

$$M = \bigcup_{k=0}^{N} \partial_k M, \tag{4.6.1}$$

of open k-dimensional submanifolds (cf. (9.3.5)). (If you are working with I^N, then this is just a decomposition of I^N into k-dimensional facets, so that $\partial_N M = M^\circ$ is the interior of the cube, $\partial_{N-1} M$ the union of the interiors of the $(N-1)$-dimensional faces, and so forth, down to $\partial_0 M$, which is made up of the 2^N vertices of the cube.)

Let

$$M_u^E(M) \triangleq \#\left(\text{extended outward maxima of } f \text{ in } M \text{ with } f_t \geq u\right).$$

In the terminology of Section 9.1, M_u^E is more precisely defined as the number of extended outward critical points $t \in \partial_k M \subset M$ of f for which the Hessian of $f|_{\partial_k M}$ has maximal index. For the example of I^N, these points are the local maxima $t \in \partial_k M$, $k = 0, \dots, N$, of $f|_{\partial_k M}$ for which $f(t_n) \uparrow f(t)$ whenever the $t_n \in \partial_{k+1} M$ converge to t monotonically (for the usual partial order on $\mathbb{R}^N$) along a direction normal to $\partial_k M$ at t.

Since the notion of extended local maxima[9] includes the 0-dimensional sets $\partial_0 M$ (i.e., the "corners" of M) it is not hard to see that if $\sup_{t\in M} f_t \geq u$, then f must have at least one extended outward maximum on M, and vice versa (see the argument in Section 14.1.1 if you want details). Consequently, we can argue as follows:

---

[9] Note that if we write

$$M_u(M) \triangleq \#(\text{local maxima of } f \text{ in } M \text{ with } f_t \geq u),$$

then one can actually carry through all of the following argument with M_u^E replaced by M_u. In fact, this was Piterbarg's original argument (cf. [126]). Since it is always true that $M_u \geq M_u^E$, this means that the upper bound given in (4.6.5) below is poorer with this change. Since the lower bound involves a difference of two terms, each of which is larger with this change, it is unclear which variation of the argument is better, although one expects the version with M_u^E rather than M_u to give the tighter bound. In some sense, the difference is of academic interest only, since neither bound can be explicitly evaluated. When we turn to the explicit computations for the Gaussian case in Chapter 14, we shall in any case adopt a somewhat different approach.

$$\mathbb{P}\Big\{\sup_{t\in M} f_t \geq u\Big\} = \mathbb{P}\{M_u^E(M) \geq 1\} \leq \sum_{k=0}^{N} \mathbb{P}\{M_u^E(\partial_k M) \geq 1\} \leq \sum_{k=0}^{N} \mathbb{E}\{M_u^E(\partial_k M)\}. \tag{4.6.2}$$

This gives a simple upper bound for excursion probabilities in terms of the mean number of outward extended maxima. For a lower bound, note that

$$\Big\{\sup_{t\in M} f_t \geq u\Big\} \iff \{M_u^E(M^\circ) \geq 1,\ M_u^E(\partial M) = 0\} \cup \{M_u^E(\partial M) \geq 1\}, \tag{4.6.3}$$

where $\partial M = \bigcup_{k=0}^{N-1} \partial_k M$. Now assume that f, as a function on the N-dimensional manifold M, does not have any critical points on ∂M. Write

$$\begin{aligned}\{M_u^E(M^\circ) \geq 1,\ M_u^E(\partial M) = 0\} &= \{M_u^E(M^\circ) = 1, M_u^E(\partial M) = 0\}\\ &\quad \cup \{M_u^E(M^\circ) \geq 2, M_u^E(\partial M) = 0\}.\end{aligned}$$

To compute the probabilities of the above two events, set

$$p_k = \mathbb{P}\{M_u^E(M^\circ) = k\},$$

and note that

$$\mathbb{E}\{M_u^E(M^\circ)\} = \mathbb{P}\{M_u^E(M^\circ) = 1\} + \sum_{k=2}^{\infty} kp_k,$$

so that

$$\begin{aligned}&\mathbb{P}\{M_u^E(M) = 1, M_u^E(\partial M) = 0\}\\ &\quad = \mathbb{P}\{M_u^E(M) = 1\} - \mathbb{P}\{M_u^E(M^\circ) = 1, M_u^E(\partial M) \geq 1\}\\ &\quad = \mathbb{E}\{M_u^E(M^\circ)\} - \sum_{k=2}^{\infty} kp_k - \mathbb{P}\{M_u^E(M^\circ) = 1, M_u^E(\partial M) \geq 1\}.\end{aligned}$$

In a similar vein,

$$\begin{aligned}&\mathbb{P}\{M_u^E(M^\circ) \geq 2, M_u^E(\partial M) = 0\}\\ &\quad = \sum_{k=2}^{\infty} p_k - \mathbb{P}\{M_u^E(M^\circ) \geq 2, M_u^E(\partial M) \geq 1\}.\end{aligned}$$

Putting the last two equalities together with (4.6.3) immediately gives us that

$$\begin{aligned}\mathbb{P}\Big\{\sup_{t\in M} f_t \geq u\Big\} &= \mathbb{E}\{M_u^E(M^\circ)\} - \sum_{k=2}^{\infty}(k-1)p_k + \mathbb{P}\{M_u^E(\partial M) \geq 1\}\\ &\quad - \mathbb{P}\{M_u^E(M^\circ) \geq 1, M_u^E(\partial M) \geq 1\}\\ &\geq \mathbb{E}\{M_u^E(M^\circ)\} - \frac{1}{2}\mathbb{E}\{M_u^E(M^\circ)[M_u^E(M^\circ) - 1]\}\end{aligned} \tag{4.6.4}$$

on noting that $k-1 < k(k-1)/2$ for $k \geq 2$. Iterate this argument through $\partial_k M$, $0 \leq k \leq N-1$, and combine it with (4.6.2) to obtain the following.

Theorem 4.6.1. *Let f be almost surely C^2 on a C^2 Whitney stratified manifold M. With the preceding notation, assume that $f|_{\partial_k M}$ has, almost surely, no critical points on $\bigcup_{j=0}^{k-1} \partial_j M$. Furthermore, assume that the random variables $M_u^E(\partial_k M)$ have finite second moments for all k and given u. Then*

$$\sum_{k=0}^{N} \mathbb{E}\{M_u^E(\partial_k M)\} \geq \mathbb{P}\Big\{ \sup_{t\in M} f_t \geq u \Big\} \tag{4.6.5}$$
$$\geq \sum_{k=0}^{N} \left[\mathbb{E}\{M_u^E(\partial_k M)\} - \frac{1}{2}\mathbb{E}\{M_u^E(\partial_k M)[M_u^E(\partial_k M) - 1]\} \right].$$

Of course, we have not really proven Theorem 4.6.1 with the rigor that it deserves, so you should feel free to call it a conjecture rather than a theorem.

What you should note about this theorem/conjecture is that it makes no distributional assumptions on f beyond ensuring that all the terms it involves are well defined and finite. It certainly does not require that f be Gaussian. Consequently, it covers a level of generality far beyond the Gaussian theory we have treated so far.

It also makes sense that it would be quite a general phenomenon for the negative terms in (4.6.5) to be of smaller order that the others. After all, for $M_u^E(\partial_k M)[M_u^E(\partial_k M) - 1]$ to be nonzero, there must be at least two extended outward maxima of $f|_{\partial_k M}$ above the level u on $\partial_k M$, and this is unlikely to occur if u is large.

Thus Theorem 4.6.1 seems to hold a lot of promise for approximating extremal probabilities in general, assuming that we could actually compute explicit expressions for the simple expectations in (4.6.5) along with some useful bounds for the product expectations.

In Chapter 11 we shall work hard to obtain generic expressions that, in principle, allow us to compute all these expectations. In particular, we shall have Theorem 11.2.1 for the simple means and Theorem 11.5.1 for the factorial moments in (4.6.5). Unfortunately, however, carrying out the computations in practice will turn out to be impossible, even in the case that f is Gaussian. Consequently, we shall be forced to take a less-direct path to approximating excursion probabilities, by evaluating mean Euler characteristics that actually approximate the first-order expectations in (4.6.5). The details of this argument will appear in Chapter 14.

First of all, however, we shall need to spend quite some time in Part II of this book on studying geometry.

5

Stationary Fields

Stationarity has always been the backbone of almost all examples in the theory of Gaussian processes for which specific computations were possible. As described in the preface, one of the main reasons we shall be studying Gaussian processes on manifolds is to get around this assumption. Nevertheless, despite the fact that we shall ultimately try to avoid it, we invest a chapter on the topic for two reasons:

- Many of the results of Part III are *significantly* easier to interpret when specialized down to cases under which stationarity holds.
- Even in the nonstationary case, many of the detailed computations of Part III can be considered as deriving from a "local conversion to pseudostationarity," or, even more so, to pseudoisotropy. This will be taken care of there via the "induced Riemannian metric" defined in Section 12.2. Knowledge of what happens under stationarity is therefore important for knowing what to do in the general case.

We imagine that many of you will be familiar with most of the material of this chapter and so will skip it and return only when specific details are required later. For the newcomers, you should be warned that our treatment is full enough only to meet our specific needs and that both style and content are occasionally a little eclectic. In other words, you should go elsewhere for fuller, more standard treatments. References will be given along the way.

The most important classic results of this chapter are the spectral distribution and spectral representation theorems for $\mathbb{R}^N$ of Section 5.4. However, the most important results for us will be some of the consequences of these theorems for relationships between spectral moments that are concentrated in Section 5.5. This is the one section of this chapter that you will almost definitely need to come back to, even if you have decided that you are familiar enough with stationarity to skip this chapter for the moment.

5.1 Basic Stationarity

Although our primary interest lies in the study of real-valued random fields, it is mathematically more convenient to discuss stationarity in the framework of complex-valued processes. Hence, unless otherwise stated, we shall assume throughout this chapter that $f(t) = (f_R(t) + i f_I(t))$ takes values in the complex plane $\mathcal{C}$ and that $\mathbb{E}\{\|f(t)\|^2\} = \mathbb{E}\{f_R^2(t) + f_I^2(t)\} < \infty$. (Both f_R and f_I are, obviously, to be real-valued.) As for a definition of normality in the complex scenario, we first define a complex random variable to be Gaussian if the vector of its two components is bivariate Gaussian.[1] A complex process f is Gaussian if $\sum_i \alpha_{t_i} f_{t_i}$ is a complex Gaussian variable for all sequences $\{t_i\}$ and complex $\{\alpha_{t_i}\}$.

We also need some additional assumptions on the parameter space T. In particular, we require that it have a group structure[2] and an operation with respect to which the field is stationary. Consequently, we now assume that T has such a structure, "+" represents the binary operation on T and "−" repesents inversion. As usual, $t - s = t + (-s)$. For the moment, we need no further assumptions on the group.

Since $f_t \in \mathcal{C}$, it follows that the mean function $m(t) = \mathbb{E}\{f(t)\}$ is also complex-valued, as is the covariance function, which we redefine for the complex case as

$$C(s,t) \stackrel{\Delta}{=} \mathbb{E}\{[f(s) - m(s)][\overline{f(t)} - \overline{m(t)}]\}, \tag{5.1.1}$$

with the bar denoting complex conjugation.

Some basic properties of covariance functions follow immediately from (5.1.1):

- $C(s,t) = \overline{C(t,s)}$, which becomes the simple symmetry $C(s,t) = C(t,s)$ if f (and so C) is real-valued.
- For any $k \geq 1$, $t_1, \dots, t_k \in T$, and $z_1, \dots, z_k \in \mathcal{C}$, the Hermitian form $\sum_{i=1}^k \sum_{j=1}^k C(t_i, t_j) z_i \overline{z}_j$ is always real and nonnegative. We summarize this, as before, by saying that C is nonnegative definite.

(The second of these properties follows from the equivalence of the double-sum to $\mathbb{E}\{\| \sum_{i=1}^k [f(t_i) - m(t_i)] z_i \|^2\}$.)

Suppose for the moment that T is Abelian. A random field f is called *strictly homogeneous* or *strictly stationary* over T, with respect to the group operation +, if its finite-dimensional distributions are invariant under this operation. That is, for any $k \geq 1$ and any set of points $\tau, t_1, \dots, t_k \in T$,

$$(f(t_1), \dots, f(t_k)) \stackrel{\mathcal{L}}{=} (f(t_1 + \tau), \dots, f(t_k + \tau)). \tag{5.1.2}$$

An immediate consequence of strict stationarity is that $m(t)$ is constant and $C(s,t)$ is a function of the difference $s - t$ only.

Going the other way, if for a random field with $\mathbb{E}\{\|f(t)\|^2\} < \infty$ for all $t \in T$ we have that $m(t)$ is constant and $C(s,t)$ is a function of the difference $s - t$ only, then we call f simply *stationary* or *homogeneous*, occasionally adding either of the two adjectives "weakly" or "second-order."

---

[1] It therefore follows that a complex Gaussian variable $X = X_R + iX_I$ is defined by five parameters: $\mathbb{E}\{X_I\}$, $\mathbb{E}\{X_R\}$, $\mathbb{E}\{X_I^2\}$, $\mathbb{E}\{X_R^2\}$, and $\mathbb{E}\{X_I X_R\}$.

[2] If stationarity is a new concept for you, you will do well by reading this section the first time taking $T = \mathbb{R}^N$ with the usual notions of addition and subtraction.

Note that none of the above required the Gaussianity we have assumed up until now on f. If, however, we do add the assumption of Gaussianity, it immediately follows from the structure (1.2.3) of the multivariate Gaussian density[3] that a weakly stationary Gaussian field will also be strictly stationary if $C'(s,t) = \mathbb{E}\{[f(s) - m(s)][f(t) - m(t)]\}$ is also a function only of $s - t$. If f is real-valued, then since $C \equiv C'$ it follows that all weakly stationary real-valued Gaussian fields are also strictly stationary, and the issue of qualifying adjectives is moot.[4]

If T is not Abelian we must distinguish between left and right stationarity. We say that a random field f on T is *right-stationary* if (5.1.2) holds and that f is *left-stationary* if $f'(t) \triangleq f(-t)$ is right-stationary. The corresponding conditions on the covariance function change accordingly.

In order to build examples of stationary processes, we need to make a brief excursion into (Gaussian) stochastic integration.

5.2 Stochastic Integration

We return to the setting of Section 1.4.3, so that we have a σ-finite[5] measure space $(T, \mathcal{T}, \nu)$, along with the Gaussian ν-noise W defined over $\mathcal{T}$. Our aim will be to establish the existence of integrals of the form

$$\int_T f(t) W(dt), \tag{5.2.1}$$

for *deterministic* $f \in L^2(\nu)$ and, eventually, complex W. Appropriate choices of f will give us examples of stationary Gaussian fields, many of which we shall meet in the following sections.

Before starting, it is only fair to note that we shall be working with two and a bit assumptions. The "bit" is that for the moment we shall treat only real f and W. We shall, painlessly, lift that assumption soon. Of the other two, one is rather restrictive and one not, but neither is of importance to us. The nonrestrictive assumption is the Gaussian nature of the process W. Indeed, since all of what follows is based only on L^2 theory, we can, and so shall, temporarily drop the assumption that W is a Gaussian

[3] Formally, (1.2.3) is not quite enough. Since we are currently treating complex-valued processes, the k-dimensional marginal distributions of f now involve $2k$-dimensional Gaussian vectors. (The real and imaginary parts each require k dimensions.)

[4] The reason for needing the additional condition in the complex case should be intuitively clear: C alone will not even determine the variance of each of f_I and f_R but only their sum. However, together with C' both of these parameters, as well as the covariance between f_I and f_R, can be computed. See [113] for further details.

[5] Note that "σ-finite" includes "countable" and "finite," in which case ν will be a discrete measure and all of the theory we are about to develop is really much simpler. As we shall see later, these are more than just important special cases and so should not be forgotten as you read this section. Sometimes it is all too easy to lose sight of the important simple cases in the generality of the treatment.

noise and replace conditions (1.4.11)–(1.4.13) with the following three requirements for all $A, B \in \mathcal{T}$:

$$\mathbb{E}\{W(A)\} = 0, \qquad \mathbb{E}\{[W(A)]^2\} = \nu(A), \tag{5.2.2}$$

$$A \cap B = \emptyset \Rightarrow W(A \cup B) = W(A) + W(B) \text{ a.s.}, \tag{5.2.3}$$

$$A \cap B = \emptyset \Rightarrow \mathbb{E}\{W(A)W(B)\} = 0. \tag{5.2.4}$$

Note that in the Gaussian case (5.2.4) is really equivalent to the seemingly stronger (1.4.13), since zero covariance and independence are then equivalent.

The second restriction is that the integrand f in (5.2.1) is deterministic. Removing this assumption would lead us to having to define the Itô integral, which is a construction for which we shall have no need.

Since, by (5.2.3), W is a finitely additive (signed) measure, (5.2.1) is evocative of Lebesgue integration. Consequently, we start by defining the stochastic version for simple functions

$$f(t) = \sum_1^n a_i \mathbb{1}_{A_i}(t), \tag{5.2.5}$$

where $A_1, \ldots, A_n$ are disjoint $\mathcal{T}$-measurable sets, by writing

$$W(f) \equiv \int_T f(t)W(dt) \triangleq \sum_1^n a_i W(A_i). \tag{5.2.6}$$

It follows immediately from (5.2.2) and (5.2.4) that in this case $W(f)$ has zero mean and variance given by $\sum a_i^2 \nu(A_i)$. Think of $W(f)$ as a mapping from simple functions in $L^2(T, \mathcal{T}, \nu)$ to random variables[6] in $L^2(\mathbb{P}) \equiv L^2(\Omega, \mathcal{F}, \mathbb{P})$. The remainder of the construction involves extending this mapping to a full isomorphism from $L^2(\nu) \equiv L^2(T, \mathcal{T}, \nu)$ onto a subspace of $L^2(\mathbb{P})$. We shall use this isomorphism to define the integral.

Let $\mathcal{S} = \mathcal{S}(T)$ denote the class of simple functions of the form (5.2.5) for some finite n. Note first there is no problem with the consistency of the definition (5.2.6) over different representations of f. Furthermore, W clearly defines a linear mapping on $\mathcal{S}$ that preserves inner products. To see this, write $f, g \in \mathcal{S}$ as

$$f(t) = \sum_1^n a_i \mathbb{1}_{A_i}(t), \qquad g(t) = \sum_1^n b_i \mathbb{1}_{A_i}(t),$$

in terms of the same partition, to see that the $L^2(\mathbb{P})$ inner product between $W(f)$ and $W(g)$ is given by

[6] Note that if W is Gaussian, then so is $W(f)$.

$$\begin{aligned}\langle W(f), W(g)\rangle_{L^2(\mathbb{P})} &= \mathbb{E}\left\{\sum_1^n a_i W(A_i)\cdot\sum_1^n b_i W(A_i)\right\} &(5.2.7)\\ &= \sum_1^n a_i b_i \mathbb{E}\left\{[W(A_i)]^2\right\}\\ &= \int_T f(t)g(t)\nu(dt)\\ &= \langle f, g\rangle_{L^2(\nu)},\end{aligned}$$

the second line following from (5.2.4) and the second-to-last from (5.2.2) and the definition of the Lebesgue integral.

Since $\mathcal{S}$ is dense in $L^2(\nu)$ (e.g., [134]) for each $f \in L^2(\nu)$ there is a sequence $\{f_n\}$ of simple functions such that $\|f_n - f\|_{L^2(\nu)} \to 0$ as $n \to \infty$. Using this, for such f we define $W(f)$ as the mean square limit

$$W(f) \stackrel{\Delta}{=} \lim_{n\to\infty} W(f_n). \tag{5.2.8}$$

Furthermore, by (5.2.7), the limit is independent of the approximating sequence $\{f_n\}$, and the mapping W so defined is linear and preserves inner products; i.e., the mapping is an isomorphism. We take this mapping as our definition of the integral and so now not only do we have existence, but from (5.2.7) we have that for all $f, g \in L^2(\nu)$,

$$\mathbb{E}\{W(f)W(g)\} = \int_T f(t)g(t)\nu(dt). \tag{5.2.9}$$

Note also that since L^2 limits of Gaussian random variables remain Gaussian (cf. (1.2.5) and the discussion above it) under the additional assumption that W is a Gaussian noise, it follows that $W(f)$ is also Gaussian.

With our integral defined, we can now start looking at some examples of what can be done with it.

5.3 Moving Averages

We now return to the main setting of this chapter, in which T is an Abelian group under the binary operation $+$ and $-$ represents inversion. Let ν be a Haar measure on $(T, \mathcal{T})$ (assumed to be σ-finite) and take $F : T \to \mathbb{R}$ in $L^2(\nu)$. If W is a ν-noise on $\mathcal{T}$, then the random process

$$f(t) \stackrel{\Delta}{=} \int_T F(t-s)W(ds) \tag{5.3.1}$$

is called a *moving average* of W, and we have the following simple result.

Lemma 5.3.1. *Under the preceding conditions, f is a stationary random field on T. Furthermore, if W is a Gaussian noise, then f is also Gaussian.*

Proof. To establish stationarity we must prove that

$$\mathbb{E}\{f(t)f(s)\} = C(t-s)$$

for some C. However, from (5.2.9) and the invariance of ν under the group operation,

$$\begin{aligned}\mathbb{E}\{f(t)f(s)\} &= \int_T F(t-u)F(s-u)\nu(du) \\ &= \int_T F(t-s+v)F(v)\nu(dv) \\ &\triangleq C(t-s),\end{aligned}$$

and we are done.

If W is Gaussian, then we have already noted when defining stochastic integrals that $f(t) = W(F(t-\cdot))$ is a real-valued Gaussian random variable for each t. The same arguments also show that f is Gaussian as a *process*. □

A similar but slightly more sophisticated construction also yields a more general class of examples, in which we think of the elements g of a group G acting on the elements t of an underlying space T. This will force us to change notation a little and, for the argument to be appreciated in full, to assume that you also know a little about manifolds. If you do not, then you can return to this example later, after having read Chapter 6, or simply take the manifold to be $\mathbb{R}^N$. In either case, you may still want to read the very concrete and quite simple examples at the end of this section now.

Thus, taking the elements g of a group G acting on the elements t of an underlying space T, we denote the identity element of G by e and the left and right multiplication maps by L_g and R_g. We also write $I_g = L_g \circ R_g^{-1}$ for the inner automorphism of G induced by g.

Since we are now working in more generality, we shall also drop the commutativity assumption that has been in force so far. This necessitates some additional definitions, since we must distinguish between left and right stationarity. We say that a random field f on G is strictly *left-stationary* if for all n, all $(g_1, \dots, g_n)$, and any g_0,

$$(f(g_1), \dots, f(g_n)) \overset{\mathcal{L}}{=} (f \circ L_{g_0}(g_1), \dots, f \circ L_{g_0}(g_n)).$$

It is called strictly *right-stationary* if $f'(g) \triangleq f(g^{-1})$ is strictly left-stationary and strictly *bistationary*, or simply strictly *stationary*, if it is both left and right strictly stationary. As before, if f is Gaussian and has constant mean and covariance function C satisfying

$$C(g_1, g_2) = C'(g_1^{-1}g_2), \tag{5.3.2}$$

for some $C' : G \to \mathbb{R}$, then f is strictly left-stationary. Similarly, if C satisfies

$$C(g_1, g_2) = C''(g_1 g_2^{-1}) \tag{5.3.3}$$

for some C'', then f is right-stationary. If f is not Gaussian, but has constant mean and (5.3.2) holds, then f is *weakly* left-stationary. Weak right-stationarity and stationarity are defined analogously.

We can now start collecting the building blocks of the construction, which will be of a left-stationary Gaussian random field on a group G. An almost identical argument will construct a right-stationary field. It is then easy to see that this construction will give a bistationary field on G only if it is unimodular, i.e., if any left Haar measure on G is also right invariant.

We first add the condition that G be a *Lie group*, i.e., a group that is also a C^∞ manifold such that the maps taking g to g^{-1} and (g_1, g_2) to $g_1 g_2$ are both C^∞. We say G has a smooth (C^∞) (left) action on a smooth (C^∞) manifold T if there exists a map $\theta : G \times T \to T$ satisfying, for all $t \in T$ and $g_1, g_2 \in G$,

$$\theta(e, t) = t,$$
$$\theta(g_2, \theta(g_1, t)) = \theta(g_2 g_1, t).$$

We write $\theta_g : T \to T$ for the partial mapping $\theta_g(t) = \theta(g, t)$. Suppose ν is a measure on T, and let $\theta_{g*}(\nu)$ be the *push-forward* of ν under the map θ_g; i.e., $\theta_{g*}(\nu)$ is given by

$$\int_A \theta_{g*}(\nu) = \int_{\theta_{g^{-1}}(A)} \nu.$$

Furthermore, we assume that $\theta_{g*}(\nu)$ is absolutely continuous with respect to ν, with Radon–Nikodym derivative

$$D(g) \triangleq \frac{d\theta_{g*}(\nu)}{d\nu}(t), \tag{5.3.4}$$

independent of t. We call such a measure ν *left relatively invariant* under G. It is easy to see that $D(g)$ is a C^∞ homomorphism from G into the multiplicative group of positive real numbers, i.e., $D(g_1 g_2) = D(g_1) D(g_2)$. We say that ν is *left invariant* with respect to G if it is left relatively invariant and $D \equiv 1$.

Here, finally, is the result.

Lemma 5.3.2. *Suppose G acts smoothly on a smooth manifold T and ν is left relatively invariant under G. Let D be as in* (5.3.4) *and let W be Gaussian ν-noise on T. Then for any $F \in L^2(T, \nu)$,*

$$f(g) = \frac{1}{\sqrt{D(g)}} W(F \circ \theta_{g^{-1}})$$

is a left stationary Gaussian random field on G.

Proof. We must prove that

$$\mathbb{E}\{f(g_1) f(g_2)\} = C(g_1^{-1} g_2)$$

for some $C : G \to \mathbb{R}$. From the definition of W, we have

$$\begin{aligned}
\mathbb{E}\{f(g_1)f(g_2)\} &= \frac{1}{\sqrt{D(g_1)D(g_2)}} \int_T F(\theta_{g_1^{-1}}(t))F(\theta_{g_2^{-1}}(t))\nu(dt) \\
&= \frac{1}{\sqrt{D(g_1)D(g_2)}} \int_T F(\theta_{g_1^{-1}}(\theta_{g_2}(t)))F(t)\theta_{g_2*}(\nu)(dt) \\
&= \frac{D(g_2)}{\sqrt{D(g_1)D(g_2)}} \int_T F(\theta_{g_1^{-1}g_2}(t))F(t)\nu(dt) \\
&= \sqrt{D(g_1^{-1}g_2)} \int_T F(\theta_{g_1^{-1}g_2}(t))F(t)\nu(dt) \\
&\triangleq C(g_1^{-1}g_2).
\end{aligned}$$

This completes the proof. □

It is easy to find simple examples to which Lemma 5.3.2 applies. The most natural generic example of a Lie group acting on a manifold is its action on itself. In particular, any right Haar measure is left relatively invariant, and this is a way to generate stationary processes. To apply Lemma 5.3.2 in this setting one needs only to start with a Gaussian noise based on a Haar measure on G. In fact, this is the example (5.3.1) with which we started this section.

A richer but still concrete example of a group G acting on a manifold T is given by the group of rigid motions $G_N = \mathrm{GL}(N, \mathbb{R}) \times \mathbb{R}^N$ acting[7] on $T = \mathbb{R}^N$. For $g = (A, t)$ and $s \in \mathbb{R}^N$, set

$$\theta(A, t)(s) = As + t.$$

In this example it is easy to see that Lebesgue measure $\lambda_N(dt) = dt$ is relatively invariant with respect to G with $D(g) = \det A$. For an even more concrete example, take compact Borel $B \subset \mathbb{R}^N$ and $F(s) = \mathbb{1}_B(s)$. It then follows from Lemma 5.3.2 that

$$f(A, t) = \frac{W\left(A^{-1}(B - t)\right)}{\sqrt{|\det(A)|}}$$

is a stationary process with variance $\lambda_N(B)$ and covariance function

$$C((A_1, t_1), (A_2, t_2)) = \frac{\lambda_N\left(A_1 A_2^{-1}(B - (t_1 - t_2))\right)}{\sqrt{|\det(A_1 A_2^{-1})|}},$$

where we have adopted the usual notation that $B + t = \{s + t : s \in B\}$ and $AB = \{At : t \in B\}$.

Examples of this kind have been used widely in practice. For examples involving the statistics of brain mapping see, for example, [142, 143].

[7] Recall that $\mathrm{GL}(N, \mathbb{R})$ is the (general linear) group of invertible transformations of $\mathbb{R}^N$.

5.4 Spectral Representations on $\mathbb{R}^N$

The moving averages of the previous section gave us examples of stationary fields that were rather easy to generate in quite general situations from Gaussian noise. Now, however, we want to look at a general way of generating *all* stationary fields, via the so-called *spectral representation*. This is quite a simple task when the parameter set is $\mathbb{R}^N$, but rather more involved when a general group is taken as the parameter space and issues of group representations arise. Thus we shall start with the Euclidean case, which we treat in detail, and then discuss some aspects of the general case in the following section. In both cases, while an understanding of the spectral representation is a powerful tool for understanding stationarity and a variety of sample path properties of stationary fields, it is not necessary for what comes later in the book.

We return to the setting of complex-valued fields, take $T = \mathbb{R}^N$, and assume, as usual, that $\mathbb{E}\{f_t\} = 0$. Furthermore, since we are now working only with stationary processes, it makes sense to abuse notation somewhat and write

$$C(t-s) = C(s,t) = \mathbb{E}\{f(s)\overline{f(t)}\}.$$

We call C, which is now a function of one parameter only, nonnegative definite if $\sum_{i,j=1}^{n} z_i C(t_i - t_j)\overline{z_j} \geq 0$ for all $n \geq 1$, $t_1, \ldots, t_n \in \mathbb{R}^N$, and $z_1, \ldots, z_n \in \mathcal{C}$. Then we have the following result, which dates back to Bochner [27], in the setting of (nonstochastic) Fourier analysis, a proof of which can be found in almost any text on Fourier analysis.

Theorem 5.4.1 (spectral distribution theorem). *A continuous function $C : \mathbb{R}^N \to \mathcal{C}$ is nonnegative definite (i.e., a covariance function) if and only if there exists a finite measure ν on the Borel σ-field $\mathcal{B}^N$ such that*

$$C(t) = \int_{\mathbb{R}^N} e^{i\langle t,\lambda\rangle}\nu(d\lambda), \tag{5.4.1}$$

for all $t \in \mathbb{R}^N$.

With randomness in mind, we write $\sigma^2 = C(0) = \nu(\mathbb{R}^N)$. The measure ν is called the *spectral measure* (for C), and the function $F : \mathbb{R}^N \to [0, \sigma^2]$ given by

$$F(\lambda) \triangleq \nu\left(\prod_{i=1}^{N}(-\infty, \lambda_i]\right), \qquad \lambda = (\lambda_1, \ldots, \lambda_N) \in \mathbb{R}^N,$$

is called the *spectral distribution function.*[8] When F is absolutely continuous the corresponding density is called the *spectral density*.

The spectral distribution theorem is a purely analytic result and would have nothing to do with random fields were it not for the fact that covariance functions are

[8] Of course, unless ν is a probability measure, so that $\sigma^2 = 1$, F is not a distribution function in the usual usage of the term.

nonnegative definite. Understanding of the result comes from the spectral representation theorem (Theorem 5.4.2), for which we need to extend somewhat the stochastic integration of Section 5.2.

As usual, we start with a measure ν on $\mathbb{R}^N$, and define a *complex* ν-noise to be a $\mathbb{C}$-valued, set-indexed process satisfying

$$\mathbb{E}\{W(A)\} = 0, \qquad \mathbb{E}\{W(A)\overline{W(A)}\} = \nu(A), \tag{5.4.2}$$

$$A \cap B = \emptyset \Rightarrow W(A \cup B) = W(A) + W(B) \text{ a.s.}, \tag{5.4.3}$$

$$A \cap B = \emptyset \Rightarrow \mathbb{E}\{W(A)\overline{W(B)}\} = 0, \tag{5.4.4}$$

where the bar denotes complex conjugation.[9]

It is then a straightforward exercise to extend the construction of Section 5.2 to obtain an L^2 stochastic integral

$$W(f) = \int_{\mathbb{R}^N} f(\lambda) W(d\lambda)$$

for $f : \mathbb{R}^N \to \mathbb{C}$ with $\|f\| \in L^2(\nu)$, and satisfying

$$\mathbb{E}\{W(f)\overline{W(g)}\} = \int_{\mathbb{R}^N} f(\lambda)\overline{g(\lambda)}\nu(d\lambda). \tag{5.4.5}$$

This construction allows us to state the following important result.

Theorem 5.4.2 (spectral representation theorem). *Let ν be a finite measure on $\mathbb{R}^N$ and W a complex ν-noise. Then the complex-valued random field*

$$f(t) = \int_{\mathbb{R}^N} e^{i\langle t,\lambda\rangle} W(d\lambda) \tag{5.4.6}$$

has covariance

$$C(s,t) = \int_{\mathbb{R}^N} e^{i\langle (s-t),\lambda\rangle} \nu(d\lambda) \tag{5.4.7}$$

and so is (weakly) stationary. If W is Gaussian, then so is f.

Furthermore, to every mean square continuous, centered (Gaussian) stationary random field f on $\mathbb{R}^N$ with covariance function C and spectral measure ν (cf. Theorem 5.4.1), there corresponds a complex (Gaussian) ν-noise W on $\mathbb{R}^N$ such that (5.4.6) holds in mean square for each $t \in \mathbb{R}^N$.

In both cases, W is called the spectral process corresponding to f.

---

[9] You should note for later reference that if we split a complex ν-noise into its real and imaginary parts, W_R and W_I say, it does not follow from (5.4.4) that $A \cap B = \emptyset$ implies any of $\mathbb{E}\{W(A)W(B)\} = 0$, $\mathbb{E}\{W_R(A)W_R(B)\} = 0$, $\mathbb{E}\{W_I(A)W_I(B)\} = 0$, or $\mathbb{E}\{W_I(A)W_R(B)\} = 0$. Indeed, this is most definitely not the case for the complex W of Theorem 5.4.2 when, for example, the stationary process f there is real-valued. See the discussion below on "real" spectral representations, in which the above problem is addressed by taking W to be defined on a half-space rather than all of $\mathbb{R}^N$.

Proof. The fact that (5.4.6) generates a stationary field with covariance (5.4.7) is an immediate consequence of (5.4.5). What needs to be proven is the statement that *all* stationary fields can be represented as in (5.4.6). We shall only sketch the basic idea of the proof, leaving the details to the reader. (They can be found in almost any book on time series—our favorite is [34]—for processes on either $\mathbb{Z}$ or $\mathbb{R}$ and the extension to $\mathbb{R}^N$ is trivial.)

For the first step, set up a mapping Θ from[10] $\mathcal{H} \stackrel{\Delta}{=} \text{span}\{f_t,\ t \in \mathbb{R}^N\} \subset L^2(\mathbb{P})$ to $\mathcal{K} \stackrel{\Delta}{=} \text{span}\{e^{it\cdot}, t \in \mathbb{R}^N\} \subset L^2(\nu)$ via

$$\Theta\left(\sum_{j=1}^{n} a_j f(t_j)\right) = \sum_{j=1}^{n} a_j e^{it_j \cdot} \tag{5.4.8}$$

for all $n \geq 1$, $a_j \in \mathcal{C}$, and $t_j \in \mathbb{R}^N$. A simple computation shows that this gives an isomorphism between $\mathcal{H}$ and $\mathcal{K}$, which can then be extended to an isomorphism between their closures $\overline{\mathcal{H}}$ and $\overline{\mathcal{K}}$.

Since indicator functions are in $\overline{\mathcal{K}}$, we can define a process W on $\mathbb{R}^N$ by setting

$$W(\lambda) = \Theta^{-1}(\mathbb{1}_{(-\infty,\lambda]}), \tag{5.4.9}$$

where $(-\infty, \lambda] = \prod_{j=1}^{N}(-\infty, \lambda_j]$. Working through the isomorphisms shows that W is a complex ν-noise and that $\int_{\mathbb{R}^N} \exp(i\langle t, \lambda\rangle) W(d\lambda)$ is ($L^2(\mathbb{P})$ indistinguishable from) f_t. □

There is also an inverse[11] to (5.4.6), expressing W as an integral involving f, but we shall have no need of it and so now turn to some consequences of Theorems 5.4.1 and 5.4.2.

When the basic field f is real, it is natural to expect a "real" spectral representation, and this is in fact the case, although notationally it is still generally more convenient to use the complex formulation. Note firstly that if f is real, then the covariance function is a symmetric function ($C(t) = C(-t)$), and so it follows from the spectral distribution theorem (cf. (5.4.1)) that the spectral measure ν must also be symmetric. Introduce three[12] new measures, on $\mathbb{R}_+ \times \mathbb{R}^{N-1}$, by

[10] Actually, we really define $\overline{\mathcal{H}}$ to be the closure in $L^2(\mathbb{P})$ of all sums of the form $\sum_{i=1}^{n} a_i f(t_i)$ for $a_i \in \mathbb{C}$ and $t_i \in T$, thinned out by identifying all elements indistinguishable in $L^2(\mathbb{P})$, i.e., elements U, V for which $\mathbb{E}\{\|(U - V)\|^2\} = 0$. Recall that we already met this space, as well as the mapping Θ, when treating orthonormal expansions in Section 3.1.

[11] Formally, the inverse relationship is based on (5.4.9), but it behaves like regular Fourier inversion. For example, if $\Delta(\lambda, \eta)$ is a rectangle in $\mathbb{R}^N$ that has a boundary of zero ν measure, then

$$W(\Delta(\lambda, \eta)) = \lim_{K\to\infty} (2\pi)^{-N} \int_{-K}^{K} \cdots \int_{-K}^{K} \prod_{k=1}^{N} \frac{e^{-i\lambda_k t_k} - e^{-i\eta_k t_k}}{-it_k} f(t)\, dt.$$

As is usual, if $\nu(\Delta(\lambda, \eta)) \neq 0$, then additional boundary terms need to be added.

[12] Note that if ν is absolutely continuous with respect to Lebesgue measure (so that there is a spectral density), one of these, ν_2, will be identically zero.

$$\nu_1(A) = \nu(A \cap \{\lambda \in \mathbb{R}^N : \lambda_1 > 0\}),$$
$$\nu_2(A) = \nu(A \cap \{\lambda \in \mathbb{R}^N : \lambda_1 = 0\}),$$
$$\mu(A) = 2\nu_1(A) + \nu_2(A).$$

We can now rewrite[13] (5.4.1) in real form as

$$C(t) = \int_{\mathbb{R}_+ \times \mathbb{R}^{N-1}} \cos(\langle \lambda, t \rangle) \mu(d\lambda). \tag{5.4.10}$$

There is also a corresponding real form of the spectral representation (5.4.6). The fact that the spectral representation yields a real-valued process also implies certain symmetries[14] on the spectral process W. In particular, it turns out that there are two independent *real*-valued μ-noises, W_1 and W_2, such that[15]

$$f_t = \int_{\mathbb{R}_+ \times \mathbb{R}^{N-1}} \cos(\langle \lambda, t \rangle) W_1(d\lambda) + \int_{\mathbb{R}_+ \times \mathbb{R}^{N-1}} \sin(\langle \lambda, t \rangle) W_2(d\lambda). \tag{5.4.11}$$

It is easy to check that f so defined has the right covariance function.

5.5 Spectral Moments

Since they will be very important later on, we now take a closer look at spectral measures and, in particular, their moments. It turns out that these contain a lot of simple, but very useful, information. Given the spectral representation (5.4.7), that is,

$$C(t) = \int_{\mathbb{R}^N} e^{i\langle t, \lambda \rangle} \nu(d\lambda), \tag{5.5.1}$$

we define the *spectral moments*

$$\lambda_{i_1 \dots i_N} \triangleq \int_{\mathbb{R}^N} \lambda_1^{i_1} \cdots \lambda_N^{i_N} \nu(d\lambda), \tag{5.5.2}$$

for all $(i_1, \dots, i_N)$ with $i_j \geq 0$. Recalling that stationarity implies that $C(t) = \overline{C(-t)}$ and $\nu(A) = \nu(-A)$, it follows that the odd-ordered spectral moments, when they exist, are zero; i.e.,

---

[13] There is nothing special about the half-space $\lambda_1 \geq 0$ taken in this representation. Any half-space will do.

[14] To establish this rigorously, we really need the inverse to (5.4.6), expressing W as an integral involving f, which we do not have.

[15] In one dimension, it is customary to take W_1 as a μ-noise and W_2 as a $(2\nu_1)$-noise, which at first glance is different from what we have. However, noting that when $N = 1$, $\sin(\lambda t) W_2(d\lambda) = 0$ when $\lambda = 0$, it is clear that the two definitions in fact coincide in this case.

$$\lambda_{i_1 \dots i_N} = 0 \quad \text{if} \sum_{j=1}^{N} i_j \text{ is odd.} \tag{5.5.3}$$

Furthermore, it is immediate from (5.5.1) that successive differentiation of both sides with respect to the t_i connects the various partial derivatives of C at zero with the spectral moments. To see why this is important, recall the definition (1.4.8) of the L^2 derivatives of a random field as the L^2 limit

$$D^k_{L^2} f(t, t') = \lim_{h \to 0} F(t, ht'),$$

for $t \in \mathbb{R}^N$ and $t' \in \otimes^k \mathbb{R}^N$.

By choosing $t' = (e_{i_1}, \dots, e_{i_k})$, where e_i is the vector with ith element 1 and all others zero, we can talk of the mean square derivatives

$$\frac{\partial^k}{\partial t_{i_1} \cdots \partial t_{i_k}} f(t) \triangleq D^k_{L^2} f(t, (e_{i_1}, \dots, e_{i_k}))$$

of f of various orders.

It is then straightforward to see that the covariance function of such partial derivatives must be given by

$$\mathbb{E}\left\{ \frac{\partial^k f(s)}{\partial s_{i_1} \partial s_{i_1} \cdots \partial s_{i_k}} \frac{\partial^k f(t)}{\partial t_{i_1} \partial t_{i_1} \cdots \partial t_{i_k}} \right\} = \frac{\partial^{2k} C(s,t)}{\partial s_{i_1} \partial t_{i_1} \cdots \partial s_{i_k} \partial t_{i_k}}. \tag{5.5.4}$$

The corresponding variances have a nice interpretation in terms of spectral moments when f is stationary. For example, if f has mean square partial derivatives of orders $\alpha + \beta$ and $\gamma + \delta$ for $\alpha, \beta, \gamma, \delta \in \{0, 1, 2, \dots\}$, then[16]

$$\begin{aligned} \mathbb{E}\left\{ \frac{\partial^{\alpha+\beta} f(t)}{\partial^\alpha t_i \partial^\beta t_j} \cdot \frac{\partial^{\gamma+\delta} f(t)}{\partial^\gamma t_k \partial^\delta t_l} \right\} &= (-1)^{\alpha+\beta} \frac{\partial^{\alpha+\beta+\gamma+\delta}}{\partial^\alpha t_i \partial^\beta t_j \partial^\gamma t_k \partial^\delta t_l} C(t)\Big|_{t=0} \\ &= (-1)^{\alpha+\beta} i^{\alpha+\beta+\gamma+\delta} \int_{\mathbb{R}^N} \lambda_i^\alpha \lambda_j^\beta \lambda_k^\gamma \lambda_l^\delta \, \nu(d\lambda). \end{aligned} \tag{5.5.5}$$

Here are some important special cases of the above, for which we adopt the shorthand $f_j = \partial f / \partial t_j$ and $f_{ij} = \partial^2 f / \partial t_i \partial t_j$ along with a corresponding shorthand for the partial derivatives of C:

(i) f_j has covariance function $-C_{jj}$ and thus variance $\lambda_{2e_j} = -C_{jj}(0)$, where e_j, as usual, is the vector with a 1 in the jth position and zero elsewhere.

---

[16] If you decide to check this for yourself using (5.4.6) and (5.4.7)—which is a worthwhile exercise—make certain that you recall the fact that the covariance function is defined as $\mathbb{E}\{f(s)\overline{f(t)}\}$, or you will make the same mistake RJA did in [2] and forget the factor of $(-1)^{\alpha+\beta}$ in the first line. Also, note that although (5.5.5) seems to have some asymmetries in the powers, these disappear due to the fact that all odd-ordered spectral moments, like all odd-ordered derivatives of C, are identically zero.

(ii) In view of (5.5.3), and taking $\beta = \gamma = \delta = 0$, $\alpha = 1$ in (5.5.5),

$$f(t) \text{ and } f_j(t) \text{ are uncorrelated,} \tag{5.5.6}$$

for all j and all t. If f is Gaussian, this is equivalent to independence. Note that (5.5.6) does *not* imply that f and f_j are uncorrelated *as processes*. In general, for $s \neq t$, we will have that $\mathbb{E}\{f(s)f_j(t)\} = -C_j(s-t) \neq 0$.

(iii) Taking $\alpha = \gamma = \delta = 1$, $\beta = 0$ in (5.5.5) gives that

$$f_i(t) \text{ and } f_{jk}(t) \text{ are uncorrelated} \tag{5.5.7}$$

for all i, j, k and all t.

This concludes our initial discussion of stationarity for random fields on $\mathbb{R}^N$. In the following section we investigate what happens under slightly weaker conditions and after that what happens when the additional assumption of isotropy is added. In Section 5.8 we shall see how all of this is really part of a far more general theory for fields defined on groups. In Section 3.1 we shall see that, at least for Gaussian fields restricted to bounded regions, that spectral theory is closely related to a far wider theory of orthogonal expansions that works for general—i.e., not necessarily Euclidean—parameter spaces

5.6 Constant Variance

It will be important for us in later chapters that some of the relationships of the previous section continue to hold under a condition weaker than stationarity. Of particular interest is knowing when (5.5.6) holds; i.e., when $f(t)$ and $f_j(t)$ are uncorrelated.

Suppose that f has constant variance, $\sigma^2 = C(t,t)$, throughout its domain of definition, and that its L^2 first-order derivatives all exist. In this case, analagously to (5.5.5), we have that

$$\mathbb{E}\{f(t)f_j(t)\} = \frac{\partial}{\partial t_j} C(t,s)\Big|_{s=t} = \frac{\partial}{\partial s_j} C(t,s)\Big|_{s=t}. \tag{5.6.1}$$

Since constant variance implies that $\partial/\partial t_j C(t,t) \equiv 0$, the above two equalities imply that both partial derivatives there must be identically zero. That is, f and its first-order derivatives are uncorrelated.

One can, of course, continue in this fashion. If first derivatives have constant variance, then they, in turn, will be uncorrelated with second derivatives, in the sense that f_i will be uncorrelated with f_{ij} for all i, j. It will not necessarily be true, however, that f_i and f_{jk} will be uncorrelated if $i \neq j$ and $i \neq k$. This will, however, be true if the covariance matrix of all first-order derivatives is constant.

5.7 Isotropy

An interesting special class of homogeneous random fields on $\mathbb{R}^N$ that often arises in applications in which there is no special meaning attached to the coordinate system being used is the class of *isotropic* fields. These are characterized[17] by the property that the covariance function depends only on the Euclidean length $|t|$ of the vector t, so that

$$C(t) = C(|t|). \tag{5.7.1}$$

Isotropy has a number of surprisingly varied implications for both the covariance and spectral distribution functions, and is actually somewhat more limiting than it might at first seem. For example, we have the following result, due to Matérn [111].

Theorem 5.7.1. *If $C(t)$ is the covariance function of a centered, isotropic random field f on $\mathbb{R}^N$, then $C(t) \geq -C(0)/N$ for all t.*

Proof. Isotropy implies that C can be written as a function on $\mathbb{R}_+$ only. Let τ be any positive real number. We shall show that $C(\tau) \geq -C(0)/N$.

Choose any $t_1, \dots, t_{N+1}$ in $\mathbb{R}^N$ for which $|t_i - t_j| = \tau$ for all $i \neq j$. Then, by (5.7.1),

$$\mathbb{E}\left\{\left|\sum_{k=1}^{N+1} f(t_k)\right|^2\right\} = (N+1)[C(0) + NC(\tau)].$$

Since this must be positive, the result follows. □

The restriction of isotropy also has significant simplifying consequences for the spectral measure ν of (5.4.1). Let $\theta : \mathbb{R}^N \to \mathbb{R}^N$ be a rotation, so that $|\theta(t)| = |t|$ for all t. Isotropy then implies $C(t) = C(\theta(t))$, and so the spectral distribution theorem implies

$$\begin{aligned} \int_{\mathbb{R}^N} e^{i\langle t,\lambda\rangle}\nu(d\lambda) &= \int_{\mathbb{R}^N} e^{i\langle\theta(t),\lambda\rangle}\nu(d\lambda) \\ &= \int_{\mathbb{R}^N} e^{i\langle t,\theta(\lambda)\rangle}\nu(d\lambda) = \int_{\mathbb{R}^N} e^{i\langle t,\lambda\rangle}\nu_\theta(d\lambda), \end{aligned} \tag{5.7.2}$$

where ν_θ is the push-forward of ν by θ defined by $\nu_\theta(A) = \nu(\theta^{-1}A)$. Since the above holds for all t, it follows that $\nu \equiv \nu_\theta$; i.e., ν, like C, is invariant under rotation. Furthermore, if ν is absolutely continuous, then its density is also dependent only on the modulus of its argument.

An interesting consequence of this symmetry is that an isotropic field cannot have all the probability of its spectral measure concentrated in one small region in

[17] Isotropy can also be defined in the nonstationary case, the defining property then being that the distribution of f is invariant under rotation. Under stationarity, this is equivalent to (5.7.1). Without stationarity, however, this definition does not imply (5.7.1). We shall treat isotropy *only* in the scenario of stationarity.

$\mathbb{R}^N$ away from the origin. In particular, it is not possible to have a spectral measure degenerate at one point, unless that point is the origin. The closest the spectral measure of an isotropic field can come to this sort of behavior is to have all its probability concentrated in an annulus of the form

$$\{\lambda \in \mathbb{R}^N : a \leq |\lambda| \leq b\}.$$

In such a case it is not hard to see that the field itself is then composed of a "sum" of waves traveling in all directions but with wavelengths between $2\pi/b$ and $2\pi/a$ only.

Another consequence of isotropy is that the spherical symmetry of the spectral measure significantly simplifies the structure of the spectral moments, and hence the correlations between various derivatives of f. In particular, it follows immediately from (5.5.5) that

$$\mathbb{E}\left\{f_i(t)f_j(t)\right\} = -\mathbb{E}\left\{f(t)f_{ij}(t)\right\} = \lambda_2\delta_{ij}, \tag{5.7.3}$$

where δ_{ij} is the Kronecker delta and $\lambda_2 \stackrel{\Delta}{=} \int_{\mathbb{R}^N} \lambda_i^2 \nu(d\lambda)$, which is independent of the value of i. Consequently, if f is Gaussian, then the first-order derivatives of f are independent of one another, as they are of f itself.

Since isotropy has such a limiting effect in the spectrum, it is natural to ask how the spectral distribution and representation theorems are affected under isotropy. The following result, due originally to Schoenberg [140] (in a somewhat different setting) and Yaglom [177], describes what happens.

Theorem 5.7.2. *For C to be the covariance function of a mean square continuous, isotropic, random field on $\mathbb{R}^N$ it is necessary and sufficient that*

$$C(t) = \int_0^\infty \frac{J_{(N-2)/2}(\lambda|t|)}{(\lambda|t|)^{(N-2)/2}}\,\mu(d\lambda), \tag{5.7.4}$$

where μ is a finite measure on $\mathbb{R}_+$ and J_m is the Bessel function of the first kind of order m, that is,

$$J_m(x) = \sum_{k=0}^{\infty}(-1)^k \frac{(x/2)^{2k+m}}{k!\Gamma(k+m+1)}.$$

Proof. The proof consists in simplifying the basic spectral representation by using the symmetry properties of ν.

We commence by converting to polar coordinates, $(\lambda, \theta_1, \dots, \theta_{N-1})$, $\lambda \geq 0$, $(\theta_1, \dots, \theta_{N-1}) \in S^{N-1}$, where S^{N-1} is the unit sphere in $\mathbb{R}^N$. Define a measure μ on $\mathbb{R}_+$ by setting $\mu([0,\lambda]) = \nu(B^N(\lambda))$, and extending as usual, where $B^N(\lambda)$ is the N-ball of radius λ and ν is the spectral measure of (5.4.1).

Then, on substituting into (5.4.1) with $t = (|t|, 0, \dots, 0)$ and performing the coordinate transformation, we obtain

$$C(|t|) = \int_0^\infty \int_{S^{N-1}} \exp(i|t|\lambda\cos\theta_{N-1})\sigma(d\theta)\mu(d\lambda),$$

where σ is surface area measure on S^{N-1}. Integrating out $\theta_1, \ldots, \theta_{N-2}$, it follows that

$$C(|t|) = s_{N-2} \int_0^\infty \int_0^\pi e^{i\lambda|t|\cos\theta_{N-1}} (\sin\theta_{N-1})^{N-2}\, d\theta_{N-1}\mu(d\lambda),$$

where

$$s_N \triangleq \frac{2\pi^{N/2}}{\Gamma(N/2)}, \qquad N \geq 1, \tag{5.7.5}$$

is the surface area[18] of S^{N-1}.

The inside integral can be evaluated in terms of Bessel functions to yield

$$\int_0^\pi e^{i\lambda|t|\cos\theta} \sin^{N-2}\theta\, d\theta = \frac{J_{(N-2)/2}(\lambda|t|)}{(\lambda|t|)^{(N-2)/2}},$$

which, on absorbing s_{N-2} into μ, completes the proof. □

For small values of the dimension N, (5.7.4) can be simplified even further. For example, substituting $N = 2$ into (5.7.4) yields that in this case,

$$C(t) = \int_0^\infty J_0(\lambda|t|)\mu(d\lambda),$$

while substituting $N = 3$ and evaluating the inner integral easily yields that in this case,

$$C(t) = \int_0^\infty \frac{\sin(\lambda|t|)}{\lambda|t|}\mu(d\lambda).$$

Given the fact that the covariance function of an isotropic field takes such a special form, it is natural to seek a corresponding form for the spectral representation of the field itself. Such a representation does in fact exist and we shall now describe it, albeit without giving any proofs. These can be found, for example, in the book by Wong [169], or as special cases in the review by Yaglom [178], which is described in Section 5.8 below. Another way to verify it would be to check that the representation given in Theorem 5.7.3 below yields the covariance structure of (5.7.4). Since this is essentially an exercise in the manipulation of special functions and not of intrinsic probabilistic interest, we shall avoid the temptation to carry it out.

The spectral representation of isotropic fields on $\mathbb{R}^N$ is based on the so-called *spherical harmonics*[19] on the $(N-1)$-sphere, which form an orthonormal basis for

---

[18] When $N = 1$, we have the "boundary" of the "unit sphere" $[-1, 1] \in \mathbb{R}$. This is made up of the two points ± 1, which, in counting measure, has measure 2. Hence $s_1 = 2$ makes sense.

[19] We shall also avoid giving details about spherical harmonics. A brief treatment would add little to understanding them. The kind of treatment required to, for example, get the code correct in programming a simulation of an isotropic field using the representations that follow will, in any case, send you back to the basic reference of Erdélyi [61] followed by some patience in sorting out a software help reference. A quick web search will yield you many interactive, colored examples of these functions within seconds.

the space of square-integrable functions on S^{N-1} equipped with the usual surface measure. We shall denote them by $\{h_{ml}^{(N-1)}, l = 1, \dots, d_m, m = 0, 1, \dots\}$, where the d_m are known combinatorial coefficents.[20]

Now use the spectral decomposition

$$f(t) = \int_{\mathbb{R}^N} e^{i\langle t,\lambda\rangle} W(d\lambda)$$

to define a family of noises on $\mathbb{R}_+$ by

$$W_{ml}(A) = \int_A \int_{S^{N-1}} h_{ml}^{(N-1)}(\theta) W(d\lambda, d\theta),$$

where, once again, we work in polar coordinates. Note that since W is a ν-noise, where ν is the spectral measure, information about the covariance of f has been coded into the W_{ml}. From this family, define a family of mutually uncorrelated, stationary, one-dimensional processes $\{f_{ml}\}$ by

$$f_{ml}(r) = \int_0^\infty \frac{J_{m+(N-2)/2}(\lambda r)}{(\lambda r)^{(N-2)/2}} W_{ml}(d\lambda),$$

where, as in the spectral representation (5.4.6), one has to justify the existence of this L^2 stochastic integral. These are all the components we need in order to state the following.

Theorem 5.7.3. *A centered, mean square continuous, isotropic random field on* $\mathbb{R}^N$ *can be represented by*

$$f(t) = f(r, \theta) = \sum_{m=0}^{\infty} \sum_{l=1}^{d_m} f_{ml}(r) h_{ml}^{(N-1)}(\theta). \tag{5.7.6}$$

In other words, isotropic random fields can be decomposed into a countable number of mutually uncorrelated stationary processes with a one-dimensional parameter, a result that one would not intuitively expect. As noted above, there is still a hidden spectral process in (5.7.6), entering via the W_{ml} and f_{ml}. This makes for an important difference between (5.7.6) and the similar looking Karhunen–Loève expansion we saw Section 3.2. Another difference lies in the fact that while it is possible to truncate the expansion (5.7.6) to a finite number of terms and retain isotropy, this is not true of the standard Karhunen–Loève expansion. In particular, isotropic fields can never have finite Karhunen–Loève expansions. For a heuristic argument as to why this is the case, recall from Section 5.7 that under isotropy the spectral measure must be invariant under rotations, and so cannot be supported on a finite, or even countable, number of points. Consequently, one also needs an uncountable number of independent variables in the spectral noise process to generate the process via (5.4.6).

[20] The spherical harmonics on S^{N-1} are often written as $\{h_{m,l_1,\dots,l_{N-2},\pm l_{N-1}}^{(N-1)}\}$, where $0 \le l_{N-1} \le \cdots \le l_1 \le m$. The constants d_m in our representation can be computed from this.

However, a process with a finite Karhunen–Loève expansion provides only a finite number of such variables, which can never be enough.

An interesting corollary of Theorem 5.7.3 is obtained by fixing r in (5.7.6). We then have a simple representation in terms of uncorrelated random coefficients $f_{ml}(r)$ and spherical harmonics of an isotropic random field on the N-sphere of radius r. If the random field is Gaussian, then the coefficients are actually independent, and we will, essentially, have generated a Karhunen–Loève expansion. You can find many more results of this kind, with much more detail, in [102].

One can keep playing these games for more and more special cases. For example, it is not uncommon in applications to find random fields that are functions of "space" x and "time" t, so that the parameter set is most conveniently written as $(t, x) \in \mathbb{R}_+ \times \mathbb{R}^N$. Such processes are often stationary in t and isotropic in x, in the sense that

$$\mathbb{E}\{f(s, u)\overline{f(s+t, u+x)}\} = C(t, |x|),$$

where C is now a function from $\mathbb{R}^2$ to $\mathbb{C}$. In such a situation the methods of the previous proof suffice to show that C can be written in the form

$$C(t, x) = \int_{-\infty}^{\infty}\int_{0}^{\infty} e^{itv} G_N(\lambda x)\mu(dv, d\lambda),$$

where

$$G_N(x) = \left(\frac{2}{x}\right)^{(N-2)/2} \Gamma\left(\frac{N}{2}\right) J_{(N-2)/2}(x)$$

and μ is a measure on the half-plane $\mathbb{R}_+ \times \mathbb{R}^N$.

By now it should be starting to become evident that all of these representations must be special cases of some general theory, which might also be able to cover non-Euclidean parameter spaces. This is indeed the case, although for reasons that will soon be explained, the general theory is such that, ultimately, each special case requires almost individual treatment.

5.8 Stationarity over Groups

We have already seen in Section 5.3 that the appropriate setting for stationarity is that in which the parameter set has a group structure. In this case it made sense, in general, to talk about left and right stationarity (cf. (5.3.2) and (5.3.3)). Simple "stationarity" requires both of these and so makes most sense if the group is abelian (commutative).

In essence, the spectral representation of a random field over a group is intimately related to the representation theory of the group. This, of course, is far from being a simple subject. Furthermore, its level of difficulty depends very much on the group in question, and so it is correspondingly not easy to give a general spectral theory for random fields over groups. The most general results in this area are in the paper by

Yaglom [178] already mentioned above, and the remainder of this section is taken from there.[21]

We shall make life simpler by assuming for the rest of this section that T is a *locally compact Abelian* (LCA) group. As before, we shall denote the binary group operation by $+$, while $-$ denotes inversion. The Fourier analysis of LCA groups is well developed (e.g., [133]) and based on *characters*. A homomorphism γ from T to the multiplicative group of complex numbers is called a character if $\|\gamma(t)\| = 1$ for all $t \in T$ and if

$$\gamma(t+s) = \gamma(t)\gamma(s), \qquad s, t \in T.$$

If $T = \mathbb{R}^N$ under the usual addition, then the characters are given by the family

$$\left\{\gamma_\lambda(t) = e^{i\langle t,\lambda\rangle}\right\}_{\lambda\in\mathbb{R}^N}$$

of complex exponential functions, which were at the core of the spectral theory of fields over $\mathbb{R}^N$. If $T = \mathbb{Z}^N$, again under addition, the characters are as for $T = \mathbb{R}^N$, but λ is restricted to $[-\pi, \pi]^N$. If $T = \mathbb{R}^N$ under rotation rather than addition, then the characters are the spherical harmonics on S^{N-1}.

The set of all continuous characters also forms a group, Γ say, called the *dual group*, with composition defined by

$$(\gamma_1 + \gamma_2)(t) = \gamma_1(t)\gamma_2(t).$$

There is also a natural topology on Γ (cf. [133]) that gives Γ an LCA structure and under which the map $(\gamma, t) \stackrel{\Delta}{=} \gamma(t) : \Gamma \times T \to \mathcal{C}$ is continuous. The spectral distribution theorem in this setting can now be written as

$$C(t) = \int_\Gamma (\gamma, t)\nu(d\gamma), \tag{5.8.1}$$

where the finite spectral measure ν is on the σ-algebra generated by the topology on Γ. The spectral representation theorem can be correspondingly written as

$$f(t) = \int_\Gamma (\gamma, t)W(d\gamma), \tag{5.8.2}$$

where W is a ν-noise on Γ.

Special cases now follow from basic group-theoretic results. For example, if T is discrete, then Γ is compact, as we noted already for the special case of $T = \mathbb{Z}^N$. Consequently, the integral in the spectral representation (5.8.2) is actually a sum and

---

[21] There is also a very readable, albeit less exhaustive, treatment in Hannan [76]. In addition, Letac [103] has an elegant exposition based on Gelfand pairs and Banach algebras for processes indexed by unimodular groups, which, in a certain sense, give a generalization of isotropic fields over $\mathbb{R}^N$. Ylinen [179] has a theory for noncommutative locally compact groups that extends the results in [178].

as such will be familiar to every student of time series. Alternatively, if T is compact, Γ is discrete. This implies that ν must be a sum of point masses and W is actually a discrete collection of independent random variables.

An interesting and important special case that we have already met is $T = S^{N-1}$. Treating T as both an N-sphere and the rotation group $O(N)$, we write the sum $\theta_1 + \theta_2$, for $\theta_1, \theta_2 \in S^{N-1}$, for the rotation of θ_1 by θ_2, the latter considered as an element of $O(N)$. In this case the characters are again the spherical harmonics, and we have that

$$f(\theta) = \sum_{m=0}^{\infty} \sum_{l=1}^{d_m} W_{ml} h_{ml}^{(N-1)}(\theta), \tag{5.8.3}$$

where the W_{ml} are uncorrelated with variance depending only on m. This, of course, is simply (5.7.6) once again, derived from a more general setting. Similarly, the covariance function can be written as

$$C(\theta_1, \theta_2) = C(\theta_{12}) = \sum_{m=0}^{\infty} \sigma_m^2 C_m^{(N-1)/2}(\cos(\theta_{12})), \tag{5.8.4}$$

where θ_{12} is the angular distance between θ_1 and θ_2, and the C_m^N are the Gegenbauer polynomials.

Other examples for the LCA situation follow in a similar fashion from (5.8.1) and (5.8.2) by knowing the structure of the dual group Γ.

The general situation is much harder and, as has already been noted, relies heavily on knowing the representation of T. In essence, given a representation of G on $\mathrm{GL}(H)$ for some Hilbert space H, one constructs a (left or right) stationary random field on G via the canonical white noise on H. The construction of Lemma 5.3.2 with $T = G$ and $H = L^2(G, \mu)$ can be thought of as a special example of this approach. For further details, you should go to the references given earlier in this section.

Part II

Geometry

If you have not yet read the preface, then please do so now.

Since you have read the preface, you already know that central to much of what we shall be looking at in Part III is the geometry of *excursion sets*:

$$A_u \equiv A_u(f, T) \stackrel{\Delta}{=} \{t \in T : f(t) \geq u\} \equiv f^{-1}([u, \infty))$$

for random fields f over general parameter spaces T.

In order to do this, we are going to need quite a bit of geometry. In fact, we are going to need a lot more than one might expect, since the answers to the relatively simple questions that we shall be asking end up involving concepts that do not, at first sight, seem to have much to do with the original questions.

In Part III we shall see that an extremely convenient description of the geometric properties of A is its *Euler*, or *Euler–Poincaré, characteristic*. In many cases, the Euler characteristic is determined by the fact that it is the unique integer-valued functional φ, defined on collections of nice A, satisfying

$$\varphi(A) = \begin{cases} 0 & \text{if } A = \emptyset, \\ 1 & \text{if } A \text{ is ball-like}, \end{cases}$$

where by "ball-like" we mean homotopically equivalent[22] to the unit N-ball, $B^N = \{t \in \mathbb{R}^N : |t| \leq 1\}$, and with the additivity property that

$$\varphi(A \cup B) = \varphi(A) + \varphi(B) - \varphi(A \cap B).$$

More global descriptions follow from this. For example, if A is a nice set in $\mathbb{R}^2$, then $\varphi(A)$ is simply the number of connected components of A minus the number of holes in it. In $\mathbb{R}^3$, $\varphi(A)$ is given by the number of connected components, minus the number of "handles," plus the number of holes. Thus, for example, the Euler characteristics of a baseball, a tennis ball, and a coffee cup are, respectively, 1, 2, and 0.

One of the basic properties of the Euler characteristic is that it is determined by the homology class of a set. That is, smooth transformations that do not change the basic "shape" of a set do not change its Euler characteristic. In Euclidean spaces finer geometric information, which *will* change under such transformations, lies in the so-called *Minkowski functionals*. However, even if we were to decide to do without this finer information, we would still arrive at a scenario involving it. We shall see, again in Part III, that the expected value of $\varphi(A_u(f, T))$ will, in general, involve the Minkowski functionals of T. This is what was meant above about simple questions leading to complicated answers.

There are basically two different approaches to developing the geometry that we shall require. The first works well for sets in $\mathbb{R}^N$ that are made up of the finite union of simple building blocks, such as convex sets. For many of our readers, we imagine that this will suffice. The basic framework here is that of integral geometry.

The second, more fundamental approach is via the differential geometry of abstract manifolds. As described in the preface, this more general setting has very concrete applications, and moreover, often provides more powerful and elegant proofs

[22] The notion of homotopic equivalence is formalized below by Definition 6.1.4. However, for the moment, "ball-like" will be just as useful a concept.

for a number of problems related to random fields even on the "flat" manifold $\mathbb{R}^N$. This approach is crucial if you want to understand the full theory. Furthermore, since some of the proofs of later results, even in the integral-geometric setting, are more natural in the manifold scenario, it is essential if you need to see full proofs. However, the jump in level of mathematical sophistication between the two approaches is significant, so that unless you feel very much at home in the world of manifolds you are best advised to read the integral-geometric story first.

Chapter 6 handles all the integral geometry that we shall need. The treatment there is detailed, complete, and fully self-contained. This is not the case when we turn to differential geometry, where the theory is too rich to treat in full. We start with a quick and nasty revision of basic differential geometry in Chapter 7, most of which is standard graduate-course material. Chapter 8 treats piecewise smooth manifolds, which provide the link between the geometry of Chapter 6, with its sharp corners and edges, and the smooth manifolds of Chapter 7. In Chapter 9 we look at Morse theory in the piecewise smooth setting. While Morse theory in the smooth setting is, once again, standard material, the piecewise smooth case is less widely known. In fact, Chapter 9 actually contains some new results that, unlike most of the rest of Part II, we were not able to find elsewhere.

In the passage from integral to differential geometry a number of things will happen. Among them, the space T (time, or multidimensional time) will become M (manifold)[23] and the Minkowski functionals will become the *Lipschitz–Killing curvatures*.

While Lipschitz–Killing curvatures are well-known objects in global differential geometry, we do not imagine that they will be too well known to the probabilist reader. Hence we devote Chapter 10 to a study of the so-called *tube formulas*, originally due to Hermann Weyl [168] in Euclidean spaces. In addition, as we have already noted in the preface, some of the proofs are different from what is found in the standard geometry literature, in that they rely on properties of Gaussian distributions. Furthermore, Chapter 10 also discusses a relatively new generalization of the Weyl results to manifolds in Gauss space, due to JET [158].

The tube formulas of Chapter 10 can also be exploited to develop a formal approximation to the excursion probability

$$\mathbb{P}\left\{ \sup_{t \in M} f(t) \geq u \right\}$$

for certain Gaussian fields with finite orthonormal expansions,[24] and we shall look at this in Sections 10.2 and 10.6.

All of Part II, old and new, is crucial for understanding the general proofs of Part III.

---

[23] A true transition from T to M would also have the elements t of T becoming elements p (points) of M. However, as we have already mentioned, this seems to be too heavy a psychological price for a probabilist to pay, so we shall remain with points $t \in M$. For this mortal sin of misnotation we beg forgiveness from our geometer colleagues.

[24] The most general case will be treated in detail in Chapter 14 via Morse-theoretic techniques.

6

Integral Geometry

Our aim in this chapter is to develop a framework for handling *excursion sets*, which we now redefine in a nonstochastic framework.

Definition 6.0.1. *Let f be a measurable real-valued function on some measurable space, and let T be a measurable subset of that space. Then, for each $u \in \mathbb{R}$,*

$$A_u \equiv A_u(f, T) \triangleq \{t \in T : f(t) \geq u\} \qquad (6.0.1)$$

is called the excursion set of f in T over the level u.

Throughout this chapter, T will be a "nice" (in a sense to be specified soon) subset of $\mathbb{R}^N$, and our tools will be those of integral geometry.

6.1 Basic Integral Geometry

Quite comprehensive studies of integral geometry are available in the monographs of Hadwiger [74] and Santaló [139], although virtually all the results we shall derive can also be found in the paper by Hadwiger [75]. Other very useful and somewhat more modern and readable references are Schneider [141] and Klain and Rota [92].

Essentially, we shall be interested in the study of a class of geometric objects known as *basic complexes*. Later on, we shall show that with probability one, the excursion sets of a wide variety of random fields belong to this class, and the concepts that we are about to discuss are relevant to random fields. We commence with some definitions and simple results, all of which are due to Hadwiger [75].

Assume that we have equipped $\mathbb{R}^N$ with a Cartesian coordinate system, so that the N vectors e_j (with 1 in the jth position and zeros elsewhere) serve as an orthonormal basis. Throughout this and the following two sections, everything that we shall have to say will be dependent on this choice of basis. The restriction to a particular coordinate system will disappear in the coordinate-free approach based on manifolds.

We call a k-dimensional affine subspace of the form

$$E = \{t \in \mathbb{R}^N : t_j = a_j,\ j \in J,\ -\infty < t_j < \infty,\ j \notin J\}$$

a (*coordinate*) *k-plane* of $\mathbb{R}^N$ if J is a subset of size $N-k$ of $\{1, \ldots, N\}$ and $a_j,\ j \in J$, are fixed.

We shall call a compact set B in $\mathbb{R}^N$ *basic* if the intersections $E \cap B$ are simply connected for every k-plane E of $\mathbb{R}^N$, $k = 1, \ldots, N$. Note that this includes the case $E = \mathbb{R}^N$. These sets, as their name implies, will form the basic building blocks from which we shall construct more complicated and interesting structures. It is obvious that the empty set $\emptyset$ is basic, as is any closed convex set. Indeed, convex sets remain basic under rotation, a property that characterizes them. Note that it follows from the definition that if B is basic, then so is $E \cap B$ for any k-plane E.

A set $A \subset \mathbb{R}^N$ is called a *basic complex* if it can be represented as the union of a finite number of basic sets, $B_1, \ldots, B_m$, for which the intersections $B_{\nu_1} \cap \cdots \cap B_{\nu_k}$ are basic for any combination of indices $\nu_1, \ldots, \nu_k$, $k = 1, \ldots, m$. The collection of basic sets

$$p = p(A) = \{B_1, \ldots, B_m\}$$

is called a *partition* of A, and their number, m, is called the *order* of the partition. Clearly, partitions are in no way unique, nor, despite the terminology, are the elements of a partition necessarily disjoint.

The class of all basic complexes, which we shall denote by $\mathcal{C}_B$, or $\mathcal{C}_B^N$ if we need to emphasize its dimension, possesses a variety of useful properties. For example, if $A \in \mathcal{C}_B$, then $E \cap A \in \mathcal{C}_B$ for every k-plane E. In fact, if E is a k-plane with $k \leq N$ and $A \in \mathcal{C}_B^N$, we have $E \cap A \in \mathcal{C}_B^k$. To prove this it suffices to note that if

$$p = \{B_1, \ldots, B_m\}$$

is a partition of A, then

$$p' = \{E \cap B_1, \ldots, E \cap B_m\}$$

is a partition of $E \cap A$, and since $E \cap B_j$ is a k-dimensional basic set for all j, $E \cap A \in \mathcal{C}_B^k$.

Another useful property of $\mathcal{C}_B$ is that it is invariant under those rotations of $\mathbb{R}^N$ that map the vectors $e_1, \ldots, e_N$ onto one another. It is not, however, invariant under all rotations, a point to which we shall return later.[1] Furthermore, $\mathcal{C}_B$ is additive, in the sense that $A, B \in \mathcal{C}_B$ and $A \cap B = \emptyset$ imply[2] $A \cup B \in \mathcal{C}_B$.

These and similar properties make the class of basic complexes quite large, and, in view of the fact that convex sets are basic, implies that they include the convex ring (i.e., the collection of all sets formed via finite union and intersection of convex sets).

With a class of sets in mind, we can now begin our search for a way to describe the shape of sets by looking for an integer-valued functional[3] with the following two basic properties:

$$\varphi(A) = \begin{cases} 0 & \text{if } A = \emptyset, \\ 1 & \text{if } A \neq \emptyset \text{ is basic,} \end{cases} \tag{6.1.1}$$

and

$$\varphi(A \cup B) = \varphi(A) + \varphi(B) - \varphi(A \cap B), \tag{6.1.2}$$

whenever $A,\ B,\ A \cup B,\ A \cap B \in \mathcal{C}_B$.

An important result of integral geometry states that not only does a functional possessing these two properties exist, but it is *uniquely* determined by them. We shall prove this by obtaining an explicit formulation for φ, which will also be useful in its own right.

Let $p = p(A)$ be a partition of order m of some $A \in \mathcal{C}_B$ into basic sets. Define the *characteristic of the partition* to be

---

[1] A note for the purist: As noted earlier, our definition of $\mathcal{C}_B$ is dependent on the choice of basis, which is what loses us rotation invariance. An easy counterexample in $\mathcal{C}_B^2$ is the descending staircase $\bigcup_{j=1}^{\infty} B_j$, where $B_j = \{(x, y) : 0 \le x \le 2^{-j}, 1 - 2^{1-j} \le y \le 1 - 2^{-j}\}$. This is actually a basic set with relation to the natural axes, but not even a basic complex if the axes are rotated 45 degrees, since then it cannot be represented as the union of a *finite* number of basic sets.

Hadwiger's [75] original definition of basic complexes was basis-independent but covered a smaller class of sets. In essence, basic sets were defined as above (but relative to a coordinate system) and basic complexes were required to have a representation as a finite union of basic sets for *every* choice of coordinate system. Thus our descending staircase is not a basic complex in his scenario.

While more restrictive, Hadwiger's approach gives a rotation-invariant theory. The reasons for our choice will become clearer later on, when we return to a stochastic setting. See, in particular, Theorem 11.3.3 and the comments following it. See also the axis-free approach of Section 9.2.

[2] Note that if $A \cap B \neq \emptyset$, then $A \cup B$ is not necessarily a basic complex. For a counterexample, take A to be the descending staircase of the previous footnote, and let B be the line segment $\{(x, y) : y = 1 - x, x \in [0, 1]\}$. There is no way to represent $A \cup B$ as the union of a *finite* number of basic sets, essentially because of the infinite number of single point intersections between A and B, or, equivalently, the infinite number of holes in $A \cup B$.

[3] A functional on a lattice of sets that satisfies (6.1.2) and $\varphi(A) = 0$ if $A \neq \emptyset$ is called an *evaluation*; cf. [92].

$$\kappa(A,p) = \sum^{(1)} \epsilon(B_j) - \sum^{(2)} \epsilon(B_{j_1} \cap B_{j_2}) + \cdots \qquad (6.1.3)$$
$$+ (-1)^{n+1} \sum^{(n)} \epsilon(B_{j_1} \cap \cdots \cap B_{j_n}) + \cdots$$
$$+ (-1)^{m+1} \epsilon(B_1 \cap \cdots \cap B_m),$$

where $\sum^{(n)}$ denotes summation over all subsets $\{j_1, \ldots, j_n\}$ of $\{1, \ldots, m\}$, $1 \le n \le m$, and ϵ is an indicator function for basic sets, defined by

$$\epsilon(A) = \begin{cases} 0 & \text{if } A = \emptyset, \\ 1 & \text{if } A \neq \emptyset \text{ is basic.} \end{cases}$$

Then, if a functional φ satisfying (6.1.1) and (6.1.2) does in fact exist, it follows iteratively from these conditions and the definition of basic complexes that for any $A \in \mathcal{C}_B$ and any partition p,

$$\varphi(A) = \kappa(A, p). \qquad (6.1.4)$$

Thus, given existence, uniqueness of φ will follow if we can show that $\kappa(A,p)$ is independent of the partition p.

Theorem 6.1.1. *Let $A \subset \mathbb{R}^N$ be a basic complex and $p = p(A)$ a partition of A. Then the quantity $\kappa(A,p)$ is independent of p. If we denote this by $\varphi(A)$, then φ satisfies* (6.1.1) *and* (6.1.2)*, and $\varphi(A)$ is called the Euler characteristic of A.*

Proof. The main issue is that of existence, which we establish by induction. When $N = 1$, basics are simply closed intervals or points, or the empty set. Thus, setting

$$\varphi(A) = \text{number of disjoint intervals and isolated points in } A$$

yields a function satisfying $\varphi(A) = \kappa(A,p)$ for every p and $A \subset \mathcal{C}_B^1$ for which (6.1.1) and (6.1.2) are clearly satisfied.

Now let $N > 1$ and assume that for all spaces of dimension k less than N we have a functional φ^k on $\mathcal{C}_B^k$ for which $\varphi^k(A) = \kappa(A,p)$ for all $A \in \mathcal{C}_B^k$ and every partition p of A. Choose one of the vectors e_j, and for $x \in (-\infty, \infty)$ let E_x (which depends on j) denote the $(N-1)$-plane

$$E_x \triangleq \{t \in \mathbb{R}^N : t_j = x\}. \qquad (6.1.5)$$

Let $A \in \mathcal{C}_B^N$ and let $p = p(A) = \{B_1, \ldots, B_m\}$ be one of its partitions. Then clearly the projections onto E_0 of the cross-sections $A \cap E_x$ are all in $\mathcal{C}_B^{N-1}$, so that there exists a partition-independent functional φ_x defined on $\{A \cap E_x, A \in \mathcal{C}_B^N\}$ determined by

$$\varphi_x(A \cap E_x) = \varphi^{N-1}(\text{projection of } A \cap E_x \text{ onto } E_0).$$

Via φ_x we can define a new partition-independent function f by

$$f(A, x) = \varphi_x(A \cap E_x). \tag{6.1.6}$$

However, by the induction hypothesis and (6.1.6), we have from (6.1.3) that

$$f(A, x) = \sum\nolimits^{(1)} \epsilon(B_{j_1} \cap E_x) - \sum\nolimits^{(2)} \epsilon(B_{j_1} \cap B_{j_2} \cap E_x) + \cdots .$$

Consider just one of the right-hand terms, writing

$$\epsilon(x) = \epsilon(B_{j_1} \cap \cdots \cap B_{j_k} \cap E_x).$$

Assume that the intersection $B_{j_1} \cap \cdots \cap B_{j_k}$ is nonempty. Otherwise, there is nothing to prove. Since $\epsilon(x)$ is zero when the intersection of the B_{j_i} with E_x is empty and one otherwise, we have for some finite a and b that $\epsilon(x) = 1$ if $a \le x \le b$ and $\epsilon(x) = 0$ otherwise. Thus $\epsilon(x)$ is a step function, taking at most two values. Hence $f(A, x)$, being the sum of a finite number of such functions, is again a step function, with a finite number of discontinuities. Thus the left-hand limits

$$f(A, x^-) = \lim_{y \downarrow 0} f(A, x - y) \tag{6.1.7}$$

always exist. Now define a function h, which is nonzero at only a finite number of points x, by

$$h(A, x) = f(A, x) - f(A, x^-) \tag{6.1.8}$$

and define

$$\varphi(A) = \sum_x h(A, x), \tag{6.1.9}$$

where the summation is over the finite number of x for which the summand is nonzero. Note that since f is independent of p, so are h and φ.

Thus we have defined a functional on $\mathcal{C}_B$, and we need only check that (6.1.1) and (6.1.2) are satisfied to complete this section of the proof. Firstly, note that if B is a basic set and $B \ne \emptyset$, and if a and b are the extreme points of the linear set formed by projecting B onto e_j, then $f(B, x) = 1$ if $a \le x \le b$ and equals zero otherwise. Thus $h(B, a) = 1$, while $h(B, x) = 0,\ x \ne a$, so that $\varphi(B) = 1$. This is (6.1.1) since $\varphi(\emptyset) = 0$ is obvious. Now let A, B, $A \cup B$, $A \cap B$ all belong to $\mathcal{C}_B$. Then the projections onto E_0 of the intersections

$$A \cap E_x, \qquad B \cap E_x, \qquad (A \cup B) \cap E_x, \qquad A \cap B \cap E_x$$

all belong to $\mathcal{C}_B^{N-1}$, and so by (6.1.6) and the induction hypothesis,

$$f(A \cup B, x) = f(A, x) + f(B, x) - f(A \cap B, x).$$

Replacing x by x^- and performing a subtraction akin to that in (6.1.8) we obtain

$$h(A \cup B, x) = h(A, x) + h(B, x) - h(A \cap B, x).$$

Summing over x gives

$$\varphi(A \cup B) = \varphi(A) + \varphi(B) - \varphi(A \cap B),$$

so that (6.1.2) is established and we have the existence of our functional φ. It may, however, depend on the partition p used in its definition.

For uniqueness, note that since by (6.1.4), $\kappa(A, p') = \varphi(A)$ for any partition p', we have that $\kappa(A, p')$ is independent of p' and so that $\varphi(A)$ is independent of p. That is, we have the claimed uniqueness of φ.

Finally, since $\kappa(A, p)$ is independent of the particular choice of the vector e_j appearing in the proof, so is φ. □

The proof of Theorem 6.1.1 actually contains more than we have claimed in the statement of the result, since in developing the proof, we actually obtained an alternative way of computing $\varphi(A)$ for any $A \in \mathcal{C}_B$. This is given explicitly in the following theorem, for which E_x is as defined in (6.1.5).

Theorem 6.1.2. *For basic complexes $A \in \mathcal{C}_B^N$, the Euler characteristic φ, as defined by* (6.1.4), *has the following equivalent iterative definition:*

$$\varphi(A) = \begin{cases} \text{number of disjoint intervals in } A & \text{if } N = 1, \\ \sum_x \{\varphi(A \cap E_x) - \varphi(A \cap E_{x^-})\} & \text{if } N > 1, \end{cases} \tag{6.1.10}$$

where

$$\varphi(A \cap E_{x^-}) = \lim_{y \downarrow 0} \varphi(A \cap E_{x-y})$$

and the summation is over all real x for which the summand is nonzero.

This theorem is a simple consequence of (6.1.9) and requires no further proof. Note that it also follows from the proof of Theorem 6.1.1 (cf. the final sentence there) that the choice of vector e_j used to define E_x is irrelevant.[4]

The importance of Theorem 6.1.2 lies in the iterative formulation it gives for φ, for using this, we shall show in a moment how to obtain yet another formulation that makes the Euler characteristic of a random excursion set amenable to probabilistic investigation.

Figure 6.1.1 shows an example of this iterative procedure in $\mathbb{R}^2$. Here the vertical axis is taken to define the horizontal 1-planes E_x. The values of $\varphi(A \cap E_x)$ appear closest to the vertical axis, with the values of h to their left. Note in particular the set with the hole "in the middle." It is on sets like this, and their counterparts in higher dimensions, that the characteristic φ and the number of connected components of the set differ. In this example they are, respectively, zero and one. For the moment, ignore the arrows and what they are pointing at.

[4] It is also not hard to show that if A is a basic complex with respect to each of two coordinate systems (which are not simple relabelings of one another) then $\varphi(A)$ will be the same for both. However, this is taking us back to the original formulation of [75], which we have already decided is beyond what we need here.

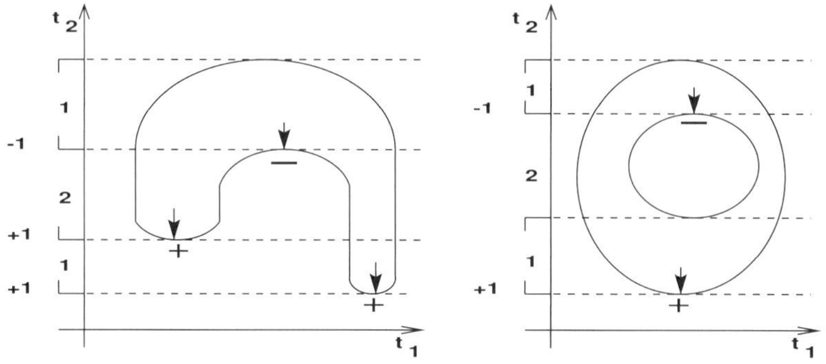

Fig. 6.1.1. Computing the Euler characteristic.

To understand how this works in higher dimensions, you should try to visualize some N-dimensional examples to convince yourself that for the N-dimensional unit ball, B^N, and the N-dimensional unit sphere, S^{N-1},

$$\varphi(B^N) = 1, \qquad \varphi(S^{N-1}) = 1 + (-1)^{N-1}. \tag{6.1.11}$$

It is somewhat less easy (and, indeed, quite deep in higher dimensions) to see that if $K_{N,k}$ denotes B^N with k nonintersecting cylindrical holes drilled through it, then, since both $K_{N,k}$ and its boundary belong to $\mathcal{C}_B^N$,

$$\varphi(K_{N,k}) = 1 + (-1)^N k,$$

while

$$\varphi(\partial K_{N,k}) = [1 + (-1)^{N-1}](1 - k).$$

Finally, if we write $\bar{K}_{N,k}$ to denote B^N with k "handles" attached, then

$$\varphi(\bar{K}_{N,k}) = 1 - k.$$

In view of (6.1.11), knowing the Euler characteristic of balls means that we know it for all basic sets. However, this invariance goes well beyond balls, since it can also be shown that the Euler characteristic is the same for all homotopically equivalent sets, i.e., sets that are geometrically "alike" in that they can be deformed into one another in a continuous fashion. Here is a precise definition of this equivalence.

Definition 6.1.3. *Let f and g be two mappings from A to B, both subsets of a Hausdorff space*[5] *X. If there exists a* continuous *mapping, $F : A \times [0,1] \to X$ with the following three properties, then we say that f is* homotopic *or* deformable *to g:*

$$\begin{aligned} F(t,0) &= f(t) \quad && \forall t \in A, \\ F(t,\tau) &\in X && \forall t \in A,\ \tau \in [0,1], \\ F(t,1) &= g(t) && \forall t \in A. \end{aligned}$$

[5] A Hausdorff space is a topological space T for which, given any two distinct points $s, t \in T$, there are open sets $U, V \subset T$ with $s \in U$, $t \in V$ and $U \cap V = \emptyset$.

Definition 6.1.4. *Let A and B be subsets of (possibly different) Hausdorff spaces. If there exist continuous $f : A \to B$ and $g : B \to A$ for which the composite maps $g \circ f : A \to A$ and $f \circ g : B \to B$ are deformable, respectively, to the identity maps on A and B, then A and B are called homotopically equivalent.*

6.2 Excursion Sets Again

We now return briefly to the setting of excursion sets. Our aim in this section will be to find a way to compute their Euler characteristics directly from the function f, without having to look at the sets themselves. In particular, we will need to find a method that depends only on *local* properties of f, since later, in the random setting, this (via finite-dimensional distributions) will be all that will be available to us for probabilistic computation.

Since the main purpose of this section is to develop understanding and since we shall ultimately use the techniques of the critical point theory developed in Chapter 9 to redo everything that will be done here in far greater generality, we shall not give all the details of all the proofs. The interested reader can find most of them either in GRF [2] or the review [4]. Furthermore, we shall restrict the parameter set T to being a bounded rectangle of the form

$$T = [s,t] \stackrel{\Delta}{=} \prod_{i=1}^{N} [s_i, t_i], \qquad -\infty < s_i < t_i < \infty. \tag{6.2.1}$$

For our first definition, we need to decompose T into a disjoint union of open sets, starting with its interior, its faces, edges, etc. More precisely, a *face* J, of dimension k, is defined by fixing a subset $\sigma(J)$ of $\{1, \dots, N\}$ of size k and a sequence of $N-k$ zeros and ones, which we write as $\varepsilon(J) = \{\varepsilon_j,\ j \notin \sigma(J)\}$, so that

$$\begin{aligned} J = \Big\{ v \in T : v_j &= (1-\varepsilon_j)s_j + \varepsilon_j t_j \text{ if } j \notin \sigma(J), \\ s_j &< v_j < t_j \text{ if } j \in \sigma(J) \Big\}. \end{aligned} \tag{6.2.2}$$

Note that faces are always open sets.

In anticipation of later notation, we write $\partial_k T$ for the collection of faces of dimension k in T. This is known as the *k-skeleton* of T. Then $\partial_N T$ contains only T°, while $\partial_0 T$ contains the 2^N vertices of the rectangle. In general, $\partial_k T$ has

$$J_k \stackrel{\Delta}{=} 2^{N-k} \binom{N}{k} \tag{6.2.3}$$

elements. Note also that the usual boundary of T, ∂T, is given by the disjoint union

$$\partial T = \bigcup_{k=0}^{N-1} \bigcup_{J \in \partial_k T} J.$$

We start with some simple assumptions on f and, as usual, write the first- and second-order partial derivatives of f as $f_j = \partial f/\partial t_j$, $f_{ij} = \partial^2 f/\partial t_i \partial t_j$.

Definition 6.2.1. *Let T be a bounded rectangle in $\mathbb{R}^N$ and let f be a real-valued function defined on an open neighborhood of T.*

Then, if for a fixed $u \in \mathbb{R}$ the following three conditions are satisfied for each face J of T for which $N \in \sigma(J)$, we say that f is suitably regular with respect to T at the level u if the following conditions hold.

(A) *f has continuous partial derivatives of up to second order in an open neighborhood of T.* (6.2.4)

(B) *$f_{|J}$ has no critical points at the level u.* (6.2.5)

(C) *If $J \in \partial_k T$ and $D_J(t)$ denotes the symmetric $(k-1) \times (k-1)$ matrix with elements f_{ij}, $i, j \in \sigma(J) \setminus \{N\}$, then there are no $t \in J$ for which $0 = f(t) - u = \det D_J(t) = f_j(t)$ for all $j \in \sigma(J) \setminus \{N\}$.* (6.2.6)

The first two conditions of suitable regularity are meant to ensure that the boundary $\partial A_u = \{t \in T : f(t) = u\}$ of the excursion set is smooth in the interior T° of T and that its intersections with ∂T also is smooth. The last condition is a little more subtle, since it relates to the curvature of ∂A_u both in the interior of T and on its boundary.

The main importance of suitable regularity is that it gives us the following theorem.

Theorem 6.2.2. *Let $f : \mathbb{R}^N \to \mathbb{R}^1$ be suitably regular with respect to a bounded rectangle T at level u. Then the excursion set $A_u(f, T)$ is a basic complex.*

The proof of Theorem 6.2.2 is not terribly long, but since it is not crucial to what will follow, it can be skipped at first reading. The reasoning behind it is all in Figure 6.2.1, and after understanding this you can skip to the examples immediately following the proof without losing much.

For those of you remaining with us, we start with a lemma.

Lemma 6.2.3. *Let $f : \mathbb{R}^N \to \mathbb{R}^1$ be suitably regular with respect to a bounded rectangle T at the level u. Then there are only finitely many $t \in T$ for which*

$$f(t) - u = f_1(t) = \cdots = f_{N-1}(t) = 0. \tag{6.2.7}$$

Proof. To start, let $g = (g^1, \ldots, g^N) : T \to \mathbb{R}^N$ be the function defined by

$$g^1(t) = f(t) - u, \quad g^j(t) = f_{j-1}(t), \quad j = 2, \ldots, N. \tag{6.2.8}$$

Let $t \in T$ satisfy (6.2.7). Then, by (6.2.5), $t \notin \partial T$; i.e., t is in the interior of T. We shall show that there is an open neighborhood of t throughout which no other point satisfies (6.2.7), which we rewrite as $g(t) = 0$. This would imply that the points in T satisfying (6.2.7) are isolated, and thus, from the compactness of T, we would have that there are only a finite number of them. This, of course, would prove the lemma.

The inverse mapping theorem[6] implies that such a neighborhood will exist if the $N \times N$ matrix $(\partial g^i/\partial t_j)$ has a nonzero determinant at t. However, this matrix has the following elements:

$$\frac{\partial g^1}{\partial t_j} = f_j(t) \qquad \text{for } j = 1, \ldots, N,$$
$$\frac{\partial g^i}{\partial t_j} = f_{i-1,j}(t) \quad \text{for } i = 2, \ldots, N,\ j = 1, \ldots, N.$$

Since t satisfies (6.2.7), all elements in the first row of this matrix, other than the Nth, are zero. Expanding the determinant along this row gives us that it is equal to

$$(-1)^{N-1} f_N(t) \det D(t), \tag{6.2.9}$$

where $D(t)$ is as defined in (6.2.6). Since (6.2.7) is satisfied, (6.2.5) and (6.2.6) imply, respectively, that neither $f_N(t)$ nor the determinant of $D(t)$ is zero, which, in view of (6.2.9), is all that is required. □

Proof of Theorem 6.2.2. When $N = 1$ we are dealing throughout with finite collections of intervals, and so the result is trivial.

Now take the case $N = 2$. We need to show how to write A_u as a finite union of basic sets.

Consider the set of points $t \in T$ satisfying either

$$f(t) - u = f_1(t) = 0 \tag{6.2.10}$$

or

$$f(t) - u = f_2(t) = 0. \tag{6.2.11}$$

For each such point draw a line containing the point and parallel to either the horizontal or vertical axis, depending, respectively, on whether (6.2.10) or (6.2.11) holds. These lines form a grid over T, and it is easy to check that the connected regions of A_u within each cell of this grid are basic. Furthermore, these sets have intersections that are either straight lines, points, or empty, and Lemma 6.2.3 (applied to the original axes and a 90° rotation of them) guarantees that there are only a finite number of them, so that they form a partition of A_u. (An example of this partitioning procedure is shown in Figure 6.2.1. The dots mark the points where either (6.2.10) or (6.2.11) holds.) This provides the required partition, and we are done.

[6] The inverse mapping theorem goes as follows: Let $U \subset \mathbb{R}^N$ be open and $g = (g^1, \ldots, g^N) : U \to \mathbb{R}^N$ a function possessing continuous first-order partial derivatives $\partial g^i/\partial t_j$, $i, j = 1, \ldots, N$. Then if the matrix $(\partial g^i/\partial t_j)$ has a nonzero determinant at some point $t \in U$, there exist open neighborhoods U_1 and V_1 of t and $g(t)$, respectively, and a function $g^{-1} : V_1 \to U_1$, for which

$$g^{-1}(g(t)) = t \quad \text{and} \quad g(g^{-1}(s)) = s$$

for all $t \in U_1$ and $s \in V_1$.

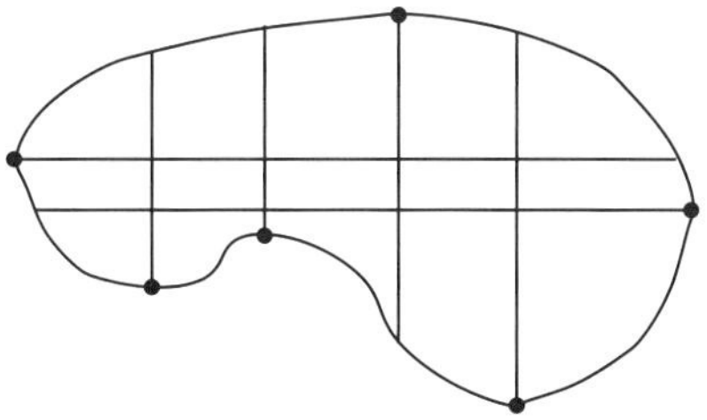

Fig. 6.2.1. Partitioning excursion sets into basic components.

For $N > 2$ essentially the same argument works, using partitioning $(N-1)$ planes passing through points at which

$$f(t) - u = f_1(t) = \cdots = f_{N-1}(t).$$

Lemma 6.2.3 again guarantees the finiteness of the partition. The details are left to you.[7] □

We now attack the problem of obtaining a simple way of computing the Euler characteristic of A_u. As you are about to see, this involves looking at each $A_u \cap J$, $J \in \partial_k T$, $0 \le k \le N$, separately. We start with the simple example $T = I^2$, in which $\partial_2 T = T^o$, $\partial_1 T$ is composed of four open intervals parallel to the axes, and $\partial_0 T$ contains the four vertices of the square. Since this is a particularly simple case, we shall pool $\partial_1 T$ and $\partial_0 T$, and handle them together as ∂T.

Thus, let $f : \mathbb{R}^2 \to \mathbb{R}^1$ be suitably regular with respect to I^2 at the level u. Consider the summation (6.1.10) defining $\varphi(A_u(f, I^2))$, that is,

$$\varphi(A_u) = \sum_{x \in (0,1]} \{\varphi(A_u \cap E_x) - \varphi(A_u \cap E_{x^-})\}, \tag{6.2.12}$$

where now E_x is simply the straight line $t_2 = x$, and so $n_x \stackrel{\Delta}{=} \varphi(A_u \cap E_x)$ counts the number of distinct intervals in the cross-section $A_u \cap E_x$. The values of x contributing to the sum correspond to values of x where n_x changes.

It is immediate from the continuity of f that contributions to $\varphi(A_u)$ can occur only when E_x is tangential to ∂A_u (Type I contributions) or when $f(0, x) = u$ or $f(1, x) = u$ (Type II contributions). Consider the former case first.

If E_x is tangential to ∂A_u at a point t, then $f_1(t) = 0$. Furthermore, since $f(t) = u$ on ∂A_u, we must have that $f_2(t) \neq 0$, as a consequence of suitable regularity. Thus, in the neighborhood of such a point and on the curve ∂A_u, t_2 can be expressed as an implicit function of t_1 by

$$f(t_1, g(t_1)) = u.$$

[7] You should at least try the three-dimensional case, to get a feel for the source of the conditions on the various faces of T in the definition of suitable regularity.

The implicit function theorem[8] gives us

$$\frac{dt_2}{dt_1} = -\frac{f_1(t)}{f_2(t)},$$

so that applying what we have just noted about the tangency of E_x to ∂A_u, we have for each contribution of Type I to (6.2.12) that there must be a point $t \in I^2$ satisfying

$$f(t) = u \tag{6.2.13}$$

and

$$f_1(t) = 0. \tag{6.2.14}$$

Furthermore, since the limit in (6.2.12) is one-sided, continuity considerations imply that contributing points must also satisfy

$$f_2(t) > 0. \tag{6.2.15}$$

Conversely, for each point satisfying (6.2.13) to (6.2.15) there is a unit contribution of Type I to $\varphi(A_u)$. Note that there is no contribution of Type I to $\varphi(A_u)$ from points on the boundary of I^2 because of the regularity condition (6.2.5). Thus we have a one-to-one correspondence between unit contributions of Type I to $\varphi(A_u)$ and points in the interior of I^2 satisfying (6.2.13) to (6.2.15). It is also easy to see that contributions of $+1$ will correspond to points for which $f_{11}(t) < 0$ and contributions of -1 to points for which $f_{11}(t) > 0$. Furthermore, because of (6.2.6) there are no contributing points for which $f_{11}(t) = 0$.

Consider now Type II contributions to $\varphi(A)$, which is best done by looking first at Figure 6.2.2.

The eight partial and complete disks there lead to a total Euler characteristic of 8. The three sets A, B, and C are accounted for by Type I contributions, since in each case the above analysis will count $+1$ at their lowest points. We need to account for the remaining five sets, which means running along ∂I^2 and counting points there. In fact, what we need to do is count $+1$ at the points marked with a •. There is never

[8] The implicit function theorem goes as follows: Let $U \subset \mathbb{R}^N$ be open and $F : U \to \mathbb{R}$ possess continuous first-order partial derivatives. Suppose at $t^* \in U$, $F(t^*) = u$ and $F_N(t^*) \neq 0$. Then the equation

$$F(t_1, \dots, t_{N-1}, g(t_1, \dots, t_{N-1})) = u$$

defines an implicit function g that possesses continuous, first-order partial derivatives in some interval containing $(t_1^*, \dots, t_{N-1}^*)$, and such that $g(t_1^*, \dots, t_{N-1}^*) = t_N^*$. The partial derivatives of g are given by

$$\frac{\partial g}{\partial t_j} = -\frac{F_j}{F_N} \quad \text{for } j = 1, \dots, N-1.$$

Furthermore, if F is k-times differentiable, so is g.

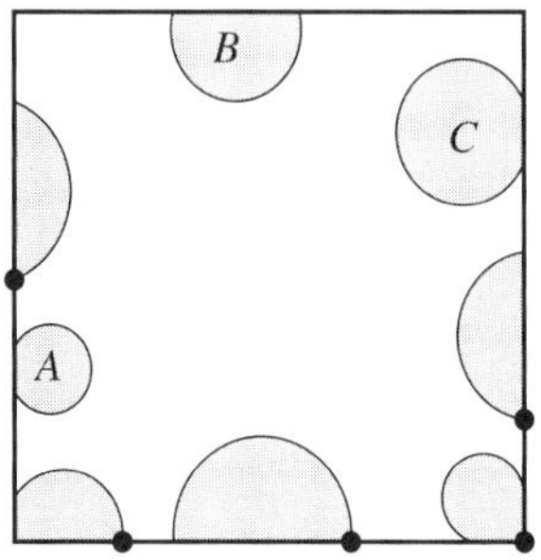

Fig. 6.2.2. Contributing points to the Euler characteristic.

a need to count -1 on the boundary. Note that on the two vertical sections of ∂I^2 we count $+1$ whenever we enter the set (at its intersection with ∂I^2) from below. There is never a contribution from the top side of I^2. For the bottom, we need to count the number of connected components of its intersection with the excursion set, which can be done either on "entering" or "leaving" the set in the positive t_1 direction. We choose the latter, and so must also count a $+1$ if the point $(1, 0)$ is covered.

Putting all this together, we have a Type II contribution to $\varphi(A)$ whenever one of the following four sets os conditions is satisfied:

$$\begin{cases} t = (t_1, 0), \ f(t) = u, \ f_1(t) < 0, \\ t = (0, t_2), \ f(t) = u, \ f_2(t) > 0, \\ t = (1, t_2), \ f(t) = u, \ f_2(t) > 0, \\ t = (1, 0), \ \ f(t) > u. \end{cases} \tag{6.2.16}$$

The above argument has established the following, which in Chapter 9, with completely different techniques and a much more sophisticated and powerful language, will be extended to parameter sets in $\mathbb{R}^N$ and on C^2 manifolds with piecewise smooth boundaries:

Theorem 6.2.4. *Let f be suitably regular with respect to I^2 at the level u. Then the Euler characteristic of its excursion set $A_u(f, I^2)$ is given by the number of points t in the interior of I^2 satisfying*

$$f(t) - u = f_1(t) = 0, \qquad f_2(t) > 0, \quad \textit{and} \quad f_{11}(t) < 0, \tag{6.2.17}$$

minus the number of points satisfying

$$f(t) - u = f_1(t) = 0, \qquad f_2(t) > 0, \quad \textit{and} \quad f_{11}(t) > 0, \tag{6.2.18}$$

plus the number of points on the boundary of I^2 satisfying one of the four sets of conditions in (6.2.16).

This is what we have been looking for in a doubly simple case: The ambient dimension was only 2, and the set T a simple square. There is another proof of

Theorem 6.2.4 in Section 9.4, built on Morse theory. There you will also find a generalization of this result to I^N for all finite N, although the final point set representation is a little different from that of (6.2.16). Morse theory is also the key to treating far more general parameter spaces. Nevertheless, what we have developed so far, along with some ingenuity, does let us treat some more general cases, for which we give an example. You should be able to fill in the details of the computation by yourself, although some hand waving may be necessary. Here is the example:

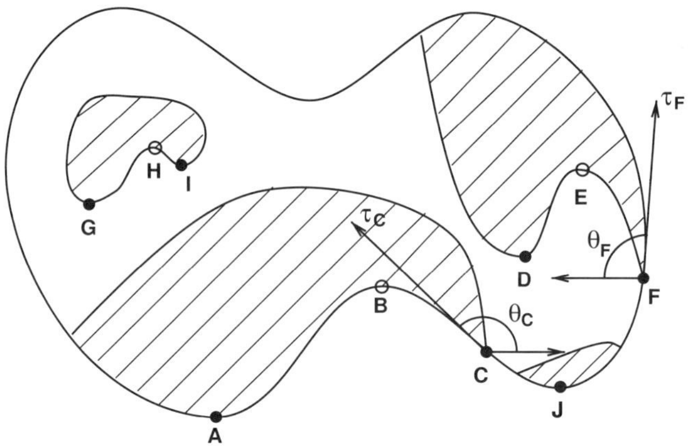

Fig. 6.2.3. Computing the Euler characteristic for a general shape.

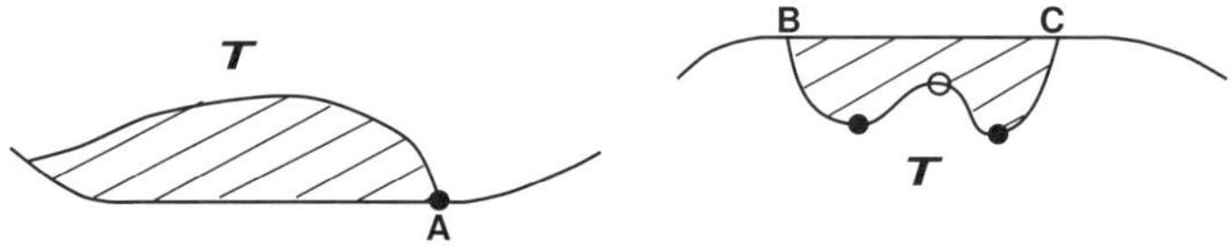

Fig. 6.2.4. Points on a horizontal part of the boundary.

Consider the parameter space T to be the surrounding dumbell shape of Figure 6.2.3. Of the four components of the excursion set (the hatched objects), three intersect ∂T. We already know how to characterize the small component in the top left: We count a $+1$ for each $\bullet$, -1 for each $\circ$, and sum. The problem is what to do with the remaining two components. Worsley [174] showed that what needs to be done is again to subtract from the number of points in the interior of T that satisfy (6.2.17) the number satisfying (6.2.18) and then to go along ∂T and add a $+1$ for each point marked with a $\bullet$.

More precisely, the rules for marking are as follows:

(a) If t is in the interior of T, then apply the criteria (6.2.17) and (6.2.18) exactly as before, marking $\bullet$ $(+1)$ and $\circ$ (-1), respectively.

(b) If $t \in \partial T \cap \partial A_u$, and the tangent to ∂T is *not* parallel to the horizontal axis, then let $f_{\rm up}(t)$ be the derivative of f in the direction of the tangent to ∂T pointing in the positive t_2 direction. (Two such tangent vectors appear as τ_C and τ_F in Figure 6.2.3.) Furthermore, take the derivative of f with respect to t_1 in the direction pointing *into* T. Call this $f_\perp$. (It will equal either f_1 or $-f_1$, depending on whether the angles θ in Figure 6.2.3 from the horizontal to the τ vector develop in a counterclockwise or clockwise direction, respectively.) Now mark t as a $\bullet$ (and so count as $+1$) if $f_\perp(t) < 0$ and $f_{\rm up}(t) > 0$. There are no $\circ$ points in this class.

(c) If $t \in \partial T \cap \partial A_u$, and the tangent to ∂T *is* parallel to the horizontal axis, but t is not included in an open interval ∂T that is parallel to this axis, then proceed as in (b), simply defining $f_\perp$ to be f_1 if the tangent is above ∂T and $-f_1$ otherwise.

(d) If $t \in \partial T \cap \partial A_u$ belongs to an open interval of ∂T that is parallel to the horizontal axis (as in Figure 6.2.4), then mark it as a $\bullet$ if T is above ∂T and $f_1(t) < 0$. (Thus, as in Figure 6.2.4, points such as B and C by which A_u "hangs" from ∂T will never be counted.)

(e) Finally, if $t \in \partial T \cap A_u$ has not already been marked, and coincides with one of the points that contribute to the Euler characteristic of T itself (e.g., A, B, and J in Figure 6.2.3), then mark it exactly as it was marked in computing $\varphi(T)$.

All told, this can be summarized as follows.

Theorem 6.2.5 (Worsley [174]). *Let $T \subset \mathbb{R}^2$ be compact with boundary ∂T that is twice differentiable everywhere except, at most, at a finite number of points. Let f be suitably regular for T at the level u. Let $\chi(A_u(f, T))$ be the number of points in the interior of T satisfying* (6.2.17) *minus the number satisfying* (6.2.18).

Denote the number of points satisfying (b)–(d) *above as $\chi_{\partial T}$, and denote the sum of the contributions to $\varphi(T)$ of those points described in* (e) *by $\varphi_{\partial T}$. Then*

$$\varphi(A) = \chi(A) + \chi_{\partial T} + \varphi_{\partial T}. \tag{6.2.19}$$

Theorem 6.2.5 can be extended to three dimensions (also in [174]). In principle, it is not too hard to guess what has to be done in higher dimensions as well. Determining whether a point in the interior of T and on ∂A_u contributes a $+1$ or -1 will depend on the curvature of ∂A_u, while if $t \in \partial T$, both this curvature and that of ∂T will have roles to play.

It is clear that these kinds of arguments are going to get rather messy very quickly, and a different approach is advisable. This is provided via the critical point theory of differential topology, which we shall develop in Chapter 9. However, before doing this we want to develop some more geometry in the still relatively simple scenario of Euclidean space and describe how all of this relates to random fields.

6.3 Intrinsic Volumes

The Euler characteristic of Section 6.1 arose as the unique additive functional (cf. (6.1.2)) that assigned the value 1 to sets homotopic to a ball, and 0 to the empty set.

It turned out to be integer-valued, although we did not demand this in the beginning, and has an interpretation in terms of "counting" the various topological components of a set. But there is more to life than mere counting, and one would also like to be able to say things about the volume of sets, the surface area of their boundaries, their curvatures, etc. In this vein, it is natural to look for a class of N additional position- and rotation-invariant functionals $\{\mathcal{L}_j\}_{j=1}^N$ that are also additive in the sense of (6.1.2) and scale with dimensionality in the sense that

$$\mathcal{L}_j(\lambda A) = \lambda^j \mathcal{L}_j(A), \quad \lambda > 0, \tag{6.3.1}$$

where $\lambda A \triangleq \{t : t = \lambda s,\ s \in A\}$.

Such functionals exist, and together with $\mathcal{L}_0 \triangleq \varphi$, the Euler characteristic itself, make up what are known as the *intrinsic volumes* defined on a suitable class of sets A. The reason why the intrinsic volumes are of crucial importance to us will become clear when we get to the discussion following Theorem 6.3.1, which shows that many probabilistic computations for random fields are intimately related to them. They can be defined in a number of ways, one of which is a consequence of *Steiner's formula* [92, 139], which, for *convex*[9] subsets of $\mathbb{R}^N$, goes as follows:

For $A \subset \mathbb{R}^N$ and $\rho > 0$, let

$$\mathrm{Tube}(A, \rho) = \{x \in \mathbb{R}^N : d(x, A) \le \rho\} \tag{6.3.2}$$

be the "tube" of radius ρ, or "ρ-tube," around A, where

$$d(x, A) \triangleq \inf_{y \in A} |x - y|$$

is the usual Euclidean distance from the point x to the set A. An example is given in Figure 6.3.1, in which A is the inner triangle and $\mathrm{Tube}(A, \rho)$ the larger triangular object with rounded-off corners.

With λ_N denoting, as usual, Lebesgue measure in $\mathbb{R}^N$, Steiner's formula states[10] that $\lambda_N(\mathrm{Tube}(A, \rho))$ has a finite expansion in powers of ρ. In particular,

$$\lambda_N\left(\mathrm{Tube}(A, \rho)\right) = \sum_{j=0}^{N} \omega_{N-j} \rho^{N-j} \mathcal{L}_j(A), \tag{6.3.3}$$

where

$$\omega_j = \lambda_j(B(0, 1)) = \frac{\pi^{j/2}}{\Gamma\left(\frac{j}{2} + 1\right)} = \frac{s_j}{j} \tag{6.3.4}$$

[9] Moving from basic complexes down to the very special case of convex sets is a severe restriction in terms of what we need, and we shall lift this restriction soon. Nevertheless, convex sets are a comfortable scenario in which to first meet intrinsic volumes.

[10] There is a more general version of Steiner's formula for the case in which T and its tube are embedded in $\mathbb{R}^{N'}$, where $N' > N = \dim(A)$. In that case (6.3.2) still holds, but with N replaced by N'. See Theorem 10.5.6 for more details in the manifold setting.

is the volume of the unit ball in $\mathbb{R}^j$. (Recall that s_j was the corresponding surface area; cf. (5.7.5).)

We shall see a proof of (6.3.3) later on, in Chapter 10, in a far more general scenario, but it is easy to see from Figure 6.3.1 from where the result comes.

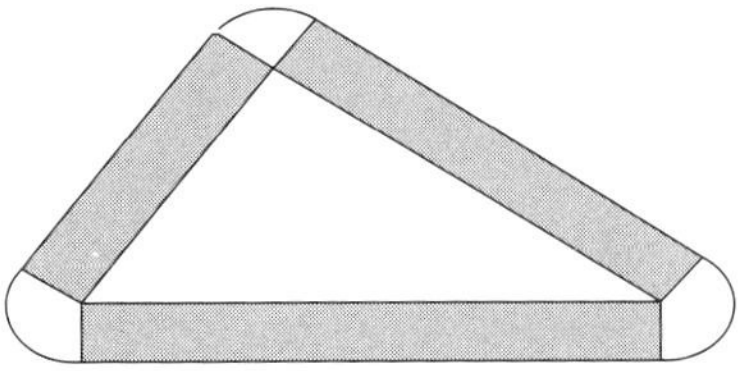

Fig. 6.3.1. The tube around a triangle.

To find the area (i.e., two-dimensional volume) of the enlarged triangle, one needs only to sum three terms:

- The area of the original, inner triangle.
- The area of the three rectangles. Note that this is the perimeter (i.e., "surface area") of the triangle multiplied by ρ.
- The area of the three corner sectors. Note that the union of these sectors will always give a disk of Euler characteristic 1 and radius ρ.

In other words,

$$\text{area}(\text{Tube}(A, \rho)) = \pi\rho^2\varphi(A) + \rho \,\text{perimeter}(A) + \text{area}(A).$$

Comparing this to (6.3.3), it now takes only a little thought to guess what the intrinsic volumes must measure. If the ambient space is $\mathbb{R}^2$, then $\mathcal{L}_2$ measures area, $\mathcal{L}_1$ measures boundary length, while $\mathcal{L}_0$ gives the Euler characteristic. In $\mathbb{R}^3$, $\mathcal{L}_3$ measures volume, $\mathcal{L}_2$ measures surface area, $\mathcal{L}_1$ is a measure of cross-sectional diameter, and $\mathcal{L}_0$ is again the Euler characteristic. In higher dimensions, it takes some imagination, but $\mathcal{L}_N$ and $\mathcal{L}_{N-1}$ are readily seen to measure volume and surface area, while $\mathcal{L}_0$ is always the Euler characteristic. Precisely why this happens, how it involves the curvature of the set and its boundary, and what happens in less-familiar spaces forms much of the content of Section 7.6 and is treated again, in fuller detail, in Chapter 10.

In the meantime, you can try checking for yourself, directly from (6.3.3) and a little first-principles geometry, that for an N-dimensional cube of side length T the intrinsic volumes are given by

$$\mathcal{L}_j\left([0, T]^N\right) = \binom{N}{j} T^j. \tag{6.3.5}$$

It is also not hard to see that for N-dimensional rectangles,

$$\mathcal{L}_j\left(\prod_{i=1}^{N}[0, T_i]\right) = \sum T_{i_1}\cdots T_{i_j}, \tag{6.3.6}$$

where the sum is taken over the $\binom{N}{j}$ different choices of subscripts $i_1, \ldots, i_j$.

For handling B_λ^N, the ball of radius λ, it useful to go beyond first principles. Noting that $\text{Tube}(B_\lambda^N, \rho) = B_{\lambda+\rho}^N$, we have

$$\begin{aligned}\lambda_N\left(\text{Tube}\left(B_\lambda^N, \rho\right)\right) &= (\lambda+\rho)^N \omega_N = \sum_{j=0}^{N}\binom{N}{j}\lambda^j \rho^{N-j}\omega_N \\ &= \sum_{j=0}^{N}\omega_{N-j}\rho^{N-j}\binom{N}{j}\lambda^j\frac{\omega_N}{\omega_{N-j}}.\end{aligned}$$

Comparing this to Steiner's formula (6.3.3), it is immediate that

$$\mathcal{L}_j\left(B_\lambda^N\right) = \binom{N}{j}\lambda^j\frac{\omega_N}{\omega_{N-j}}. \tag{6.3.7}$$

For S_λ^{N-1} the sphere of radius λ in $\mathbb{R}^N$, a similar argument, using the fact that

$$\text{Tube}(S_\lambda^{N-1}, \rho) = B_{\lambda+\rho}^N - B_{\lambda-\rho}^N,$$

yields

$$\mathcal{L}_j(S_\lambda^{N-1}) = 2\binom{N}{j}\frac{\omega_N}{\omega_{N-j}}\lambda^j = 2\binom{N-1}{j}\frac{s_N}{s_{N-j}}\lambda^j \tag{6.3.8}$$

if $N-1-j$ is even and 0 otherwise.

Further examples can be found, for example, in [139].

A useful normalization of the intrinsic volumes are the so-called *Minkowski functionals*, defined as

$$\mathcal{M}_j(A) = (j!\omega_j)\mathcal{L}_{N-j}(A), \tag{6.3.9}$$

so that, when expressed in terms of the $\mathcal{M}_j$, Steiner's formula now reads like a Taylor series expansion

$$\lambda_N(\text{Tube}(A, \rho)) = \sum_{j=0}^{N}\frac{\rho^j}{j!}\mathcal{M}_j(A). \tag{6.3.10}$$

It is an important and rather deep fact, due to Weyl [168] in the manifold setting,[11] that the $\mathcal{L}_j$ are independent of the ambient space in which sets sit. Because of the reversed numbering system and the choice of constants, this is not true of the Minkowski functionals.[12]

---

[11] See also [73, 92] and Chapter 10, where we develop this in detail.

[12] To see why the $\mathcal{M}_j$ are dependent on the ambient space, let i_{NM} be the standard inclusion of $\mathbb{R}^N$ into $\mathbb{R}^M$, $M \geq N$, defined by $i_{NM}(x) = (x_1, \ldots, x_N, 0, \ldots, 0) \in \mathbb{R}^M$

There is another way to define intrinsic volumes that works just as well for our main scenario of basic complexes as it does for convex sets, based on the idea of *kinematic density*.[13] Let $G_N = \mathbb{R}^N$ be the isometry group (of rigid motions) on $\mathbb{R}^N$, and equip it with the Haar measure μ_N, normalized to be Lebesgue measure on $\mathbb{R}^N$ and the invariant probability measure on the rotation group $O(N)$. A formula of Hadwiger states that

$$\mathcal{L}_j(A) = \begin{bmatrix} N \\ j \end{bmatrix} \int_{G_N} \varphi(A \cap gE_{N-j}) \mu_N(dg), \tag{6.3.11}$$

where E_j is any j-dimensional affine subspace of $\mathbb{R}^N$,

$$\begin{bmatrix} N \\ j \end{bmatrix} = \frac{[N]!}{[N-j]![j]!} = \binom{N}{j} \frac{\omega_N}{\omega_{N-j}\omega_j}, \tag{6.3.12}$$

where $[N]! = N!\omega_N$ and φ is our old friend the Euler characteristic. The numbers $\left[\begin{smallmatrix} n \\ k \end{smallmatrix}\right]$ are known as *flag coefficients* [92]. Hadwiger's formula is important, and we shall return to it later.

Note, however, that although over the class of convex sets the two definitions of intrinsic volumes are equivalent, Steiner's formula no longer holds for these intrinsic volumes over the class of basic complexes, or even over the convex ring. For a counterexample, take the union of a vertical and horizontal line as in Figure 6.3.2 and consider the tube around it. Blindly applying Steiner's formula, one finds that there are two problems: The area of the quarter disk on the lower left corner has been forgotten, and the area marked with a black square has been double counted. Since these two areas are different, they do not cancel to give the right volume for the tube.

Fig. 6.3.2. A tube around the letter L.

The way around this will be to develop a somewhat different notion, that of Lipschitz–Killing curvatures, which we shall do in Chapters 7 and 8 in the setting on

for $x \in \mathbb{R}^N$. Consider $M > N$, and note that the polynomials $\lambda_M(\text{Tube}(A, \rho))$ and $\lambda_M(\text{Tube}(i_{NM}(A), \rho))$ lead to different geometric interpretations. For example, for a curve C in $\mathbb{R}^2$, $\mathcal{M}_1(C)$ will be proportional to the arc length of C and $\mathcal{M}_2(C) = 0$, while $\mathcal{M}_2(i_{2,3}(C))$, rather than $\mathcal{M}_1(i_{2,3}(C))$, is proportional to arc length.

[13] We shall meet this again, in a slightly different format and in far more detail, as a special case of Crofton's formula, Theorem 13.1.1. In Chapter 13 we shall also develop a significant extension of the Hadwiger formula (6.3.11), and use it to lift results about Euler characteristics to results about Lipschitz–Killing curvatures.

manifolds, and for which we shall develop in Chapter 10 an extension of Steiner's formula. One of the key differences between Lipschitz–Killing curvatures and the intrinsic volumes that we have looked at so far will be that the former will also be allowed to take negative values. This will allow, for example, for the subtraction of the area of the double-counted square in Figure 6.3.2.

The time has now come to explain what all of the above has to do with random fields. For this we need one more result from integral geometry, due to Hadwiger [74], which also seems to be the only place to find a proof. However, unless you have a weakness for classical German, you should turn to Schneider [141] to read more about it. In Klain and Rota [92], a proof is given for continuous, invariant functionals on the convex ring, which, as noted previously, is a subset of the basic complexes. Zahle [182] has the most general result in terms of the classes of sets covered, although with continuity replacing monotonicity among the conditions of the theorem.

Theorem 6.3.1. *Let ψ be a real-valued function on basic complexes in $\mathbb{R}^N$, invariant under rigid motions, additive (in the sense of* (6.1.2)*) and monotone, in the sense that for all pairs A, B, either $A \subseteq B \Rightarrow \psi(A) \le \psi(B)$ or $A \subseteq B \Rightarrow \psi(A) \ge \psi(B)$. Then*

$$\psi(A) = \sum_{j=0}^{N} c_j \mathcal{L}_j(A), \tag{6.3.13}$$

where $c_0, \dots, c_N$ are nonnegative (ψ-dependent) constants.

Now take an isotropic field f on $\mathbb{R}^N$ and consider the set-indexed functional

$$\psi(A) \triangleq \mathbb{P}\left\{ \sup_{t \in A} f(t) > u \right\}. \tag{6.3.14}$$

Then ψ is clearly monotone and rigid-motion-invariant. Unfortunately, it is not quite additive, since even if A and B are disjoint,

$$\begin{aligned}\psi(A \cup B) &= \psi(A) + \psi(B) \\ &\quad - \mathbb{P}\left\{ \left(\sup_{t \in A} f(t) > u \right) \cap \left(\sup_{t \in B} f(t) > u \right) \right\}.\end{aligned} \tag{6.3.15}$$

Consider the last term here for large u. In that case both $\psi(A)$ and $\psi(B)$ will be small. If A and B are disjoint, then it is reasonable to expect that the events $\{\sup_A f > u\}$ and $\{\sup_B f >\}$ will be close to independent,[14] and so the final term in (6.3.15) would be a product of two very small terms, and so of smaller order than either.

[14] This would happen, for example, if A and B were sufficiently distant for the values of f in A and B to be close to independent. It turns out that at least for Gaussian f, these heuristics are true even if A and B are close, as long as u is large enough. We shall not treat these results, but you can find examples in [17] and [97].

On the other hand, if $A \cap B \neq \emptyset$ then one could argue as follows: Since it is difficult for $\sup_A f$ to be greater than u, and it is difficult for $\sup_B f$ to be greater than u, if both of these events are to happen, then it is most likely that the supremum of the process will occur somewhere in $A \cap B$, making the final term in (6.3.15) approximately $\psi(A \cap B)$.

Putting the two cases together, it is therefore not unreasonable to expect that there might be an invariant, additive, *and* monotone functional $\widetilde{\psi}_u$ for which

$$\mathbb{P}\Big\{\sup_{t\in A} f(t) > u\Big\} = \widetilde{\psi}_u(A) + o(\widetilde{\psi}_u(A)), \tag{6.3.16}$$

in which case, by Theorem 6.3.1, we would be able to conclude that

$$\mathbb{P}\left\{\sup_{t\in A} f(t) > u\right\} = \sum_{j=0}^{N} c_j(u)\mathcal{L}_j(A) + \text{lower-order terms} \tag{6.3.17}$$

for some functions $c_j(u)$.

The fact that such a function does indeed exist, and is given by

$$\widetilde{\psi}_u = \mathbb{E}\{\varphi(A_u(f, A))\},$$

where φ is the Euler characteristic and A_u an excursion set, is one of the main punch lines of this book and of the last few years of research in Gaussian random fields. Proving this, in wide generality, and computing the coefficients c_j in (6.3.17) as functions of u is what much of Part III of this book is about.

If you are interested mainly in nice Euclidean parameter spaces, and do not need to see the details of all the proofs, you can now comfortably skip the rest of Part II, with the exception of Section 9.4, which gives a version of Theorem 6.2.4 in general dimensions. The same is true if you have a solid background in differential geometry, even if you plan to follow all the proofs later on. You can return later when the need to confirm notation arises.

7

Differential Geometry

As we have said more than once, this chapter is intended to serve as a rapid and noncomprehensive introduction to differential geometry, basically in the format of a "glossary of terms." Most will be familiar to those who have taken a couple courses in differential geometry, and hopefully informative enough to allow the uninitiated[1] to follow the calculations in later chapters. However, to go beyond merely following the arguments there and to reach the level of a real understanding of what is going on, it will be necessary to learn the material from its classical sources.[2]

Essentially all that we have to say is "well known," in the sense that it appears in textbooks of differential geometry. Thus, the reader familiar with these books will be able to skim the remainder of this chapter, needing only to pick up our notation and emphases.

7.1 Manifolds

A differentiable manifold is a mathematical generalization, or abstraction, of objects such as curves and surfaces in $\mathbb{R}^N$. Intuitively, it is a smooth set with a locally defined, Euclidean, coordinate system.

Manifolds without boundary

We call M a *topological* N*-manifold* if M is a locally compact Hausdorff space such that for every $t \in M$, there exist an open $U \subset M$ containing t, an open $\tilde{U} \subset \mathbb{R}^N$, and a homeomorphism $\varphi : U \to \tilde{U}$.

[1] For first-timers we have added numerous footnotes and simple examples along the way that are meant to help them through the general theory. In general, we shall be satisfied with the exposition as long as we have not achieved the double failure of both boring the expert and bamboozling the novice. We also apologize to the experts for attempting the impossible: to reduce their beautiful subject matter to a "glossary."

[2] For the record, the books that we found most useful were Boothby [29], Jost [88], Millman and Parker [114], Morita [116], and O'Neill [121]. The two recent books by Lee [100, 101] stand out from the pack as being particularly comprehensive and easy to read, and we highly recommend them as the right place to start.

A *(coordinate) chart* on M is a pair (φ, U), where, as above, $U \subset M$ is open and $\varphi : U \to \varphi(U) \subset \mathbb{R}^N$ is a homeomorphism. A collection $\{\varphi_i : U_i \to \mathbb{R}^N\}_{i\in I}$ of charts is called C^k *compatible* if

$$\varphi_i \circ \varphi_j^{-1} : \varphi_j(U_i \cap U_j) \to \varphi_i(U_i \cap U_j) \tag{7.1.1}$$

is a C^k diffeomorphism[3] for every $i, j \in I$ for which $U_i \cap U_j \neq \emptyset$.

If a collection of charts gives a covering of M, i.e., $\bigcup_{i\in I} U_i = M$, then it is called a (C^k) *atlas* for M. An atlas is called *maximal* if it is not contained in any strictly larger atlas with homeomorphisms satisfying (7.1.1). A topological N manifold, together with a C^k maximal atlas, is called a C^k *differential manifold*. The maximal atlas is usually referred to as the *differential structure* of M.

The component functions $x_1, \dots, x_N$ of a coordinate chart (φ, U), defined by $\varphi(t) = (x_1(t), \dots, x_N(t))$, are called the *local coordinates* on U.

Manifolds with boundary

If some of the U_i map to subsets of the half-space $\mathbb{H}^N \triangleq \mathbb{R}^{N-1} \times \mathbb{R}_+$ in $\mathbb{R}^N$, then we talk about a *manifold with boundary*, rather than a simple manifold. A manifold with boundary can be thought of as a disjoint union of two manifolds: ∂M, its boundary, an $(N-1)$-dimensional manifold, and M°, its interior, an N-dimensional manifold.

Continuous and differentiable functions on manifolds

The notions of continuous and differentiable functions on a C^k manifold M are straightforward. A function $f : M \to \mathbb{R}$ is said to be of class C^k if $f \circ \varphi^{-1}$ is of class C^k, in the usual Euclidean sense, for every chart in the atlas.

Tangent spaces

For a manifold M embedded in $\mathbb{R}^N$, such as a curve or a surface, it is straightforward to envision what is meant by a *tangent vector* at a point $t \in M$. It is no more than a vector with origin at t, sitting in the tangent plane to M at t. Given such a vector, v, one can differentiate functions $f : M \to \mathbb{R}$ along the direction v. Thus, to each v there corresponds a local derivative. For abstract manifolds, the basic notion is not that of a vector sitting in a tangent plane, but rather that of a differential operator.

To see how this works, we start with the simplest case, in which $M = \mathbb{R}^N$ and everything reduces to little more than renaming familiar objects. Here we can manage with the atlas containing the single chart (M, i_{NN}), where i_{NN} is the inclusion map. We change notation after the inclusion, writing $x = i_{NN}(t)$ and $M' = i_{NN}(M)$ $(= \mathbb{R}^N)$. To every vector X_x with origin $x \in M'$, we can assign a linear map from $C^1(M')$ to $\mathbb{R}$ as follows:

If $f \in C^1(M)$ and X_x is a vector of the form $X_x = x + \sum_{i=1}^N a_i e_i$, where $\{e_i\}_{1\le i\le N}$ is the standard basis for $\mathbb{R}^N$, we define the first-order differential operator[4]

[3] That is, both $\varphi_i \circ \varphi_j^{-1}$ and its inverse $\varphi_j \circ \varphi_i^{-1}$ are k-times differentiable as functions from subsets of $\mathbb{R}^N$ to subsets of $\mathbb{R}^N$.

[4] Hopefully, the standard usage of X_x to denote both a vector and a differential operator will not cause too much confusion. In this simple case, they clearly amount to essentially the same thing. In general, they are the same by definition.

X_x by its action on f:

$$X_x f = \sum_{i=1}^{N} a_i \frac{\partial f}{\partial x_i}\bigg|_x . \tag{7.1.2}$$

It is elementary calculus that X_x satisfies the product rule $X_x(fg) = fX_x g + gX_x f$, and that for any two functions f, g that agree on a neighborhood of x we have $X_x f = X_x g$.

For each $x \in U$, identifying $\frac{\partial}{\partial x_i}\big|_x$ with e_i, we see that

$$\left\{ \frac{\partial}{\partial x_i}\bigg|_x \right\}_{1 \le i \le N}$$

forms a basis for an N-dimensional space of first-order differential operators, which we call the *tangent space* at x and denote by $T_x\mathbb{R}^N$. Returning to M, and identifying $T_t\mathbb{R}^N$ with $T_x\mathbb{R}^N$, we have now defined the tangent space here as well.

Before turning to the case of an abstract manifold M, we need to be a little more precise about the term "first-order differential operator" used above. While it may be obvious what is meant in $\mathbb{R}^N$, we need to be a little more careful in general. Loosely speaking, given an $x \in \mathbb{R}^N$ and a neighborhood U of x, a *first-order differential operator* is a linear operator on $C^1(U)$ that satisfies the product rule and depends only on the derivatives of a function when evaluated at x. The looseness in the term "depends only on the derivatives of a function when evaluated at x" can be made precise by requiring that the operator be actually defined on the *germ*[5] of C^1 functions at x. Since this definition is local, it extends easily to manifolds. Thus, choosing $t \in M$ and a chart (U, φ) around t, we can define similarly a first-order differential operator at $t \in M$ as a linear operator (on the germs of C^1 functions) that satisfies the product rule.

Formally, if $x \in \varphi(U) \subset \mathbb{R}^N$ for some chart (U, φ), then we can lift the basis of $T_x\mathbb{R}^N$ (built as in the Euclidean case) to the point $t = \varphi^{-1}(x)$ via $\varphi^{-1}_{*,t}$, the so-called differential or *push-forward* of φ^{-1} at t. The map $\varphi^{-1}_{*,t}$ is a linear map that maps the first-order differential operators $(\partial/\partial x_i)|_{\varphi(t)}$, $1 \le i \le N$, acting on $C^1(\varphi(U))$ to first-order differential operators on $C^1(U)$, and we define it by

$$\left(\varphi^{-1}_{*,t}(X_{\varphi(t)})\right) f = X_{\varphi(t)} \left(f \circ \varphi^{-1}\right) \tag{7.1.3}$$

for any $f \in C^1(M)$. Since $\varphi^{-1}_{*,t}$ is linear, it follows that the set

$$\left\{ \varphi^{-1}_{*,t} \left(\frac{\partial}{\partial x_i}\bigg|_{\varphi(t)} \right) \right\}_{1 \le i \le N} \tag{7.1.4}$$

[5] The germs of f at x are the (equivalence) class of functions g for which $f \equiv g$ over some open neighborhood U of x. All local properties of f at x depend only on the germs to which f belongs.

is linearly independent. We define the space it spans to be T_tM, the *tangent space of M at t*. Its elements X_t, while being differential operators, are called the *tangent vectors* at t.

With some abuse of notation, in a chart (U, φ) with a coordinate system $(x_1, \dots, x_N)$ for $\varphi(U)$ the basis (7.1.4) is usually written as

$$\left\{ \frac{\partial}{\partial x_i}\bigg|_t \right\}_{1 \le i \le N}, \tag{7.1.5}$$

and is referred to as the *natural basis* for T_tM in the chart (U, φ).

Chain rule: The push-forward

The chain rule of vector calculus is a formula for differentiating the composition $h \circ g$ for $g \in C^1(\mathbb{R}^k; \mathbb{R}^j)$, $h \in C^1(\mathbb{R}^j; \mathbb{R}^l)$. In differential geometry, the chain rule is expressed through the push-forward of differentiable functions. Specifically, given two manifolds M and N and any $g \in C^1(M; N)$,[6] we define its *push-forward* g_* by defining it on charts (U, φ) and (V, ψ) of M and N, for which $g(U) \cap V \ne \emptyset$, as

$$(g_* X_t)h = X_t(h \circ g),$$

for any $h \in C^1(N)$.

Vector fields

Since each $X_t \in T_tM$ is a linear map on $C^1(M)$ satisfying the product rule, we can construct a first-order differential operator X, called a *vector field*, that takes C^k functions ($k \ge 1$) to real-valued functions, as follows:

$$(Xf)_t = X_t f \quad \text{for some } X_t \in T_tM.$$

In other words, a vector field is a map that assigns, to each $t \in M$, a tangent vector $X_t \in T_tM$. Thus, in a chart (U, φ) with coordinates $(x_1, \dots, x_N)$, a vector field can be represented (cf. (7.1.5)) as

$$X_t = \sum_{i=1}^{N} a_i(t) \frac{\partial}{\partial x_i}\bigg|_t. \tag{7.1.6}$$

If the a_i are C^k functions, we can talk about C^k vector fields. Note that for $j \ge 1$, a C^k vector field maps $C^j(M)$ to $C^{\min(j-1,k)}(M)$.

[6] $C^k(M; N)$, the space of "k-times differentiable functions from M to N," is defined analogously to $C^1(M)$. Thus $f \in C^k(M; N)$ if for all $t \in M$, there are a chart (U_M, φ_M) for M and a neighborhood V of t with $V \subset U_M$ such that $f(V) \subset U_N$ for some chart (U_N, φ_N) for N for which the composite map $\varphi_N \circ f \circ (\varphi_M)^{-1} : \varphi_M(V) \to \varphi_N(f(V))$ is C^k in the usual, Euclidean, sense.

Vector bundles

A C^l *vector bundle* is a triple (E, M, F), along with a map $\pi : E \to M$, where E and M are, respectively, $(N+q)$- and N-dimensional manifolds of class at least C^l and F is a q-dimensional vector space. The manifold E is locally a product. That is, every $t \in M$ has a neighborhood U such that there is a homeomorphism $\varphi_U : U \times F \to \pi^{-1}(U)$ satisfying $(\pi \circ \varphi_U)(t, f_U) = t$, for all $f_U \in F$. Furthermore, for any two such overlapping neighborhoods U, V with $t \in U \cap V$, we require that

$$\varphi_U(t, f_U) = \varphi_V(t, f_V) \Longleftrightarrow f_U = g_{UV}(t)(f_V),$$

where $g_{UV}(t) : F \to F$ is a nondegenerate linear transformation. The functions $g_{UV} : U \cap V \to \mathrm{GL}(F)$ are called the *transition functions* of the bundle (E, M, F). The manifold M is called the *base manifold*, and the vector space $\pi^{-1}(t)$, which is isomorphic to F (as a vector space), is called the *fiber* over t. Usually, it is clear from the context what M and F are, so we refer to the bundle as E. A C^j, $j \leq l$, *section* of a vector bundle is a C^j map $s : M \to E$ such that $\pi \circ s$ is the identity map on M. In other words, if one thinks of a vector bundle as assigning, to each point of $t \in M$, the entire vector space $\pi^{-1}(t)$, then a C^j section is a rule for choosing from this space in a smooth (C^j) fashion.

We denote the space of C^k sections of a vector bundle E by $\Gamma^k(E)$.

In a similar fashion, one can define *fiber bundles* when the fibers F are not vector spaces, although we shall not go through such a construction yet. Two such fiber bundles that will be of interest to us are the *sphere bundle* $S(M)$ and the *orthonormal frame bundle* $\mathcal{O}(M)$ of a Riemannian manifold (M, g). These will be described below once we have the definition of a Riemannian manifold.

Tangent bundles

Perhaps the most natural example of a vector bundle is obtained from the collection of all tangent spaces T_tM, $t \in M$. This can be parameterized in a natural way into a manifold, $T(M)$, called the *tangent bundle*.

In a chart (U, φ), any tangent vector X_t, $t \in U$, can be represented as in (7.1.6), so the set of all tangent vectors at points $t \in U$ is a $2N$-dimensional space, with local coordinates $\widetilde{\varphi}(X_t) = (x_1(t), \dots, x_N(t); a_1(t), \dots, a_N(t))$. Call this E_t, and call the projection of E_t onto its last N coordinates F_t. Denote the union (over $t \in M$) of the E_t by E, and of the F_t by F, and define the natural projection $\pi : E \to M$ given by $\pi(X_t) = t$, for $X_t \in E_t$. The triple (E, M, F), along with π, defines a vector bundle, which we call the *tangent bundle* of M and denote by $T(M)$.

The tangent bundle of a manifold is itself a manifold, and as such can be given a differential structure in the same way that we did for M. To see this, suppose M is a C^k manifold and note that an atlas $\{U_i, \varphi_i\}_{i \in I}$ on M determines a covering on $T(M)$, the charts $\{\pi^{-1}(U_i), \widetilde{\varphi}_i\}_{i \in I}$ of which determine a topology on $T(M)$, the smallest topology such that $\widetilde{\varphi}_{|\pi^{-1}(U)}$ are homeomorphisms. In any two charts (U, φ_U) with local coordinates $(x_1, \dots, x_N)$ and (V, φ_V) with local coordinates $(y_1, \dots, y_N)$ with $U \cap V \neq \emptyset$, a vector $X_t \in T(M)$ is represented by

$$\begin{aligned}X_t &= (x_1(t), \ldots, x_N(t); a_1(t), \ldots, a_N(t)) \\ &= (y_1(t), \ldots, y_N(t); b_1(t), \ldots, b_N(t)),\end{aligned}$$

where we have the relation

$$b_i(t) = \sum_{j=1}^{N} a_j(t) \frac{\partial y_i}{\partial x_j}, \tag{7.1.7}$$

and $\partial y_i / \partial x_j$ is the first-order partial derivative of the real-valued function y_i with respect to x_j.

Since $\varphi_U \circ \varphi_V^{-1}$ is a C^k diffeomorphism, we have that $\widetilde{\varphi}_U \circ \widetilde{\varphi}_V^{-1}$ is a C^{k-1} diffeomorphism, having lost one order of differentiation because of the partial derivatives in (7.1.7). In summary, the atlas $\{U_i, \varphi_i\}_{i \in I}$ determines a C^{k-1} differential structure on $T(M)$ as claimed.

7.2 Tensor Calculus

Tensors are basically linear operators on vector spaces. They have a multitude of uses, but there are essentially only two main approaches to setting up the (somewhat heavy) definitions and notation that go along with them. When tensors appear in applied mathematics or physics, the definitions involve high-dimensional arrays that transform (as the ambient space is transformed) according to certain definite rules. The approach we shall adopt, however, will be the more modern differential-geometric one, in which the transformation rules result from an underlying algebraic structure via which tensors are defined. This approach is neater and serves two of our purposes: We can fit everything we need into only six pages, and the approach is essentially coordinate-free.[7] The latter is of obvious importance for the manifold setting. The downside of this approach is that if this is the first time you see this material, it is very easy to get lost among the trees without seeing the forest. Thus, since one of our main uses of tensors will be for volume (and, later, curvature) computations on manifolds, we shall accompany the definitions with a series of footnotes showing how all of this relates to simple volume computations in Euclidean space. Since manifolds are locally Euclidean, this might help a first-timer get some feeling for what is going on.

Tensors and exterior algebras

Given an N-dimensional vector space V and its dual V^*, a map ω is called a *tensor* of order (n, m) if

$$\omega \in L(\underbrace{V \oplus \cdots \oplus V}_{n \text{ times}} \oplus \underbrace{V^* \oplus \cdots \oplus V^*}_{m \text{ times}} \mathbb{R}),$$

[7] Ultimately, however, we shall need to use the array notation when we come to handling specific examples. It appears, for example, in the connection forms of (7.3.14) and the specific computations of Section 7.7.

where $L(E; F)$ denotes the set of (multi)linear[8] mappings between two vector spaces E and F. We denote the space of tensors of order (n, m) by $\mathcal{T}_m^n$, where n is said to be the *covariant* order and m the *contravariant* order.[9]

Let $\mathcal{T}(V) = \bigoplus_{i,j=1}^{\infty} \mathcal{T}_i^j(V)$ be the direct sum of all the tensor spaces $\mathcal{T}_j^i(V)$. The bilinear associative *tensor product* $\otimes : \mathcal{T}(V) \times \mathcal{T}(V) \to \mathcal{T}(V)$ is defined on $\mathcal{T}_j^i \times \mathcal{T}_l^k$ as

$$\begin{aligned}&(\alpha \otimes \beta)(v_1, \ldots, v_{i+k}, v_1^*, \ldots, v_{j+l}^*)\\ &\quad \stackrel{\Delta}{=} \alpha(v_1, \ldots, v_i, v_1^*, \ldots, v_j^*) \times \beta(v_{i+1}, \ldots, v_{i+k}, v_{j+1}^*, \ldots, v_{j+l}^*).\end{aligned}$$

The algebra $(\mathcal{T}(V), \otimes)$ can be split into two subalgebras, the covariant and contravariant tensors $\mathcal{T}^*(V) = \bigoplus_{i=1}^{\infty} \mathcal{T}_0^i(V)$ and $\mathcal{T}_*(V) = \bigoplus_{i=1}^{\infty} \mathcal{T}_i^0(V)$.

Alternating and symmetric covariant tensors

A covariant tensor γ of order k (a $(k, 0)$-tensor) is said to be *alternating* if

$$\gamma(v_{\sigma(1)}, \ldots, v_{\sigma(k)}) = \varepsilon_\sigma \gamma(v_1, \ldots, v_k) \quad \text{for all } \sigma \in S(k),$$

where $S(k)$ is the symmetric group of permutations of k letters and ε_σ is the sign of the permutation σ. It is called *symmetric* if

$$\gamma(v_{\sigma(1)}, \ldots, v_{\sigma(k)}) = \gamma(v_1, \ldots, v_k) \quad \text{for all } \sigma \in S(k).$$

For example, computing the determinant of the matrix formed from N-vectors gives an alternating covariant tensor of order N on $\mathbb{R}^N$.

For $k \geq 0$, we denote by $\Lambda^k(V)$ (respectively, $\mathrm{Sym}(\mathcal{T}_0^k(V))$) the space of alternating (symmetric) covariant k-tensors on V, and by

$$\Lambda^*(V) = \bigoplus_{k=0}^{\infty} \Lambda^k(V), \qquad \mathrm{Sym}^*(V) = \bigoplus_{k=0}^{\infty} \mathrm{Sym}(\mathcal{T}_0^k(V)),$$

their direct sums. If $k = 0$ then both $\Lambda^0(V)$ and $\mathrm{Sym}(\mathcal{T}_0^0(V))$ are isomorphic to $\mathbb{R}$. Note that $\Lambda^*(V)$ is actually a finite-dimensional vector space, since $\Lambda^k(V) \equiv \{0\}$ if $k > N$.

There exist natural projections $\mathcal{A} : \mathcal{T}^*(V) \to \Lambda^*(V)$ and $\mathcal{S} : \mathcal{T}^*(V) \to \mathrm{Sym}^*(V)$ defined on each $\mathcal{T}^k(V)$ by

---

[8] A multilinear mapping is linear, separately, in each variable. In general, we shall not bother with the prefix.

[9] For a basic example, let $V = \mathbb{R}^3$. There are three, very elementary, covariant tensors of order 1 operating on vectors $v = (v_1, v_2, v_3)$. These are $\theta_1(v) = v_1$, $\theta_2(v) = v_2$, and $\theta_3(v) = v_3$. Thus θ_j measures the "length" of v in direction j, or, equivalently, the "length" of the projection of v onto the jth axis, where "length" is signed. The fact that the length may be signed is crucial to all that follows. Not also that since $\mathbb{R}^3$ is its own dual, we could also treat the same θ_i as contravariant tensors of order 1.

$$\mathcal{A}\gamma(v_1, \dots, v_k) = \frac{1}{k!} \sum_{\sigma \in S(k)} \varepsilon_\sigma \gamma(v_{\sigma(1)}, \dots, v_{\sigma(k)}),$$

$$\mathcal{S}\gamma(v_1, \dots, v_k) = \frac{1}{k!} \sum_{\sigma \in S(k)} \gamma(v_{\sigma(1)}, \dots, v_{\sigma(k)}).$$

Grassmann algebra

The bilinear, associative *wedge product*,[10] or *exterior product*, $\wedge : \Lambda^*(V) \times \Lambda^*(V) \to \Lambda^*(V)$ is defined on $\Lambda^r(V) \times \Lambda^s(V)$ by

$$\alpha \wedge \beta = \frac{(r+s)!}{r!s!} \mathcal{A}(\alpha \otimes \beta). \tag{7.2.1}$$

The algebra $(\Lambda^*(V), \wedge)$ is called the *Grassmann* or *exterior algebra* of V.

Induced inner products on $\mathcal{T}^*(V)$

Any inner product on V induces, in a natural way, an inner product on $\mathcal{T}^*(V)$, and correspondingly on any subspace of $\mathcal{T}^*(V)$. We shall be particularly interested in $\Lambda^*(V)$ and so give the details for this case.

Specifically, given an inner product on V and an orthonormal basis $B_{V^*} = \{\theta_1, \dots, \theta_N\}$ for V^*, then

$$B_{\Lambda^*(V)} = \{\theta_0\} \cup \bigcup_{j=1}^{N} \left\{\theta_{i_1} \wedge \cdots \wedge \theta_{i_j} : i_1 < i_2 < \cdots < i_j\right\}, \tag{7.2.2}$$

---

[10] Continuing the example of footnote 9, take $u, v \in \mathbb{R}^3$ and check that this definition gives $(\theta_1 \wedge \theta_2)(u, v) = (u_1 v_2 - v_1 u_2)$, which is, up to a sign, the area of the parallelogram with corners 0, $\pi(u)$, $\pi(v)$, and $\pi(u) + \pi(v)$, where $\pi(u)$ is the projection of u onto the plane spanned by the first two coordinate axes. That is, this particular alternating covariant tensor of order 2 performs an area computation, where "areas" may be negative. Now take $u, v, w \in \mathbb{R}^3$, and take the wedge product of $\theta_1 \wedge \theta_2$ with θ_3. A little algebra gives that $(\theta_1 \wedge \theta_2 \wedge \theta_3)(u, v, w) = \det\langle u, v, w\rangle$, where $\langle u, v, w\rangle$ is the matrix with u in the first column, etc. In other words, it is the (signed) volume of the parallelepiped three of whose edges are u, v, and w. Extending this example to N dimensions, you should already be able to guess a number of important facts, including the following:

- (i) Alternating covariant tensors of order n have a lot to do with computing n-dimensional (signed) volumes.
- (ii) The wedge product is a way of combining lower-dimensional tensors to generate higher-dimensional ones.
- (iii) Our earlier observation that $\Lambda^k(V) \equiv \{0\}$ if $k > N$ translates, in terms of volumes, to the trivial observation that the k-volume of an N-dimensional object is zero if $k > N$.
- (iv) Since area computations and determinants are intimately related, so will be tensors and determinants.

where θ_0 is a 0-form, is a basis[11] for $\Lambda^*(V)$, which itself is a vector space of dimension $\sum_{k=0}^{N}\binom{N}{k} = 2^N$.

The unique inner product on $\Lambda^*(V)$ that makes this basis orthogonal is the "corresponding" inner product to which we referred above.

Algebra of double forms

Define $\Lambda^{n,m}(V)$ to be the linear span of the image of $\Lambda^n(V) \times \Lambda^m(V)$ under the map $\otimes$, and define, with some abuse of notation, $\Lambda^*(V) \otimes \Lambda^*(V) = \bigoplus_{n,m=0}^{\infty} \Lambda^{n,m}(V)$. An element of $\Lambda^{n,m}(V)$ is called a *double form of type* (n, m). Note that an (n, m) double form is alternating in its first n and last m variables.

The *double wedge product* $\cdot$ on $\Lambda^*(V) \otimes \Lambda^*(V)$ is defined on tensors of the form $\gamma = \alpha \otimes \beta \in \Lambda^*(V) \times \Lambda^*(V)$ by

$$(\alpha \otimes \beta) \cdot (\alpha' \otimes \beta') = (\alpha \wedge \alpha') \otimes (\beta \wedge \beta'), \tag{7.2.3}$$

and then extended by linearity. For $\gamma \in \Lambda^{n,m}(V)$ and $\theta \in \Lambda^{p,q}(V)$ this gives $\gamma \cdot \theta \in \Lambda^{n+p,m+q}(V)$, for which

$$\begin{aligned}
&(\gamma \cdot \theta)\left((u_1, \ldots, u_{n+p}), (v_1, \ldots, v_{m+q})\right) \qquad (7.2.4)\\
&\quad = \frac{1}{n!m!p!q!} \sum_{\substack{\sigma \in S(n+p)\\ \rho \in S(m+q)}} \varepsilon_\sigma \varepsilon_\rho \Big[\gamma\left((u_{\sigma_1}, \ldots, u_{\sigma_n}), (v_{\rho_1}, \ldots, v_{\rho_m})\right)\\
&\qquad \times \theta\left((u_{\sigma_{n+1}}, \ldots, u_{\sigma_{n+p}}), (v_{\rho_{m+1}}, \ldots, v_{\rho_{m+q}})\right)\Big].
\end{aligned}$$

Note that a double form of type $(n, 0)$ is simply an alternating covariant tensor, so that comparing (7.2.4) with (7.2.1), it is clear that the restriction of $\cdot$ to $\bigoplus_{j=0}^{N} \Lambda^{j,0}(V)$ is just the usual wedge product.

We shall be most interested in the restriction of this product to

$$\Lambda^{*,*}(V) \stackrel{\Delta}{=} \bigoplus_{j=0}^{\infty} \Lambda^{j,j}(V).$$

The pair $(\Lambda^{*,*}(V), \cdot)$ is then a commutative (sub)algebra.[12]

For a double form $\gamma \in \Lambda^{*,*}(V)$, we define the polynomial γ^j as the product of γ with itself j times for $j \geq 1$, and set $\gamma^0 = 1$. If γ is of type (k, k) then γ^j is of type (jk, jk). Furthermore, for powers, (7.2.4) simplifies somewhat. In particular, if γ is a $(1, 1)$ double form, then it is easy to check from (7.2.4) that

$$\gamma^j\left((u_1, \ldots, u_j), (v_1, \ldots, v_j)\right) = j! \det\left(\gamma\,(u_k, v_\ell)_{k,\ell=1,\ldots,j}\right). \tag{7.2.5}$$

Note that if γ is of the simple form $\alpha \otimes \beta$, then it follows from either the above or directly from (7.2.3) that $\gamma^j \equiv 0$ for all $j > 1$.

[11] The notation of the example of footnote 9 should now be clear. The 1-tensor θ_j defined by $\theta_j(v) = v_j$ is in fact the jth basis element of V^* if V is given the usual basis $\{e_j\}_{j=1}^N$. Thus, in view of the fact that $B_{\Lambda^*(V)}$ is a basis for $\Lambda^*(V)$, the examples of tensors of order 2 and 3 that we built are actually archetypal.

[12] The product is not commutative on all of $\Lambda^*(V) \otimes \Lambda^*(V)$, since in general, for $\alpha \in \Lambda^{n,m}(V)$ and $\beta \in \Lambda^{p,q}(V)$ we have $\alpha \cdot \beta = (-1)^{np+mq} \beta \cdot \alpha$.

Trace of a double form

As described above, any inner product on V induces a corresponding inner product on $\Lambda^*(V)$. Therefore, any inner product on V also induces a real-valued map on $\Lambda^{*,*}(V)$, called the *trace*, defined on tensors of the form $\gamma = \alpha \otimes \beta \in \Lambda^{k,k}$ by

$$\mathrm{Tr}(\gamma) = \langle \alpha, \beta \rangle$$

and then extended by linearity to the remainder of $\Lambda^{*,*}(V)$. Given an orthonormal basis $(v_1, \ldots, v_N)$ of V, a simple calculation shows that for $\gamma \in \Lambda^{k,k}(V)$, $k \geq 1$, we have the following rather useful expression for Tr:

$$\mathrm{Tr}(\gamma) = \frac{1}{k!} \sum_{a_1,\ldots,a_k=1}^{N} \gamma\big((v_{a_1}, \ldots, v_{a_k}), (v_{a_1}, \ldots, v_{a_k})\big). \tag{7.2.6}$$

If $\gamma \in \Lambda^{0,0}(V)$, then $\gamma \in \mathbb{R}$ and we define $\mathrm{Tr}(\gamma) = \gamma$. One can also check that while the above seemingly depends on the choice of basis, the trace operator is, in fact, basis-independent, a property generally referred to as *invariance*.

There is also a useful extension to (7.2.6) for powers of symmetric forms in $\gamma \in \Lambda^{1,1}$. Using (7.2.5) to compute γ^j, and (7.2.6) to compute its trace, it is easy to check that

$$\mathrm{Tr}\left(\gamma^j\right) = j!\,\mathrm{detr}_j\left(\gamma\left(v_i, v_j\right)\right)_{i,j=1,\ldots,N}, \tag{7.2.7}$$

where for a matrix A,

$$\mathrm{detr}_j(A) \triangleq \text{sum over all } j \times j \text{ principal minors of } A. \tag{7.2.8}$$

When there is more than one inner product space under consideration, say V_1 and V_2, we shall denote their traces by Tr^{V_1} and Tr^{V_2}.

For a more complete description of the properties of traces, none of which we shall need, you could try [64, Section 2]. This is not easy reading, but is worth the effort.

A useful trace formula

To each $\gamma \in \Lambda^{k,k}(V)$ there corresponds a linear map $T_\gamma : \Lambda^k(V) \to \Lambda^k(V)$. The correspondence is unique, and so one can think of γ and T_γ as equivalent objects. To define T_γ, take a basis element of $\Lambda^k(V)$ of the form $\theta_{i_1} \wedge \cdots \wedge \theta_{i_k}$ (cf. (7.2.2)) and define its image under T_γ by setting

$$T_\gamma(\theta_{i_1} \wedge \cdots \wedge \theta_{i_k})(w_1, \ldots, w_k) = \gamma(v_{i_1}, \ldots, v_{i_k}, w_1, \ldots, w_k) \tag{7.2.9}$$

and extending linearly to all of $\Lambda^k(V)$.

It is a simple computation to check that if $k = 1$ and we write

$$I = \sum_{i=1}^{N} \theta_i \otimes \theta_i, \tag{7.2.10}$$

then T_I is the identity from $\Lambda^1(V)$ to $\Lambda^1(V)$. In general, T_{I^k} is the identity from $\Lambda^k(V)$ to $\Lambda^k(V)$.

We shall have need of the following useful formula (cf. [64, p. 425]) in Part III when we calculate the expected Euler characteristic of a Gaussian random field on M. For $\gamma \in \Lambda^{k,k}(V)$, and $0 \le j \le N-k$,

$$\mathrm{Tr}(\gamma I^j) = \frac{(N-k)!}{(N-k-j)!}\,\mathrm{Tr}(\gamma). \tag{7.2.11}$$

Tensor bundles, differential forms, and Grassmann bundles

Having defined the tangent bundle $T(M)$ and, for a fixed vector space V, the tensor spaces $\mathcal{T}^n_m(V)$, $\Lambda^*(V)$, and $\Lambda^{*,*}(V)$, it is a simple matter to define corresponding tensor bundles over M.

The basic observation is that given a C^k manifold M and its tangent space T_tM with dual T_t^*M, we can carry out constructions of tensors on $V = T_tM$, at every $t \in M$, exactly as we did in the previous section. Then, just as we defined vector fields as maps $t \mapsto X_t \in T_tM$, $t \in M$, we can define tensor fields, covariant tensor fields, alternating covariant tensor fields, etc.

It is also quite simple to associate a differential structure with these objects, following the argument at the end of Section 7.1. In particular, since tensor fields of order (n,m) are determined, locally, in the same way that vector fields are defined, any atlas $\{U_i, \varphi_i\}_{i\in I}$ for M determines[13] a C^{k-1} differential structure[14] on the collection

$$\mathcal{T}^n_m(M) \stackrel{\Delta}{=} \bigcup_{t\in M} \mathcal{T}^n_m(T_tM),$$

which itself is called the *tensor bundle of order* (n,m) over M.

Constructing the Grassmann algebra on each T_tM and taking the union

$$\Lambda^*(M) \stackrel{\Delta}{=} \bigcup_{t\in M} \Lambda^*(T_tM)$$

gives what is known as the *Grassmann bundle* of M. As was the case for vector fields, one can define C^j, $j \le k-1$, sections of $\Lambda^*(M)$, and these are called the C^j *differential forms*[15] of mixed degree.

[13] The finite dimensionality of spaces $\mathcal{T}^n_m(T_tM)$, noted earlier, is crucial to make this construction work.

[14] Included in this "differential structure" is a set of rules describing how tensor fields transform under transformations of the local coordinate systems, akin to what we had for simple vector fields in (7.1.7).

In fact, there is a lot hidden in this seemingly simple sequence of constructions. In particular, recall that at the beginning of our discussion of tensors we mentioned that the tensors of applied mathematics and physics are defined via arrays that transform according to very specific rules. However, nowhere in the path we have chosen have these demands explicitly appeared. They are, however, implicit in the constructions that we have just carried out.

[15] The reason for the addition of the adjective "differential" will be explained later; cf. (7.3.16) and the discussion following it.

Similarly, carrying out the same construction over the $\Lambda^k(T_t M)$ gives the *bundle of (differential) k-forms* on M, while carrying it out for $\Lambda^{*,*}(T_t M)$ yields the *bundle of double (differential) forms* on M.

7.3 Riemannian Manifolds

If you followed the footnotes while we were developing the notion of tensors, you will have noted that we related them to the elementary notions of area and volume in the simple Euclidean setting of $M = \mathbb{R}^N$. In general, of course, they are somewhat more complicated. In fact, if you think back, we do not as yet even have a notion of simple distance on M, let alone notions of volume. For this, we need to add a Riemannian structure to the manifold.

Riemannian manifolds

Formally, a C^{k-1} *Riemannian metric* g on a C^k manifold M is a C^{k-1} section of $\mathrm{Sym}(\mathcal{T}_0^2(M))$ such that for each t in M, g_t is positive definite; that is, $g_t(X_t, X_t) \geq 0$ for all $t \in M$ and $X_t \in T_t M$, with equality if and only if $X_t = 0$. Thus a C^{k-1} Riemannian metric determines a family of smoothly (C^{k-1}) varying inner products on the tangent spaces $T_t M$.

A C^k manifold M together with a C^{k-1} Riemannian metric g is called a C^k *Riemannian manifold* (M, g).

Sphere and orthonormal frame bundles

The first thing that a Riemannian metric gives us is the length $g_t(X_t, X_t)$ of tangent vectors $X_t \in T_t M$, and this allows us to define two rather important fiber bundles.

The *sphere bundle* $S(M)$ of (M, g) is the set of all unit tangent vectors of M, i.e., elements $X_t \in T(M)$, for some $t \in M$, with $g_t(X_t, X_t) = 1$. This is an example of a bundle with fibers that are not vector spaces, since the fibers are the spheres $S(T_t M)$.

Another example is the *orthonormal frame bundle*, $\mathcal{O}(M)$, the set of all sets of N unit tangent vectors $(X_t^1, \ldots, X_t^N)$ of M such that $(X_t^1, \ldots, X_t^N)$ form an orthonormal basis for $T_t M$.

Geodesic metric induced by a Riemannian metric

Despite its name, a Riemannian metric is not a true metric on M. However, it does induce a metric τ_g on M. Since g determines the length of a tangent vector, we can define the length of a C^1 curve $c : [0, 1] \to M$ by

$$L(c) = \int_{[0,1]} \sqrt{g_t(c_t', c_t')}\, dt$$

and define the metric τ_g by

$$\tau_g(s, t) = \inf_{c \in D^1([0,1];M)_{(s,t)}} L(c), \tag{7.3.1}$$

where $D^1([0,1]; M)_{(s,t)}$ is the set of all piecewise C^1 maps $c : [0,1] \to M$ with $c(0) = s$, $c(1) = t$. When the Riemannian metric g is unambiguous, we write τ_g as τ.

A curve in M connecting two points $s, t \in M$, along which the infimum in (7.3.1) is achieved, is called a *geodesic* connecting them. Geodesics need not be unique.

Gradient of $f \in C^1((M, g))$

The *gradient* of f is the unique continuous vector field on M such that

$$g_t(\nabla f_t, X_t) = X_t f \tag{7.3.2}$$

for every vector field X.

Differentiating vector fields: Connections on vector bundles

Differentiating vector fields is considerably more subtle on manifolds than in simple Euclidean space, and since it is crucial to all that follows, we shall spend some time discussing it. In fact, it is already a delicate issue if the manifold M is a simple surface in $\mathbb{R}^3$, and so we look at this case first.

Suppose X and Y are two vector fields, and we want to "differentiate" Y, at the point $t \in M$, in the direction X_t. Following the usual procedure, we need to "move" a distance δ in the direction X_t and compute the limit of the ratios $(Y_t - Y_{t+\delta X_t})/\delta$. There are three problems here. The first is that there is no guarantee (and it will not generally be the case) that $t + \delta X_t$ lies on M, so that $Y_{t+\delta X_t}$ is not well defined. Thus we have to find a way of "moving in the direction X_t" while staying on M. This in itself is not a problem, and can be achieved by moving along any C^1 curve c with $c(0) = t$, $\dot{c}(0) = X_t$. Then, instead of $Y_{t+\delta X_t}$, which may not be well defined, we consider $Y_{c(\delta)}$, and so need to evaluate the limit

$$\lim_{\delta \to 0} \frac{Y_{c(\delta)} - Y_t}{\delta}. \tag{7.3.3}$$

The second problem is that $Y_{c(\delta)} \in T_{c(\delta)}\mathbb{R}^3$ and $Y_t \in T_t\mathbb{R}^3$, so that the above difference is also not well defined. However, fixing a basis for $\mathbb{R}^3$, we can use the natural identifications of $T_t\mathbb{R}^3$ and $T_{c(\delta)}\mathbb{R}^3$ with $\mathbb{R}^3$ to define the difference $Y_{c(\delta)} - Y_t$ and so, under appropriate conditions on Y, to also define the limit in (7.3.3).

The third problem is clearest in the case in which Y is also a tangent vector field, and for reasons that will become clearer later we decide that we would like a notion of derivative that also yields a tangent vector in T_tM. The above construction will not necessarily do this, even when (7.3.3) exists and yields a vector in $T_t\mathbb{R}^3$. This last problem can, however, be easily solved by projecting (using the natural basis for $T_t\mathbb{R}^3$) the limit (7.3.3) onto $T_tM \subset T_t\mathbb{R}^3$.

This solves the problem of differentiating vector fields when M is embedded in a general Euclidean space, since there is really nothing special about dimension three in the above construction. We shall denote the result of the above construction by

$$\nabla_X Y \equiv (\nabla_X Y)(t),$$

and call it the *covariant derivative*[16] of Y in the direction X. Note that unless the surface M is flat, $\nabla_X Y(t)$ is quite different from the usual derivative $X_t Y_t$. In particular, while $\nabla_X Y(t) \in T_t M$, the same is not generally true of $X_t Y_t$. Furthermore, in general, $\nabla_X Y \neq \nabla_Y X$.

It is not trivial to extend the above construction to general manifolds, and so we shall adopt a path of definition by characterization, rather than by construction. As a first step, we define the notion of a (*linear*) *connection* on the tangent bundle of a manifold M as a bilinear mapping

$$\nabla : \Gamma^q(T(M)) \times \Gamma^k(T(M)) \to \Gamma^{\min(q,k-1)}(T(M))$$

satisfying the following three properties:

$$\nabla \text{ is linear over } \mathbb{R} \text{ in } Y: \qquad (7.3.4)$$
$$\nabla_X(aY_1 + bY_2) = a\nabla_X Y + b\nabla_X Y \quad \text{for } a, b \in \mathbb{R},$$

$$\nabla \text{ is linear over } \mathbb{R} \text{ in Y}: \qquad (7.3.5)$$
$$\nabla_{fX_1+gX_2} Y = f\nabla_{X_1} Y + g\nabla_{X_2} Y \quad \text{for } f, g \in C^1(M),$$

$$\nabla \text{ satisfies the product rule}: \qquad (7.3.6)$$
$$\nabla_X(fY) = (Xf)Y + f\nabla_X Y \quad \text{for } f \in C^1(M).$$

It is easy to check that the covariant derivative we defined above for the Euclidean setting satisfies all of the above three requirements. Furthermore, it is quite easy to generate connections on a general manifold.

In particular, given a chart (U, φ) on an N-dimensional manifold M, suppose that we have N^3 functions $\{\Gamma_{ij}^k : U \to \mathbb{R}\}_{1\le i,j,k\le N}$. If, using the natural bases for the $T_t M$, we define

$$\nabla_{\frac{\partial}{\partial x_i}} \frac{\partial}{\partial x_j} = \sum_{k=1}^{N} \Gamma_{ij}^k \frac{\partial}{\partial x_k}, \qquad (7.3.7)$$

and extend ∇ to $T(M) \times T(M)$ by linearity, then it is easy to check that the three conditions required of a connection are satisfied on U. It is then standard fare to patch charts together to construct a connection on the full manifold, given enough functions Γ_{ij}^k. In other words, if we can determine the Γ_{ij}^k for a given connection, then we have determined the connection. Note, however, that there is uniqueness here in only one direction. It is clear that the Γ_{ij}^k determine a connection. It is neither clear nor in general true that on a given Riemannian manifold there is only one connection.

For uniqueness, we need to demand a little more. We start by noticing that the construction we gave of the Euclidean connection, with its projection onto tangent spaces, actually determines a unique set of Γ_{ij}^k. This follows from properties (7.3.4)–(7.3.6), the representation (7.3.7), and the following two consequences of the construction:

[16] Hopefully, the double usage of ∇ for both gradient and the covariant derivative (and, soon, the Riemannian connection) will not cause too many difficulties. Note that, like the gradient, the connnection "knows" about the Riemannian metric g. The reason for this will become clear soon.

$$\nabla_{\frac{\partial}{\partial x_i}} \frac{\partial}{\partial x_j} = \nabla_{\frac{\partial}{\partial x_j}} \frac{\partial}{\partial x_i}, \tag{7.3.8}$$

$$\frac{\partial}{\partial x_i}\left\langle \frac{\partial}{\partial x_j}, \frac{\partial}{\partial x_k} \right\rangle = \left\langle \nabla_{\frac{\partial}{\partial x_i}} \frac{\partial}{\partial x_j}, \frac{\partial}{\partial x_k} \right\rangle + \left\langle \nabla_{\frac{\partial}{\partial x_i}} \frac{\partial}{\partial x_k}, \frac{\partial}{\partial x_j} \right\rangle, \tag{7.3.9}$$

where $\langle \, , \rangle$ is the usual Euclidean inner product.

Indeed, the above two equations actually determine the connection that we constructed, without any need to go back to the construction. The connection resulting from these equations is now called the *Levi-Civitá connection* for M. The advantage of using these equations over our previous construction is that they do not depend on the embedding of M in $\mathbb{R}^l$ and so present a purely "intrinsic" (to M) approach.

This approach also works for general C^3 Riemannian manifolds (M, g), and yields the Levi-Civitá connection determined by the following two conditions, which extend (7.3.8) and (7.3.9) to the non-Euclidean scenario:

$$\begin{aligned} &\nabla \text{ is } \textit{torsion-free}, \text{ i.e.,} \\ &\qquad \nabla_X Y - \nabla_Y X - [X, Y] = 0; \end{aligned} \tag{7.3.10}$$

$$\begin{aligned} &\nabla \text{ is } \textit{compatible} \text{ with the metric } g, \text{ i.e.,} \\ &\qquad Xg(Y, Z) = g(\nabla_X Y, Z) + g(\nabla_X Z, Y), \end{aligned} \tag{7.3.11}$$

where $[X, Y]f = XYf - YXf$ is the so-called *Lie bracket*[17] of the vector fields X and Y.[18]

It is a matter of simple algebra to derive from (7.3.10) and (7.3.11) that for C^1 vector fields X, Y, Z,

$$\begin{aligned} 2g(\nabla_X Y, Z) = {} & Xg(Y, Z) + Yg(X, Z) - Zg(X, Y) \\ & + g(Z, [X, Y]) + g(Y, [Z, X]) + g(X, [Z, Y]). \end{aligned} \tag{7.3.12}$$

This equation is known as *Koszul's formula*. Its importance lies in the fact that the right-hand side depends only on the metric g and differential-topological notions. Consequently, it gives a coordinate-free formula that actually determines ∇.

Another way to determine the Levi-Civitá connection is via an extension of (7.3.7), which was written in terms of the natural bases of the $T_t M$. For this, we need the notion of a C^k *orthonormal frame field* $\{E_i\}_{1 \le i \le N}$, which is a C^k section of the orthonormal (with respect to g) frame bundle $\mathcal{O}(M)$. Then, given a connection ∇, an orthonormal frame field E on M, and defining N^3 functions Γ_{ij}^k by the relationships

$$\nabla_{E_i} E_j = \Gamma_{ij}^k E_k, \tag{7.3.13}$$

---

[17] The Lie bracket measures the failure of partial derivatives to commute, and is always zero in the Euclidean case.

[18] Actually, (7.3.11) is a little misleading, and would be more consistent with what has gone before if it were written as

$$X_t g(Y_t, Z_t) = g_t((\nabla_X Y)_t, Z_t) + g_t(Y_t, (\nabla_X Z)_t).$$

The preponderence of t's here is what leads to the shorthand of (7.3.11).

determines the connection by setting

$$\nabla_{E_i} E_j = \sum_{k=1}^{N} \Gamma_{ij}^{k} E_k,$$

and extending by linearity. The functions Γ_{ij}^{k} are known as the *Christoffel symbols* (of the second kind) of the connection ∇ (with respect to the given orthonormal frame field).

Yet another way to determine the Levi-Civitá connection, which is essentially a rewriting of the previous paragraph, is to define a collection of N^2 1-forms $\{\theta_i^j\}_{i,j=1}^{N}$, known as *connection forms*, via the requirement that

$$\nabla_X E_i = \sum_{j=1}^{N} \theta_i^j(X) E_j. \tag{7.3.14}$$

Since the connection forms must satisfy

$$\theta_i^j(X) = g\left(\nabla_X E_i, E_j\right),$$

the importance of Koszul's formula for determining the connection is now clear. We shall see in detail how to compute the θ_i^j for some examples in Section 7.7. In general, they are determined by (7.3.19) below, in which $\{\theta_i\}_{1\le i\le N}$ denotes the orthonormal *dual frame* corresponding to $\{E_i\}_{1\le i\le N}$. To understand (7.3.19) we need one more concept, that of the (*exterior*) *differential* of a k-form.

If $f : M \to \mathbb{R}$ is C^1, then its *exterior derivative* or *differential*, df, is the 1-form defined by

$$df(E_{it}) = f_i(t) \stackrel{\Delta}{=} E_{it} f. \tag{7.3.15}$$

Alternatively, df is just the push-forward of $f : M \to \mathbb{R}$. Consequently, we can write df, using the dual frame, as

$$df = \sum_{i=1}^{N} f_i \theta_i.$$

If $\theta = \sum_{i=1}^{N} h_i \theta_i$ is a 1-form, then its exterior derivative is the 2-form

$$d\theta \stackrel{\Delta}{=} \sum_{i=1}^{N} dh_i \wedge \theta_i. \tag{7.3.16}$$

Note that the exterior derivative of a 0-form (i.e., a function) gave a 1-form, and that of a 1-form gave a 2-form. There is a general notion of exterior differentiation, which in general takes k-forms to $(k+1)$-forms,[19] but which we shall not need.

[19] This is why we used the terminology of "differential forms" when discussing Grassmann bundles at the end of Section 7.1.

We do now, however, have enough to define the 1-forms θ_i^j of (7.3.14). They are the unique set of N^2 1-forms satisfying the following two requirements:

$$d\theta^i - \sum_{j=1}^{N} \theta^j \wedge \theta_j^i = 0, \tag{7.3.17}$$

$$\theta_i^j + \theta_j^i = 0. \tag{7.3.18}$$

The Riemannian metric g is implicit in these equations in that, as usual, it determines the notion of orthonormality and so the choice of the θ_i. In fact, (7.3.17) is a consequence of the compatibility requirement (7.3.11), while (7.3.18) is a consequence of the requirement (7.3.10) that the connection be torsion-free.

Since it will occasionally be useful, we note that (7.3.18) takes the following form if the tangent frame and its dual are not taken to be orthonormal:

$$dg_{ij} = \sum_{k=1}^{N} \left(\theta_k^i g_{kj} + \theta_k^j g_{ki} \right), \tag{7.3.19}$$

where $g_{ij} = (E_i, E_j) = g(E_i, E_j)$.

Finally, we note that if $M = \mathbb{R}^N$ and g is the usual Euclidean inner product, then it is easy to check that $\nabla_X Y$ is no more than the directional derivative of the vector field Y in the directions given by X. In this case $\nabla_X Y$ is known as the *flat connection*, and the Christoffel symbols are all identically zero.

Exponential map and geodesics

As we noted earlier, a geodesic from t to s on (M, g) is a curve that locally minimizes arc length among all curves between t and s. It is straightforward to show that a geodesic $c : (-\delta, \delta) \to M$ satisfies

$$\nabla_{\dot{c}} \dot{c} = 0.$$

Hence, for every $\dot{c}(0) \in T_t M$, there is a unique geodesic leaving t in the direction $\dot{c}(0)/|\dot{c}(0)|$, obtained by solving the above second-order differential equation.

This procedure is formalized by the exponential map

$$\exp \equiv \exp_{(M,g)} : T_t M \to M \tag{7.3.20}$$

of a Riemannian manifold (M, g), which sends a tangent vector $X_t \in T_t M$ to the geodesic of length $|X_t|$ in the direction $X_t/|X_t|$.

The exponential map also allows us to choose a special coordinate system for charts. An orthonormal basis $\{E_i\}$ for $T_t M$ gives an isomorphism $E : \mathbb{R}^N \to T_t M$ by $E(x_1, \ldots, x_N) = \sum_i x_i E_i$. If U is a small enough neighborhood of t in M we can use E to define the coordinate chart

$$\varphi_t \stackrel{\Delta}{=} E^{-1} \circ \exp^{-1} : U \to \mathbb{R}^N. \tag{7.3.21}$$

Any such coordinates are called *(Riemannian) normal coordinates* centered at t.

Hessian of $f \in C^2((M, g))$

The (*covariant*) *Hessian* $\nabla^2 f$ of a function $f \in C^2(M)$ on a Riemannian manifold (M, g) is the bilinear symmetric map from $C^1(T(M)) \times C^1(T(M))$ to $C^0(M)$ (i.e., it is a double differential form) defined by[20]

$$\nabla^2 f(X, Y) \triangleq XYf - \nabla_X Yf = g(\nabla_X \nabla f, Y), \tag{7.3.22}$$

where ∇_X is the Levi-Civitá connection of (M, g). The equality here is a consequence of (7.3.11) and (7.3.2).

Note the obvious but important point that if t is such that $\nabla f(t) = 0$,[21] then $Xf(t) = 0$ for all $X \in T_t M$ (cf. (7.3.2)), and so by (7.3.22) it follows that $\nabla^2 f(X, Y)(t) = XYf(t)$. Consequently, at these points the Hessian is independent of the metric g.

7.4 Integration on Manifolds

We now have the tools and vocabulary to start making mathematical sense out of our earlier comments linking tensors and differential forms to volume computations. However, rather than computing only volumes, we shall need general measures over manifolds. Since a manifold M is a locally compact topological space, there is no problem defining measures over it, and by Riesz's theorem, these are positive, bounded linear functionals on $C_c^0(M)$, the c denoting compact support. This description, however, does not have much of a geometric flavor to it, and so we shall take a different approach. The branch of mathematics of "adding a geometric flavor" to this description is known, appropriately, as *geometric measure theory*. A user-friendly introduction to this beautiful area of mathematics can be found in Morgan's book [115]. The definitive treatment, however, is still to be found in Federer's treatise [65]. We shall not, however, have need for much of this rather heavy machinery.

Integration of differential forms

Recall that for any two manifolds M and N, a C^1 map $g : M \to N$ induces a mapping $g_* : T(M) \to T(N)$, referred to as the push-forward (cf. (7.1.3)). Just as g_* replaces the chain rule of vector calculus, another related mapping $g^* : \Lambda^*(T(N)) \to \Lambda^*(T(M))$, called the *pullback*, replaces the change of variables formula. Actually, the pullback is more than just a generalization of the change of variables formula, although for our purposes it will suffice to think of it that way. We define g^* on $(k, 0)$-tensors by

[20] $\nabla^2 f$ could also have been defined to be $\nabla(\nabla f)$, in which case the first relationship in (7.3.22) becomes a consequence rather than a definition. Recall that in the simple Euclidean case the Hessian is defined to the $N \times N$ matrix $H_f = (\partial^2 f / \partial x_i \partial x_j)_{i,j=1}^N$. Using H_f to define the two-form $\nabla^2 f(X, Y) = X H_f Y'$, (7.3.22) follows from simple calculus.

[21] These are the *critical points* of f, and will play a central role in the Morse theory of Chapter 9 and throughout Part III.

$$(g^*\alpha)(X_1, \ldots, X_k) \stackrel{\Delta}{=} \alpha(g_* X_1, \ldots, g_* X_k). \tag{7.4.1}$$

The pullback has many desirable properties, the main ones being that it commutes with both the wedge product of forms and the exterior derivative d (cf. (7.3.16)).

With the notion of a pullback defined, we can add one more small piece of notation. Take a chart (U, φ) of M, and recall the notation of (7.1.5) in which we used $\{\partial/\partial x_i\}_{1 \le i \le N}$ to denote both the natural Euclidean basis of $\varphi(U)$ and its push-forward to $T(U)$. We now do the same with the notation

$$\{dx_1, \ldots, dx_N\}, \tag{7.4.2}$$

which we use to denote both the natural dual coordinate basis in $\mathbb{R}^N$ (so that $dx_i(v) = v_i$) and its pullback under φ.

Now we can start defining integration, all of which hinges on a single definition: If $A \subset \mathbb{R}^N$, and $f : A \to \mathbb{R}$ is Lebesgue integrable over A, then we define

$$\int_A f\, dx_1 \wedge \cdots \wedge dx_N \stackrel{\Delta}{=} \int_A f(x)\, dx, \tag{7.4.3}$$

where $\{dx_i\}_{1 \le i \le N}$, as above, is the natural dual coordinate basis in $\mathbb{R}^N$, and the right-hand side is simply Lebesgue integration.

Since the wedge products in the left-hand side of (7.4.3) generate all N-forms on $\mathbb{R}^N$ (cf. (7.2.2)), this and additivity solves the problem of integrating N-forms on $\mathbb{R}^N$ in full generality.

Now we turn to manifolds. For a given chart, (U, φ), and an N-form α on U, we define the integral of α over $V \subset U$ as

$$\int_V \alpha \stackrel{\Delta}{=} \int_{\varphi(V)} \left(\varphi^{-1}\right)^* \alpha, \tag{7.4.4}$$

where the right-hand side is defined by virtue of (7.4.3).

To extend the integral beyond single charts, we require a new condition, that of *orientability*. A C^k manifold M is said to be *orientable* if there is there is an atlas $\{U_i, \varphi_i\}_{i \in I}$ for M such that for any pair of charts (U_i, φ_i), (U_j, φ_j) with $U_i \cap U_j \ne \emptyset$, the Jacobian of the map $\varphi_i \circ \varphi_j^{-1}$ has a positive determinant. For orientable manifolds it is straightforward to extend the integral (7.4.4) to general domains by additivity. This leads to the notion of an *integrable N-form*, essentially any N-form for which (7.4.4) can be extended to all of M as a σ-finite signed measure. In particular, every integrable form α determines such a measure, μ_α say, by setting

$$\mu_\alpha(A) = \int_A \alpha. \tag{7.4.5}$$

Given an oriented manifold M with atlas as above, one can also define the notion of an *oriented (orthonormal) frame field* over M. This is a C^0 (orthonormal) frame field $\{E_{1t}, \ldots, E_{Nt}\}$ over M for which, for each chart (U, φ) in the atlas, and at each

$t \in U$, the push-forward $\{\varphi_* E_{1t}, \dots, \varphi_* E_{Nt}\}$ can be transformed to the standard basis of $\mathbb{R}^N$ via a transformation with positive determinant.

Given an oriented orthonormal frame field, there is a unique volume form, which we denote either by Ω or by Vol_g, the latter if we want to emphasize the dependence on the metric g, which plays the role of Lebesgue measure on M and which is defined by the requirement that

$$(\mathrm{Vol}_g)_t(E_{1t}, \dots, E_{Nt}) \equiv \Omega_t(E_{1t}, \dots, E_{Nt}) = +1. \tag{7.4.6}$$

From this, and the fact that Ω is a differential form, it follows that for $X_i \in T_t M$,

$$(\mathrm{Vol}_g)_t(X_1, \dots, X_n) = \det(\{g_t(X_i, E_{jt})\}_{1 \le i,j \le N}). \tag{7.4.7}$$

The integral of Ω is comparatively easy to compute. For a fixed (oriented) chart (U, φ), with natural basis $\{\partial/\partial x_i\}_{1 \le i \le N}$, write

$$g_{ij}(t) = g_t\left(\frac{\partial}{\partial x_i}, \frac{\partial}{\partial x_j}\right), \tag{7.4.8}$$

where g is the Riemannian metric. This determines, for each t, a positive definite matrix $(g_{ij}(t))_{1 \le i,j \le N}$. Then, for $A \subset U$,

$$\begin{aligned} \int_A \Omega &\equiv \int_{\varphi(A)} (\varphi^{-1})^* \Omega \\ &= \int_{\varphi(A)} \sqrt{\det(g_{ij} \circ \varphi^{-1})}\, dx_1 \wedge \cdots \wedge dx_N \\ &= \int_{\varphi(A)} \sqrt{\det(g_{ij} \circ \varphi^{-1})}\, dx, \end{aligned} \tag{7.4.9}$$

where the crucial intermediate term comes from (7.4.3)–(7.4.8) and some algebra.

Under orientability, both this integral and that in (7.4.4) can be extended to larger subsets of M by additivity.[22] This yields a σ-finite measure μ_Ω associated to Ω, called *Riemannian measure*, which, with some doubling up of notation, we shall also write as Vol_g when convenient.

For obvious reasons, we shall often write the volume form Ω as $dx_1 \wedge \cdots \wedge dx_N$, where the 1-forms dx_i are the (dual) basis of $T^*(M)$.

An important point to note is that Vol_g agrees with the N-dimensional Hausdorff measure[23] associated with the metric τ induced by g, and so we shall also often write

[22] The last line of (7.4.9), when written out in longhand, should be familiar to most readers as an extension of the formula giving the "surface area" of a regular N-dimensional surface. The extension lies in the fact that an arbitrary Riemannian metric now appears, whereas in the familiar case there is only one natural candidate for g.

[23] If M is an N-dimensional manifold, treat (M, τ) as a metric space, where τ is the geodesic metric given by (7.3.1). The diameter of a set $S \subset M$ is then $\mathrm{diam}(S) = \sup\{\tau(s,t) : s, t \in S\}$, and for integral n, the Hausdorff n-measure of $A \subset M$ is defined by

it as $\mathcal{H}_N$. In this case we shall usually write integrals as $\int_M h(t)\, d\mathcal{H}_N(t)$ rather than $\int_M h\mathcal{H}_N$, which would be more consistent with our notation so far.

We now return to the issue of orientability. In setting up the volume form Ω, we first fixed an orthonormal frame field and then demanded that $\Omega_t(E_{1t}, \dots, E_{Nt}) = +1$ for all $t \in M$ (cf. (7.4.6)). We shall denote M, along with this orientation, by M^+. However, there is another orientation of M for which $\Omega = -1$ when evaluated on the orthonormal basis. We write this manifold as M^-. On an orientable manifold, there are only two such possibilities.

In fact, it is not just Ω that can be used to determine an orientation. Any nonvanishing continuous N-form α on an orientable manifold can be used to determine one of two orientations, with the orientation being determined by the sign of α on the orthonormal basis at any point on M. We can thus talk about the "orientation induced by α."

With an analogue for Lebesgue measure in hand, we can set up the analogues of Borel measurability and (Lebesgue) integrability in the usual ways. Furthermore, it follows from (7.4.3) that there is an inherent smoothness[24] in the above construction. In particular, if M is compact, then for any continuous, nonvanishing N-form α that induces the same orientation on M as does Ω, there is an Ω-integrable function $d\alpha/d\Omega$ for which

$$\int_M \alpha = \int_M \frac{d\alpha}{d\Omega}\Omega.$$

For obvious reasons, $d\alpha/d\Omega$ is called the *Radon–Nikodym derivative* of α.

From now on, and without further comment, *we shall assume that all manifolds in this book are orientable, and the frame fields and orientation chosen so as the make the volume form positive.*

Coarea formula

The following important result, due to Federer [64] and known as his *coarea formula*, allows us to break up integrals over manifolds into iterated integrals over submanifolds of lower dimension. Consider a differentiable map $f : M \to N$ between two

$$\mathcal{H}_n(A) \overset{\Delta}{=} \omega_n 2^{-n} \liminf_{\varepsilon \downarrow 0} \sum_i (\text{diam } B_i)^n,$$

where for each $\varepsilon > 0$, the infimum is taken over all collections $\{B_i\}$ of open τ-balls in M whose union covers A and for which $\text{diam}(B_i) \le \varepsilon$. As usual, ω_n is the volume of the unit ball in $\mathbb{R}^n$. For the moment, we need only the case $n = N$. When both are defined, and the underlying metric is Euclidean, Hausdorff and Lebesgue measures agree.

Later we shall need a related concept, that of *Hausdorff dimension*. If $A \subset M$, the Hausdorff dimension of A is defined as

$$\dim(A) \overset{\Delta}{=} \inf\left\{\alpha : \liminf_{\varepsilon \downarrow 0} \sum_i (\text{diam } B_i)^\alpha = 0\right\},$$

where the conditions on the infimum and the B_i are as above.

[24] This smoothness implies, for example, that N-forms cannot be supported, as measures, on lower-dimensional submanifolds of M.

Riemannian manifolds, with $m = \dim(M) \geq n = \dim(N)$. At each $t \in M$ its push-forward, $f_*(t)$, is a linear map between two Hilbert spaces and can be written as

$$\sum_{j=1}^{n} \tilde{\theta}_j(f(t)) \sum_{i=1}^{m} A_{ij}^{f}(t)\theta_i(t) \tag{7.4.10}$$

for orthonormal bases $\{\theta_j(t)\}_{i=1}^m$ of $(T_tM)^*$ and $\{\tilde{\theta}_j(f(t))\}_{j=1}^n$ of $T_{f(t)}N$.

Writing $\{\partial/\partial t_1, \dots, \partial/\partial t_m\}$ for the corresponding basis of T_tM, the A_{ij}^f are given by

$$A_{ij}^{f}(t) = (\partial/\partial t_i) f_j(t). \tag{7.4.11}$$

Similarly, the pullback f^*, evaluated at $t \in M$, is a linear map from $\Lambda^*(T_{f(t)}N)$ to $\Lambda^*(T_tM)$. If we restrict $f^*(t)$ to $\Lambda^n(T_{f(t)}N)$, it can be identified with an element in the linear span of $\Lambda^n(T_{f(t)}N) \times \Lambda^n(T_tM)$.

Both the push-forward and the pullback share the same norm, which we write as $Jf(t)$ and which can be calculated by taking the square root of the sum of squares of the determinants of the $n \times n$ minors of the matrix A^f of (7.4.10). Alternatively, it is given by

$$Jf(t) \stackrel{\Delta}{=} \sqrt{\det\left(\langle \nabla f_i(t), \nabla f_j(t)\rangle\right)}. \tag{7.4.12}$$

Federer's coarea formula [64] states that for differentiable[25] f and for $g \in L^1(M, \mathcal{B}(M), \mathcal{H}_m)$,

$$\int_M g(t) Jf(t)\, d\mathcal{H}_m(t) = \int_N d\mathcal{H}_n(u) \int_{f^{-1}(u)} g(s)\, d\mathcal{H}_{m-n}(s). \tag{7.4.13}$$

Simplifying matters a little bit, consider two special cases. If $M = \mathbb{R}^N$ and $N = \mathbb{R}$, it is easy to see that $Jf = |\nabla f|$, so that

$$\int_{\mathbb{R}^N} g(t)|\nabla f(t)|\, dt = \int_{\mathbb{R}} du \int_{f^{-1}\{u\}} g(s)\, d\mathcal{H}_{N-1}(s). \tag{7.4.14}$$

There is not a great deal of simplification in this case beyond the fact that it is easy to see what the functional J is. On the other hand, if $M = N = \mathbb{R}^N$, then $Jf = |\det \nabla f|$, and

$$\begin{aligned} \int_{\mathbb{R}^N} g(t)|\det \nabla f(t)|\, dt &= \int_{\mathbb{R}^N} du \int_{f^{-1}\{u\}} g(s)\, d\mathcal{H}_0(s) \\ &= \int_{\mathbb{R}^N} \left(\sum_{t: f(t)=u} g(t) \right) du. \end{aligned} \tag{7.4.15}$$

We shall return to (7.4.15) in Section 11.4 and to the coarea formula in general in Chapter 15, where it will play a very important role in our calculations.

[25] Federer's setting is actually somewhat more general than this, since he works with Lipschitz mappings. In doing so he replaces derivatives with "approximate derivatives" throughout.

7.5 Curvature

We now come to what is probably the most central of all concepts in differential geometry, that of curvature. In essence, much of what we have done so far in developing the calculus of manifolds can be seen as no more than setting up the basic tools for handling the ideas to follow.

Curvature is the essence that makes the local properties of manifolds inherently different from simple Euclidean space, where curvature is always zero. Since there are many very different manifolds, and many different Riemannian metrics, there are a number of ways to measure curvature. In particular, curvature can be measured in a somewhat richer fashion for manifolds embedded in ambient spaces of higher dimension than it can for manifolds for which no such embedding is given.[26] A simple example of this is given in Figure 7.5.1, where you should think of the left-hand circle S^1 as being embedded in the plane, while the right-hand circle exists without any embedding. In the embedded case there are notions of "up" and "down," with the two arrows at the top and bottom of the circle pointing "up." In one case the circle curves "away" from the arrow, in the other, "toward" it, so that any reasonable treatment of curvature has to be able to handle this difference. However, for the nonembedded case, in which there is nothing external to the circle, the curvature must be the same everywhere. In what follows, we shall capture the first, richer, notion of curvature via the *second fundamental form* of the manifold, and the second via its *curvature tensor*, which is related to the second fundamental form, when both are defined, via the Gauss equation (7.5.9) below.

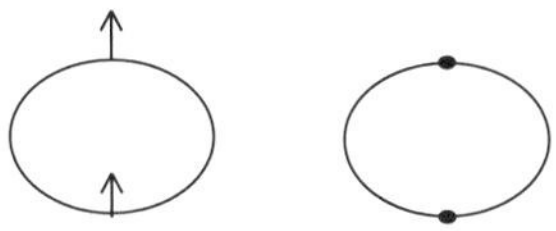

Fig. 7.5.1. Embedded and intrinsic circles.

---

[26] We have used this term often already, albeit in a descriptive sense. The time has come to define it properly: Suppose $f : M \to \widehat{M}$ is C^1. Take $t \in M$ and charts (U, φ) and (V, ψ) containing t and $f(t)$, respectively. The *rank* of f at t is defined to be the rank of the mapping $\psi \circ f \circ \varphi^{-1} : \varphi(U) \to \psi(V)$ between Euclidean spaces. If f is everywhere of rank $\dim M$, then it is called an *immersion*. If $\dim M = \dim \widehat{M}$, then it is called a *submersion*. Note that this is a purely local property of M.

If, furthermore, f is a one-to-one homeomorphism of M onto its image $f(M)$ (with its topology as a subset of $\widehat{M}$), then we call f an *embedding* of M in $\widehat{M}$ and refer to M as an *embedded (sub)manifold* and to $\widehat{M}$ as the *ambient* manifold. This is a global property, and amounts to the fact that M cannot "intersect" itself on $\widehat{M}$.

Finally, let M and $\widehat{M}$ be Riemannian manifolds with metrics g and $\widehat{g}$, respectively. Then we say that (M, g) is an *isometrically embedded Riemannian manifold* of $(\widehat{M}, \widehat{g})$ if, in addition to the above, $g = f^*\widehat{g}$, where $f^*\widehat{g}$ is the pullback of $\widehat{g}$ (cf. (7.4.1)).

Riemannian curvature tensor

We start with the (Riemannian) curvature tensor. While less informative than the second fundamental form, the fact that it is *intrinsic* (i.e., it is uniquely determined by the Riemannian metric) actually makes it a more central concept to Riemannian geometry.

Much in the same way that the Lie bracket $[X, Y] = XY - YX$ measures the failure of partial derivatives to commute, the Riemannian curvature tensor, R, measures the failure of covariant derivatives to commute. A relatively simple computation shows that for vector fields X, Y it is not generally true that $\nabla_X \nabla_Y - \nabla_Y \nabla_X = 0$. However, rather than taking this difference as a measure of curvature, to ensure linearity, it is convenient to define the *(Riemannian) curvature operator*[27] as

$$R(X, Y) \stackrel{\Delta}{=} \nabla_X \nabla_Y - \nabla_Y \nabla_X - \nabla_{[X,Y]}. \tag{7.5.1}$$

Note that if $[X, Y] = 0$, as is the case when X_t and Y_t are coordinate vectors in the natural basis of some chart, then $R(X, Y) = \nabla_X \nabla_Y - \nabla_Y \nabla_X$, and so the operator R is the first measure of lack of commutativity of ∇_X and ∇_Y mentioned above.

The *(Riemannian) curvature tensor*, also denoted by R, is defined by

$$\begin{aligned} R(X, Y, Z, W) &\stackrel{\Delta}{=} g\left(\nabla_X \nabla_Y Z - \nabla_Y \nabla_X Z - \nabla_{[X,Y]} Z,\ W\right) \\ &= g(R(X, Y)Z,\ W), \end{aligned} \tag{7.5.2}$$

where the R in the second line is, obviously, the curvature operator. It is easy to check that for $\mathbb{R}^N$, equipped with the standard Euclidean metric, $R \equiv 0$.[28]

The definition (7.5.2) of R is not terribly illuminating, although one can read it as "the amount, in terms of g and in the direction W, by which ∇_X and ∇_Y fail to commute when applied to Z." To get a better idea of what is going on, it is useful to think of planar sections of $T_t M$.

For any $t \in M$, we call the span of two linearly independent vectors $X_t, Y_t \in T_t M$ the *planar section* spanned by X_t and Y_t, and denote it by $\pi(X_t, Y_t)$. Such a planar section is determined by any pair of orthonormal vectors E_{1t}, E_{2t} in $\pi(X_t, Y_t)$, and

$$\kappa(\pi) \stackrel{\Delta}{=} -R\left(E_{1t}, E_{2t}, E_{1t}, E_{2t}\right) = R\left(E_{1t}, E_{2t}, E_{2t}, E_{1t}\right) \tag{7.5.3}$$

is called the *sectional curvature* of the planar section. It is independent of the choice of basis. Sectional curvatures are somewhat easier to understand than the curvature tensor, but essentially equivalent, since it is easy to check from the symmetry properties of the curvature tensor that it is uniquely determined by the sectional curvatures.

We shall later need a further representation of R, somewhat reminiscent of the representation (7.3.14) for the Riemannian connection. The way that R was defined

---

[27] Note that the curvature operator depends on the underlying Riemannian metric g via the dependence of the connection on g.

[28] Manifolds for which $R \equiv 0$ are called *flat spaces*. However, not only Euclidean space is flat. It is easy to check, for example, that the cylinder $S^1 \times \mathbb{R}$ is also flat when considered as a manifold embedded in $\mathbb{R}^3$ with the usual Euclidean metric.

in (7.5.2) shows that it is clearly a covariant tensor of order 4. However, it is not a difficult computation, based on (7.3.14), to see that it can also be expressed as a symmetric double form of order (2, 2). In particular, if g is C^2 and $\{\theta_i\}_{1\le i\le N}$ is the dual of a C^2 orthonormal frame field, then $R \in C^0(\Lambda^{2,2}(M))$, and we can write

$$R = \frac{1}{2}\sum_{i,j=1}^{N} \Omega_{ij} \otimes \left(\theta_i \wedge \theta_j\right), \tag{7.5.4}$$

where the Ω_{ij} are skew-symmetric C^0 differential 2-forms ($\Omega_{ij} = -\Omega_{ji}$) known as the *curvature forms* for the section $\{E_i\}_{1\le i\le N}$ and are defined by

$$\Omega_{ij}(X, Y) = R\left(E_i, E_j, X, Y\right) \tag{7.5.5}$$

for vector fields X, Y.

Projection formula

Although it is more a result from linear algebra than differential geometry, since we are about to start talking about projections, and these will appear frequently throughout the book, we recall the following:

Suppose v is an element in a vector space V with inner product $\langle \cdot, \cdot \rangle$, and $w_1, \dots, w_n$ are vectors in V. Then the projection of v onto the span of the w_j is given by

$$P_{\text{span}(w_1,\dots,w_n)} v = \sum_{i,j=1}^{n} \langle w_i, v\rangle g^{ij} w_j, \tag{7.5.6}$$

where the g^{ij} are the elements of G^{-1}, and G is the matrix with elements $g_{ij} = \langle w_i, w_j\rangle$.

Second fundamental form

Let (M, g) be a Riemannian manifold embedded in an ambient Riemannian manifold $(\widehat{M}, \widehat{g})$ of codimension[29] at least one. Write ∇ for the Levi-Civitá connection on M and $\widehat{\nabla}$ for the Levi-Civitá connection on $\widehat{M}$. If $t \in M$, then the *normal space* to M in $\widehat{M}$ at t is

$$T_t^{\perp} M \stackrel{\Delta}{=} \{X_t \in T_t\widehat{M} : \widehat{g}_t(X_t, Y_t) = 0 \text{ for all } Y_t \in T_t M\}, \tag{7.5.7}$$

and we also write $T^{\perp}(M) = \bigcup_t T_t^{\perp} M$. Note that since $T_t\widehat{M} = T_t M \oplus T_t^{\perp} M$ for each $t \in M$, it makes sense to talk about tangential and normal components of an element of $T_t\widehat{M}$ and so of the orthogonal projections

$$P_{T(M)} : \pi_{\widehat{M}}^{-1} M \to T(M), \qquad P_{T(M)}^{\perp} : \pi_{\widehat{M}}^{-1} M \to T^{\perp}(M),$$

[29] The *codimension* of M in $\widehat{M}$ is $\dim(\widehat{M}) - \dim(M)$.

where $\pi_{\widehat{M}}$ is the canonical projection on the tangent bundle $T(\widehat{M})$, and $T(M)$ and $T^{\perp}(M)$ are embedded subbundles of $T(\widehat{M})$.

The *second fundamental form* of M in $\widehat{M}$ can now be defined to be the operator S from $T(M) \times T(M)$ to $T^{\perp}(M)$ satisfying

$$S(X, Y) \triangleq P_{T(M)}^{\perp}\left(\widehat{\nabla}_X Y\right) = \widehat{\nabla}_X Y - \nabla_X Y, \tag{7.5.8}$$

where the equality here is known as *Gauss's formula*.

When (M, g) isometrically embedded in $(\widehat{M}, \widehat{g})$ there is a useful relation between the second fundamental form and the curvature tensors R of (M, g) and $\widehat{R}$ of $(\widehat{M}, \widehat{g})$, known as the *Gauss equation*, and given by

$$\begin{aligned} &\widehat{g}(S(X, Z), S(Y, W)) - \widehat{g}(S(X, W), S(Y, W)) \\ &\quad = 2\left(\widehat{R}(X, Y, Z, W) - R(X, Y, Z, W)\right). \end{aligned} \tag{7.5.9}$$

In the special case that $\widehat{M}$ is a manifold with smooth boundary $\partial\widehat{M} = M$, the above simplifies to

$$S^2((X, Y), (Z, W)) = -2\left(R^{\partial\widehat{M}}(X, Y, Z, W) - R^{\widehat{M}}(X, Y, Z, W)\right). \tag{7.5.10}$$

Now let ν denote a unit normal vector field on M, so that $\nu_t \in T_t^{\perp}M$ for all $t \in M$. Then the *scalar second fundamental form* of M in $\widehat{M}$ for ν is defined, for $X, Y \in T(M)$, by

$$S_\nu(X, Y) \triangleq \widehat{g}\left(S(X, Y), \nu\right), \tag{7.5.11}$$

where the internal S on the right-hand side refers to the second fundamental form (7.5.8). Note that despite its name, the scalar fundamental form is not a differential form, since it is symmetric (rather than alternating) in its arguments. When there is no possibility of confusion we shall drop the qualifier "scalar" and refer also to S_ν as the second fundamental form.

In view of the fact that $\widehat{\nabla}$ is compatible with the metric g (cf. (7.3.11)), we also have the so-called *Weingarten equation* that for $X, Y \in T(M)$ is given by

$$S_\nu(X, Y) = \widehat{g}(\widehat{\nabla}_X Y, \nu) = -\widehat{g}(Y, \widehat{\nabla}_X \nu). \tag{7.5.12}$$

As we already noted, S_ν is a symmetric 2-tensor, so that as for the curvature tensor, we can view it as a symmetric section of $\Lambda^{*,*}(M)$. As such, it contains a lot of information about the embedding of M in $\widehat{M}$ and thus curvature information about M itself. For example, fix ν and, doubling up a little on notation, use the second fundamental form to define an operator $\mathcal{S}_\nu : T(M) \to T(M)$ by

$$\widehat{g}(\mathcal{S}_\nu(X), Y) \triangleq S_\nu(X, Y),$$

for all $Y \in T(M)$. Then $\mathcal{S}_\nu$ is known as the *shape operator*. It has N real eigenvalues, known as the *principal curvatures* of M in the direction ν, and the corresponding eigenvectors are known as the *principal curvature vectors*.

All of the above becomes quite familiar and particularly useful if M is a simple surface determined by the graph of $f : \mathbb{R}^N \to \mathbb{R}$, with the usual Euclidean metric. In this case, the principal curvatures are simply the eigenvalues of the Hessian $(\partial^2 f/\partial x_i \partial x_j)_{i,j=1}^N$ (cf. (7.3.22)). In particular, if (M, g) is a surface in $\mathbb{R}^3$ with the induced Euclidean metric, then the product of these eigenvalues is known as *Gaussian curvature* and it is Gauss's *Theorema Egregium* that it is an isometry invariant of (M, g), i.e., it is independent of the embedding of M in $\mathbb{R}^3$. In the next section we shall see that the Gaussian curvature is not unique as far as isometric invariance goes and that there are also other invariants, which can be obtained as integrals of mixed powers of curvature and second fundamental forms over manifolds.

7.6 Intrinsic Volumes for Riemannian Manifolds

In Section 6.3, in the context of integral geometry, we described the notions of *intrinsic volumes* and *Minkowski functionals* and how they could be used, via Steiner's formula (6.3.3), to find an expression for the tube around a convex Euclidean set. In Chapter 10 we shall extend these results considerably, looking at tubes around submanifolds of some larger manifold. In doing so we shall encounter extensions of intrinsic volumes known as Lipschitz–Killing curvatures, and study them in some depth.

However, while the detailed calculations of Chapter 10 are not needed for Part III, the same is not true of Lipschitz–Killing curvatures, which will have a crucial role to play there. Hence we define them now, with no further motivation.

Let (M, g) be a C^2 Riemannian manifold. For each $t \in M$, the Riemannian curvature tensor R_t given by (7.5.2) is in $\Lambda^{2,2}(T_tM)$, so that for $j \le \dim(M)/2$ it makes sense to talk about the jth power R_t^j of R_t. We can also take the trace

$$\mathrm{Tr}^M(R^j)(t) \triangleq \mathrm{Tr}^{T_tM}(R_t^j).$$

Integrating these powers over M gives the *Lipschitz–Killing curvature measures* of (M, g) defined, for measurable subsets U of M, by

$$\mathcal{L}_j(M, U) = \begin{cases} \frac{(-2\pi)^{-(N-j)/2}}{\left(\frac{N-j}{2}\right)!} \int_U \mathrm{Tr}^M(R^{(N-j)/2})\,\mathrm{Vol}_g & \text{if } N-j \text{ is even}, \\ 0 & \text{if } N-j \text{ is odd}. \end{cases} \tag{7.6.1}$$

The *Lipschitz–Killing curvature*, or *intrinsic volume*, of (M, g) is defined as

$$\mathcal{L}_j(M) = \mathcal{L}_j(M, M). \tag{7.6.2}$$

While (7.6.1) is a tidy formula, the integral is not always easy to compute. Perhaps the easiest example is given by the Lipschitz–Killing curvatures of S^N, which you should be able to compute directly from (7.6.1). We shall carry out the computation at the end of the following section, after having looked a little more carefully at some of the components making up the trace in the integrand.

7.7 A Euclidean Example

We close this chapter with what might be considered a "concrete" example, taking as our manifold M a compact C^2 domain in $\mathbb{R}^N$ endowed with a Riemannian metric g. We shall show how to explicitly compute both the curvature tensor R and the second fundamental form S, as well as traces of their powers. This section can be skipped on first reading, although you will probably want to come back to it once we start looking at specific examples in later chapters.

Christoffel symbols

We first met Christoffel symbols back in Section 7.3, where we used them to motivate the construction of Riemannian connections. Now, however, we want to work in the opposite direction. That is, given M and g, which uniquely define a Levi-Civitá connection, we want to represent this connection in a form conducive to performing computations. In doing this we shall recover the Christoffel symbols of Section 7.3 as well as develop a related class of symbols.

We start with $\{E_i\}_{1\le i\le N}$, the standard[30] coordinate vector fields on $\mathbb{R}^N$. This also gives the natural basis in the global chart $(\mathbb{R}^N, i)$, where i is the inclusion map. We now define[31] the so-called *Christoffel symbols of the first kind* of ∇,

$$\Gamma_{ijk} \stackrel{\Delta}{=} g(\nabla_{E_i} E_j, E_k), \quad 1 \le i, j, k \le N. \tag{7.7.1}$$

We further define

$$g_{ij} = g(E_i, E_j). \tag{7.7.2}$$

Despite the possibility of some confusion, we also denote the corresponding matrix function by $g = (g_{ij})_{i,j=1}^N$, doubling up on the notation for the metric.

With this notation it now follows via a number of successive applications of (7.3.10) and (7.3.11) that

$$\Gamma_{ijk} = \left(E_j g_{ik} - E_k g_{ij} + E_i g_{jk}\right)/2. \tag{7.7.3}$$

We need two more pieces of notation, the elements g^{ij} of the inverse matrix g^{-1} and the *Christoffel symbols of the second kind* of ∇, defined by

$$\Gamma_{ij}^k = \sum_{s=1}^N g^{ks}\Gamma_{ijs}.$$

[30] Note that while the E_i might be "standard," there is no reason why they should be the "right" coordinate system to use for a given g. In particular, they are no longer orthonormal, since g_{ij} of (7.7.2) need not be a Kronecker delta. Thus, although we start here, we shall soon leave this choice of basis for an orthonormal one.

[31] An alternative, and somewhat better motivated, definition of the Γ_{ijk} comes by taking the vector fields $\{E_i\}$ to be orthonormal with respect to the metric g. In that case, they can be defined via their role in determining the Riemannian connection through the set of N^2 equations $\nabla_{E_i} E_j = \sum_{k=1}^N \Gamma_{ijk} E_k$. Taking this as a definition, it is easy to see that (7.7.1) must also hold. In general, the Christoffel symbols are dependent on the choice of basis.

Riemannian curvature

With the definitions above, it is now an easy and standard exercise to show that if the metric g is C^2, then

$$R^E_{ijkl} \stackrel{\Delta}{=} R((E_i, E_j), (E_k, E_l)) \tag{7.7.4}$$
$$= \sum_{s=1}^{N} \left[g_{sl} \left(E_i \left(\Gamma^s_{jk} \right) - E_j \left(\Gamma^s_{ik} \right) \right) + \Gamma_{isl} \Gamma^s_{jk} - \Gamma_{jsl} \Gamma^s_{ik} \right]$$
$$= E_i \Gamma_{jkl} - E_j \Gamma_{ikl} + \sum_{s,t=1}^{N} \left(\Gamma_{iks} g^{st} \Gamma_{jlt} - \Gamma_{jks} g^{st} \Gamma_{ilt} \right).$$

Returning to the definition of the curvature tensor, and writing $\{de_i\}_{1 \le i \le N}$ for the dual basis of $\{E_i\}_{1 \le i \le N}$, it now follows (after some algebra) that the curvature tensor itself can be written as

$$R = \frac{1}{4} \sum_{i,j,k,l=1}^{N} R^E_{ijkl} (de_i \wedge de_j) \otimes (de_k \wedge de_l). \tag{7.7.5}$$

Next, we express the curvature tensor in an orthonormal frame field. To this end, let $X = \{X_i\}_{1 \le i \le N}$ be a section of the orthonormal frame bundle $\mathcal{O}(M)$, having dual frames $\{\theta_i\}_{1 \le i \le N}$, so that

$$\theta_i = \sum_{i'=1}^{N} g^{\frac{1}{2}}_{ii'} \, de_{i'}, \tag{7.7.6}$$

where $g^{\frac{1}{2}}$ is given by

$$(g^{\frac{1}{2}})_{ij} = g(E_i, X_j)$$

and the notation comes from the easily verified fact that $g^{\frac{1}{2}} (g^{\frac{1}{2}})' = g$, so that $g^{\frac{1}{2}}$ is a square root of g.

It follows that

$$R = \frac{1}{4} \sum_{i,j,k,l=1}^{N} R^X_{ijkl} (\theta_i \wedge \theta_j) \otimes (\theta_k \wedge \theta_l), \tag{7.7.7}$$

where

$$R^X_{ijkl} = \sum_{i',j',k',l'=1}^{N} R^E_{i'j'k'l'} g^{-\frac{1}{2}}_{ii'} g^{-\frac{1}{2}}_{jj'} g^{-\frac{1}{2}}_{kk'} g^{-\frac{1}{2}}_{ll'} = R\left((X_i, X_j), (X_k, X_l) \right),$$

and you are free to interpret the $g^{-\frac{1}{2}}_{ij}$ as either the elements of $(g^{\frac{1}{2}})^{-1}$ or of a square root of g^{-1}.

In (7.7.7) we now have a quite computable representation of the curvature tensor for any orthonormal basis. Given this, we also have the curvature forms $\Omega_{ij}(X,Y) = R\left(E_i, E_j, X, Y\right)$ of (7.5.4), and so, via (7.5.5), we can rewrite the curvature tensor as

$$R = \frac{1}{2}\,\Omega_{ij} \otimes \left(\theta_i \wedge \theta_j\right).$$

With the product and general notation of (7.2.4) we can thus write R^k as

$$R^k = \frac{1}{2^k} \sum_{i_1,\dots,i_{2k}=1}^{N} \left(\bigwedge_{l=1}^{k} \Omega_{i_{2l-1}i_{2l}}\right) \otimes \left(\bigwedge_{l=1}^{k} (\theta_{i_{2l-1}} \wedge \theta_{i_{2l}})\right),$$

where

$$\begin{aligned} &\bigwedge_{l=1}^{k} \Omega_{i_{2l-1}i_{2l}}(X_{a_1},\dots,X_{a_{2k}}) \\ &\quad = \frac{1}{2^k} \sum_{\sigma\in S(2k)} \varepsilon_\sigma \prod_{l=1}^{k} \Omega_{i_{2l}i_{2l-1}}(X_{a_{\sigma(2l-1)}}, X_{a_{\sigma(2l)}}) \\ &\quad = \frac{1}{2^k} \sum_{\sigma\in S(2k)} \varepsilon_\sigma \prod_{l=1}^{k} R^X_{i_{2l-1}i_{2l}a_{\sigma(2l-1)}a_{\sigma(2l)}}. \end{aligned}$$

It follows that

$$\begin{aligned} &R^k\left((X_{a_1},\dots,X_{a_{2k}}),(X_{a_1},\dots,X_{a_{2k}})\right) \\ &\quad = \frac{1}{2^{2k}} \sum_{i_1,\dots,i_{2k}=1}^{N} \delta^{(a_1,\dots,a_{2k})}_{(i_1,\dots,i_{2k})} \left(\sum_{\sigma\in S(2k)} \varepsilon_\sigma \prod_{l=1}^{k} R^X_{i_{2l-1}i_{2l}a_{\sigma(2l-1)}a_{\sigma(2l)}} \right), \end{aligned}$$

where for all m,

$$\delta^{(c_1,\dots,c_m)}_{(b_1,\dots,b_m)} = \begin{cases} \varepsilon_\sigma & \text{if } c = \sigma(b), \text{ for some } \sigma \in S(m), \\ 0 & \text{otherwise.} \end{cases}$$

We are finally in a position to write down an expression for the trace of R^k, as defined by (7.2.6):

$$\begin{aligned} \operatorname{Tr}(R^k) &= \frac{1}{(2k)!} \sum_{a_1,\dots,a_{2k}=1}^{N} R^k\left((X_{a_1},\dots,X_{a_{2k}}),(X_{a_1},\dots,X_{a_{2k}})\right) \qquad (7.7.8) \\ &= \frac{1}{2^{2k}} \sum_{a_1,\dots,a_{2k}=1}^{n} \left(\sum_{\sigma\in S(2k)} \varepsilon_\sigma \prod_{l=1}^{k} R^X_{a_{2l-1}a_{2l}a_{\sigma(2l-1)}a_{\sigma(2l)}} \right). \end{aligned}$$

This is the equation we have been searching for to give a “concrete” example of the general theory.

A little thought will show that while the above was presented as an example of a computation on $\mathbb{R}^N$, it is, in fact, far more general. Indeed, you can reread the above, replacing $\mathbb{R}^N$ by a general manifold and the E_i by a family of local coordinate systems that are in some sense "natural" for computations. Then (7.7.8) still holds, as do all the equations leading up to it. Thus the title of this section is somewhat of a misnomer, since the computations actually have nothing to do with Euclidean spaces!

The Lipschitz–Killing curvatures of S_λ^{N-1}

We can now take a moment to look at the very familiar example of S_λ^{N-1}, the sphere of radius λ in $\mathbb{R}^N$, and compute its Lipschitz–Killing curvatures from what we have just established.

In Chapter 6 we already computed what we called there the intrinsic volumes[32] S_λ^{N-1}. These were given by (6.3.8) as

$$\mathcal{L}_j\left(S_\lambda^{N-1}\right) = 2\binom{N-1}{j}\frac{s_N}{s_{N-j}}\lambda^j \tag{7.7.9}$$

if $N-1-j$ is even, and 0 otherwise. The argument that led to this was integral-geometric, and involved no more than knowing how to compute the volume of N balls and some trivial algebra.

We shall now rederive this from the integral (7.6.1) but shall have to work a little harder. The advantage, of course, is that while the harder calculation generalizes to more complex cases, the simple one does not.

Since it well known that S_λ^{N-1} has constant curvature, the trace in (7.6.1) is constant. Furthermore, since the volume element there is simply spherical measure, it is immediate that (7.6.1) reduces to

$$\mathcal{L}_j\left(S_\lambda^{N-1}\right) = \frac{(-2\pi)^{-(N-1-j)/2}}{\left(\frac{N-1-j}{2}\right)!}\lambda^{N-1}s_N \operatorname{Tr}^{S_\lambda^{N-1}}\left(R_t^{(N-1-j)/2}\right)$$

when $N-1-j=0$, and zero otherwise. Here t is any point on S_λ^{N-1}. Thus, all we need to do is compute the trace. We shall indicate how to do this, and leave the remaining algebra needed to obtain (7.7.9) to you.

The key point to note is that if X, Y, U, V are unit vectors in $T_t S_\lambda^{N-1}$, then

$$R(X, Y, U, V) = \begin{cases} -\lambda^{-2}, & (X,Y) = (U,V), \\ \lambda^{-2}, & (X,Y) = (V,U), \\ 0 & \text{otherwise.} \end{cases} \tag{7.7.10}$$

This can be checked, for example, by taking a parametric representation of S_λ^{N-1} to compute the Christoffel symbols of the first kind of (7.7.1) as the first step of the

[32] We have not really shown that the integral-geometric intrinsic volumes of Chapter 6 are the same as those that arise from differential-geometric considerations, of this and later chapters. That this in indeed the case is nontrivial to establish in general. The current calculation at least shows equivalence for S_λ^{N-1}.

computations described above. Alternatively, at least for the nonzero cases, you can use the "well-known" (and easily checkable) result that the curvature of S^2_λ is λ^{-2} and then use (7.5.3) to go from two to general dimensions.

Once we have (7.7.10), then (7.7.8) and a little counting gives

$$\mathrm{Tr}\left(R^k\right) = (-1)^k \lambda^{-2k} 2^{-k} \frac{(N-1)!}{(N-1-2k)!}.$$

This, along with some more algebra, will yield (7.7.9), as required.

Second fundamental form

We close this chapter by doing for the scalar second fundamental form S_ν what we did for Riemannian curvature above and also going a little further, deriving an expression for the trace of mixed powers of the form $R^k S^j_\nu$, $(j \le N-2k)$ on ∂M. We shall need these soon for computing the Lipschitz–Killing curvatures of manifolds with boundaries (cf. (10.7.1)).

We start much as we did for the curvature tensor, by choosing a convenient set of bases. However, this time the "natural" Euclidean basis is no longer natural, since our primary task is to parameterize the surface ∂M in a convenient fashion.

Thus, this time we start with $\{E^*_i\}_{1\le i\le N-1}$, the natural basis determined by some atlas on ∂M. This generates $T(\partial M)$. It is then straightforward to enlarge this to a section $E^* = \{E^*_i\}_{1\le i\le N}$ of $S(M)$, the sphere bundle of M, in such a way that on ∂M, we have $E^*_N = \nu$, the inward-pointing unit normal vector field on ∂M.

It then follows from the definition of the scalar second fundamental form (cf. (7.5.11)) that

$$S_\nu = \sum_{i=1}^{N-1} \gamma_{iN} \otimes de^*_i, \tag{7.7.11}$$

where, by (7.5.12), the connection forms γ_{ij} (on $\mathbb{R}^N$) satisfy

$$\gamma_{ij}(Y) = g(\nabla_Y E^*_i, E^*_j). \tag{7.7.12}$$

If we now define $\Gamma^*_{ijk} = g(\nabla_{E^*_i} E^*_j, E^*_k)$, then the connection forms γ_{iN} can be expressed[33] as

[33] It is often possible to write things in a format that is computationally more convenient. In particular, if the metric is Euclidean and if it is possible to explicitly determine functions a_{ij}, such that

$$E^*_{it} = \sum_{k=1}^{N} a_{ik}(t) E_{kt},$$

then it follows trivially from the definition of the Γ^*_{jiN} that

$$\Gamma^*_{jiN}(t) = \sum_{k,l,m=1}^{N} a_{jk}(t) \frac{\partial}{\partial t_k}\big(a_{Nl}(t) a_{im}(t) g_{ml}(t)\big).$$

$$\gamma_{iN} = \sum_{j=1}^{N-1} \Gamma^*_{jiN} de^*_j. \tag{7.7.13}$$

If, as for the curvature tensor, we now choose a smooth section X of $\mathcal{O}(M)$ with dual frames θ such that on ∂M, $X_N = \nu$, similar calculations yield that

$$S_\nu = \sum_{i,j=1}^{N-1} S^X_{ij} \theta_i \otimes \theta_j = \sum_{i=1}^{N-1} \theta_{iN} \otimes \theta_i,$$

where

$$S^X_{ij} \triangleq \sum_{i',j'=1}^{N-1} g_{ii'}^{-\frac{1}{2}} g_{jj'}^{-\frac{1}{2}} \Gamma_{j'i'N}, \qquad \theta_{iN} \triangleq \sum_{i=1}^{N-1} S^X_{ij} \theta_j.$$

Finally, on setting $p = 2k + j$, it follows that

$$R^k S^j_\nu = \frac{1}{2^k} \sum_{a_1,\dots,a_p=1}^{N-1} \left(\bigwedge_{l=1}^{k} \Omega_{a_{2l-1},a_{2l}} \right) \wedge \left(\bigwedge_{m=1}^{j} \theta_{a_{2k+m}N} \right) \otimes \left(\bigwedge_{l=1}^{p} \theta_{a_l} \right), \tag{7.7.14}$$

a formula we shall need in Chapter 12.

8

Piecewise Smooth Manifolds

So far, all that we have had to say about manifolds and calculus on manifolds has been of a local nature; i.e., it depended only on what was happening in individual charts. However, looking back at what we did in Sections 6.1–6.3 in the setting of integral geometry, we see that this is not going to solve our main problem, which is understanding the global structure of excursion sets of random fields now defined over manifolds.

In order to handle this, we are going to need a reasonably heavy investment in notation leading to various notions of *piecewise smooth spaces*.[1] The investment will be justified by ultimately producing results, in Part III, that are both elegant and applicable. To understand the need for piecewise smooth spaces, two simple examples should suffice: the sphere S^2, which is a C^∞ manifold without boundary, and the unit cube I^3, a flat manifold with a boundary made up of six faces that intersect at twelve edges, themselves intersecting at eight vertices. The cube, faces, edges, and vertices are themselves flat C^∞ manifolds, of dimensions 3, 2, 1, and 0, respectively.

In the first case, if $f \in C^k(S^2)$, the excursion set $A_u(S^2, f)$ is made of smooth subsets of S^2, each one bounded by a C^k curve. In the second case, for $f \in C^k(I^3)$, while the individual components of $A_u(I^3, f)$ will have a C^k boundary away from ∂I^3, their boundaries will also have faces, edges, and vertices where they intersect with ∂I^3. We already know from Section 6.2 that when we attempt to find point set representations for the Euler characteristics of excursion sets, these boundary intersections are important (e.g., (6.2.16) for the case of I^2). This is even the case if the boundary of the parameter set is itself smooth (e.g., Theorem 6.2.5). Consequently, as soon as we permit as parameter spaces manifolds with boundaries, we are going to require techniques to understand how these boundaries intersect with excursion sets.

[1] We shall soon meet many types of piecewise smooth spaces, including stratified manifolds, Whitney stratified manifolds, all in tame and locally convex (or not) versions. We shall therefore use the term "piecewise smooth" in a loose generic sense, referring to any or all of these examples. In the formal statements of results, however, we shall be careful about specifying which case is under consideration. Unfortunately—since one would prefer a general theory—this is necessary, since very often, results that appear at first as if they should hold in wider generality than stated do not.

Thus, what we are searching for is a framework that will cover basically smooth spaces, which, however, are allowed to have edges, corners, etc. This will ultimately involve blending both integral and differential geometry. There are many ways to do this, none of which could be considered canonical. The two most popular approaches are based on the related, but nonequivalent, theories of Whitney stratified manifolds, for which the standard reference is the monograph by Goresky and MacPherson [72], and sets of finite reach, as developed by Federer [64, 65] with extensions by Zähle [182] and others. In this book we shall adopt an approach based on Whitney stratified manifolds, and in the remainder of the chapter we shall define carefully what these are, and then proceed, both here and in Chapter 9, to add some additional restrictions needed to make the results of Part III work.

The resulting treatment is therefore occasionally unmotivated, and it would only be natural for you to ask why we suddenly add one side condition or another, and if this one, why not another. The rather unsatisfactory answer is that we add conditions that we require to make later proofs work. Indeed, it is probably instructional for you to know that although this and the following chapter are only the eighth and ninth out of fifteen, they were the last to be finished, since we had to return here time and again to tailor conditions to match what we could prove. Our only consolation is that this phenomenon seems to be endemic to the integral/differential geometry interface in general; i.e., one chooses a basic framework, and then appends conditions to generate proofs that morally should not require the additional conditions but in practice, do.[2]

Note, however, that if you are interested only in parameter spaces that are manifolds without boundary then you can go directly to Chapter 9 and skip the current one. However, you will then have to forgo fully understanding how to handle excursion sets over parameter spaces as simple as cubes, which, from the point of view of applications, is a rather significant loss.

Finally, we note that as much of the material of this chapter is not completely standard material in differential geometry, and even those readers comfortable with Chapter 7 may find it useful, at least for establishing notation.

8.1 Whitney Stratified Spaces

Our first step toward classifying piecewise smooth spaces will be what are known as Whitney stratified spaces. The basic reference for these is the monograph by Goresky and MacPherson [72], although we shall occasionally also have need of material that can be found in Pflaum [122].

We start with a topological subspace M of a C^k ambient manifold $\widetilde{M}$. The term "stratified" refers to a decomposition of M into strata, which we shall take to be smooth manifolds. A Whitney stratified space is a stratified subspace[3] $M \subset \widetilde{M}$

[2] Although, as noted above, we shall adopt an approach based on Whitney stratified manifolds, it is clear that the same fine-scale detailing would have arisen had we adopted an approach based on sets of finite reach. In addition, while this might have given an approach both powerful and mathematically elegant, it involves a heavier background investment than is justified for our purposes.

[3] We shall henceforth drop the term "subspace" in favor of space, with the understanding that for our purposes, there is *always* some ambient $\widetilde{M}$ in which M is embedded.

with some additional regularity constraints imposed. These regularity conditions are essential in setting up Morse theory for stratified spaces, which we describe in Chapter 9.

More formally, a C^l, $l \leq k$, *stratified space*, or *decomposed space*, $M \subset \widetilde{M}$ is a subspace of $\widetilde{M}$ along with a partition $\mathcal{Z}$ of M such that the following conditions are satisfied:

1. Each piece, or *stratum*, $S \in \mathcal{Z}$ is an embedded C^l submanifold of $\widetilde{M}$, without boundary.
2. For $R, S \in \mathcal{Z}$, if $R \cap \overline{S} \neq \emptyset$ then $R \subset \overline{S}$, and R is said to be *incident* to S.

We shall use the terminologies "stratified space" and *stratified manifold* interchangeably, and denote such spaces by $(M, \mathcal{Z})$.

The partition $\mathcal{Z}$ has a natural partial order, namely $R \preceq S$ if $R \subset \overline{S}$. Thus we can treat $\mathcal{Z}$ either as a partition of M, or as an abstract partially ordered set.

To fix a concrete example of a stratified space take $M = I^3$. The natural decomposition is to decompose I^3 into the eight zero-dimensional vertices, the eight one-dimensional edges, the six two-dimensional faces, and the single three-dimensional interior. Similarly, if $M = \partial I^3$, the natural stratification is to decompose ∂I^3 into the vertices, the edges, and the faces.

In general, a stratified space $M \subset \widetilde{M}$ with decomposition $\mathcal{Z}$ can be written as

$$M = \bigcup_{l=0}^{\dim M} \partial_l M,$$

where $\partial_l M$ is the *l-dimensional boundary* of M made up of the disjoint union of a finite number of l-dimensional manifolds. Given $0 \leq j \leq \dim M$, the space

$$\bigcup_{l=0}^{j} \partial_l M$$

is again a decomposed space, with partition $\{S \in \mathcal{Z} : \dim(S) \leq j\}$.

Although there is a natural decomposition of I^3, there are also many others, some of which are similar and some of which are not. For example, in the "natural" decomposition we made above, we could further decompose the three-dimensional interior of I^3 into the interior minus the point $(\frac{1}{2}, \frac{1}{2}, \frac{1}{2})$, and this point. For an even simpler example note that $\mathbb{R}$ can be decomposed in many ways, three of which are

$$\begin{aligned}
\mathcal{Z}_1 &= \{(-\infty, 0), \{0\}, (0, \infty)\}, \\
\mathcal{Z}_2 &= \{\mathbb{R} \setminus \{0\}, \{0\}\}, \\
\mathcal{Z}_3 &= \mathbb{R}.
\end{aligned}$$

For the first two, the decompositions look locally the same at every point in $\mathbb{R}$. However, $\mathcal{Z}_3$ is fundamentally different at 0. A priori, this means that some of the expressions for the intrinsic volumes described in Section 10.7 seem as if they may depend on the decomposition of M. That this is not the case is a somewhat subtle issue, to which we shall return when the need arises.[4]

So far, we have not imposed any regularity on how the respective l-dimensional boundaries $\partial_l M$ are "glued" together. In the case of I^3, and the natural decomposition described above, the pieces all fit together in a nice fashion. For more general parameter spaces, we shall have to impose some further regularity on M.

Returning to I^3 for motivation, note that every point $t \in I^3$ has a neighborhood that is isomorphic[5] to the product of an open neighborhood of the stratum that contains t and a cone. For example, each vertex of I^3 has a neighborhood that is isomorphic to a closed octant in $\mathbb{R}^3$, and every point t in an edge of I^3 has a neighborhood isomorphic to the product of an open interval around 0 and a closed quadrant in $\mathbb{R}^2$.

This "locally conic" property of I^3 will be essential to the Morse theory developed in Chapter 9, and that for general stratified spaces follows from Whitney's conditions (A) and (B) below. These conditions are regularity conditions imposed on a stratified space that, in particular, imply that each point $t \in M$ has a neighborhood isomorphic to the product of an open subset of the stratum S containing t and a cone[6] $\mathrm{Cone}(L_S)$ with base L_S over a stratified space L_S, the *link* of M at t.

A stratified space $(M, \mathcal{Z})$, is said to satisfy *Whitney condition (A)* at $t \in S$ if the following holds for every $S \preceq \widetilde{S}$:

(A) If $t_n \to t \in S$ such that $t_n \in \widetilde{S}$ for all n and the sequence of tangent spaces $T_{t_n}\widetilde{S}$ converges in the Grassmannian[7] bundle of $\dim(\widetilde{S})$-dimensional tangent (sub)spaces of $\widetilde{M}$ to some $\tau \subset T_t\widetilde{M}$, then $\tau \supset T_t S$. A limit of such tangent spaces is called a *generalized tangent space* of M at t.

---

[4] This point is treated in detail in [122], where a distinction is also made between a decomposition and a stratification, where the latter are equivalence classes of decompositions. However, since any decomposition uniquely determines a stratification, for our purposes we can assume that we are given a decomposition $\mathcal{Z}$ of M and talk about the stratification induced by $\mathcal{Z}$. Reference [122] also raises the possibility of parameterizing strata by something known as "depth" rather than by dimension.

[5] Two stratified spaces $(M_1, \mathcal{Z}_1)$ and $(M_2, \mathcal{Z}_2)$ of class C^l are said to be *isomorphic* if there exists a map $H : M_1 \to M_2$ that is an isomorphism of the partially ordered sets $\mathcal{Z}_1$ and $\mathcal{Z}_2$. That is, for each $S \in \mathcal{Z}_1$, $H(S) \in \mathcal{Z}_2$, and this map is an isomorphism when S and $H(S)$ are considered as elements of the partially ordered sets $\mathcal{Z}_1$ and $\mathcal{Z}_2$, respectively. Furthermore, H is such that for each $S \in \mathcal{Z}_1$, $H_{|S} : S \to H(S)$ is a diffeomorphism of class C^l.

[6] Recall that the cone $\mathrm{Cone}(L)$ over a topological space L is defined as the quotient space $Y = L \times [0, 1)$, where $y_1 \sim y_2 \iff y_1 = (x_1, 0), y_2 = (x_2, 0)$ for $x_1, x_2 \in L$.

If the topological space is also a real vector space V, then a cone is a subset K of V for which $\lambda K = K$ for all $\lambda > 0$. (Technically, this is a *positive* cone, but all our cones will be of this form, so we shall drop the qualifier.) Then K can be written in the form

$$K = \{\lambda x : x \in K_{\mathrm{base}},\ \lambda \geq 0\}.$$

We refer to K as the *cone over* the *base* K_{base} and denote it by $\mathrm{Cone}(K_{\mathrm{base}})$.

[7] The Grassmannian bundle referred to here is the natural one, namely the union of the Grassmannian manifolds of $\dim(\widetilde{S})$-dimensional subspaces of $T_t\widetilde{M}$, where the union is taken over all $t \in \widetilde{M}$, and the collection is parameterized similarly to the tangent bundle.

In a chart (U, φ) on $\widetilde{M}$ containing t, a stratified space $(M, \mathcal{Z})$ is said to satisfy *Whitney condition (B)* at $t \in S$ if the following is satisfied for every $S \preceq \widetilde{S}$:

(B) Let $t_n \to t$ and $s_n \to t$ be such that $t_n \in S, s_n \in \widetilde{S}$, for all n. Furthermore, suppose that the sequence of line segments $\overline{\varphi(t_n)\varphi(s_n)}$ converges in projective space to a line ℓ and the sequence of tangent spaces $T_{s_n}\widetilde{S}$ converges in the Grassmannian to a subspace $\tau \subset T_t\widetilde{M}$. Then $\varphi_*^{-1}(\ell) \subset \tau$.

It is easy to show that condition (B) implies condition (A), and it is not too much more difficult to show that whether condition (B) is satisfied at t is independent of the chart (cf. [122]). A stratified space $(M, \mathcal{Z})$ is called a *Whitney stratified space* if it satisfies Whitney condition (B) (and hence condition (A)) at every $t \in M$.

Examples of Whitney stratified spaces abound, and include the following:

- piecewise linear sets;
- finite simplicial complexes;
- Riemannian polyhedra;
- Riemannian manifolds (with boundary);
- closed semialgebraic (subanalytic) subsets of Euclidean spaces (i.e., sets that are finite Boolean combinations of excursion sets of algebraic (analytic) functions);
- basic complexes with C^2 boundary.

The last of these, of course, provides the link between this chapter and the integral geometry of Chapter 6.

Furthermore, Whitney stratified spaces have many desirable properties, among them these:

- Whitney stratified spaces can be triangulated.
- Whitney stratified spaces have a well-defined Euler characteristic.
- The intersection of two Whitney stratified spaces is generally a Whitney stratified space, whose strata are the intersections of the strata of the two spaces.

These three properties will all be of key importance to us, so we shall discuss them a little, including giving rough definitions of the terms we have used in stating them.

By a *triangulation* of a Whitney stratified space M we mean a covering of M by diffeomorphic images of simplices of dimension no more than $\dim(M)$ such that if two such images have a nonempty intersection, then the preimages of the intersection must be full facets (subsimplices) of each of the original simplices.

The simplices of the triangulation can be joined to form a single simplicial complex by joining them along the facets where their images intersected on M. We call this a *simplicial complex generated by* M, and denote it by $\mathcal{S}_M$. There is, of course, no uniqueness in such a triangulation. We can then define the *Euler*, or *Euler–Poincaré, characteristic* of M as the alternating sum

$$\varphi(M) \triangleq \sum_{j=0}^{\dim M} (-1)^{(N-j)} \alpha_j\left(\mathcal{S}_M\right), \tag{8.1.1}$$

where $\alpha_j(\mathcal{S}_M)$ is the number of j-dimensional facets in $\mathcal{S}_M$. Despite the fact that there is no uniqueness for triangulations and so the right-hand side here would seem to depend on $\mathcal{S}_M$, it is a basic theory of algebraic topology that the Euler characteristic is well defined and independent of the triangulation.

Finally, we turn to the issue of tranverse intersections. We say that two Whitney stratified submanifolds, M_1 and M_2, subsets of the same ambient N-dimensional manifold $\widetilde{M}$, *intersect transversally* if for each pair (j, k) with $0 \leq j \leq \dim(M_1)$ and $0 \leq k \leq \dim(M_2)$, and each $t \in \partial_j M_1 \cap \partial_k M_2$, the dimension of

$$\operatorname{span}\left\{X_t + Y_t : X_t \in T_t^{\perp}\partial_j M_1,\ Y_t \in T_t^{\perp}\partial_k M_2\right\} \tag{8.1.2}$$

is equal to $2N - j - k \geq 0$.

Now suppose that $f \in C^k(\widetilde{M})$ is such that its excursion set over $\widetilde{M}$, $A_u(f, \widetilde{M}) = f^{-1}[u, \infty)$ is a Whitney stratified manifold. If M is a Whitney stratified submanifold of $\widetilde{M}$, and if $A_u(f, \widetilde{M})$ and M intersect transversally, then $A_u(f, M)$ will also be a Whitney stratified manifold with a stratification inherited from those of $A_u(f, \widetilde{M})$ and M.

The punch line of the above paragraph is that if we start with Whitney stratified manifolds as the parameter spaces of random fields, and if our random fields behave well, then excursion sets will also be Whitney stratified manifolds. This closure, together with the fact that Whitney stratified manifolds provide a natural setting for both smooth and angular parameter spaces, is one of the main reasons that this class of sets will be, for us, the right choice for Part III.

8.2 Locally Convex Spaces

For much of what follows, we are going to have to slightly limit Whitney stratified manifolds to those that are, in an appropriate sense, locally convex.

To make this precise, we need the notion of the tangent to a curve c in a manifold M. Take a chart (U, φ) in the atlas of $\widetilde{M}$, a point $t \in U \cap M$, and a curve $c : [-1, 1] \to M$ for which $c(0) = t$. Then, at least for s sufficiently close to 0, we can write the coordinate representation of c as $c(s) = (c_1(s), \ldots, c_N(s))$, and the formula for the push-forward in the natural coordinates corresponding to φ gives us the natural definition

$$\dot{c}(0) \stackrel{\Delta}{=} \sum_{i=1}^{N} \dot{c}_i(0) \frac{\partial}{\partial x_i}\bigg|_{c(0)}$$

for the tangent vector to c at 0.

If, for a manifold M, we now collect all limiting directions

$$\begin{aligned} \mathcal{S}_t M \stackrel{\Delta}{=} \big\{ & X_t \in T_t\widetilde{M} : \exists \delta > 0,\ c \in C^1((-\delta, \delta), \widetilde{M}), \\ & c(0) = t,\ \dot{c}(0) = X_t,\ c(s) \in M \text{ for all } s \in [0, \delta)\big\}, \end{aligned} \tag{8.2.1}$$

we obtain a cone in $T_t\widetilde{M}$ known as the *support cone* of M at t. Support cones behave predictably under intersections, in that

$$\mathcal{S}_t(M_1 \cap M_2) = \mathcal{S}_t M_1 \cap \mathcal{S}_t M_2. \tag{8.2.2}$$

This follows from the simple observation that the support cone at $t \in M_1 \cap M_2$ is the set of all directions in which one can leave $t \in M_1 \cap M_2$ while remaining in $M_1 \cap M_2$. Thus each such direction must be contained in the support cones of both M_1 and M_2, i.e., it must lie in the intersection of the two support cones.

Figure 8.2.1 shows a part of the (shaded) support cones for two domains in $\mathbb{R}^2$, where the bases of the cones are the points at the base of the concavity in each domain. Note that while neither of the domains is itself convex, the smooth domain always has convex support cones, while the domain with the concave cusp does not.

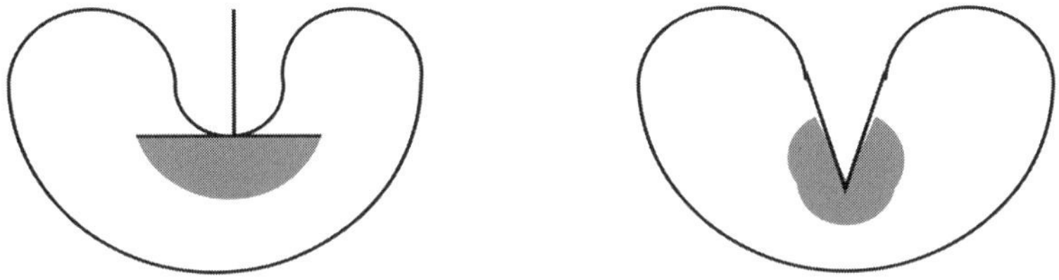

Fig. 8.2.1. Convex and nonconvex support cones.

We now have all we need to define a family of Whitney stratified spaces that will be important for us later.

Definition 8.2.1. *A Whitney stratified space* $(M, \mathcal{Z})$ *is called locally convex if the support cone* $\mathcal{S}_t M$ *is convex for every* $t \in M$.

Under this definition, the smooth domain of Figure 8.2.1 is locally convex, while the domain with concave cusp is not. These examples are actually quite generic, since the main import of the convex support cone assumption is to exclude sharp, concave cusps. Similarly, while the N-cube I^N is locally convex (and indeed, convex), its boundary ∂I^N is not.

With the notion of support cones fresh in our minds, this is probably a good place to define their duals, for which we need to assume that $\widetilde{M}$ has a Riemannian structure. This we do, writing $\widetilde{g}$ for the Riemannian metric on $\widetilde{M}$ and g or $\widetilde{g}_{|\partial_j M}$ for the metric it induces on M or $\partial_j M$.

This allows us to define a dual to each support cone $\mathcal{S}_t M$, known as the *normal cone* of M at t and defined by

$$N_t M \stackrel{\Delta}{=} \{X_t \in T_t\widetilde{M} : \widetilde{g}(X_t, Y_t) \leq 0 \text{ for all } Y_t \in \mathcal{S}_t M\}. \tag{8.2.3}$$

In the two domains in Figure 8.2.1, assuming the usual Euclidean metric, the normal cones at the base of the concavity are the outward normal in the first case, and empty in the other.

A more interesting example is given in Figure 8.2.2 for the tetrahedron. In this case we have shown the (truncated) normal cones at (a) a vertex (b) a point on an edge, and (c) a point on a face. What remains is a point in the interior of the tetrahedron, for which the normal cone is empty. Note that in each case the dimension of the normal cone is the codimension of the stratum in which the base point sits, with respect to that of the ambient manifold (in this case, $\mathbb{R}^3$).

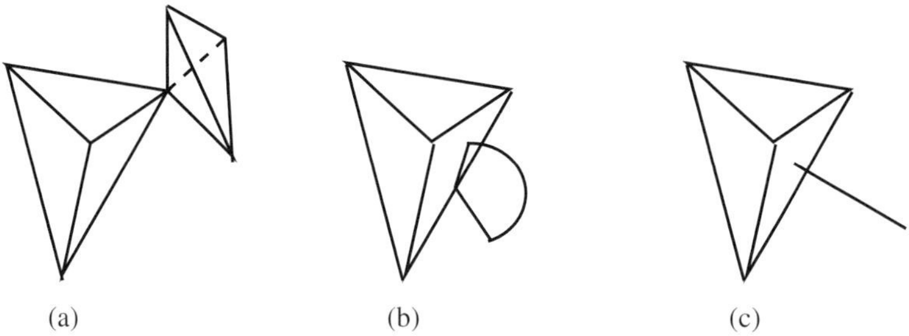

Fig. 8.2.2. Normal cones for a tetrahedron.

8.3 Cone Spaces

For technical reasons[8] we shall soon have to place a further restriction on the Whitney stratified manifolds with which we shall work, this being that they be C^2 *cone spaces*.

To define these, recall that every point t in a stratum S of a Whitney stratified space has a neighborhood U in $\widetilde{M}$ homeomorphic to the product $(S \cap U) \times \text{Cone}(L_S)$, where L_S is its link in M. However, this homeomorphism is not necessarily differentiable. (See the example of Neil's parabola at the end of this section.) Requiring differentiability is what leads to the notion of $C^{l,m}$ cone spaces, which we define following [122].[9]

Definition 8.3.1. *Let $M \subset \widetilde{M}$ be a C^l stratified space with stratification $(\mathcal{Z}, S)$, with l a nonnegative integer. Then M is said to be a cone space of class C^l and depth 0 if it is the topological sum of countably many connected C^l manifolds, the strata S of which are the unions of connected components of equal dimension.*[10]

A space M is said to be a cone space of class $C^{l,m}$ $(m \geq 0)$ and depth $d+1$ $(d \geq 0)$ if every $t \in S \subset M$ has a neighborhood $U \subset \widetilde{M}$ such that $U \cap M$ is C^m diffeomorphic to $(U \cap S) \times \text{Cone}(L_S)$, where L_S is a compact C^l cone space of

[8] The point at which we shall need C^2 cone spaces is in the proof of Theorem 9.2.6 in the following chapter. It is not clear to us at this stage whether the assumption that the manifolds there are also cone spaces is necessary for the result to hold, or merely a requirement of our method of proof. However, we were unable to find a proof without this assumption.

[9] In [122], cone spaces are defined only for $m = 0$, i.e., works only for homeomorphisms. Unfortunately, we shall need a little more for the proof of Theorem 9.2.6.

[10] Note that these are true manifolds, and not manifolds with boundary.

depth d, and $\mathrm{Cone}(L_S)$ *denotes the cone. When* $m = 0$, *"*C^m *diffeomorphic" is to be understood as "homeomorphic."*

In general, the set $(U \cap S) \times \mathrm{Cone}(L_S)$ can be thought of as a subset of $\mathcal{S}_t M$ if we choose an appropriate chart (φ, U) on $\widetilde{M}$. In this sense, the set M is locally approximated by (is homeomorphic to) the disjoint union

$$\bigcup_{t \in S} \mathcal{S}_t M$$

of support cones.

For an example, consider the important special case of convex simplicial complexes in Euclidean spaces. Such spaces have canonical stratifications given by the facets. In this case, for a given $t \in M$, the "link" L_S is naturally identified with the unit normal vectors of the supporting hyperplanes at t, chosen to point toward the interior of M. Furthermore, if we denote the span of a facet S by $[S]$, it is clear that $[S] \oplus \mathrm{Cone}(L_S)$ is $\mathcal{S}_t M$, where $\oplus$ denotes Minkowski addition

$$A \oplus B = \{x + y : x \in A, y \in B\} \tag{8.3.1}$$

(cf. [141]).

For a simple example of a $C^{\infty,0}$ cone space that is not $C^{\infty,m}$ for any $m \geq 1$, consider the so-called Neil's parabola (cf. [122]) given by

$$M_{\mathrm{Neil}} = \{(s, t) \in \mathbb{R}^2 : s^3 = t^2\}.$$

Its stratification is given by the origin and the two legs $\{(s, t) \in M_{\mathrm{Neil}} : t > 0\}$ and $\{(s, t) \in M_{\mathrm{Neil}} : t < 0\}$.

M_{Neil} is clearly homeomorphic to $\mathrm{Cone}(\{-1, 1\})$, which we can identify with the graph

$$\{(s, t) \in \mathbb{R}^2 : s = |t|\}.$$

However, there is no diffeomorphism of the plane that maps M_{Neil} to $\mathrm{Cone}(\{-1, 1\})$.

9

Critical Point Theory

In the preceding chapter we set up the two main geometric tools that we shall need in Part III of the book. The first of these are piecewise smooth manifolds of one kind or another, which will serve there as parameter spaces for our random fields, as well as appearing in the proofs. The second are the Lipschitz–Killing curvatures that we met briefly in Chapter 7 and shall look at far more closely, in the piecewise smooth scenario, in Chapter 10. These will appear in the answers to the questions we shall ask. Between the questions and the answers will lie considerable computation, and the main geometric tool that we shall need there is the topic of this short chapter.

Critical point theory, also known as Morse theory, is a technique for describing various global topological characteristics of manifolds via the local behavior, at critical points, of functions defined over the sets. We have already seen a version of this back in Section 6.2, where we obtained point set representations for the Euler characteristic of excursion sets (cf. Theorems 6.2.4 and 6.2.5, which gave point set representations for excursion sets in $\mathbb{R}^2$, over squares and over bounded sets with C^2 boundaries). Our aim now is to set up an analogous set of results for the excursion sets of C^2 functions defined over C^3 piecewise smooth spaces. We shall show in Section 9.4 how to specialize these back down to the known, Euclidean, examples of Section 6.2.

A full development of this theory for manifolds, which goes well beyond what we shall need, is in the classic treatise of Morse and Cairns [117], but you can also find a very readable introduction to this theory in the recent monograph of Matsumoto [112]. The standard theory, however, concentrates on smooth manifolds, as opposed to the piecewise smooth case that we need. The standard reference in this case is the excellent monograph of Goresky and MacPherson [72].

9.1 Critical Points

We begin with a general definition of critical points. Let $(M, \mathcal{Z})$ be a C^l Whitney stratified space embedded in a C^k ambient N-manifold $\widetilde{M}$. For $\widetilde{f} \in C^2(\widetilde{M})$, a *critical point* of $\widetilde{f}$ is a point $t \in \widetilde{M}$ such that $\nabla \widetilde{f}_t = 0$. Points that are not critical are called *regular*.

Now take M to be compact, N-dimensional, and C^2 piecewise smooth, embedded in a C^3 ambient manifold $(\widetilde{M}, \widetilde{g})$, writing, as usual, g for the induced metric on M. Extending the notion of critical points to $f = \widetilde{f}_{|M}$ requires taking note of the fact that the various boundaries in M are of different dimensions and so, in essence, involves repeating the above definition for each $f_{|\partial_j M}$. However, our heavy investment in notation now starts to pay dividends, since it is easy to see from the general definition that a point $t \in \partial_j M$, for some $0 \le j \le N$, is a critical point if and only if

$$\nabla \widetilde{f}_t \in T_t^{\perp} \partial_j M \tag{9.1.1}$$

(cf. (7.5.7) for notation). Thus we need work only with the single function f and not, explicitly at least, with its various restrictions.[1]

This definition implies that all points in $\partial_0 M$ are to be considered as critical points, and, when $\dim(M) = \dim(\widetilde{M})$, that critical points of $f_{|\partial_N M} \equiv \widetilde{f}_{|M^\circ}$ are just critical points of $\widetilde{f}$ in the sense of the initial definition.

We call the set

$$\bigcup_{j=0}^{N} \{t \in \partial_j M : \nabla \widetilde{f}_t \in T_t^{\perp} \partial_j M\}$$

the *set of critical points of* $f_{|M}$. All other points are known as *regular* points.

A critical point $t \in \partial_j M$ of $f_{|M}$ is called *nondegenerate* if the covariant Hessian $\nabla^2 f_{|T_t \partial_j M}$ is nondegenerate, when considered as a bilinear mapping. A function $f \in C^2(\widetilde{M})$ is said to be nondegenerate on M if all the critical points of $f_{|M}$ are nondegenerate. The *tangential Morse index*

$$\iota_{f,\partial_j M}(t) \tag{9.1.2}$$

of a nondegenerate critical point $t \in \partial_j M$ of $f_{|M}$ is the dimension of the largest subspace L of $T_t \partial_j M$ such that $\nabla^2 f(t)\big|_L$ is negative definite. Thus, a point of tangential Morse index zero is a local minimum of f on $\partial_j M$, while a point of index j is a local maximum. Other indices correspond to saddle points of various kinds. Before defining Morse functions on stratified Whitney spaces, we need one further concept: the normal Morse index, which, as its name suggests, complements the tangential Morse index.

[1] This assumes, however, that one remembers where all these spaces are sitting, or (9.1.1) makes little sense. Assuming that the ambient space $\widetilde{M}$ is N-dimensional ($N = \dim(M)$), we have, on the one hand, that $\nabla \widetilde{f}$ is also N-dimensional, whereas $T_t^{\perp} \partial_j M$ is $(N - j)$-dimensional. It is important, therefore, to think of $T_t^{\perp} \partial_j M$ as a subspace of $T_t^{\perp} M$ for the inclusion to make sense. Overall, all of these spaces are dependent on $\widetilde{M}$ and its Riemannian metric.

9.2 The Normal Morse Index

The *normal Morse index*, α, is a measure of local change in the topology of the manifold. In that sense it is very much like the Euler–Poincaré functional φ of (6.1.10). However, unlike φ, it records this change for any normal direction, and later we shall take averages over all directions. In this way we overcome the drawback of the theory of Chapter 6, which was axis-dependent (cf. footnote 1 there).

Beyond its use in Morse theory, α will also play a key role in the definition of Lipschitz–Killing curvature measures for Whitney stratified spaces in Chapter 10.

9.2.1 The Index

We start with another item of notation. If $(M, \mathcal{Z})$ is a Whitney stratified submanifold of an ambient Riemannian manifold $(\widetilde{M}, \widetilde{g})$, we write

$$T^{\perp}M = \bigcup_{S\in\mathcal{Z}} T^{\perp}S$$

for the *stratified normal bundle* of $(M, \mathcal{Z})$ (cf. (7.5.7)).

The next step in defining the normal Morse index involves moving everything to Euclidean space. Thus, fixing a $t \in M$, take normal coordinates (U_t, φ_t) (cf. (7.3.21)) on $\widetilde{M}$ at t. Now fix a unit vector $\nu \in T_t^{\perp}S$, and choose $(\delta_0, \varepsilon_0)$ small enough so that the topology of

$$\varphi_t(U_t \cap M) \cap \left\{x \in \mathbb{R}^{\dim(\widetilde{M})} : \langle x, \varphi_* \nu\rangle = -\varepsilon\right\} \cap B_\delta^{\dim(\widetilde{M})} \tag{9.2.1}$$

is unchanged for $0 < \delta < \delta_0$ and $0 < \varepsilon < \varepsilon_0$. This set is easily seen to be a Whitney stratified manifold embedded in $\mathbb{R}^{\dim(\widetilde{M})}$, and so has a well-defined Euler characteristic given either via (8.1.1) or via the integral geometry of Chapter 6 if the setup there happens to apply here as well. For $0 < \delta < \delta_0$ and $0 < \varepsilon < \varepsilon_0$, denote it by $\chi(\nu)$.

We then define the (*normal*) *Morse index* of M at t in the direction ν to be

$$\alpha(\nu) \equiv \alpha(\nu; M) \stackrel{\Delta}{=} 1 - \chi(\nu). \tag{9.2.2}$$

In fact, the normal Morse index is really dependent only on the structure of the support cone $\mathcal{S}_t M$, which, recall, we can express as $T_t S \oplus K_t$ for some cone $K_t \subset T_t S^{\perp}$. Moreover, a little thought shows that it is actually the structure of the cone K_t that is important, and so it therefore makes sense to work with

$$\alpha(\nu_t; K_t) \stackrel{\Delta}{=} \alpha(\nu_t) = \alpha(\nu_t; M).$$

Actually, it is not immediately clear that $\alpha(\nu)$ is well defined, i.e., that the above Euler characteristic depends only on the vector $\nu \in T^{\perp}S$ (more precisely on its identification with a covector canonically determined by $\widetilde{g}$)[2] and not on the choice of chart φ_t. However, this is indeed the case, as is established in [72, Theorem 7.5.1].

The above definition of the Morse index is not as complex as it may at first seem, as is seen from some simple examples. For example, if $M = ([-1,1] \times \{0\}) \cup (\{0\} \times [-1,1])$ is a $+$ in the plane, then $\alpha(\nu) = -1$ for all $\nu \in T_0\mathbb{R}^2$ unless ν is parallel to one of the axes. In this case $\alpha(\nu) = 0$. If, on the other hand, $M = ([0,1] \times \{0\}) \cup (\{0\} \times [0,1])$, then $\alpha(\nu)$ can take only one of the values $-1, 0, 1$, depending on the angle $\theta = \theta(\nu)$ that ν makes with the x-axis. Specifically,

$$\alpha(\nu) = \begin{cases} 1, & \theta \in \left(0, \frac{\pi}{2}\right), \\ 0, & \theta \in \left[\frac{\pi}{2}, \pi\right] \cup \left[\frac{3\pi}{2}, 2\pi\right], \\ -1, & \theta \in \left(\pi, \frac{3\pi}{2}\right). \end{cases}$$

The Morse index takes a particularly simple form for convex polytopes. In this case it is easy to check that

$$\alpha(\nu) = \begin{cases} 1, & -\nu \in (N_t M)^\circ, \\ 0 & \text{otherwise}, \end{cases} \tag{9.2.3}$$

where $N_t M$ is the normal cone of (8.2.3) under the Euclidean metric.

In fact, (9.2.3) also holds if M is a Whitney stratified manifold with convex support cones, that is, if M is a locally convex space. We shall use this fact heavily in Part III.

It is also not difficult to prove directly from the definition that $\alpha(\nu_t, \cdot)$ is additive, in the sense that for each $\nu_t \in T_t S^\perp$,

$$\alpha(\nu_t; K_t \cup \widetilde{K}_t) = \alpha(\nu_t; K_t) + \alpha(\nu_t; \widetilde{K}_t) - \alpha(\nu_t; K_t \cap \widetilde{K}_t). \tag{9.2.4}$$

Furthermore, since the support cones of $M_1 \cap M_2$ for two Whitney stratified subspaces are the intersections of the corresponding support cones of M_1 and M_2 (cf. (8.2.2)), we can rewrite (9.2.4) as

$$\alpha(\nu_t; M_1 \cup M_2) = \alpha(\nu_t; M_1) + \alpha(\nu_t; M_2) - \alpha(\nu_t; M_1 \cap M_2). \tag{9.2.5}$$

9.2.2 Generalized Tangent Spaces and Tame Manifolds

We have avoided one technicality in the above discussion, related to what are known as generalized tangent spaces. To define these, let $S_1 \subset S_2$ ($S_1 \neq S_2$) be two strata of a Whitney stratified manifold, and take $t \in \overline{S_1} \subset S_2$ and $\{t_n\}$ a sequence of points in S_1 converging to t. Then a *generalized tangent space* at t is any limit

$$\lim_{t_n \to t} T_{t_n} S_1.$$

---

[2] Note that the Morse index can actually be defined independently of the Riemannian metric, although then we would need to define it on the stratified conormal bundle of M. Since all our examples are Riemannian, we prefer to think of α as an integer-valued map on the stratified normal bundle instead.

We shall say that a vector ν *annihilates* a generalized tangent space of M at t if $\widetilde{g}(\nu, X) = 0$ for all X in the space.

Vectors ν for which there exist such generalized tangent spaces are referred to as *degenerate tangent vectors*[3] and all others as *nondegenerate*.

A little thought shows that there can be problems in defining the normal Morse index for degenerate vectors. We shall therefore (almost surely) assume these problems away.

Definition 9.2.1. *If C is a positive integer, then a closed stratified manifold M embedded in an ambient manifold $\widetilde{M}$ is said to be "C-tame," or simply "tame,"*[4] *if it satisfies the Whitney condition* (B) *as well as the following two conditions:*

(i) *If S is a stratum of M, then the collection*

$$\left\{ \lim_{t_n \to t} T_{t_n} S : t \in \partial S \right\}$$

of all generalized tangent spaces coming from S has Hausdorff dimension less than $\dim(S)$ *in the appropriate Grassmannian.*[5]

(ii) *Wherever the normal Morse index is defined, we have*

$$|\alpha(\nu_t; M)| \leq C.$$

The requirement that a manifold be tame is not, in general, a serious restriction. In particular, all the examples of Whitney stratified manifolds that we gave in Section 8.1 are tame.

However, tameness has an important consequence for the degenerate tangent vectors of a manifold. In particular, note that for fixed t, the set of degenerate tangent vectors is a closed cone in $T_t\widetilde{M}$ and the collection of all degenerate tangent vectors,

$$\bigcup_{t \in M} \left\{ \nu_t \in T_t\widetilde{M} : \nu_t \text{ is a degenerate tangent vector of } M \text{ at } t \right\}, \tag{9.2.6}$$

glued together across each $t \in M$ is a locally conic space. If M is tame, then it is immediate that this space has Hausdorff dimension strictly less than $\dim(\widetilde{M}) - 1$ (cf. [33]).

This property will be important when we come to discuss the Morse theory of excursion sets of random fields on manifolds in Chapter 12, as well as for defining the Lipschitz–Killing curvatures of Whitney stratified spaces in Chapter 10.

---

[3] For example, if we decompose $\mathbb{R} \subset \mathbb{R}^2$ as $(-\infty, 0) \times \{0\}$, $\{(0, 0)\}$, $(0, \infty) \times \{0\}$, then the set of degenerate tangent vectors at 0 corresponds to the y-axis, since there is only one generalized tangent space, the x-axis.

[4] The classic use of the adjective "tame," without the C, actually relates to stratified manifolds satisfying a slightly milder (integral) condition on the normal indices (cf. [33]) than the condition (ii) that we assume. However, since we shall later need to assume boundedness in any case, we shall assume it already now and adopt the same term.

[5] Cf. footnote 7 of Chapter 8.

Another consequence of nondegeneracy, which we note now but, again, is for later use, is the fact that if ν is nondegenerate then

$$\begin{aligned}\{X_t \in \mathcal{S}_t M : \langle X_t, \nu\rangle = -\varepsilon\} \cap B_{T_t\widetilde{M}}(0, \delta) \\ \simeq \{X_t \in K_t : \langle X_t, \nu\rangle \leq 0\} \cap S(T_t\widetilde{M}),\end{aligned} \tag{9.2.7}$$

where here, and in the future, $\simeq$ refers to homotopy equivalence.

9.2.3 Regular Stratified Manifolds

We now come to a simple definition, to which, since it is so central to all that follows, we have devoted an entire subsection. Recall that cone spaces are defined in Definition 8.3.1, local convexity at Definition 8.2.1, and tame spaces were defined above.

Definition 9.2.2. *Let M be a C^2 Whitney stratified manifold, embedded in an ambient manifold $\widetilde{M}$. Assume that M is also a $C^{2,1}$ cone space of arbitrary depth and that M is C-tame for some finite C. Then M is called a regular stratified manifold.*

If, in addition, M is locally convex, then, not surprisingly, it is called a locally convex, regular manifold.

In general, we shall not require that our manifolds be locally convex, although many formulas (and some proofs) become easier in this case. However, this will be a crucial assumption for Chapter 14, where we shall prove what is our main result about excursion probabilities for smooth Gaussian fields.

9.2.4 The Index on Intersections of Sets

An issue that will recur often throughout Part III of the book will be the geometry of a set $M_1 \cap M_2$, where both M_1 and M_2 are stratified spaces. The simplest and almost ubiquitous example will arise when we look at excursion sets, which we can write as

$$A_u(f, M) = M \cap f^{-1}[u, \infty),$$

where we think of both M and $f^{-1}[u, \infty)$ as submanifolds of an ambient space $\widetilde{M}$.

Another class of examples, which will be at the core of the generalized Crofton formulas of Chapter 13, is provided by sets of the form

$$M \cap f^{-1}(u),$$

which are boundaries of excursion sets, but still stratified manifolds.

In all these cases we shall need to relate the normal Morse index $\alpha(\cdot;\, M_1 \cap M_2)$ of the intersection to those of the individual M_j, that is, to the $\alpha(\cdot;\, M_j)$. The final result is given in Theorem 9.2.6 below, and is simple to understand without the technicalities leading to its proof.

While it is logically consistent to present and prove these results now, we do warn you that the arguments are rather technical and of no interest to the rest of the book beyond their application in establishing Theorem 9.2.6. Consequently, we recommend that you pass them on first (at least) reading, and return only when you have seen why, and how, we use the result.[6] However, despite all of this, because Theorem 9.2.6 itself will be very important for us, since we could not find anything quite like it anywhere in the literature, and since the proof is definitely not trivial, we give some details.

To start, we note that since normal Morse indices are always computed on the Euclidean image of a stratified manifold under its coordinate charts (7.3.21) (cf. (9.2.1)), it actually suffices to leave the setting of general manifolds and work rather with cones in Euclidean[7] space.

In particular, we shall start with two Euclidean cones $\widetilde{K}_1$ and $\widetilde{K}_2$, which may contain subspaces[8] and which we therefore write as

$$\widetilde{K}_1 = K_1 \oplus L_1, \qquad \widetilde{K}_2 = K_2 \oplus L_2, \tag{9.2.8}$$

where each L_j is the largest proper subspace contained in $\widetilde{K}_j$.

We begin with the case that K_1 and K_2 are simplicial cones, for which we have the following.

Theorem 9.2.3. *With the above setup, suppose*[9] *that*

$$K_1 = \left\{ \sum_{i=1}^{n} a_i v_i : a_i \geq 0,\ 1 \leq i \leq n \right\},$$

$$K_2 = \left\{ \sum_{i=1}^{m} a_i w_i : a_i \geq 0,\ 1 \leq i \leq m \right\},$$

for two sets of linearly independent vectors $V = \{v_1, \ldots, v_n\}$ *and* $W = \{w_1, \ldots, w_m\}$, *where* $0 \leq n \leq \mathrm{Codim}(L_1)$ *and* $0 \leq m \leq \mathrm{Codim}(L_2)$.

Suppose that

---

[6] To be honest, we should point out that this indeed was how this section was written. We completed the book presuming that Theorem 9.2.6 was true, and then returned to prove it only when, to our surprise, we could not find it in the literature.

[7] Actually, Theorem 9.2.3 and its proof could be stated in any Hilbert space, but we gain little by doing so.

[8] Note that we assume nothing about the dimensions of any of the sets here. All we shall need, assumed implicitly in what follows, is that each $\mathrm{Codim}(L_j)$ is finite. Thus, in terms of the Hilbert space example of the previous footnote, all sets could be infinite dimensional, as long as the required codimensions were finite.

Note also that one or both of the K_j could be equal to 0, the zero vector.

[9] In view of the previous footnote, we could replace the conditions defining K_1 and K_2 by the equivalent but notationally simpler requirement that they be cones over simplices. However, since we need the additional notation for the definition of the operator $\mathcal{T}_{L_2}$ below, we take the longer definition already here.

$$\text{Codim}\,(L_1 \cap L_2) = \text{Codim}\,(L_1) + \text{Codim}\,(L_2)\,. \tag{9.2.9}$$

Then, for almost[10] *every*[11] $\nu \in (L_1 \cap L_2)^\perp$,

$$\alpha(\nu;\, \widetilde{K}_1 \cap \widetilde{K}_2) = \alpha(\nu;\, P_{L_2}\widetilde{K}_1) \cdot \alpha(\nu;\, P_{L_1}\widetilde{K}_2). \tag{9.2.10}$$

Furthermore, suppose we enlarge W *to a set*

$$\widetilde{W} = \{w_1, \dots, w_m, \dots, w_{\text{Codim}(L_2)}\}$$

that $\widetilde{W}$ *spans* $L_2^\perp$, *and then define*[12] *a linear transformation* $\mathcal{T}_{L_2} : (L_1 \cap L_2)^\perp \to L_1^\perp$ *by*

$$\mathcal{T}_{L_2}\nu \stackrel{\Delta}{=} P_{L_1}^\perp \nu - \sum_{i,j=1}^{\text{Codim}(L_2)} \langle P_{L_1}\nu, P_{L_1}w_i\rangle \widetilde{g}^{ij} P_{L_1}^\perp w_j, \tag{9.2.11}$$

where the $\widetilde{g}^{ij}$ *are the entries of the inverse of the matrix with elements*

$$\widetilde{g}_{ij} = \langle P_{L_1}w_i, P_{L_1}w_j\rangle, \quad w_i, w_j \in \widetilde{W}. \tag{9.2.12}$$

Then, again for almost every $\nu \in (L_1 \cap L_2)^\perp$,

$$\alpha(\nu;\, \widetilde{K}_1 \cap \widetilde{K}_2) = \alpha(\mathcal{T}_{L_2}\nu;\, \widetilde{K}_1) \cdot \alpha(\nu;\, P_{L_1}\widetilde{K}_2). \tag{9.2.13}$$

Proof. While (9.2.10) may seem like a tidier and more natural result than (9.2.13), the latter result will actually be rather important for us, and we shall also prove it first.

We start with a little notation, for which we fix a point $t \in \widetilde{L}_1 \cap \widetilde{L}_2$ and look at a variety of vectors, matrices, and spaces at t. However, since the notation is heavy enough with carrying the dependence on t, it will not appear explicitly in any of what follows. Note, however, that as far as the cones K_1 and K_2 are concerned, t is always at the apex of the cone.

Enlarge the set V to

$$\widetilde{V} = \left\{v_1, \dots, v_n, v_{n+1}, \dots, v_{\text{Codim}(L_1)}\right\}$$

in such a way that $\text{span}(\widetilde{V}) = L_1^\perp$ and

$$v_{n+i} \in \text{span}(\widetilde{K}_1)^\perp, \quad 1 \le i \le \text{Codim}(L_1) - n.$$

[10] The vectors that cause us problems at this stage are those that may annihilate generalized tangent spaces, which in the current simple scenario have measure zero.

[11] Since we normally write the Morse index as depending on a vector ν_t emanating from a point t, and here there is no reference to t, note that we are implicitly assuming that given a $\nu \in (L_1 \cap L_2)^\perp$, the relevant t is the point in $L_1 \cap L_2$ from which it emanates.

[12] The fact that $w_j \in L_2^\perp$ for all $1 \le j \le m$ comes from the orthogonal structure in (9.2.8). It is also straightforward to check that the transformation $\mathcal{T}_{L_2}$ is independent of the choice of extension of W to $\widetilde{W}$.

Next, let $\widehat{G}$ be the $\mathrm{Codim}(L_1) \times \mathrm{Codim}(L_1)$ matrix with elements

$$\widehat{g}_{ij} = \langle v_i, v_j \rangle$$

and $\widehat{g}^{ij}$ the elements of $\widehat{G}^{-1}$. Define the dual vectors, $v_i^* \in L_1^{\perp}$, by

$$v_i^* = \sum_{j=1}^{\mathrm{Codim}(L_1)} \widehat{g}^{ij} v_j, \quad 1 \le i \le n.$$

It is straightforward to check that this set of vectors forms a basis for the normal cone $\widetilde{K}_1^*$, which we can now write as

$$\begin{aligned}\widetilde{K}_1^* &= \left\{ \sum_{i=1}^{n} a_i v_i^* : a_i \le 0,\ 1 \le i \le n \right\} \oplus \left\{ v : P_{\mathrm{span}(\widetilde{K}_1)} v = 0 \right\} \\ &= \left\{ \sum_{i=1}^{\mathrm{Codim}(L_1)} a_i v_i^* : a_i \le 0,\ 1 \le i \le n \right\}.\end{aligned}$$

Similarly, define a set of vectors $w_i^* \in L_2^{\perp}$, starting with the w_i rather than with the v_i, and use these to write an analogous representation for the normal cone $\widetilde{K}_2^*$ of $\widetilde{K}_2$.

Given this notation, we are now in a position to start the proof.

Since intersections of simplicial cones are simplicial cones and simplicial cones are convex, it follows from (9.2.3) that their normal Morse index is actually an indicator function. Specifically, for $\nu \in (L_1 \cap L_2)^{\perp}$,

$$\alpha(\nu; \widetilde{K}_1 \cap \widetilde{K}_2) = \mathbb{1}_{(\widetilde{K}_1 \cap \widetilde{K}_2)^*}(-\nu), \tag{9.2.14}$$

where

$$(\widetilde{K}_1 \cap \widetilde{K}_2)^* = \widetilde{K}_1^* \oplus \widetilde{K}_2^*.$$

In particular, for $\nu \in (L_1 \cap L_2)^{\perp}$,

$$-\nu \in \widetilde{K}_1^* \oplus \widetilde{K}_2^*$$

if and only if the coefficients of ν in the basis[13]

$$B = \left\{ b_1, \dots, b_{\mathrm{Codim}(L_1 \cap L_2)} \right\} \stackrel{\Delta}{=} \left\{ v_1^*, \dots, v_{\mathrm{Codim}(L_1)}^*, w_1^*, \dots, w_{\mathrm{Codim}(L_2)}^* \right\}$$

corresponding to $\{v_1^*, \dots, v_n^*, w_1^*, \dots, w_m^*\}$ are nonnegative.

To determine whether a vector is a nonnegative combination of these particular elements of B, it suffices to compute the coefficients of ν in the basis B. However, these coefficients can also be written as $\langle \nu, b_i^* \rangle$, where

[13] The linear independence follows from the assumption that $\mathrm{Codim}(L_1 \cap L_2) = \mathrm{Codim}(L_1) + \mathrm{Codim}(L_2)$.

$$b_i^* = \sum_{j=1}^{\mathrm{Codim}(L_1 \cap L_2)} g^{ij} b_j \tag{9.2.15}$$

is the dual basis of B and

$$g_{ij} = \langle b_i, b_j \rangle$$

is the matrix of cross products of the b_i's with $\widehat{g}$ in its upper left corner and

$$\langle w_i, w_j \rangle, \quad 1 \le i, j \le \mathrm{Codim}(L_2),$$

in its lower right corner. As usual, the g^{ij} are the elements of the inverse of the matrix of g_{ij}.

Note, for later use, that the above construction implies that for $1 \le i \le \mathrm{Codim}(L_1)$ and $\mathrm{Codim}(L_1) + 1 \le j \le \mathrm{Codim}(L_1 \cap L_2)$,

$$\langle v_i, b_j^* \rangle = \langle b_i^*, w_j \rangle = 0. \tag{9.2.16}$$

With the above notation, we can now write the normal Morse index of $\widetilde{K}_1 \cap \widetilde{K}_2$ as

$$\alpha(\nu; \widetilde{K}_1 \cap \widetilde{K}_2) = \left(\prod_{j=1}^{n} \mathbb{1}_{[0,\infty)}(\langle \nu, b_j^* \rangle) \right) \cdot \left(\prod_{j=\mathrm{Codim}(L_1)+1}^{\mathrm{Codim}(L_1)+m} \mathbb{1}_{[0,\infty)}(\langle \nu, b_j^* \rangle) \right). \tag{9.2.17}$$

Our goal now is to relate each factor in the product to the factors on the right-hand sides of (9.2.10) and (9.2.13). We start with (9.2.13).

Recall that the dual basis (9.2.15) is also uniquely determined by the orthonormality relationship

$$\langle b_i^*, b_j \rangle = \delta_{ij}$$

for all $b_i \in B$. Therefore, for any sequence of reals, c_j,

$$\left\langle \nu + \sum_{l \ne j} c_l b_l, b_j^* \right\rangle = \left\langle \nu, b_j^* \right\rangle.$$

In particular, taking $\nu \in (L_1 \cap L_2)^\perp$ and $1 \le l \le n$, and noting that the w_j are then orthogonal to b_l^*, we have

$$\begin{aligned} \left\langle \nu, b_l^* \right\rangle &= \left\langle \nu - \sum_{i,j=1}^{\mathrm{Codim}(L_2)} \langle P_{L_1}\nu, P_{L_1} w_i \rangle \widetilde{g}^{ij} w_j, b_l^* \right\rangle \\ &= \left\langle P_{L_1}^\perp \nu - \sum_{i,j=1}^{\mathrm{Codim}(L_2)} \langle P_{L_1}\nu, P_{L_1} w_i \rangle \widetilde{g}^{ij} P_{L_1}^\perp w_j, b_l^* \right\rangle. \end{aligned} \tag{9.2.18}$$

The last line follows from expressing $\nu \in (L_1 \cap L_2)^\perp$ as

$$\nu = P_{L_1}^{\perp}\nu + P_{L_1}\nu = P_{L_1}^{\perp}\nu + \sum_{i,j=1}^{\mathrm{Codim}(L_2)} \langle P_{L_1}\nu, P_{L_1}w_i\rangle \widetilde{g}^{ij} P_{L_1}w_j,$$

where the matrix $\widetilde{g}$ was defined in (9.2.12), and the representation of $P_{L_1}\nu$ follows from the fact that the codimension assumptions of the theorem ensure that the $\{P_{L_1}w_j\}_{j=1,\dots,\mathrm{Codim}(L_2)}$ are linearly independent and span the orthogonal complement of $L_1^{\perp}$ in $(L_1 \cap L_2)^{\perp}$.

Returning to (9.2.18), noting the definition (9.2.11) of the mapping $\mathcal{T}_{L_2}$ and the fact that $\mathcal{T}_{L_2}\nu \in L_1^{\perp}$, we now have

$$\left\langle \nu, b_l^* \right\rangle = \langle \mathcal{T}_{L_2}\nu, b_l^*\rangle = \langle \mathcal{T}_{L_2}\nu, P_{L_1}^{\perp}b_l^*\rangle = \langle \mathcal{T}_{L_2}\nu, v_l^*\rangle, \tag{9.2.19}$$

where the final equality follows from the observation that for $1 \le i, l \le n$,

$$\langle v_i, P_{L_1}^{\perp}b_l^*\rangle = \langle v_i, b_l^*\rangle = \delta_{il}$$

(cf. (9.2.16)). By the uniqueness of the dual basis this implies that $P_{L_1}^{\perp}b_l^* = v_l^*$.

Substituting (9.2.19) into (9.2.17), we find that the first term on the right-hand side there, and so also in (9.2.13), is given by

$$\prod_{l=1}^{n} \mathbb{1}_{[0,\infty)}(\langle \mathcal{T}_{T_2}\nu, v_l^*\rangle) = \mathbb{1}_{K_1^*}(-\mathcal{T}_{L_2}\nu) = \alpha(\mathcal{T}_{L_2}\nu, \widetilde{K}_1),$$

and so we are half-done.

We turn now to the second term. Note first that it follows from (9.2.16) that $b_j^* \in L_1$ for $\mathrm{Codim}(L_1) + 1 \le j \le \mathrm{Codim}(L_1 \cap L_2)$.

Therefore, for $1 \le i, j \le \mathrm{Codim}(L_2)$,

$$\left\langle P_{L_1}w_i, b^*_{\mathrm{Codim}(L_1)+j} \right\rangle = \left\langle w_i, b^*_{\mathrm{Codim}(L_1)+j} \right\rangle = \delta_{ij},$$

and so the vectors $\{b^*_{\mathrm{Codim}(L_1)+j}\}$ are dual to the $\{P_{L_1}w_i\}$. If we choose the extension of W to $\widetilde{W}$ in such a way that

$$w_{m+i} \in \mathrm{span}(K_2)^{\perp}, \quad 1 \le i \le \mathrm{Codim}(L_2) - m,$$

then the vectors $\{b^*_{\mathrm{Codim}(L_1)+i}, 1 \le i \le m\}$ are dual to $\{P_{L_1}w_i, 1 \le i \le m\}$, and so we can write

$$\prod_{i=1}^{m} \mathbb{1}_{[0,\infty)}(\langle \nu, b^*_{\mathrm{Codim}(L_1)+i}\rangle) = \mathbb{1}_{P_{L_1}K_2^*}(-\nu) = \alpha(\nu, P_{L_1}\widetilde{K}_2).$$

This takes care of the second term in (9.2.13).

To complete the proof, it remains only to check that (9.2.10) also holds. But this is now easy, since by symmetry with respect to $\widetilde{K}_1$ and $\widetilde{K}_2$, we have

$$\alpha\left(\nu; P_{L_1}\widetilde{K}_2\right) = \alpha\left(\mathcal{T}_{L_1}\nu; \widetilde{K}_2\right),$$
$$\alpha\left(\nu; P_{L_2}\widetilde{K}_1\right) = \alpha\left(\mathcal{T}_{L_2}\nu; \widetilde{K}_1\right).$$

Substituting into (9.2.13) immediately yields (9.2.10) and we are done. □

The following two corollaries are now reasonably straightforward.

Corollary 9.2.4. *Suppose that*

$$\widetilde{K}_1 = K_1 \oplus L_1,$$
$$\widetilde{K}_2 = K_2 \oplus L_2,$$

where K_1 and K_2 are cones in $\mathbb{R}^N$ for which the following conditions hold:

(i) *There exist sequences $K_{1,n}$ and $K_{2,n}$ of cones over simplicial complexes with $K_{j,n} \to K_j$ in the Hausdorff metric.*
(ii) *For each j and almost all ν we have $\alpha(\nu; K_{j,n} \oplus L_j) \to \alpha(\nu; K_j \oplus L_j)$ as $n \to \infty$.*

Let Codim $(L_1 \cap L_2)$ = Codim (L_1) + Codim (L_2). *Then, for almost every $\nu \in (L_1 \cap L_2)^\perp$,*

$$\begin{aligned}\alpha(\nu; \widetilde{K}_1 \cap \widetilde{K}_2) &= \alpha(\mathcal{T}_{L_2}\nu; \widetilde{K}_1) \cdot \alpha(\nu, P_{L_1}\widetilde{K}_2) \\ &= \alpha(\nu; P_{L_2}\widetilde{K}_1) \cdot \alpha(\nu, P_{L_1}\widetilde{K}_2).\end{aligned}$$

Proof. Note first that the result of Theorem 9.2.3 extends trivially to the case in which the simplicial cones K_1 and K_2 are replaced by cones over simplicial complexes. This follows from footnote 9, on applying Theorem 9.2.3 to the individual simplices making up the complex and then exploiting the additivity (9.2.4).

Thus the result holds when each $\widetilde{K}_j$ is replaced by a $K_{j,n} \oplus L_j$. Conditions (i) and (ii) of the theorem now imply the result in general. □

Before turning to our main result, we note the following corollary to the preceding results.

Corollary 9.2.5. *Suppose that*

$$\widetilde{K}_1 = K_1 \oplus L_1,$$

where K_1 satisfies the conditions of Theorem 9.2.3 *or Corollary* 9.2.4 *and*

$$\widetilde{K}_2 = L_2,$$

where L_1 and L_2 are subspaces of $\mathbb{R}^N$. If Codim$(L_1 \cap L_2)$ = Codim(L_1) + Codim(L_2), *then for almost every $\nu \in (L_1 \cap L_2)^\perp$,*

$$\alpha(\nu; \widetilde{K}_1 \cap \widetilde{K}_2) = \alpha(\nu; \widetilde{K}_1 \cap L_2) = \alpha(\mathcal{T}_{L_2}\nu; \widetilde{K}_1).$$

Proof. First note that $P_{L_1}\widetilde{K}_2 = P_{L_1}L_2 \simeq L_1 \cap L_2$ is a subspace of less than full dimension in $\mathbb{R}^N$ and so has normal Morse index one at every point, and in all normal directions.

Note also that $P_{L_2}\widehat{K}_1 \simeq K_1 \cap L_2$. Now apply the appropriate theorem or corollary. □

Now that we have a set of basic results for cones, we can turn to the manifold version of these results, which is what we shall actually need, but for which we offer only an outline of a proof.

Theorem 9.2.6. *Let M_1 and M_2 be regular stratified manifolds with stratifications*

$$M_1 = \bigcup_{j=0}^{\dim M_1} M_{1j}, \qquad M_2 = \bigcup_{k=0}^{\dim M_2} M_{2k}.$$

Suppose that for each j and k, M_{1j} and M_{2k} intersect transversally and

$$\text{Codim}\left(M_{1j} \cap M_{2k}\right) = \text{Codim}\left(M_{1j}\right) + \text{Codim}\left(M_{2k}\right). \tag{9.2.20}$$

Fix a $t \in M_{1j} \cap M_{2k}$. Then, for every such t and almost every $\nu_t \in T_t(M_{1j} \cap M_{2k})^{\perp}$,

$$\begin{aligned} \alpha\left(\nu_t; \mathcal{S}_t(M_1 \cap M_2)\right) &= \alpha\left(\nu_t; \mathcal{S}_t M_1 \cap \mathcal{S}_t M_2\right) \\ &= \alpha\left(\nu_t; P_{T_t M_{2k}} \mathcal{S}_t M_1\right) \cdot \alpha\left(\nu_t; P_{T_t M_{1j}} \mathcal{S}_t M_2\right) \\ &= \alpha\left(\mathcal{T}_{jk} \nu_t; \mathcal{S}_t M_1\right) \cdot \alpha\left(\nu_t; P_{T_t M_{1j}} \mathcal{S}_t M_2\right), \end{aligned} \tag{9.2.21}$$

where $\mathcal{T}_{jk} : T_t(M_{1j} \cap M_{2k})^{\perp} \to T_t M_{1j}^{\perp}$ is defined by

$$\mathcal{T}_{jk} \nu = P^{\perp}_{T_t M_{1j}} \nu - \sum_{r,s=1}^{\text{Codim}(M_{2k})} \langle P_{T_t M_{1j}} \nu, P_{T_t M_{1j}} w_r \rangle \widetilde{g}^{rs} P^{\perp}_{T_t M_{1j}} w_s,$$

where $\{w_1, \dots, w_{\text{Codim}(M_{2k})}\}$ is a collection of vectors spanning $T_t M_{2k}^{\perp}$ and the $\widetilde{g}^{rs}$ are the entries of the inverse of the matrix with elements

$$\widetilde{g}_{rs} = \langle P_{T_t M_{1j}} w_r, P_{T_t M_{1j}} w_s \rangle, \quad 1 \le r, s \le \text{Codim}(M_{2k}).$$

Furthermore, if $\mathcal{S}_t(M_2)$ is a half-space, then for almost every $\nu_t \in T_t(M_{1,j} \cap M_{2,k})^{\perp}$,

$$\begin{aligned} \alpha\left(\nu_t; \mathcal{S}_t(M_1 \cap M_2)\right) &= \alpha\left(\nu_t; P_{T_t M_{2,k}} \mathcal{S}_t M_1\right) \\ &= \alpha\left(\mathcal{T}_{T_{jk}} \nu_t; \mathcal{S}_t M_1\right). \end{aligned} \tag{9.2.22}$$

Proof. Starting the proof is easy; finishing it is, to say the least, "tedious."

To start it, note that the fact that $\mathcal{S}_t(M_1 \cap M_2) = \mathcal{S}_t M_1 \cap \mathcal{S}_t M_2$ is an immediate consequence of the definition of support cones, and so the first equality in (9.2.21) is trivial.

To continue, note that $\alpha(\cdot, M)$ was actually defined on the Euclidean image of M under the exponential map. Thus we can assume that we are in Euclidean space.

The rest (i.e., both (9.2.21) and (9.2.22)) would now follow with very little work from Theorem 9.2.3 and Corollary 9.2.5 if it were true that M_1 and M_2 were simplicial complexes. Unfortunately, this, in general, is not the case.

There are, however, a number of standard techniques for approximating tame Whitney stratified manifolds by simplicial complexes.[14] Furthermore, there are many techniques[15] for showing that results that hold for the simplicial approximations carry over to the manifolds, as long as they are not too badly behaved.

In essence, one needs results akin to our Corollary 9.2.4, but whereas there we were able to assume much about the convergence of approximations to make things work, now we have to show that these assumptions are justified. Among other things, one needs to show that the Morse indices of the approximations converge to those of the limit, and it is precisely here that we need the condition that M_1 and M_2 be $C^{2,1}$ cone spaces.

All of this is, while standard fare in integrodifferential geometry, would take us far beyond the level of this book, and so we shall not attempt a detailed proof.[16] □

9.3 Morse's Theorem for Stratified Spaces

In this section, we give an extension of the point set representation of the Euler characteristic developed in Sections 6.1 and 6.2 to Whitney stratified spaces. The extension is one of the central results of differential geometry and is known as Morse's theorem.

9.3.1 Morse Functions

The last definition we need before stating Morse's theorem is that of a Morse function.

Definition 9.3.1. *A function $f \in C^2(\widetilde{M})$, where M is a C^2 Whitney stratified manifold embedded in a C^3 ambient manifold $\widetilde{M}$, is called a* Morse function *on M if it satisfies the following two conditions on each stratum $\partial_k M$, $k = 0, \dots, \dim(M)$:*

(i) *$f_{|\partial_k M}$ it is nondegenerate on $\partial_k M$.*
(ii) *The restriction of f to $\overline{\partial_k M} = \bigcup_{j=0}^{k} \partial_j M$ has no critical points on $\bigcup_{j=0}^{k-1} \partial_j M$.*

Note that (ii) *is equivalent to requiring the following:*

(iii) *At each critical point t of $f_{|\partial_k M}$, $\nabla f_{|\partial_k M, t}$ is a nondegenerate tangent vector.*[17]

[14] There are many references that could be given here, but perhaps the most appropriate approximation is due to Cheeger et al. [39]. To see how to use it in the setting of tame manifolds, see the papers by Zähle [182] and Bröcker and Kuppe [33].

[15] In our case the approach of Zähle [182] is probably the most appropriate.

[16] You might ask, why, if we had not planned to prove Theorem 9.2.6, we bothered with the proof of Theorem 9.2.3 and its corollaries.

The reason is that we could not find anything like Theorem 9.2.3 in the literature, and, since the product result here for the Morse index is crucial for later parts of the book, we felt duty bound to prove it. On the other hand, the move from simplicial complexes to tame manifolds, while certainly not easy, is more standard fare, and so a description of the proof as "tedious but straightforward" is not unjustified.

[17] Cf. (9.2.6) and the discussion preceding it.

9.3.2 Morse's Theorem

With all the definitions cleared up, we have the necessary ingredients to state the following version of Morse's theorem, due to Goresky and MacPherson [72].

Theorem 9.3.2 (Morse's theorem). *Let $(M, \mathcal{Z})$ be a compact C^2 Whitney stratified space embedded in a C^3 Riemannian manifold $(\widetilde{M}, \widetilde{g})$ and let $\widetilde{f} \in C^2(\widetilde{M})$ be a Morse function on M. Then, setting $f = \widetilde{f}_{|M}$,*

$$\varphi(M) = \sum_{j=0}^{N} \sum_{\{t \in \partial_j M : \nabla f_t \in T_t^{\perp} \partial_j M\}} (-1)^{\iota_{f,\partial_j M}(t)} \alpha(P^{\perp}_{T_t \partial_j M} \nabla f_t; M), \tag{9.3.1}$$

where $P^{\perp}_{T_t \partial_j M}$ is the orthogonal projection onto $(T_t \partial_j M)^{\perp}$, $\varphi(M)$ is the Euler characteristic of M, and the $\iota_{f,\partial_j M}(t)$ are the tangential Morse indices of (9.1.2).

If the support cones $\mathcal{S}_t M$ are convex for each $t \in M$, i.e., M is locally convex, then (cf. (9.2.3)) the above theorem reads as follows.

Corollary 9.3.3 (Morse's theorem for locally convex manifolds). *Let $(M, \mathcal{Z})$ be a compact C^2, locally convex, Whitney stratified space embedded in a C^3 Riemannian manifold $(\widetilde{M}, \widetilde{g})$ and let $\widetilde{f} \in C^2(\widetilde{M})$ be a Morse function on M. Then, setting $f = \widetilde{f}_{|M}$,*

$$\varphi(M) = \sum_{j=0}^{N} \sum_{\{t \in \partial_j M : \nabla f_t \in T_t^{\perp} \partial_j M\}} (-1)^{\iota_{f,\partial_j M}(t)} \mathbb{1}_{\{-\nabla f_t \in N_t M\}}, \tag{9.3.2}$$

where $\varphi(M)$ is the Euler characteristic of M.

The points counted in the above corollary are given a special name.

Definition 9.3.4. *In the above setup, we call a point $t \in M$ an extended inward critical point of $f = \widetilde{f}_{|M}$ if*

$$-\nabla f_t \in N_t(M),$$

where $N_t(M)$ is the normal cone of (8.2.3). *Similarly, t is an extended outward critical point if $\nabla f_t \in N_t(M)$.*

Morse's theorem is a deep and important result in differential topology and is actually somewhat more general than as stated here, since in its full form it also gives a series of inequalities linking the Betti numbers[18] of M to the critical points of Morse functions on M. We shall make no attempt to prove Morse's theorem, which, given

[18] Betti numbers are additional geometrical quantifiers of M of somewhat less straightforward interpretation than the Euler characteristic, and not unrelated to the Lipshitz–Killing curvatures of Section 7.6.

our definition of the Euler characteristic, relies on arguments of homology theory and algebraic geometry. As mentioned above, [112] and [117] have all the details. Theorem 9.3.2, as presented here, is essentially proven in [148], although the notation and terminology there are a little different from ours.

Soon, in Section 9.4, we shall see how all of this affects simple Euclidean examples, in which case we shall recover the integral-geometric results of Section 6.2 for which we *did* give full proofs.

What we shall prove now is the following corollary, which is actually what we shall be using in the future. The proof is included, since, unlike that for Morse's theorem itself, it does not seem to appear in the literature. Note that we shall have to add tameness to the list of conditions (by demanding regularity) since the proof depends, in its last line, on Theorem 9.2.6.

Corollary 9.3.5. *Let M be a regular stratified manifold embedded in a C^3 manifold $\widetilde{M}$. Let $\widetilde{f} \in C^2(\widetilde{M})$ be a Morse function on M, and let $u \in \mathbb{R}$ be a regular value of $f_{|\partial_j M}$ for all $j = 0, \dots, N$. Then, writing $f = \widetilde{f}_{|M}$,*

$$\begin{aligned}\varphi(M \cap f^{-1}[u, \infty)) \qquad (9.3.3)\\ = \sum_{j=0}^{N} \sum_{\{t \in \partial_j M : f_t \geq u, \nabla f_t \in T_t^{\perp} \partial_j M\}} (-1)^{\iota_{-f, \partial_j M}(t)} \alpha(-P^{\perp}_{T_t \partial_j M} \nabla f_t; M).\end{aligned}$$

If M is also locally convex, then

$$\begin{aligned}\varphi(M \cap f^{-1}[u, \infty)) \qquad (9.3.4)\\ = \sum_{j=0}^{N} \sum_{\{t \in \partial_j M : f_t > u, \nabla f_t \in T_t^{\perp} \partial_j M\}} (-1)^{\iota_{-f, \partial_j M}(t)} \mathbb{1}_{\{\nabla f_t \in N_t M\}}.\end{aligned}$$

Proof. As usual, write $A_u = M \cap \widetilde{f}^{-1}[u, \infty)$. If $-\widetilde{f}$ were a Morse function on A_u, and if we changed the condition $\widetilde{f}_t > u$ to $\widetilde{f}_t \geq u$ on the right-hand side of (9.3.3), then the corollary would merely be a restatement of Morse's theorem, and there would be nothing to prove. However, the change is not obvious, and $-\widetilde{f}$ is not a Morse function on A_u, since it is constant on the strata of A_u that are subsets of $M \cap \widetilde{f}^{-1}\{u\}$.

The bulk of the proof involves finding a Morse function $\widehat{f}$ on A_u that agrees with $-\widetilde{f}$ on "most" of this set (thus solving the problem of the "non-Morseness" of $\widetilde{f}$) and that, at the same time, has critical points in a one-to-one correspondence with the critical points of f on M above the level u. (This allows us to replace $\widetilde{f}_t \geq u$ with $\widetilde{f}_t > u$).

The space $M \cap f^{-1}[u, \infty)$ can be decomposed into j-dimensional strata of the form $\partial_j M \cap f^{-1}(u, \infty)$ and $\partial_{j+1} M \cap f^{-1}\{u\}$. More formally, consider $\widetilde{f}$ as a function on $\widetilde{M}$, which, since it is a manifold, has no boundary. Then

$$\partial_N \left(\widetilde{f}^{-1}[u, \infty) \right) = \widetilde{f}^{-1}(u, \infty),$$
$$\partial_{N-1} \left(\widetilde{f}^{-1}[u, \infty) \right) = \widetilde{f}^{-1}\{u\},$$
$$\partial_j \left(\widetilde{f}^{-1}[u, \infty) \right) = \emptyset \quad \text{for } j = 0, \ldots, N-2.$$

Since $\widetilde{f}$ is a Morse function and u is a regular point for $f_{|\partial_k M}$ for all k, it follows that M and $f^{-1}[u, \infty)$ intersect transversally as subsets of $\widetilde{M}$. Therefore A_u is a (locally convex) regular subspace of $\widetilde{M}$ that can be decomposed as

$$A_u = \bigcup_{j=0}^{N} \left(\partial_j M \cap \widetilde{f}^{-1}(u, \infty) \right) \cup \left(\partial_{j+1} M \cap \widetilde{f}^{-1}\{u\} \right) \tag{9.3.5}$$
$$= \left(\bigcup_{j=0}^{N} \partial_j M \cap \widetilde{f}^{-1}(u, \infty) \right) \cup \left(\bigcup_{j=0}^{N-1} \partial_{j+1} M \cap \widetilde{f}^{-1}\{u\} \right).$$

It is the last term here that gives rise to problems, since it is here that $\widetilde{f}_{|A_u}$ loses its property of being a Morse function (on A_u). Thus we search for a replacement to f that is well behaved on this boundary set.

Since $\widetilde{f}$ is a Morse function on M, it has only finitely many critical points inside a relatively compact neighborhood V of M. Furthermore, exploiting the fact that u is a regular value of $f_{|\partial_k M}$ for every k, there exists an $\varepsilon > 0$ such that

$$U_\varepsilon = \widetilde{f}^{-1}(u - \varepsilon, u + \varepsilon) \cap M \cap V$$

contains no critical points of $f_{|M}$. It is standard fare that there exists an $h \in C^2(\widetilde{M})$ that is a Morse function on $f^{-1}\{u\}$ and that is zero outside of U_ε. Furthermore, since $\overline{V}$ is compact, there exist K_f and K_h such that $|\nabla h| < K_h$ and

$$|P_{T_t \partial_j M} \nabla \widetilde{f}| > K_f,$$

for all $t \in \partial_j M \cap U_\varepsilon$, $1 \le j \le N$.

It then follows that the function

$$\widehat{f} \triangleq -\widetilde{f} + \frac{K_f}{3K_h} h$$

is a Morse function on A_u. By our choice of h, the critical points of $\widehat{f}_{|A_u}$ agree with those of f on $M \cap U_\varepsilon^c$. Furthermore, $\nabla f_{|\partial_j M} \equiv \pi_{T_t \partial_j M} \nabla f$ can never be zero on $U_\varepsilon \cap \partial_k M$, and so there are no critical points of $\widehat{f}$ at all in this region. Consequently, the critical points of $\widehat{f}$ on $M \cap \widetilde{f}^{-1}[u, \infty)$ are in one-to-one correspondence with those of $\widetilde{f}$ on $M \cap \widetilde{f}^{-1}(u, \infty)$.

The result then follows from the fact that by Theorem 9.2.6 (cf. (9.2.22)) the normal Morse indices at these points are the same for M as they are for $M \cap f^{-1}[u, \infty)$. $\square$

9.4 The Euclidean Case

With basic Morse theory for piecewise smooth spaces under our belt, it is now time to look at one rather important example for which everything becomes quite simple. The example is that of the N-dimensional cube $I^N = [0, 1]^N$, and the ambient space is $\mathbb{R}^N$ with the usual Euclidean metric. In particular, we want to recover Theorem 6.2.4, which gave a point set representation for the Euler characteristic of the excursion set of smooth functions over the square.

To recover Theorem 6.2.4 for the unit square we use Morse's theorem 9.3.2 in its original version. Reserving the notation f for the function of interest that generates the excursion sets, write the f of Morse's theorem as f_m. We are interested in computing the Euler characteristic of the set

$$A_u = \{t \in I^2 : f(t) \geq u\}$$

and will take as our Morse function the "height function"

$$f_m(t) = f_m(t_1, t_2) = t_2.$$

Now assume that f is "suitably regular" in the sense of Definition 6.2.1. This is almost enough to guarantee that f_m is a Morse function over I^2 for the ambient manifold $\mathbb{R}^2$. Unfortunately, however, all the points along the top and bottom boundaries of I^2 are degenerate critical points for f_m. We get around this by replacing I^2 with a tilted version, I^2_ε, obtained by rotating the square through ε degrees, as in Figure 9.4.1.

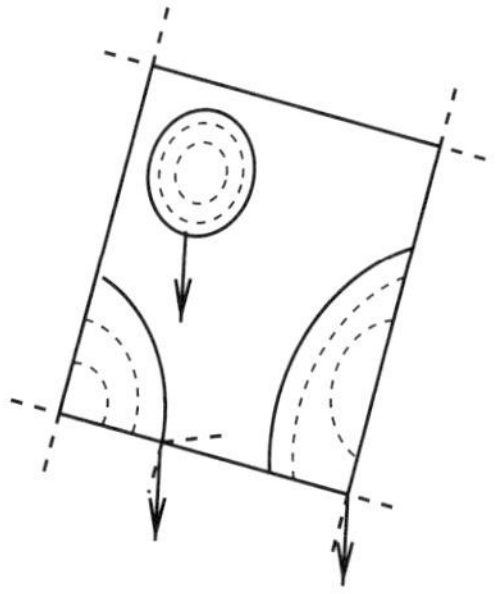

Fig. 9.4.1. The Euler characteristic via Theorem 9.3.2.

To compute $\varphi(A_u)$ we now apply (9.3.1), and so need to characterize the various critical points and their indices. The first fact to note is that, using the usual coordinate system, we have $\nabla f_m = (0, 1)$, and so there are no critical points of f_m in A°_u. Thus we can restrict interest to the boundary ∂A_u which we break into three parts:

(i) points $t \in (I^2_\varepsilon)^\circ \cap \partial A_u$;
(ii) points $t \in \partial I^2_\varepsilon \cap A_u$, but not vertices of the square;
(iii) the four vertices of the square.

An example of each of these three classes appears in Figure 9.4.1, where the excursion set of f appears along with contour lines in the interiors of the various components.

At points of type (i), $f(t) = u$. Furthermore, since the normal cone $N_t(A_u)$ is then the one-dimensional vector space normal to ∂A_u, $-\nabla f_m = (0, -1) \in N_t(A_u)$ at points for which $\partial f/\partial t_1 = 0$ and $\partial f/\partial t_2 > 0$. Such a point is at the base of the arrow coming out of the disk in Figure 9.4.1. Differentiating between points that contribute $+1$ and -1 to the Euler characteristic involves looking at $\partial^2 f/\partial t_1^2$. Comparing with Theorem 6.2.4, we see that we have characterized the contributions of (6.2.17) and (6.2.18) to the Euler characteristic.

We now turn to points of type (ii). Again due constancy of ∇f_m, this time on ∂I_ε^2, there are no critical points to be counted on $(\partial I_\varepsilon^2 \cap A_u)^\circ$. We can therefore add the endpoints of the intervals making up $\partial I_\varepsilon^2 \cap A_u$ to those of type (iii). One of these appears as the base of the leftmost arrow on the base of Figure 9.4.1. The rightmost arrow extends from such a vertex.

For points of these kinds, the normal cone is a closed wedge in $\mathbb{R}^2$, and it is left to you to check that the contributions of these points correspond (on taking $\varepsilon \to 0$) to those of those described by (6.2.16).

This gives us Theorem 6.2.4, which trivially extends to any rectangle in $\mathbb{R}^2$. You should now check that Theorem 6.2.5, which computed the Euler characteristic for a subset of $\mathbb{R}^2$ with piecewise C^2 boundary, also follows from Morse's theorem, using the same f_m.

The above argument was really unnecessarily complicated, since it did not use Corollary 9.3.5, which we built specifically for the purpose of handling excursion sets. Nevertheless, it did have the value of connecting the integral-geometric and differential-geometric approaches.

Now we apply Corollary 9.3.5. We again assume that f is suitably regular at the level u in the sense of Definition 6.2.1, which suffices to guarantee that it is a Morse function over the C^2 piecewise smooth space I^N for the ambient manifold $\mathbb{R}^N$ and that the conditions of the corollary apply.

Write $\mathcal{J}_k \equiv \partial_k I^N$ for the collection of faces of dimension k in I^N (cf. (6.2.2)). With this notation, we can rewrite the sum (9.3.3) as

$$\varphi(A_u(f, I^N)) = \sum_{k=0}^{N} \sum_{J \in \mathcal{J}_k} \sum_{i=0}^{k} (-1)^i \mu_i(J), \tag{9.4.1}$$

where, for $i \leq \dim(J)$,

$$\mu_i(J) \stackrel{\Delta}{=} \#\{t \in J : \nabla f_{|J}(t) = 0, \iota_{-f,J}(t) = i\}.$$

Recall that to each face $J \in \mathcal{J}_k$ there corresponds a subset $\sigma(J)$ of $\{1, \ldots, N\}$, of size k, and a sequence of $N-k$ zeros and ones $\varepsilon(J) = \{\varepsilon_1, \ldots, \varepsilon_{N-k}\}$ such that

$$J = \{t \in I^N : t_j = \varepsilon_j \text{ if } j \notin \sigma(J),\ 0 < t_j < 1 \text{ if } j \in \sigma(J)\}.$$

Set $\varepsilon_j^* = 2\varepsilon_j - 1$. Working with the definition of the C_i, it is then not hard to see that $\mu_i(J)$ is given by the number of points $t \in J$ satisfying the following four conditions:

$$f(t) \geq u, \tag{9.4.2}$$
$$f_j(t) = 0, \qquad j \in \sigma(J), \tag{9.4.3}$$
$$\varepsilon_j^* f_j(t) > 0, \qquad j \notin \sigma(J), \tag{9.4.4}$$
$$\text{index}(f_{mn}(t))_{(m,n\in\sigma(J))} = k - i, \tag{9.4.5}$$

where, as usual, subscripts denote partial differentiation, and, consistent with the definition of the index of a critical point, we define the *index of a matrix* to be the number of its negative eigenvalues.

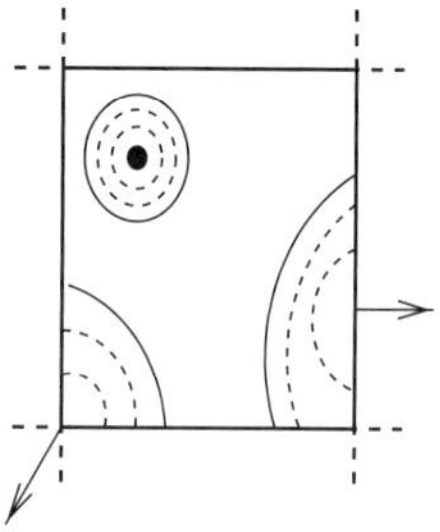

Fig. 9.4.2. The Euler characteristic via Corollary 9.3.5.

In Figure 9.4.2 there are three points that contribute to $\varphi(A_u(f, I^2))$. One, in the center of the upper left disk, contributes via $J = (I^2)^\circ = \mathcal{J}_2$. That on the right side contributes via $J =$ "right side" $\in \mathcal{J}_1$, and that on the lower left corner via $J = \{0\} \in \mathcal{J}_0$.

The representation (9.4.1) of the Euler characteristic of an excursion set, along with the prescription in (9.4.2)–(9.4.5) as to how to count the contributions of various points to the sum, is clearly a tidier way of writing things than what we obtained via integral-geometric methods. Nevertheless, it is now clear that the two are essentially different versions of the same basic result. However, it is the compactness of (9.4.1) that will be of importance to us in the upcoming computations for random excursion sets in Part III of the book.

10

Volume of Tubes

In Chapters 7 and 8 we invested a good deal of time and energy in developing the many results we need from differential geometry. The time has now come to begin to reap the benefits of our investment, while at the same time developing some themes a little further for later exploitation. This chapter focuses on the celebrated volume-of-tubes formula of Weyl [73, 168], which expresses the Lebesgue volume of a tube of radius ρ around a set M embedded in $\mathbb{R}^l$ or $S(\mathbb{R}^l)$ in terms of the radius of the tube[1] and the Lipschitz–Killing curvatures of M (see Theorem 10.5.6). It is an interesting fact, particularly in view of the fact that[2] this is a book about probability that is claimed to have applications to statistics, and despite the fact that Weyl's formula is today the basis of a large literature in geometry, that the origins of the volume-of-tubes formulas were inspired by a statistical problem. This problem, along with its solution due to Hotelling [79], were related to regression analysis and involved the one-dimensional volume-of-tubes problem on a sphere, not unrelated to the computation we shall do in a moment.

Much of the work in using Weyl's tube formula and its generalizations lies in deriving explicit expressions for Lipschitz–Killing curvature measures. This being the case, even if you are comfortable with the material of Chapter 7 and are not interested in the volume-of-tubes approximation per se, you will find it useful to read Sections 10.7 and 10.9 of this chapter. These describe the Lipschitz–Killing curvature measures of stratified Whitney spaces and the generalized Lipschitz–Killing curvature measures that will be needed in Chapter 15.

Beginning in the late 1980s and early 1990s, the usefulness of the tube formula in statistics was rediscovered, and there were a number of works applying these formulas to statistical problems, as in [87, 93]. A particularly interesting paper for us was a 1993 paper by Jiayang Sun [147], which came out of a thesis under David Siegmund. It had an early version (albeit not easily recognizable as such) of the

---

[1] A word on notation: If (T, d) is a metric space, then the sphere of radius λ in T is denoted by $S_\lambda(T)$, with $S(T) \equiv S_1(T)$. When $T = \mathbb{R}^N$ we continue, when convenient, to use the notation S_λ^{N-1} and S^{N-1} for $S_\lambda(\mathbb{R}^N)$ and $S(\mathbb{R}^N)$, respectively.

[2] Despite current appearances!

simple[3] approximation

$$\mathbb{P}\left\{\sup_{t\in M} f_t \geq u\right\} \approx \sum_{j=0}^{N} \rho_j(u)\mathcal{L}_j(\psi(M)). \tag{10.0.1}$$

Here f is a sufficently nice, centered, unit-variance Gaussian field over a sufficiently nice N-dimensional parameter space M. The functionals $\mathcal{L}_j$ are Lipschitz–Killing curvatures on a Euclidean sphere equipped with the Euclidean geodesic metric and are defined in (10.5.4). The functions ρ_j are defined as

$$\rho_j(u) \triangleq \begin{cases} 1-\Phi(u), & j=0, \\ (2\pi)^{-(j+1)/2} H_{j-1}(u)e^{-u^2/2}, & j\geq 1, \end{cases}$$

where H_j is the jth Hermite polynomial, which we shall meet in more detail later in Chapter 11 (cf. (11.6.9)). Finally, ψ is a mapping from the parameter space M to a unit sphere. Exactly which mapping and a sphere of which dimension will be clear after you have read Sections 10.2 and 10.6.

As we shall soon see, the above approximation, given enough side conditions, is actually rather simple to derive using volume-of-tube techniques. Furthermore, when both are defined, it agrees with the approximation that we shall derive in Chapter 14, which is based on the expected Euler characteristic and which is much harder to establish. Unfortunately, however, the volume-of-tubes approach has the disadvantage of being restricted to a somewhat small class of Gaussian processes, in that they are required to possess a finite Karhunen–Loève expansion. Specifically, these are processes defined on a manifold M that can be expressed as

$$f(t,\omega) = \langle \psi_t, \xi(\omega)\rangle_{\mathbb{R}^l} = \sum_{j=1}^{l} \xi_j(\omega)\psi_j(t), \tag{10.0.2}$$

for some smooth mapping $\psi : M \to S(\mathbb{R}^l)$,[4] where the ξ_j are independent, standard Gaussians.

The volume-of-tubes-based derivation of the approximation (10.0.1) also has the disadvantage that all results on its accuracy are very distributional in nature and there is no known way to even begin thinking about how to extend them to non-Gaussian processes. In any case, we shall not address the accuracy of the approximation in this chapter, since it will follow as a corollary to the accuracy of the expected Euler characteristic approach in Chapter 14 and the equivalence of the two approaches [148] for finite Karhunen–Loève processes.

[3] This is "simple" as long as you are also comfortable applying the same adjective to Lipschitz–Killing curvatures, defined below in (10.5.4).

[4] While if $M = I^N$, this expansion may indeed arise from the $L^2(I^N)$ expansion of a covariance function as in Section 3.2, this is not necessary, the main issue being the existence of a finite orthogonal expansion of some kind. Consequently, these processes are more appropriately referred to as *finite orthogonal expansion processes*, although we shall retain the nomenclature "finite Karhunen–Loève expansion processes" for historical reasons.

What we shall do in this chapter is describe the volume-of-tubes approximation to (10.0.1) in detail and also derive Weyl's tube formula for locally convex subspaces of $\mathbb{R}^l$ and $S(\mathbb{R}^l)$. In our development of this approximation, we also consider the volume-of-tubes problem in a slightly more general setting, necessary for the calculations of Chapter 15.

In the following section we shall give a very brief description of the volume-of-tubes problem in a general setting, specializing only later to the case of locally convex spaces embedded in $\mathbb{R}^l$ or $S(\mathbb{R}^l)$, where one can develop very precise results. Before getting into details, however, we shall describe in Section 10.2 the connection between tubes and (10.0.1) so that you can see why all of this might be related to probability in general and Gaussian processes in particular. However, if you do not require the motivation, feel free to skip Section 10.2.

10.1 The Volume-of-Tubes Problem

For a metric space (T, τ) the tube of radius ρ around $A \subset T$ is, as we have already seen in the Euclidean setting, defined as

$$\text{Tube}(A, \rho) \overset{\Delta}{=} \{x \in T : \tau(x, A) \leq \rho\} = \bigcup_{y \in A} B_\tau(y, \rho), \tag{10.1.1}$$

where

$$\tau(x, A) \overset{\Delta}{=} \inf_{y \in A} \tau(x, y) \tag{10.1.2}$$

is the usual distance function of the set A.

The volume-of-tubes problem[5] seeks to compute, when possible,

$$\mu\left(\text{Tube}(A, \rho)\right)$$

for some measure μ, as a function of ρ. When this is not possible, it seeks approximations to $\mu(\text{Tube}(A, \rho))$.

Of course, the possibility of such a computation, or the accuracy and validity of such an approximation, will depend on the properties of the measure μ as well as the set A. The simplest spaces to work on are the metric spaces $\mathbb{R}^l$ with its natural metric structure and $S(\mathbb{R}^l)$ with the geodesic metric from the induced Riemannian structure on $S(\mathbb{R}^l)$. The natural measures to consider in these examples are Lebesgue measure on $\mathbb{R}^l$ and surface measure on $S(\mathbb{R}^l)$, which agree with the Riemannian measures induced by the canonical Riemannian structures. These are the two main examples that we shall consider. In addition, in Section 10.9 we shall also investigate the Gaussian volume of tubes, where the metric space is $\mathbb{R}^l$ with its standard metric

[5] Since μ is a general measure, it would really be more appropriate to talk of the *measure* of tubes rather than their "volumes." However, we shall stay with the more standard "volume" terminology."

and the measure[6] $\mu = \gamma_{\mathbb{R}^l}$ the distribution of a random vector $W \sim N(0, I_{l\times l})$. This example will play a crucial role in the important computations of Chapter 15.

As for what type of sets A we need to consider, in order to derive explicit formulas for $\mu(\text{Tube}(A, \rho))$, we shall have to restrict ourselves to sets A that are embedded piecewise C^2 submanifolds of $\mathbb{R}^l$ or $S(\mathbb{R}^l)$. For much of the discussion we shall also limit ourselves to Lebesgue measure in the first case and surface measure in the second. Both of these, of course, are the volume, or Hausdorff, measure induced by the standard Euclidean metric in $\mathbb{R}^l$ (cf. footnote 23 of Chapter 7). In the final section of the chapter we shall see what happens when μ is Gauss measure on $\mathbb{R}^l$.

10.2 Volume of Tubes and Gaussian Processes

We now start a discussion of the connection between the excursion probability (10.0.1) and the volume of tubes for finite Karhunen–Loève processes. We shall conclude it only in Section 10.6, when we shall have more information on tube formulas.

Take a Gaussian process f, which is a restriction to a locally convex submanifold M of a process $\widetilde{f}$ on an ambient manifold $\widetilde{M}$ that has the representation (10.0.2), that is,

$$\widetilde{f}(t, \omega) = \langle \psi_t, \xi(\omega)\rangle_{\mathbb{R}^l} = \sum_{j=1}^{l} \xi_j(\omega)\psi_j(t). \tag{10.2.1}$$

Note that we have introduced a new parameter here. Whereas we still use N to denote $\dim(M) = \dim(\widetilde{M})$, the new parameter l denotes the order of the orthogonal expansion.

Note also that the vector $\xi/|\xi|$ is uniformly distributed on $S(\mathbb{R}^l)$, independently of $|\xi|$, which is distributed as the square root of a χ^2_l random variable. We shall write η_l to denote the distribution of $\xi/|\xi|$, that is, the uniform measure over $S(\mathbb{R}^l)$.

Finally, we shall assume that $\widetilde{f}$ has constant unit variance, so that (10.2.1) immediately implies

$$|\psi(t)|^2 \stackrel{\Delta}{=} \sum_{j=1}^{l} \psi_j^2(t) = 1, \tag{10.2.2}$$

for all $t \in \widetilde{M}$. This being the case, we can define the map $\psi : t \to (\psi_i(t), \dots, \psi_l(t))$, an embedding of M in $S(\mathbb{R}^l)$. More significantly, we can define a random field f on $\psi(M) \in S(\mathbb{R}^l)$ by setting

[6] For a finite-dimensional Hilbert space, H, we shall use γ_H to denote the canonical Gaussian random vector on H. That is, if $X \sim \gamma_H$, then for any orthonormal basis $\zeta_1, \dots, \zeta_{\dim(H)}$, the sequence

$$\left(\langle\zeta_1, X\rangle_H, \dots, \langle\zeta_{\dim(H)}, X\rangle_H\right)$$

is a sequence of independent, standard Gaussians.

$$f(x) \triangleq \widetilde{f}\left(\psi^{-1}(x)\right), \tag{10.2.3}$$

for all $x \in \psi(M)$. Note that f has the simple covariance function

$$C(x, y) = \mathbb{E}\{f(x)f(y)\} = \langle x, y\rangle, \tag{10.2.4}$$

and thus there is no problem taking a version of it on all of $S(\mathbb{R}^l)$ with this covariance function. This process is known as the *canonical (isotropic) Gausssian field on* $S(\mathbb{R}^l)$. Apart from our need of it now, it will play a central role in the calculations of Chapter 15, and more details on it can be found in Section 15.6.

For our current purposes, we note that since it is trivial that

$$\sup_{t\in M} \widetilde{f}(t) \equiv \sup_{x\in\psi(M)} f(x),$$

it is clear that in computing excursion probabilities for unit-variance finite Karhunen–Loève expansion fields, we lose no generality by treating only the canonical process over subsets of $S(\mathbb{R}^l)$.

Returning to tubes, when the metric space (T, τ) is $S(\mathbb{R}^l)$ with the geodesic metric, i.e., when

$$\tau(x, y) = \cos^{-1}\left(\langle x, y\rangle\right),$$

then a tube of radius ρ around a closed set A can be expressed as

$$\begin{aligned} \text{Tube}(A, \rho) &= \left\{x \in S(\mathbb{R}^l) : \tau(x, A) \le \rho\right\} \\ &= \left\{x \in S(\mathbb{R}^l) : \exists y \in A \text{ such that} \langle x, y\rangle \ge \cos(\rho)\right\} \\ &= \left\{x \in S(\mathbb{R}^l) : \sup_{y\in A}\langle x, y\rangle \ge \cos(\rho)\right\}. \end{aligned} \tag{10.2.5}$$

If we think of X being a random vector uniformly distributed on $S(\mathbb{R}^l)$ with distribution η_l, then (10.2.5) implies that

$$\begin{aligned} \mathbb{P}\left\{\sup_{y\in A}\langle X, y\rangle \ge \cos(\rho)\right\} &= \eta_l\left(\left\{x \in S(\mathbb{R}^l) : \sup_{y\in A}\langle x, y\rangle \ge \cos(\rho)\right\}\right) \\ &= \eta_l\left(\text{Tube}(A, \rho)\right). \end{aligned}$$

We now have enough to write out the basic equation in the volume-of-tubes approach for such processes:

$$
\begin{aligned}
\mathbb{P}\left\{\sup_{t\in M} f_t \geq u\right\} &= \int_0^\infty \mathbb{P}\left\{\sup_{t\in M} f_t \geq u \,\middle|\, |\xi| = r\right\} \mathbb{P}_{|\xi|}(dr) \\
&= \int_0^\infty \mathbb{P}\left\{\sup_{t\in M} \langle \psi_t, \xi\rangle \geq u \,\middle|\, |\xi| = r\right\} \mathbb{P}_{|\xi|}(dr) \\
&= \int_u^\infty \mathbb{P}\left\{\sup_{t\in M} \langle \psi_t, \xi\rangle \geq u \,\middle|\, |\xi| = r\right\} \mathbb{P}_{|\xi|}(dr) \\
&= \int_u^\infty \mathbb{P}\left\{\sup_{t\in M} \langle \psi_t, \xi/r\rangle \geq u/r \,\middle|\, |\xi| = r\right\} \mathbb{P}_{|\xi|}(dr) \\
&= \int_u^\infty \mathbb{P}\left\{\sup_{s\in \psi(M)} \langle s, \xi/r\rangle \geq u/r \,\middle|\, |\xi| = r\right\} \mathbb{P}_{|\xi|}(dr) \\
&= \int_u^\infty \eta_l\left(\mathrm{Tube}(\psi(M), \cos^{-1}(u/r))\right) \mathbb{P}_{|\xi|}(dr) \\
&= \frac{\Gamma\left(\frac{l}{2}\right)}{2\pi^{l/2}} \int_u^\infty \mathcal{H}_{l-1}\left(\mathrm{Tube}(\psi(M), \cos^{-1}(u/r))\right) \mathbb{P}_{|\xi|}(dr) \\
&= \frac{\Gamma\left(\frac{l}{2}\right)}{2\pi^{l/2}} \mathbb{E}\left\{\mathcal{H}_{l-1}\left(\mathrm{Tube}(\psi(M), \cos^{-1}(u/|\xi|))\right) \mathbb{1}_{\{|\xi|\geq u\}}\right\}.
\end{aligned}
\tag{10.2.6}
$$

Therefore, the excursion probability (10.0.1) is a weighted average of the volume of tubes around $\psi(M)$ of varying radii, and being able to compute

$$
\mathcal{H}_{l-1}\left(\mathrm{Tube}(\psi(M), \rho)\right)
$$

will go a long way to computing (10.0.1). In particular, if it were true that, in some approximate sense,

$$
\mathcal{H}_{l-1}\left(\mathrm{Tube}(\psi(M), \rho)\right) \approx \sum_j \widetilde{G}_{j,l}(\rho)\mathcal{L}_j(\psi(M)) \tag{10.2.7}
$$

for some functions $\widetilde{G}_{j,l}$, then the right-hand side of (10.2.6), and hence (10.0.1), would be approximately

$$
\frac{\Gamma\left(\frac{l}{2}\right)}{2\pi^{l/2}} \sum_j \mathbb{E}\left\{\widetilde{G}_{j,l}(\cos^{-1}(u/|\xi|))\mathbb{1}_{\{|\xi|\geq u\}}\right\} \mathcal{L}_j(\psi(M)). \tag{10.2.8}
$$

Much of the rest of this chapter is devoted to justifying this approximation[7] and obtaining explicit representations for $\widetilde{G}_{j,l}$. The justification is based on Weyl's tube formula on $S_\lambda(\mathbb{R}^l)$, given below in Theorem 10.5.7. Weyl's formula states that (10.2.7) is exact for ρ small enough if $\psi(M)$ is a locally convex space. Furthermore, it identifies the functions $\widetilde{G}_{j,l}(\rho)$.

[7] Note that we can really expect the approximation to work only for large u. The issue is that we expect tube expansion (10.2.7) to hold only for small ρ. Since ρ is replaced by $\cos^{-1}(u/|\xi|)$ in (10.2.8), we can therefore expect the tube approximation to work well only if, given $|\xi| \geq u$, $|\xi|$ is close to u with high probability. This will happen only if u is large.

Deriving Weyl's formula requires an understanding of the local properties of tubes, to which we devote the following section.

10.3 Local Geometry of Tube(M, ρ)

This section, as its title suggests, describes the local geometry of Tube(M, ρ), where throughout, we shall assume that M is a locally convex, C^2, Whitney stratified manifold. To formally define the local geometry of the tube, we first have to specify the ambient space in which M is assumed to be embedded and the metric space in which the tube lives. Although this sounds rather pedantic, it is important in appreciating the *intrinsic* nature of the final form of the volume-of-tubes formula. As a locally convex space, M by definition needs an ambient space $\widetilde{M}$ in which to be embedded, but $\widetilde{M}$ itself might be embedded in a Riemannian manifold $(\widehat{M}, \widehat{g})$, so that we have the inclusion[8]

$$M \subset \widetilde{M} \subset \widehat{M}.$$

In this case,

$$\text{Tube}(M, \rho) = \left\{x \in \widehat{M} : d_{\widehat{M}}(x, M) \leq \rho\right\}, \tag{10.3.1}$$

where $d_{\widehat{M}}$ is geodesic distance on $\widehat{M}$.

One of the main conclusions of Weyl's formula is that if $\widehat{M}$ is either $\mathbb{R}^l$ or $S(\mathbb{R}^l)$, then the volume of Tube(M, ρ) is intrinsic in the sense of the following definition.

Definition 10.3.1. *A quantity $Q(M)$ is intrinsic to a stratified manifold M embedded in $(\widetilde{M}, \widetilde{g})$ if it can be computed based solely on the Riemannian metric of $(\widetilde{M}, \widetilde{g})$. That is, if $\widetilde{M}$ is isometrically embedded in $(\widehat{M}, \widehat{g})$, and $i(M)$ is the image of M under the inclusion $i : \widetilde{M} \to \widehat{M}$, then $Q(M)$ is intrinsic if and only if*

$$Q(M) = Q(i(M)).$$

A particularly important example of intrinsic quantities is that of the Lipschitz–Killing curvature measures $\mathcal{L}_i(M; \cdot)$, defined in (10.5.4).

The property of being intrinsic is arguably what has made Weyl's tube formula so useful, because it implies that it is enough to know only local properties of M as it sits in $\widetilde{M}$, without needing to know anything about how it sits in $\widehat{M}$.

With the triple $M \subset \widetilde{M} \subset \widehat{M}$ in mind, we establish the following convention about notation: In general, $\widehat{\cdot}$ will denote an object when we are considering M as being embedded in $\widehat{M}$. For instance, $\widehat{\mathcal{S}_t M}$ is the support cone of M at $t \in \widehat{M}$, as opposed to $\mathcal{S}_t M$, the support cone of M at $t \in \widetilde{M}$. Sometimes, there is no distinction to be made, as in fact is the case when one is dealing with support cones, since it is

[8] To clarify things, think of $M = S^1 = S(\mathbb{R}^2)$, a one-dimensional circle embedded in the two-dimensional plane $\widetilde{M} = \mathbb{R}^2$. If $\widehat{M} = \mathbb{R}^3$, then the tube around M, as defined by (10.3.1), is a solid torus.

easy to see that $\mathcal{S}_t M$ and $\widehat{\mathcal{S}_t M}$ are isomorphic. However, there *is* a difference between $N_t M$ and $\widehat{N_t M}$. Specifically,

$$\widehat{N_t M} = N_t M \oplus T_t \widetilde{M}^{\perp} = \left\{ X_t \in T_t \widehat{M} : X_t = V_t + Y_t,\ V_t \in N_t M,\ Y_t \in T_t \widetilde{M}^{\perp} \right\},$$

where $T_t \widetilde{M}^{\perp}$ is the orthogonal complement of $T_t \widetilde{M}$ in $T_t \widehat{M}$.

10.3.1 Basic Structure of Tubes

We now begin to describe the local structure of Tube(M, ρ), beginning with a linearization of $\widehat{M}$ near $t \in M$, which we use as motivation for an explicit description of Tube(M, ρ). Our ultimate goal is to come up with an explicit parameterization of the tube that we can use to compute its volume via the differential-geometric tools of Chapter 7.

As with all linearizations, there will be a region over which it works well, and so we shall need to define various notions of *critical radius* for Tube(M, ρ). As we shall see, Weyl's tube formula is valid only for values of ρ smaller than the critical radius of M, since for larger ρ the explicit parameterization of the tube breaks down.

The linearization we shall use is essentially a linearization of the metric projection $\xi_M : \widehat{M} \to M$ given by

$$\xi_M(s) \triangleq \mathrm{argmin}_{t \in M}\, \widehat{d}(s, t), \tag{10.3.2}$$

which parameterizes the tube as

$$\mathrm{Tube}(M, \rho) = \bigcup_{t \in M} \left\{ s \in \widehat{M} : \xi_M(s) = t,\, \widehat{d}(s, t) \le \rho \right\},$$

where the union is disjoint if ρ is small enough.

Actually, we shall need more than this, since we shall also need an explicit description of the level sets

$$\left\{ s \in \widehat{M} : \xi_M(s) = t \right\}$$

of ξ_M. To do this, we start by approximating ξ_M with the projection onto the convex cone $\widehat{\mathcal{S}_t M}$, defined on $T_t \widehat{M}$ by

$$P_{\widehat{\mathcal{S}_t M}}(X_t) = \mathrm{argmin}_{Y_t \in \widehat{\mathcal{S}_t M}} |X_t - Y_t| = \mathrm{argmin}_{Y_t \in \mathcal{S}_t M} |P_{T_t \widetilde{M}} X_t - Y_t|. \tag{10.3.3}$$

A simple convexity argument then shows that X_t projects to the origin if and only if $P_{T_t \widetilde{M}} X_t \in N_t M$. Alternatively, if $P_{T_t \widetilde{M}} X_t \notin N_t M$, then there exists some point $Y_t \neq 0 \in \mathcal{S}_t M$ such that $|P_{T_t \widetilde{M}} X_t - Y_t| < |X_t|$.

This argument is based on a local linearization of $\widehat{M}$ and M at $t \in M$; hence it holds only locally. However, using the fact that near t, M can be approximated by $\mathcal{S}_t M$, it is sufficient to establish that

$$\left\{s \in \widehat{M} : \xi_M(s) = t,\ \widehat{d}(s,t) \le \rho\right\}$$
$$= \exp_{\widehat{M}}\left(\left\{X_t \in T_t\widehat{M} : P_{T_t\widetilde{M}} X_t \in N_t M, |X_t| \le \rho\right\}\right).$$

Therefore, for ρ small enough,

$$\begin{aligned} \text{Tube}(M,\rho) &= \bigcup_{t\in M} \bigcup_{\{X_t \in T_t\widehat{M} : |X_t| \le \rho\}} \exp_{\widehat{M}}(t, X_t) \qquad (10.3.4) \\ &= \bigcup_{t\in M} \bigcup_{\{X_t \in \widehat{N_t M} : |X_t| \le \rho\}} \exp_{\widehat{M}}(t, X_t). \end{aligned}$$

The second equality is based on the local linearization described above, i.e., if $X_t \notin \widehat{N_t M}$ then there exists a point $s \in M$ such that

$$\widehat{d}(\exp_{\widehat{M}}(t, X_t), s) < \widehat{d}(\exp_{\widehat{M}}(t, X_t), t).$$

Therefore, we can leave out all vectors X_t such that $X_t \notin \widehat{N_t M}$.

At this point, we have not yet quantified how small ρ should be for the above argument to work. First of all, ρ should be small enough so that

$$\exp_{\widehat{M}} : B(T_t\widehat{M}, \rho) \to \widehat{M}$$

is a diffeomorphism onto its range. This means that ρ should be less than

$$\widehat{\rho}_c(\widehat{M}) \stackrel{\Delta}{=} \sup\left\{r : \exp_{\widehat{M}} : B(T_t\widehat{M}, r) \to \widehat{M} \text{ is a diffeomorphism for all } t \in \widehat{M}\right\}, \qquad (10.3.5)$$

the *radius of injectivity* of $\widehat{M}$.

In general, we shall have to take ρ even smaller, since we want the final expression in (10.3.4) to be a *disjoint* union. Suppose, then, that $\rho < \widehat{\rho}_c(\widehat{M})$. The geodesic with origin t and direction $X_t \in S(T_t\widehat{M})$ can be extended only so far without overlapping itself or a geodesic emanating from another point. We define $\rho_{c,l} : M \times (\widehat{N_t M} \cap S(T_t\widehat{M})) \to \mathbb{R}$, the *local critical radius of M in $\widehat{M}$ at t in the direction X_t*, as

$$\rho_{c,l}(t, X_t) \stackrel{\Delta}{=} \sup\left\{r : \widehat{d}(\exp_{\widehat{M}}(t, rX_t), M) = r\right\}. \qquad (10.3.6)$$

For $r < \rho_{c,l}(t, X_t)$ the point $\exp_{\widehat{M}}(t, rX_t)$ projects uniquely to the point t.

Taking the infimum over $N_t M$, we define $\rho_c : M \to \mathbb{R}$, the *local critical radius of M in $\widehat{M}$ at t*, as

$$\rho_c(t) \stackrel{\Delta}{=} \inf_{X_t \in \widehat{N_t M} \cap S(T_t\widehat{M})} \rho_{c,l}(t, X_t), \qquad (10.3.7)$$

and finally, the *critical radius of M in $\widehat{M}$* as

$$\rho_c(M, \widehat{M}) \stackrel{\Delta}{=} \inf_{t\in M} \rho_c(t). \qquad (10.3.8)$$

Therefore, for $\rho \le \min(\rho_c(M, \widehat{M}), \widehat{\rho_c}(\widehat{M}))$, the final expression in (10.3.4) is a disjoint union. Consequently, for such ρ, Tube(M, ρ) is the image of the region

$$\left\{(t, X_t, r) : X_t \in \widehat{N_t M} \cap S(T_t \widehat{M}),\ 0 \le r < \rho\right\}$$

under the bijection (for such ρ) $\widehat{F}$ defined by

$$\widehat{F}(t, X_t, r) \triangleq \exp_{\widehat{M}}(t, rX_t).$$

For values of ρ larger than $\min(\rho_c(M, \widehat{M}), \widehat{\rho_c}(\widehat{M}))$, the map $\widehat{F}$ may not be one-to-one. However, it is always one-to-one when restricted to the region

$$\left\{(t, X_t, r) : X_t \in \widehat{N_t M} \cap S(T_t \widehat{M}),\ 0 \le r < \min\left(\rho, \rho_{c,l}(t, X_t)\right)\right\}. \tag{10.3.9}$$

10.3.2 Stratifying the Tube

We now turn to describing the region (10.3.9) in more detail, essentially stratifying Tube(M, ρ). After this, computing the volume of Tube(M, ρ) reduces to computing an integral over each of its strata.

As usual, we assume that M is an N-dimensional, stratified Whitney space $(M, \mathcal{Z})$, decomposed as a disjoint union

$$M = \bigcup_{j=0}^{N} \partial_j M,$$

where the manifolds $\partial_j M$ are (possibly empty) j-dimensional, C^2, submanifolds of a Riemannian N-manifold $(\widetilde{M}, \widetilde{g})$. Recall that each $\partial_j M$ will in general contain many connected components, namely, all those sets in the partition $\mathcal{Z}$ of M of dimension j.

For each stratum $\partial_j M$ for $0 \le j \le l$ fix an orthonormal frame field

$$\eta = \left(\eta_1, \dots, \eta_{l-j}\right)$$

for $\widehat{T_t \partial_j M}^{\perp}$ as t varies over $\partial_j M$. The region (10.3.9) can then be expressed as the image of the disjoint union of the regions

$$D_j(\rho) \triangleq \bigcup_{t \in \partial_j M} \bigcup_{s \in S(\mathbb{R}^{l-j})} \bigcup_{0 \le r \le \min\left(\rho, \rho_{c,l}(t, \sum_i s_i \eta_i(t))\right)} \left\{(t, s, r) : \sum_i s_i \eta_i(t) \in \widehat{N_t M}\right\} \tag{10.3.10}$$

under the maps

$$G_j : (t, s, r) \to \left(t, \sum_i s_i \eta_i(t), r\right).$$

Finally, $\mathrm{Tube}(M, \rho)$ is the disjoint union of the sets $D_j(\rho)$ under the maps $F_j \triangleq \widehat{F} \circ G_j$, so that we have

$$F_j : (t, s, r) \to \exp_{\widehat{M}}\left(t, r \sum_i s_i \eta_i(t)\right). \tag{10.3.11}$$

That is,

$$\mathrm{Tube}(M, \rho) = \bigcup_j F_j(D_j(\rho)). \tag{10.3.12}$$

Note that while this gives the tube in $\widehat{M}$ around M, the same construction gives the tube in $\widetilde{M}$ around M if we merely replace $\exp_{\widehat{M}}$ in (10.3.11) by $\exp_{\widetilde{M}}$, although the critical radius of M in $\widetilde{M}$ is in general different from the critical radius of M in $\widehat{M}$.

10.4 Computing the Volume of a Tube

10.4.1 First Steps

In the previous section, we decomposed $\mathrm{Tube}(M, \rho)$ into strata corresponding to the strata of M. The strata of $\mathrm{Tube}(M, \rho)$ are open sets in $\widehat{M}$, and to compute $\mu(\mathrm{Tube}(M, \rho))$ we are left with an integral over each stratum of $\mathrm{Tube}(M, \rho)$. That is,

$$\mu(\mathrm{Tube}(M, \rho)) = \sum_j \mu(F_j(D_j(\rho))). \tag{10.4.1}$$

Assume now that μ is a measure associated to an integrable l-form α_μ (see (7.4.5)). Then it can be expressed as $f_\mu \cdot \mathrm{Vol}_{\widehat{M}}$, where $\mathrm{Vol}_{\widehat{M}}$ is the Riemannian volume form on $\widehat{M}$ and f_μ is bounded. Applying the coarea formula (7.4.13) to the distance function of M (which is almost everywhere[9] C^1 with Lipschitz gradient of norm 1) on $F_j(D_j(\rho))$, we find that

$$\mu(F_j(D_j(\rho))) = \int_0^\rho \int_{S_j(r)} \mathbb{1}_{F_j(D_j(\rho))}(x) f_\mu(x) \Omega_{j,r}\, dr, \tag{10.4.2}$$

where

$$S_j(r) \triangleq \{s \in \widehat{M} : d_{\widehat{M}}(s, \partial_j M) = r\}, \tag{10.4.3}$$

and $\Omega_{j,r}$ is the volume form induced on $S_j(r)$ by α_μ.

Alternatively, we can pull back (cf. (7.4.1)) $\Omega_{j,r}$ to the level sets making up $D_j(\rho)$ to obtain

[9] Note that whereas we stated the coarea theorem for differentiable functions, the distance function to $\partial_j M$ can be nondifferentiable on a set of dimension up to $l-1$. This, however, is a set of too small dimension to affect the integral in (10.4.2). See footnote 25 in Chapter 7.

$$\mu(F_j(D_j(\rho))) = \int_0^\rho \int_{\partial_j M \times S(\mathbb{R}^{l-j})} \mathbb{1}_{D_j(\rho)}(t,s,r) f_\mu(F_{j,r}(t,s)) F^*_{j,r}(\Omega_{j,r})\, dr, \tag{10.4.4}$$

where $F_{j,r}$ is the partial map

$$F_{j,r}(t,s) \triangleq F_j(t,s,r).$$

We are left with computing the pullback $F^*_{j,r}(\Omega_{j,r})$, as well as the integration over $D_j(\rho)$.

10.4.2 An Intermediate Computation

Although the details on how best to compute $F^*_{j,r}(\Omega_{j,r})$ tend to be problem-specific, there are some general preliminaries that will help for the two cases that we wish to consider in detail, in which $\widehat{M}$ is either $\mathbb{R}^l$ or $S(\mathbb{R}^l)$.

Fixing a point $(t,s) \in \partial_j M \times S(\mathbb{R}^{l-j})$, recall that by the definition of the pullback (7.4.1) and from (7.4.7),

$$\begin{aligned} F^*_{j,r}(\Omega_{j,r})&\left(X_{1,(t,s)}, \ldots, X_{l-1,(t,s)}\right) \\ &= \Omega_{j,r}\left(F_{j,r*}X_{1,(t,s)}, \ldots, F_{j,r*}X_{l-1,(t,s)}\right) \\ &= \det\left(\left\langle F_{j,r*}X_{k,(t,s)}, W_{m,F_{j,r}(t,s)}\right\rangle\right)_{1\le k,m\le l-1}, \end{aligned} \tag{10.4.5}$$

where $X_{j,(t,s)} \in T_{(t,s)}\partial_j M \times S(\mathbb{R}^{l-j})$ and $(W_{1,F_{j,r}(t,s)}, \ldots, W_{l-1,F_{j,r}(t,s)})$ is any (suitably oriented) orthonormal basis of the tangent space at $F_{j,r}(t,s)$ to the level set (10.4.3), a hypersurface in $\widehat{M}$. The inner product here is, of course, that of $\widehat{M}$.

Therefore, in order to evaluate $F^*_{j,r}(\Omega_{j,r})$ it will, in general, be necessary to choose a suitably oriented orthonormal basis $(W_{1,F_{j,r}(t,s)}, \ldots, W_{l-1,F_{j,r}(t,s)})$ of $T_{F_{j,r}(t,s)}S_j(r)$.

We now treat two specific cases, in both of which the exponential map has an explicit form that makes computations feasible.[10]

Furthermore, we shall ultimately want to replace the integral in (10.4.4) by an integral with respect to the natural volume form on $\partial_j M \times S(\mathbb{R}^{l-j})$, and so we note here, for later use, the fact that if α is an n-form on an n-dimensional Riemannian manifold with volume form Ω, and $X_1, \ldots, X_n$ is an orthonormal basis for $T_t M$, then

$$\alpha(V_1, \ldots, V_n) = \Omega(V_1, \ldots, V_n) \cdot \alpha(X_1, \ldots, X_n) \tag{10.4.6}$$

for any collection of vectors $V_j \in T_t M$.

[10] Indeed, this is true of many spaces of constant curvature. In general, when $\widehat{M}$ is an arbitrary manifold, the Jacobian of F_j, which is needed to continue from (10.4.4), is difficult to compute, since the map $\exp_{\widehat{M}}$ has no simple closed form. In [73] the Jacobian is expressed in terms of the solution to a Riccati differential equation, but we shall not pursue this approach here.

10.4.3 Subsets of $\mathbb{R}^l$

We now treat the case in which the ambient space is $\widehat{M} = \mathbb{R}^l$. Our aim is to compute the determinant in (10.4.5).

In this case, since the exponential map is the identity, $F_{j,r}$ is particularly simple and is given by

$$F_{j,r}(t,s) = t + r\sum_{i=1}^{l-j} s_i \eta_i(t). \tag{10.4.7}$$

This allows us to establish the following lemma, which will be a major step toward computing (10.4.5).

Lemma 10.4.1. *For any coordinate systems u on $\partial_j M$ and w on $S(\mathbb{R}^{l-j})$ and any orthonormal frames $(\eta_i)_{1\le i\le l-j}$ normal to $\partial_j M$, the following relations hold:*

$$F_{j,r*}\left(\frac{\partial}{\partial u_k}\right) = \frac{\partial}{\partial u_k} + r\sum_{i=1}^{l-j} s_i \nabla_{\frac{\partial}{\partial u_k}} \eta_i(t(u)), \tag{10.4.8}$$

$$F_{j,r*}\left(\frac{\partial}{\partial w_k}\right) = 0 + r\sum_{i=1}^{l-j} \frac{\partial s_i}{\partial w_k} \eta_i(t(u)). \tag{10.4.9}$$

As a consequence, we have

$$\left\langle F_{j,r*}\left(\frac{\partial}{\partial u_k}\right), \sum_{i=1}^{l-j} s_i(w)\eta_i(t(u)) \right\rangle = 0, \tag{10.4.10}$$

$$\left\langle F_{j,r*}\left(\frac{\partial}{\partial w_k}\right), \sum_{i=1}^{l-j} s_i(w)\eta_i(t(u)) \right\rangle = 0. \tag{10.4.11}$$

Proof. Note first that throughout the proof, hopefully without too much ambiguity, we shall identify tangent vectors $X_y \in T_y\mathbb{R}^l$ with vectors in $\mathbb{R}^l$. Not doing so would lead to even more unwieldy notation.

To verify (10.4.8) and (10.4.9), let (U,φ) be a chart on M, in which case we can rewrite (10.4.7) locally as a mapping from $U \times S(\mathbb{R}^{l-j})$ by defining

$$\widetilde{F}_{j,r}(u,w) = \varphi(u) + r\sum_{i}^{l-j} s_i(w)\eta_i(\varphi(u)). \tag{10.4.12}$$

With the usual abuse of notation that identifies $\partial/\partial u_k$ and $\varphi_*(\partial/\partial u_k)$ we can compute $F_{j,r*}(\partial/\partial u_k)$ from $\widetilde{F}_{j,r*}(\partial/\partial u_k) \equiv \widetilde{F}_{j,r*}(\varphi_*(\partial/\partial u_k))$. To see how to do this, take the standard basis $(e_1,\dots,e_l)$ on $\mathbb{R}^l$ and start by noting that for $1 \le m \le l$, $1 \le k \le l-j$,

$$
\begin{aligned}
\left(\frac{\partial}{\partial u_k}\widetilde{F}_{j,r}(u,s)\right)_m
&= \left(\frac{\partial \varphi(u)}{\partial u_k}\right)_m + r\sum_i^{l-j} s_i \frac{\partial \langle \eta_i(\varphi(u)), e_m\rangle}{\partial u_k} \\
&= \left(\varphi_*\left(\frac{\partial}{\partial u_k}\right)\right)_m + r\sum_i^{l-j} s_i \left\langle\left(\varphi_*\left(\frac{\partial}{\partial u_k}\right)\eta_i\right)(\varphi(u)), e_m\right\rangle \\
&= \left(\frac{\partial}{\partial u_k}\right)_m + r\sum_i^{l-j} s_i \frac{\partial}{\partial u_k}\left\langle \eta_i(\varphi(u)), e_m\right\rangle,
\end{aligned}
$$

where in this case the notational abuse appears in the passage between the last two lines. Note, however, that from the "compatability" property of the Levi-Civitá connection (cf. (7.3.11)), we have

$$
\begin{aligned}
\frac{\partial}{\partial u_k}\langle \eta_i(\varphi(u)), e_m\rangle &= \langle \nabla_{\partial/\partial u_k}\eta_i(\varphi(u)), e_m\rangle + \langle \nabla_{\partial/\partial u_k} e_m, \eta_i(\varphi(u)),\rangle \frac{\partial}{\partial e_m} \\
&= \langle \nabla_{\partial/\partial u_k}\eta_i(\varphi(u)), e_m\rangle,
\end{aligned}
$$

since $\nabla_{\partial/\partial u_k} e_m \equiv 0$.

Substituting this in the above leads to

$$
\begin{aligned}
F_{j,r*}\left(\frac{\partial}{\partial u_k}\right) &= \widetilde{F}_{j,r*}\left(\frac{\partial}{\partial u_k}\right) \\
&= \sum_{m=1}^{l}\left(\frac{\partial}{\partial u_k}\widetilde{F}_{j,r}(u,s)\right)_m \frac{\partial}{\partial e_m} \\
&= \partial/\partial u_k + r\sum_i s_i \sum_m \frac{\partial}{\partial u_k}\langle \eta_i(\varphi(u)), e_m\rangle \frac{\partial}{\partial e_m} \\
&= \partial/\partial u_k + r\sum_i s_i \sum_m \left(\langle \nabla_{\partial/\partial u_k}\eta_i(\varphi(u)), e_m\rangle\right)\frac{\partial}{\partial e_m} \\
&= \partial/\partial u_k + r\sum_i s_i \nabla_{\partial/\partial u_k}\eta_j(\varphi(u)),
\end{aligned}
$$

which gives (10.4.8). A similar argument gives (10.4.9).

We now turn to (10.4.10) and (10.4.11). To verify (10.4.10), first note that by the definition of the pullback,

$$
F_{j,r*}\left(\frac{\partial}{\partial u_k}\right) \in T_{F_{j,r}(t,s)} S_j(r),
$$

while by (10.4.9),

$$
\sum_{i=1}^{l-j} s_i(w)\eta_i(t(u)) \in T^{\perp}_{F_{j,r}(t,s)} S_j(r).
$$

Equation (10.4.10) is now immediate.

To verify (10.4.11) note that we can express the first term there as

$$F_{j,r*}\left(\frac{\partial}{\partial w_k}\right) = \sum_{i=1}^{l-j} a_{ki}(s(w))\eta_i(t(u)),$$

where the a_{ki} are given by

$$a_{ki}(s(w)) = \frac{\partial s_i(w)}{\partial w_k}, \tag{10.4.13}$$

and so for all $1 \le k \le l-j-1$ satisfy

$$\sum_{i=1}^{l-j} a_{ki}(s(w))s_i(w) = 0. \tag{10.4.14}$$

Therefore,

$$\begin{aligned}
&\left\langle F_{j,r*}\left(\frac{\partial}{\partial w_k}\right), \sum_{i=1}^{l-j} s_i(w)\eta_i(t(u))\right\rangle \\
&\quad = \left\langle \sum_{n=1}^{l-j} a_{kn}(s(w))\eta_n(t(u)), \sum_{i=1}^{l-j} s_i(w)\eta_i(t(u))\right\rangle \\
&\quad = \sum_{i=1}^{l-j} a_{ki}(s(w))s_i(w) \\
&\quad = 0.
\end{aligned}$$

□

We now return to our main task, the computation of the pullback and, specifically, the determinant in (10.4.5). By (10.4.6), it will suffice to do this for an orthonormal basis of $T_{(t,s)}\partial_j M \times S(\mathbb{R}^{l-j})$. To this end, let $(X_{1,t}, \dots, X_{j,t})$ be an orthonormal basis of $T_t\partial_j M$, and for $1 \le m \le l-j-1$, let

$$\widetilde{X}_{m,s} = \sum_{i=1}^{l-j} a_{mi}(s)\frac{\partial}{\partial s_i},$$

with the a_{mi} as in (10.4.13), be an orthonormal basis for $T_s S(\mathbb{R}^{l-j})$.

We now need to choose an orthonormal basis

$$\left(W_{1,F_{j,r}(t,s)}, \dots, W_{l-1,F_{j,r}(t,s)}\right)$$

for $T_{F_{j,r}(t.s)}S_j(r)$. By Lemma 10.4.1, it is clear that such a basis is an orthonormal basis for the orthogonal complement, in $T_{F_{j,r}(t,s)}\mathbb{R}^l$, of the vector

$$\eta(t,s)=\sum_{i=1}^{l-j} s_i\eta_i(t), \tag{10.4.15}$$

with base at $t\in\partial_j M$ and going through the point $F_{j,r}(t,s)$.[11]

This motivates the following choice, with the $X_{j,t}$ and $\widetilde{X}_{m,s}$ as above: For $1\le m\le j$, let

$$W_{m,F_{j,r}(t,s)}=X_{m,t}$$

and, for $1\le m\le l-j-1$, define

$$W_{j+m,F_{j,r}(t,s)}=\sum_{i=1}^{l-j} a_{mi}(s)\eta_i(t).$$

It is not difficult to see that the set $\{W_{1,F_{j,r}(t,s)},\dots,W_{l-1,F_{j,r}(t,s)}\}$ is an orthonormal basis of $T_{F_{j,r}(t,s)}S_j(r)$ since it is an orthonormal set and all elements are orthogonal to $\eta(t,s)$.

Returning to (10.4.5), we can now compute

$$\det\left(\left\langle F_{j,r*}X_{k,(t,s)},\,W_{m,F_{j,r}(t,s)}\right\rangle\right)_{1\le k,m\le l-1}, \tag{10.4.16}$$

for the $W_{m,F_{j,r}(t,s)}$ just constructed, and taking $X_{1,t},\dots,X_{j,t},\widetilde{X}_{1,s},\dots,\widetilde{X}_{l-j-1,s}$ as our choice for the vectors $X_{k,(t,s)}$. In view of (10.4.6), this will suffice.

The matrix in (10.4.16) can be broken into four blocks. For the first, we take $1\le k,m\le j$, so that in the notation on Lemma 10.4.1, and for some collection of coefficients $b_{ik}(t)$, it has elements

$$\begin{aligned}
&\left\langle F_{j,r*}X_{k,t},\,W_{m,F_{j,r}(t,s)}\right\rangle\\
&\quad=\left\langle F_{j,r*}\left(\sum_{i=1}^{j} b_{ik}(t)\frac{\partial}{\partial u_i}\right),\,X_{m,t}\right\rangle\\
&\quad=\sum_{i=1}^{j} b_{ik}(t)\left\langle\left(\frac{\partial}{\partial u_i}+r\sum_{n=1}^{l-j} s_n\nabla_{\frac{\partial}{\partial u_i}}\eta_n\right),\,X_{m,t}\right\rangle\\
&\quad=\left\langle X_{k,t}+r\sum_{n=1}^{l-j} s_n\nabla_{X_{k,t}}\eta_n,\,X_{m,t}\right\rangle,
\end{aligned}$$

the second equality here following from (10.4.8) of Lemma 10.4.1 and where $\eta=\eta(s,t)$ is the normal vector of (10.4.15).

However, noting the orthonormality of the $X_{k,t}$ and applying the Weingarten equation (7.5.12), it is immediate that the above is equal to

[11] Recall our convention of identifying tangent vectors $X_y\in T_y\mathbb{R}^l$ with vectors in $\mathbb{R}^l$.

$$\delta_{km} - r \cdot S_\eta(X_{k,t}, X_{m,t}),$$

where S is the scalar second fundamental form of $\partial_j M$ in $\mathbb{R}^l$ (cf. (7.5.11)).

Applying similar arguments based on Lemma 10.4.1 to the remaining terms in (10.4.16), it is now easy to check that the determinant there has the form

$$\det \begin{pmatrix} I_{j\times j} - r \cdot \left\{S_\eta(X_{k,t}, X_{m,t})\right\}_{k,m=1,\dots,j} & \cdots \\ 0 & r \cdot I_{(l-j-1)\times(l-j-1)} \end{pmatrix}. \tag{10.4.17}$$

To compute this determinant recall first the standard formula

$$\det \begin{pmatrix} A & B \\ 0 & C \end{pmatrix} = \det(A)\det(C)$$

for A and C square matrices, so that (10.4.17) simplifies to

$$r^{l-j-1} \det\left(I_{j\times j} - r\left\{S_\eta(X_{k,t}, X_{m,t})\right\}\right). \tag{10.4.18}$$

To evaluate the remaining determinant, we can either use standard, matrix-theoretic, formulas for evaluating $\det(A + \lambda I)$, or, more formally, we can view S as a double form and use (7.2.11). Since we shall use the latter approach for more difficult calculations later, we use it here as well. Therefore,

$$\begin{aligned}
&\det \begin{pmatrix} I_{j\times j} - r\left\{S_\eta(X_{k,t}, X_{m,t})\right\} & \cdots \\ 0 & rI_{(l-j-1)\times(l-j-1)} \end{pmatrix} \\
&\quad = r^{l-j-1}\frac{1}{j!}\operatorname{Tr}^{T_t\partial_j M}\left((I - rS_\eta)^j\right) \\
&\quad = \sum_{i=0}^{j} r^{l-j+i-1}\frac{1}{i!(j-i)!}\operatorname{Tr}^{T_t\partial_j M}\left(I^{j-i}S_{-\eta}^i\right) \\
&\quad = \sum_{i=0}^{j} r^{l-j+i-1}\frac{1}{i!}\operatorname{Tr}^{T_t\partial_j M}\left(S_{-\eta}^i\right),
\end{aligned}$$

where we have used (7.2.11) in going from the third to the fourth line.

Putting the above together, we have proven the following theorem, which is one of the two main results of this section.

Theorem 10.4.2. *Suppose that M is a C^2, locally convex, Whitney stratified manifold embedded in the N-manifold $(\widetilde{M}, \widetilde{g})$ that is isometrically embedded in $\widehat{M} = \mathbb{R}^l$. Then*

$$\begin{aligned}
&F_{j,r}^*(\Omega_{j,r})\left(X_{1,t}, \dots, X_{j,t}, \widetilde{X}_{1,s}, \dots, \widetilde{X}_{l-j-1,s}\right) \\
&\quad = \sum_{i=0}^{j} r^{l-j-1+i}\frac{1}{i!}\operatorname{Tr}^{T_t\partial_j M}(S_{-\eta}^i),
\end{aligned}$$

where S is the scalar second fundamental form of $\partial_j M$ in $\mathbb{R}^l$.

In particular, if μ is the measure associated to an integrable $(l-1)$-form α_μ on $S(\mathbb{R}^l)$, then

$$\mu\left(\operatorname{Tube}(M,\rho)\right) = \sum_{j=0}^{l}\sum_{i=0}^{j} \int_{\partial_j M}\int_{S(\mathbb{R}^{l-j})}\int_0^\rho \mathbb{1}_{D_j(\rho)}(t,s,r) f_\mu(F_{j,r}(t,s)) \\ \times r^{l-j-1+i}\frac{1}{i!}\operatorname{Tr}^{T_t\partial_j M}(S_{-\eta}^i)\, dr\, \mathcal{H}_{l-j-1}(ds)\mathcal{H}_j(dt), \tag{10.4.19}$$

where $\mathcal{H}_{l-j-1}$ is standard surface measure on $S(\mathbb{R}^{l-j})$ and $\mathcal{H}_j$ is the volume measure induced on $\partial_j M$ by the Riemannian metric $\widetilde{g}$.

10.4.4 Subsets of Spheres

In this section, we treat the case $\widehat{M} = S_\lambda(\mathbb{R}^l)$, the sphere of radius λ in $\mathbb{R}^l$. This case is very similar to the previous one, so we need only go over the major differences.

The geodesic metric on $S_\lambda(\mathbb{R}^l)$ is given by

$$\widehat{d}_\lambda(t,s) = \lambda\cos^{-1}(\lambda^{-2}\langle t,s\rangle).$$

The exponential mapping $\exp_{S_\lambda(\mathbb{R}^l)}$ is also explicitly computable in this case. For a unit vector $X_t \in T_t S_\lambda(\mathbb{R}^l)$ and $r > 0$,

$$\exp_{S_\lambda(\mathbb{R}^l)}(t, rX_t) = \cos(r/\lambda)t + \lambda\sin(r/\lambda)X_t.$$

The maps $F_{j,r}$ are of the form

$$F_{j,r}(t,s) = \cos(r/\lambda)t + \lambda\sin(r/\lambda)\sum_{i=1}^{l-1-j} s_i\eta_i(t)$$

for some orthonormal frame $(\eta_1(t),\dots,\eta_{l-1-j}(t))$ spanning $\widehat{T_t\partial_j M^\perp}$, the orthogonal complement of $T_t\partial_j M$ in $T_t\widehat{M} = T_t S_\lambda(\mathbb{R}^l)$.

The orthogonality relations (10.4.10) and (10.4.11) still hold in this case, with the minor change that the W_i of the previous subsection now give a coordinate system on $S(\mathbb{R}^{l-1-j})$ instead of $S(\mathbb{R}^{l-j})$.

To be more specific, to construct an orthonormal basis on $T_{F_{j,r}(t,s)}S_j(r)$, we set

$$W_{m,F_{j,r}(t,s)} = X_{j,t}$$

as before, for $1 \le m \le j$, and for $1 \le m \le l-2-j$ we set

$$W_{j+m,F_{j,r}(t,s)} = \sum_{i=1}^{l-j-1} a_{mi}(s)\eta_i(t),$$

where

$$\widetilde{X}_m(s) = \sum_{i=1}^{l-j-1} a_{mi}(s)\frac{\partial}{\partial s_i}, \qquad 1 \le m \le l-2-j,$$

is an orthonormal basis of $T_s S(\mathbb{R}^{l-1-j})$.

Turning to the determinant (10.4.5) in $F^*_{j,r}(\Omega_{j,r})$, we have that it is now given by

$$\det\begin{pmatrix} \cos(r/\lambda)I + \lambda\sin(r/\lambda)S_{-\eta}(X_{k,t}, X_{m,t}) & \dots \\ 0 & \lambda\sin(r/\lambda)I \end{pmatrix}$$
$$= \sum_{i=0}^{j} \lambda^{l-2-j+i}\cos(r/\lambda)^{j-i}\sin(r/\lambda)^{l-j-2+i}\,\mathrm{Tr}^{T_t\partial_j M}(S^i_{-\eta}), \tag{10.4.20}$$

where the first unit matrix is of size j and the second of size $l-j-1$. The vector $\eta = \eta(s,t)$ is as in the previous subsection and S is now the scalar second fundamental form of $\partial_j M$ in $S_\lambda(\mathbb{R}^l)$ (cf. (7.5.11)).

We thus have the following result.

Theorem 10.4.3. *Suppose that M is a C^2, locally convex, Whitney stratified manifold embedded in the N-manifold $(\widetilde{M},\widetilde{g})$ that is isometrically embedded in $\widehat{M} = S_\lambda(\mathbb{R}^l)$. Then*

$$F^*_{j,r}(\Omega_{j,r})\left(X_{1,t},\dots,X_{j,t},\widetilde{X}_{1,s},\dots,\widetilde{X}_{l-j-1,s}\right)$$
$$= \sum_{i=0}^{j} \lambda^{l-2-j+i}\cos(r/\lambda)^{j-i}\sin(r/\lambda)^{l-j-2+i}\,\mathrm{Tr}^{T_t\partial_j M}(S^i_{-\eta}),$$

where $S_{-\eta}$ is the scalar second fundamental form of $\partial_j M$ in $S(\mathbb{R}^l)$.

In particular, if μ is the measure associated to an integrable $(l-1)$-form α_μ on $S(\mathbb{R}^l)$, then

$$\mu\left(\mathrm{Tube}(M,\rho)\right) = \sum_{j=0}^{l-1}\sum_{i=0}^{j}\int_{\partial_j M}\int_{S(\mathbb{R}^{l-1-j})}\int_0^\rho \mathbb{1}_{D_j(\rho)}(t,s,r)\,f_\mu\left(F_{j,r}(t,s)\right)$$
$$\times\,\lambda^{l-2-j+i}\cos(r/\lambda)^{j-i}\sin(r/\lambda)^{l-j-2+i}\frac{1}{i!}$$
$$\times\,\mathrm{Tr}^{T_t\partial_j M}(S^i_{-\eta})\,dr\,\mathcal{H}_{l-j-2}(ds)\mathcal{H}_j(dt),$$

where $\mathcal{H}_{l-j-2}$ and $\mathcal{H}_j$ are as in (10.4.19).

10.5 Weyl's Tube Formula

As we have written them, Theorems 10.4.2 and 10.4.3 are not in the form of classic Weyl's tube formulas, since the remaining integrals there, since they depend on S_η, are not intrinsic. On the other hand, they are a little more general than the classic Weyl's tube formulas, for which μ is either $\mathcal{H}_l$ or $\mathcal{H}_{l-1}$.

In order to obtain an intrinsic formulation of these results, and so rederive the classic Weyl's tube formulas, we need to specialize down to $\mathcal{H}_l$ or $\mathcal{H}_{l-1}$ and prove that only intrinsic quantities arise when the integrals in Theorems 10.4.2 and 10.4.3 are evaluated. This is the *deep* part of Weyl's tube formula,[12] and it is one of the first appearances of Lipschitz–Killing curvatures.

The derivation that follows is a little unusual, since it involves expectations of random Gaussian forms, and so is not what you will find in a standard textbook. Nevertheless, it does fit in nicely with the theme of this book, which attempts to integrate both geometry and probability.

The drawback of this approach is that the following argument depends on Lemma 12.3.1, which is still two chapters into the future. With only minor effort, we could rearrange things so that the lemma appeared here rather than there, but it is in Chapter 12, where it logically belongs. To make things a little easier for you, we shall "precall" one definition now.

Let V be a vector space and $\mu \in \Lambda^{1,1}(V)$ a double form on V. Furthermore, let $\mathrm{Cov} : (V \otimes V) \times (V \otimes V) \to \mathbb{R}$ be bilinear, symmetric, and nonnegative definite. We think of μ as a mean function and Cov as a covariance function, and call W a random, Gaussian 2-form on $V \otimes V$ with mean function μ and covariance function Cov if for all finite collections of pairs $(v_{i_1}, v_{i_2}) \in V \otimes V$, the $W(v_{i_1}, v_{i_2})$ have a joint Gaussian distribution with means

$$\mathbb{E}\left\{W(v_{i_1}, v_{i_2})\right\} = \mu(v_{i_1}, v_{i_2})$$

and covariances

$$\begin{aligned}\mathbb{E}&\left\{\left(W(v_{i_1}, v_{i_2}) - \mu(v_{i_1}, v_{i_2})\right) \cdot \left(W(v_{j_1}, v_{j_2}) - \mu(v_{j_1}, v_{j_2})\right)\right\} \\ &= \mathrm{Cov}\left((v_{i_1}, v_{i_2}), (v_{j_1}, v_{j_2})\right).\end{aligned}$$

Lemma 12.3.1 is about computing expectations of powers of such Gaussian forms, where the powers are computed via the double wedge tensor product of (7.2.4).

To return to our main task, what remains in order to derive Weyl's tube formula is proving that for ρ small enough,

$$\int_{\partial_j M} \int_{S(T_t \partial_j M^{\perp})} \int_0^{\rho} \mathbb{1}_{D_j(\rho)}(t, s, r) \, \mathrm{Tr}^{\partial_j M}(S^i_{-\eta}) \, dr \, ds \, dt = \widetilde{C}(j, i, l) Q(\partial_j M),$$

for some constant $\widetilde{C}$ and some intrinsic quantity $Q(\partial_j M)$, which will turn out to be related to the intrinsic volumes of M.[13]

---

[12] In fact, of the computations of Sections 10.3 and 10.4, albeit in the case of manifolds in $\mathbb{R}^l$ and $S(\mathbb{R}^l)$ without boundary, and for $\mathcal{H}_l$ and $\mathcal{H}_{l-1}$, Weyl wrote that "so far we have hardly done more than what could have been accomplished by any student in a course of calculus."

[13] Read $\mathrm{Vol}_{\widetilde{g}|S(T_t \partial_j M^{\perp})}(ds)$ and $\mathrm{Vol}_{\widetilde{g}|\partial_j M}(dt)$ for ds and dt here.

Recall that in the Euclidean case, we have $S(T_t\partial_j M^\perp) = S(\mathbb{R}^{l-j})$, while for the spherical case, $S(T_t\partial_j M^\perp) = S(\mathbb{R}^{l-j-1})$.

As a first step, define the constants

$$C(m,i) \triangleq \begin{cases} \frac{(2\pi)^{i/2}}{s_{m+i}}, & m+i>0, \\ 1, & m=0, \end{cases} \tag{10.5.1}$$

where as usual, $s_m = 2\pi^{m/2}/\Gamma\left(\frac{m}{2}\right)$ is the surface area of the unit sphere in $\mathbb{R}^m$.

Our first result relating the quantities in Theorem 10.4.2 to intrinsic quantities is the following.

Lemma 10.5.1. *Suppose that M is a C^2, locally convex, Whitney stratified manifold embedded in the N-manifold $(\widetilde{M},\widetilde{g})$ that is isometrically embedded in $\mathbb{R}^l$. Then, for $j<l$ the following quantities are intrinsic, for each $t\in M$:*

$$\begin{aligned} C(l-j,i) &\int_{S(\mathbb{R}^{l-j})} \mathbb{1}_{\widehat{N_t M}}(\eta)\frac{1}{i!}\operatorname{Tr}^{T_t\partial_j M}(S^i_{-\eta})\mathcal{H}_{l-j-1}(ds) \\ &= \sum_{m=0}^{\lfloor \frac{i}{2}\rfloor} C(N-j,i-2m)(-1)^i\frac{(-1)^m}{m!(i-2m)!} \\ &\quad\times \int_{S(T_t\partial_j M^\perp)} \operatorname{Tr}^{T_t\partial_j M}\left(\widetilde{R}^m\widetilde{S}^{i-2m}_{-\nu_{N-j}}\right)\mathbb{1}_{N_t M}(\nu_{N-j})\mathcal{H}_{N-j-1}(d\nu_{N-j}), \end{aligned} \tag{10.5.2}$$

where S is the scalar second fundamental form of $\partial_j M$ as it sits in $\mathbb{R}^l$, $T_t\partial_j M^\perp$ is the orthogonal complement of $T_t\partial_j M$ in $T_t\widetilde{M}$, $\widetilde{R}$ is the curvature tensor of $(\widetilde{M},\widetilde{g})$, $\widetilde{S}$ is the scalar second fundamental form $\partial_j M$ as it sits in $\widetilde{M}$, and

$$\eta = \eta(s,t) = \sum_i s_i\eta_i(t),$$

with $\{\eta_i(t)\}_{1\le i\le l-j}$ an orthonormal basis of $\widehat{T_t\partial_j M}^\perp$, the orthogonal complement of $T_t\partial_j M$ in $\mathbb{R}^l$.

Remark 10.5.2. *Above, if $j=l$, then $\widetilde{M}$ is essentially an open subset of $\mathbb{R}^l$, so that $N=l$ and the left- and right-hand sides of (10.5.2) are not defined. This of course plays havoc with our notation, since we would be integrating over the sphere in the zero subspace of $T_t\widetilde{M}$. We therefore adopt the following somewhat strange but convenient convention: for any zero-dimensional space $\mathbb{0}$ and any function h on $\mathbb{0}$,*

$$\int_{S(\mathbb{0})} h(\nu)\,d\nu \triangleq h(\mathbb{0}).$$

This is well defined since $\mathbb{0}$ has only one element and is effectively equivalent to defining $S(\mathbb{0})=\mathbb{0}$. We also adopt the convention, for the scalar second fundamental form, that

$$S_{\mathbb{0}}^j = \begin{cases} 1, & j = 0, \\ 0 & \textit{otherwise}. \end{cases}$$

Finally, we define the volume of the unit ball in $\mathbb{0}$ *to be* 1, *i.e.,*

$$\omega_0 \triangleq 1.$$

These issues will, thankfully, arise only when we define the Lipschitz–Killing curvatures. The conventions are pedantic, but are important to ensure that the Lipschitz–Killing curvature measures for locally convex spaces, defined in (10.5.4), *agree with the Lipschitz–Killing curvature measures for smooth manifolds defined in* (7.6.1).

The corresponding version of Lemma 10.5.1 for subsets of a hypersphere is as follows.

Lemma 10.5.3. *Suppose that M is a C^2, locally convex, Whitney stratified manifold embedded in an N-manifold $(\widetilde{M}, \widetilde{g})$ that is isometrically embedded in $S_\lambda(\mathbb{R}^l)$. Then, for $j < l$, the following quantities are intrinsic, for each $t \in M$:*

$$\begin{aligned} C(l-1-j, i) &\int_{S(\mathbb{R}^{l-1-j})} \mathbb{1}_{\widehat{N_t M}}(\eta) \frac{1}{i!} \operatorname{Tr}^{T_t \partial_j M}(S^i_{-\eta}) \mathcal{H}_{l-j-2}(ds) \qquad (10.5.3) \\ &= \sum_{m=0}^{\lfloor \frac{i}{2} \rfloor} C(N-j, i-2m)(-1)^i \frac{(-1)^m}{m!(i-2m)!} \\ &\quad \times \int_{S(T_t \partial_j M^\perp)} \operatorname{Tr}^{T_t \partial_j M} \left(\left(\frac{\lambda^{-2}}{2} I^2 + \widetilde{R} \right)^m \widetilde{S}^{i-2m}_{-\nu_{N-j}} \right) \\ &\quad \times \mathbb{1}_{N_t M}(\nu_{N-j}) \mathcal{H}_{N-j-1}(d\nu_{N-j}), \end{aligned}$$

where S is the scalar second fundamental form of $\partial_j M$ as it sits in $S_\lambda(\mathbb{R}^l)$ and everything else is as in Lemma 10.5.1.

Remark 10.5.4. *Above, if $j = l - 1$, then $\widetilde{M}$ is essentially an open subset of $S_\lambda(\mathbb{R}^l)$, so that $N = l - 1$ and the left- and right-hand sides of* (10.5.3) *are not defined. We therefore again adopt the conventions of Remark* 10.5.2.

The proofs of Lemmas 10.5.1 and 10.5.3 are virtually identical and are based on the following lemma, for which we recall that $\gamma_{\mathbb{R}^m}$ denotes Gauss measure on $\mathbb{R}^m$ and η_m denotes uniform measure over $S(\mathbb{R}^m)$.

Lemma 10.5.5. *Let $W \sim \gamma_{\mathbb{R}^m}$, let P_i be a homogeneous function of order i on $\mathbb{R}^m$, and let A be a cone in $\mathbb{R}^m$. Then*

$$\mathbb{E}\{P_i(W)\mathbb{1}_A(W)\} = C(m, i) \int_{S(\mathbb{R}^m)} P_i(\nu)\mathbb{1}_A(\nu)\mathcal{H}_{m-1}(d\nu).$$

Proof.

$$
\begin{aligned}
&\mathbb{E}\left\{P_i(W)\mathbb{1}_A(W)\right\} \\
&= \mathbb{E}\left\{|W|^i P_i(W/|W|)\mathbb{1}_A(W/|W|)\right\} \\
&= \int_0^\infty r^i \int_{S(\mathbb{R}^m)} P_i(\nu)\mathbb{1}_A(\nu)\eta_m(d\nu)\mathbb{P}_{|W|}(dr) \\
&= \frac{\Gamma\left(\frac{m}{2}\right)}{2\pi^{m/2}} \int_0^\infty r^i \mathbb{P}_{|W|}(dr) \int_{S(\mathbb{R}^m)} P_i(\nu)\mathbb{1}_A(\nu)\mathcal{H}_{m-1}(d\nu) \\
&= \frac{1}{2}(2\pi)^{-m/2}2^{(i+m)/2}\Gamma\left(\frac{m+i}{2}\right)\int_{S(\mathbb{R}^m)} P_i(\nu)\mathbb{1}_A(\nu)\mathcal{H}_{m-1}(d\nu),
\end{aligned}
$$

which, on comparing with the definition of the $C(m,i)$ at (10.5.1), proves the lemma. □

Proof of Lemma 10.5.1. Our proof is somewhat indirect and, as mentioned above, proceeds via Gaussian random forms, which are not generally found in geometric arguments. Rather than taking one expression in (10.5.2) and showing that it is equal to the other, we shall show that both are given by a certain expectation.

One of the nice aspects of this approach is that it immediately shows that both sides of (10.5.2) are truly intrinsic,[14] since the expectation will, a priori, involve only intrinsic quantities.

We start with the case $j < N$, leaving the easier case of $j = N$ for later.

Therefore, let $X_{l-j} \sim \gamma_{\widehat{T_t\partial_j M^\perp}}$ be a canonically distributed Gaussian random vector on $\widehat{T_t\partial_j M^\perp}$, the orthogonal complement of $T_t\partial_j M$ in $T_t\widehat{M}$. For a fixed orthonormal basis of $\widehat{T_t\partial_j M^\perp}$, the random variable

$$
\frac{1}{i!}\operatorname{Tr}^{T_t\partial_j M}(S^i_{-X_{l-j}})
$$

is a homogeneous polynomial of degree i in the components of X_{l-j} in this basis. Hence, by Lemma 10.5.5,

$$
\begin{aligned}
&\frac{1}{i!}\mathbb{E}\left\{\operatorname{Tr}^{T_t\partial_j M}(S^i_{-X_{l-j}})\mathbb{1}_{\widehat{N_t M}}(X_{l-j})\right\} \\
&= C(l-j,i)\int_{S(\widehat{T_t\partial_j M})} \mathbb{1}_{\widehat{N_t M}}(\nu_{l-j})\frac{1}{i!}\operatorname{Tr}^{T_t\partial_j M}(S^i_{-\nu_{l-j}})\mathcal{H}_{l-j-1}(\nu_{l-j}) \\
&= C(l-j,i)\int_{S(\mathbb{R}^{l-j})} \mathbb{1}_{\widehat{N_t M}}(\eta)\frac{1}{i!}\operatorname{Tr}^{T_t\partial_j M}(S^i_{-\eta})\mathcal{H}_{l-j-1}(ds),
\end{aligned}
$$

which is the left-hand side of (10.5.2).

[14] Of course, this would also follow from *any* proof that (10.5.2) held, since while the left-hand side—which seems to depend on the nonintrinsic parameter l—is not obviously intrinsic, the right-hand side depends only on intrinsic quantities.

On the other hand, X_{l-j} can be decomposed as

$$X_{l-j} = P_{T_t\widetilde{M}} X_{l-j} + P^{\perp}_{T_t\widetilde{M}} X_{l-j} \triangleq X_{N-j} + X_{l-N}.$$

With this notation, and noting that the scalar fundamental form S_η is linear in η, we can write

$$\frac{1}{i!}\,\mathrm{Tr}^{T_t\partial_j M}(S^i_{-X_{l-j}}) = \sum_{m=0}^{i} \frac{1}{m!(i-m)!}\,\mathrm{Tr}^{T_t\partial_j M}\left(S^m_{-X_{l-N}} S^{i-m}_{-X_{N-j}}\right),$$

where the double forms $S_{-X_{N-j}}$ and $S_{-X_{l-N}}$ are independent, since X_{N-j} and X_{l-N} are.

Furthermore,

$$\mathbb{1}_{\widehat{N_t M}}(X_{l-j}) = \mathbb{1}_{N_t M}(X_{N-j}).$$

Consequently, we have

$$\begin{aligned}
&\mathbb{E}\left\{\frac{1}{i!}\,\mathrm{Tr}^{T_t\partial_j M}(S^i_{-X_{l-j}})\mathbb{1}_{\widehat{N_t M}}(X_{l-j})\right\}\\
&\quad= \sum_{m=0}^{i} \frac{1}{m!(i-m)!}\,\mathrm{Tr}^{T_t\partial_j M}\left(\mathbb{E}\left\{S^m_{-X_{l-N}} S^{i-m}_{-X_{N-j}} \mathbb{1}_{N_t M}(X_{N-j})\right\}\right)\\
&\quad= \sum_{m=0}^{i} \frac{1}{m!(i-m)!}\,\mathrm{Tr}^{T_t\partial_j M}\left(\mathbb{E}\left\{S^m_{-X_{l-N}}\right\}\mathbb{E}\left\{\widetilde{S}^{i-m}_{-X_{N-j}} \mathbb{1}_{N_t M}(X_{N-j})\right\}\right),
\end{aligned}$$

where replacing S by $\widetilde{S}$ is justified by the Weingarten equation (7.5.12).

The second expectation here is easily computed, as above, via Lemma 10.5.5. For the first, we shall apply Lemma 12.3.1, but to do this we need to know the "means" and "covariances" of S. From symmetry arguments, the expected value is clearly 0. That is, for any $X_t, Y_t, Z_t, W_t \in T_t\widetilde{M}$,

$$\mathbb{E}\left\{S_{-X_{l-N}}(X_t, Y_t)\right\} = 0.$$

As for the "variance" term, by (7.5.8),

$$\begin{aligned}
&\mathbb{E}\left\{\frac{1}{2} S^2_{-X_{l-N}}((X_t, Y_t), (Z_t, W_t))\right\}\\
&\quad= \mathbb{E}\left\{\widetilde{S}_{-X_{l-N}}(X_t, Z_t)\widetilde{S}_{-X_{l-N}}(Y_t, W_t)\right.\\
&\qquad\left. - \widetilde{S}_{-X_{l-N}}(X_t, W_t)\widetilde{S}_{-X_{l-N}}(Y_t, Z_t)\right\}\\
&\quad= \mathbb{E}\left\{\left\langle P_{T_t\widetilde{M}^\perp}\widehat{\nabla}_{X_t} Z_t, X_{l-N}\right\rangle\left\langle P_{T_t\widetilde{M}^\perp}\widehat{\nabla}_{Y_t} W_t, X_{l-N}\right\rangle\right.\\
&\qquad\left. - \left\langle P_{T_t\widetilde{M}^\perp}\widehat{\nabla}_{X_t} W_t, X_{l-N}\right\rangle\left\langle P_{T_t\widetilde{M}^\perp}\widehat{\nabla}_{Y_t} Z_t, X_{l-N}\right\rangle\right\}\\
&\quad= \left\langle P_{T_t\widetilde{M}^\perp}\widehat{\nabla}_{X_t} Z_t, P_{T_t\widetilde{M}^\perp}\widehat{\nabla}_{Y_t} W_t\right\rangle - \left\langle P_{T_t\widetilde{M}^\perp}\widehat{\nabla}_{X_t} W_t, P_{T_t\widetilde{M}^\perp}\widehat{\nabla}_{Y_t} Z_t\right\rangle\\
&\quad= \widehat{R}(X_t, Y_t, Z_t, W_t) - \widetilde{R}(X_t, Y_t, Z_t, W_t)\\
&\quad= -\widetilde{R}(X_t, Y_t, Z_t, W_t),
\end{aligned}$$

where the second-to-last line follows from the Gauss equation (7.5.9) and the last from the fact that $\widehat{M} = \mathbb{R}^l$, and $\mathbb{R}^l$ is flat.

Consequently, applying Lemmas 10.5.5 and 12.3.1, we have

$$\begin{aligned}
&\mathbb{E}\left\{\frac{1}{i!}\operatorname{Tr}^{T_t\partial_j M}(S^i_{-X_{l-j}})\mathbb{1}_{\widehat{N_t M}}(X_{l-j})\right\} \\
&\quad = \sum_{m=0}^{\lfloor \frac{i}{2} \rfloor} \frac{(-1)^m}{m!(i-2m)!}\operatorname{Tr}^{T_t\partial_j M}\left(\widetilde{R}^m \mathbb{E}\left\{\widetilde{S}^{i-2m}_{-X_{N-j}}\mathbb{1}_{N_t M}(X_{N-j})\right\}\right) \\
&\quad = \sum_{m=0}^{\lfloor \frac{i}{2} \rfloor} \frac{(-1)^m}{m!(i-2m)!}\mathbb{E}\left\{\operatorname{Tr}^{T_t\partial_j M}\left(\widetilde{R}^m \widetilde{S}^{i-2m}_{-X_{N-j}}\right)\mathbb{1}_{N_t M}(X_{N-j})\right\} \\
&\quad = \sum_{m=0}^{\lfloor \frac{i}{2} \rfloor} C(N-j, i-2m)\frac{(-1)^m}{m!(i-2m)!} \\
&\qquad \times \int_{S(T_t\partial_j M^\perp)} \operatorname{Tr}^{T_t\partial_j M}\left(\widetilde{R}^m \widetilde{S}^{i-2m}_{-\nu_{N-j}}\right)\mathbb{1}_{N_t M}(\nu_{N-j})\mathcal{H}_{N-j-1}(d\nu_{N-j}),
\end{aligned}$$

which is the right-hand side of (10.5.2), and so we are done when $j < N$.

When $j = N$ we need the conventions that we adopted in Remark 10.5.2 to sort things out. If $j = N < l$ then $N_t M = \mathbb{0} \subset T_t\widetilde{M}$, and by our conventions,

$$\begin{aligned}
&\sum_{m=0}^{\lfloor \frac{i}{2} \rfloor} C(N-j, i-2m)\frac{(-1)^m}{m!(i-2m)!} \\
&\qquad \times \int_{S(T_t\partial_j M^\perp)} \operatorname{Tr}^{T_t\partial_j M}\left(\widetilde{R}^m \widetilde{S}^{i-2m}_{-\nu_{N-j}}\right)\mathbb{1}_{N_t M}(\nu_{N-j})\mathcal{H}_{N-j-1}(d\nu_{N-j}) \\
&\quad = \begin{cases} \operatorname{Tr}^{T_t\widetilde{M}}(\widetilde{R}^k), & i = 2k \text{ is even}, \\ 0, & i \text{ is odd}. \end{cases}
\end{aligned}$$

On the other hand, if $j = N = l$, then, again by our conventions,

$$\mathbb{E}\left\{\frac{1}{i!}\operatorname{Tr}^{T_t\partial_j M}(S^i_{-X_{l-j}})\mathbb{1}_{\widehat{N_t M}}(X_{l-j})\right\} = \begin{cases} 1, & i = 0, \\ 0 & \text{otherwise}. \end{cases}$$

The right-hand side of (10.5.2) is also 0 in this case, by our conventions and the fact that $\widetilde{R} = 0$, and so we are done. □

The proof of Lemma 10.5.3 is virtually identical and so we omit it. As in the previous proof, there are some special cases ($j = N = l - 1$ and $j < N = l$) to be checked, but the conventions established in Remark 10.5.4 again suffice to handle these.

We are now almost ready to complete the derivation of Weyl's tube formula on $\mathbb{R}^l$ and $S_\lambda(\mathbb{R}^l)$. For $0 \le j \le k$ and $l - k \le i \le l$ define the (signed) Lipschitz–Killing curvature measures [64] of a locally convex space M for $0 \le i \le N$ by

$$\mathcal{L}_i(M, A) = \sum_{j=i}^{N} (2\pi)^{-(j-i)/2} \sum_{m=0}^{\lfloor \frac{j-i}{2} \rfloor} C(N-j, j-i-2m) \frac{(-1)^m}{m!(j-i-2m)!} \tag{10.5.4}$$

$$\times \int_{\partial_j M \cap A} \int_{S(T_t \partial_j M^{\perp})} \mathrm{Tr}^{T_t \partial_j M} \left(\widetilde{R}^m \widetilde{S}_{\nu_{N-j}}^{j-i-2m} \right)$$

$$\times \mathbb{1}_{N_t M}(-\nu_{N-j}) \mathcal{H}_{N-j-1}(d\nu_{N-j}) \mathcal{H}_j(dt),$$

where as usual, N is the dimension of the ambient manifold. We define $\mathcal{L}_i(M;)$ to be 0 if $i > \dim(M)$.

The above formula is somewhat simpler if M is embedded in $\mathbb{R}^l$ and endowed with the canonical Riemannian structure on $\mathbb{R}^l$. Then, by Lemma 10.5.1 and the flatness of $\mathbb{R}^l$ (which implies that $R \equiv 0$ and so only the terms with $m = 0$ remain in (10.5.4)), the curvature measures (10.5.4) can be written as

$$\mathcal{L}_i(M, A) = \sum_{j=i}^{N} (2\pi)^{-(j-i)/2} C(l-j, j-i) \tag{10.5.5}$$

$$\times \int_{\partial_j M \cap A} \int_{S(\mathbb{R}^{l-j})} \frac{1}{(j-i)!} \mathrm{Tr}^{T_t \partial_j M} (S_\eta^{j-i})$$

$$\times \mathbb{1}_{\widehat{N_t M}}(-\eta) \mathcal{H}_{l-j-1}(d\eta) \mathcal{H}_j(dt),$$

where S is the scalar second fundamental form of $\partial_j M$ as it sits in $\mathbb{R}^l$, $T_t \partial_j M^{\perp}$ is the orthogonal complement of $T_t \partial_j M$ in $T_t \mathbb{R}^l$, and $\widehat{N_t M}$ is the normal cone at $t \in \partial_j M$ as it sits in $T_t \mathbb{R}^l$.

As before, the (signed) total masses of the curvature measures give the intrinsic volumes

$$\mathcal{L}_i(M) \triangleq \mathcal{L}_i(M, M). \tag{10.5.6}$$

Theorem 10.5.6 (Weyl's tube formula on $\mathbb{R}^l$). *Suppose $M \subset \mathbb{R}^l$ is a C^2, locally convex, Whitney stratified manifold. For $\rho < \rho_c(M, \mathbb{R}^l)$,*

$$\mathcal{H}_l\left(\mathrm{Tube}(M, \rho)\right) = \sum_{i=0}^{N} \rho^{l-i} \omega_{l-i} \mathcal{L}_i(M). \tag{10.5.7}$$

The upper limit of the sum can be taken as infinity, since $\mathcal{L}_i(M) \equiv 0$ for $i > N$.

Proof. Changing the second index of summation in (10.4.19) and setting $f_\mu \equiv 1$, we have that

$$\mathcal{H}_l\left(\mathrm{Tube}(M, r)\right)$$
$$= \sum_{j=0}^{N} \sum_{i=0}^{j} \int_{\partial_j M} \int_{S(\mathbb{R}^{l-j})} \int_0^{\rho} \mathbb{1}_{D_j(\rho)}(t, s, r) \frac{r^{l-i-1}}{(j-i)!} \mathrm{Tr}^{T_t \partial_j M} (S_{-\eta}^{j-i}) \, dr \, ds \, dt.$$

By assumption, $\rho < \rho_c(M, \mathbb{R}^l)$, so that

$$D_j(\rho) = \bigcup_{t\in\partial_j M} \bigcup_{s\in S(\mathbb{R}^{l-j})} \bigcup_{0\le r\le\rho} \left\{(t, s, r) : \sum_i s_i\eta_i(t) \in N_t M, r < \rho\right\}.$$

Applying Lemma 10.5.1 to each term in the above summation, followed by integration with respect to r on $[0, \rho]$, yields the desired conclusion. □

In preparation for Weyl's tube formula on spheres, we need to extend the Lipschitz–Killing measures slightly to obtain a one-parameter family of measures. As we shall see in a moment, the κth such measure is particularly useful for generating a comparatively simple formula for tube volumes on a sphere of radius $1/\sqrt{\kappa}$. The definition, for $\kappa \in \mathbb{R}$, is given by

$$\begin{aligned} \mathcal{L}_i^\kappa(M, A) \triangleq \sum_{j=i}^N (2\pi)^{-(j-i)/2} \sum_{m=0}^{\lfloor\frac{j-i}{2}\rfloor} \frac{(-1)^m C(N-j, j-i-2m)}{m!(j-i-2m)!} \\ \times \int_{\partial_j M\cap A}\int_{S(T_t\partial_j M^\perp)} \mathrm{Tr}^{T_t\partial_j M}\left(\left(\widetilde{R} + \frac{\kappa}{2}I^2\right)^m \widetilde{S}^{j-i-2m}_{\nu_{N-j}}\right) \\ \times \mathbb{1}_{N_t M}(-\nu_{N-j})\mathcal{H}_{N-j-1}(d\nu_{N-j})\mathcal{H}_j(dt). \end{aligned} \tag{10.5.8}$$

Note that $\mathcal{L}_i^0(M, \cdot) \equiv \mathcal{L}_i(M, \cdot)$.

As for $\mathcal{L}_i(M)$, we define the one-parameter family of intrinsic volumes

$$\mathcal{L}_i^\kappa(M) \triangleq \mathcal{L}_i^\kappa(M, M). \tag{10.5.9}$$

Furthermore, if $\kappa > 0$ and $M \subset S_{\kappa^{-1/2}}(\mathbb{R}^l)$, then, as in (10.5.5), there is a simplification. This time it is due to the fact that for vectors E_j in an orthonormal frame field, $\widetilde{R}(E_i, E_j, E_i, E_j) = -\kappa$ (cf. (7.7.10)), which implies that $(\widetilde{R} + \frac{\kappa}{2}I^2)(E_i, E_j, E_i, E_j) = 0$, so that on taking a trace, only the terms in (10.5.8) with $m = 0$ remain. The result is

$$\begin{aligned} \mathcal{L}_i^\kappa(M, A) = \sum_{j=i}^N (2\pi)^{-(j-i)/2} C(l-1-j, j-i) \\ \times \int_{\partial_j M\cap A}\int_{S(T_t\partial_j M^\perp)} \frac{1}{(j-i)!}\mathrm{Tr}^{T_t\partial_j M}(S_\eta^{j-i}) \\ \times \mathbb{1}_{\widehat{N_t M}}(-\eta)\mathcal{H}_{l-j-2}(d\eta), \end{aligned} \tag{10.5.10}$$

where S is the scalar second fundamental form of $\partial_j M$ as it sits in $S_{\kappa^{-1/2}}(\mathbb{R}^l)$, $T_t\partial_j M^\perp$ is the orthogonal complement of $T_t\partial_j M$ in $T_t S_{\kappa^{-1/2}}(\mathbb{R}^l)$, and $\widehat{N_t M}$ is the normal cone at $t \in \partial_j M$ as it sits in $T_t S_{\kappa^{-1/2}}(\mathbb{R}^l)$.

A slight modification of the proof of Theorem 10.5.6 is enough to complete the derivation of Weyl's tube formula on $S_\lambda(\mathbb{R}^l)$.

Theorem 10.5.7 (Weyl's tube formula on $S_\lambda(\mathbb{R}^l)$). *Suppose M is a C^2, locally convex, Whitney stratified submanifold of $S_\lambda(\mathbb{R}^l)$. For $\rho < \rho_c(M, S_\lambda(\mathbb{R}^l))$,*

$$\begin{aligned}
&\mathcal{H}_{l-1}\left(\mathrm{Tube}(M,\rho)\right)\\
&\quad = \sum_{i=0}^{\infty} \lambda^{l-1-i} G_{i,l-1-i}(\rho/\lambda)\mathcal{L}_i^{\lambda^{-2}}(M)\\
&\quad = \sum_{j=0}^{\infty}\left(\sum_{n=0}^{\lfloor \frac{j}{2}\rfloor}(-4\pi)^{-n}\frac{\lambda^{l-1+j}}{n!}\frac{j!}{(j-2n)!}G_{j-2n,l-1+2n-j}(\rho/\lambda)\right)\mathcal{L}_j(M),
\end{aligned}$$

where

$$\begin{aligned}
G_{a,b}(\rho) &\triangleq \frac{\pi^{b/2}}{b\Gamma\left(\frac{b}{2}+1\right)}\int_0^{\rho}\cos^a(r)\sin^{b-1}(r)\,dr\\
&= \frac{\pi^{b/2}}{\Gamma\left(\frac{b}{2}\right)}\overline{IB}_{(a+1)/2,b/2}(\cos^2\rho),
\end{aligned}$$

with

$$\overline{IB}_{(a+1)/2,b/2}(x) \triangleq \int_x^1 x^{(a-1)/2}(1-x)^{(b-2)/2}\,dx$$

the tail of the incomplete beta function.

Finally, we note the following result for use in Chapter 15.

Lemma 10.5.8. *The following equivalences hold between the standard and extended Lipschitz–Killing curvatures:*

$$\mathcal{L}_i^{\kappa}(\cdot) = \sum_{n=0}^{\infty}\frac{(-\kappa)^n}{(4\pi)^n}\frac{(i+2n)!}{n!i!}\mathcal{L}_{i+2n}(\cdot) \tag{10.5.11}$$

and

$$\mathcal{L}_i(\cdot) = \sum_{n=0}^{\infty}\frac{\kappa^n}{(4\pi)^n}\frac{(i+2n)!}{n!i!}\mathcal{L}_{i+2n}^{\kappa}(\cdot). \tag{10.5.12}$$

Before starting the proof, we note that, at least for some examples, it is easier to compute the $\mathcal{L}_j^{\kappa}$ than the $\mathcal{L}_j$. For example, suppose we use (10.5.10) to compute $\mathcal{L}_j^1(S(\mathbb{R}^l))$. Then, since the scalar second fundamental form is zero for $S(\mathbb{R}^l)$ as it sits in itself, the only nonzero curvature is the highest-order one, and this is given by

$$\mathcal{L}_{l-1}^1(S(\mathbb{R}^l)) = s_l.$$

From this and (10.5.12) it is now just a little algebra to check that

$$\mathcal{L}_j(S(\mathbb{R}^l)) = 2\binom{l-1}{j}\frac{s_l}{s_{l-j}}$$

for $l - 1 - j$ even, and 0 otherwise, a result that we already computed from first principles back in Chapter 6 (cf. (6.3.8)).

Proof. We shall prove only (10.5.11), since (10.5.12) follows from an almost identical argument. In both cases the argument is primarily algebraic and rather elementary, but since keeping track of combinatorial coefficients is not always trivial, and this lemma will be rather important later, we give the details.

By the definition (10.5.8), $\mathcal{L}_i^\kappa(M, A)$ is given by

$$\begin{aligned}
&\sum_{j=i}^{N}(2\pi)^{-(j-i)/2}\sum_{m=0}^{\lfloor\frac{j-i}{2}\rfloor} C(N-j, j-i-2m)\frac{(-1)^m}{m!(j-i-2m)!} \\
&\quad\times\int_{\partial_j M\cap A}\int_{S(T_t\partial_j M^\perp)} \mathrm{Tr}^{T_t\partial_j M}\left(\left(\widetilde{R}+\frac{\kappa}{2}I^2\right)^m \widetilde{S}_{\nu_{N-j}}^{j-i-2m}\right) \\
&\quad\times \mathbb{1}_{N_tM}(-\nu_{N-j})\mathcal{H}_{N-j-1}(d\nu_{N-j})\mathcal{H}_j(dt) \\
&= \sum_{j=i}^{N}(2\pi)^{-(j-i)/2}\sum_{m=0}^{\lfloor\frac{j-i}{2}\rfloor}\sum_{n=0}^{m}(-\kappa/2)^n\binom{m}{n}\frac{C(N-j, j-i-2m)}{m!(j-i-2m)!} \\
&\quad\times\iint \mathrm{Tr}^{T_t\partial_j M}\left(I^{2n}(-\widetilde{R})^{m-n}\widetilde{S}_{\nu_{N-j}}^{j-i-2m}\mathbb{1}_{N_tM}(\nu_{N-j})\right) \\
&\quad\times \mathcal{H}_{N-j-1}(d\nu_{N-j})\mathcal{H}_j(dt) \\
&= \sum_{j=i}^{N}(2\pi)^{-(j-i)/2}\sum_{n=0}^{\lfloor\frac{j-i}{2}\rfloor}\sum_{m=n}^{\lfloor\frac{j-i}{2}\rfloor}(-\kappa/2)^n\binom{m}{n}\frac{C(N-j, j-i-2m)}{m!(j-i-2m)!} \\
&\quad\times\iint \mathrm{Tr}^{T_t\partial_j M}\left(I^{2n}(-\widetilde{R})^{m-n}\widetilde{S}_{\nu_{N-j}}^{j-i-2m}\mathbb{1}_{N_tM}(\nu_{N-j})\right) \\
&\quad\times \mathcal{H}_{N-j-1}(d\nu_{N-j})\mathcal{H}_j(dt) \\
&= \sum_{j=i}^{N}(2\pi)^{-(j-i)/2}\sum_{n=0}^{\lfloor\frac{j-i}{2}\rfloor}\sum_{m=0}^{\lfloor\frac{j-i-2n}{2}\rfloor}\frac{(-\kappa/2)^n}{n!}\frac{C(N-j, j-i-2m-2n)}{m!(j-i-2m-2n)!} \\
&\quad\times\iint \mathrm{Tr}^{T_t\partial_j M}\left(I^{2n}(-\widetilde{R})^{m}\widetilde{S}_{\nu_{N-j}}^{j-i-2m-2n}\mathbb{1}_{N_tM}(\nu_{N-j})\right) \\
&\quad\times \mathcal{H}_{N-j-1}(d\nu_{N-j})\mathcal{H}_j(dt),
\end{aligned}$$

where we have used (7.2.11) in obtaining the last line.

Continuing, the last line can be rewritten as

$$\sum_{n=0}^{\lfloor\frac{N-i}{2}\rfloor} \frac{(4\pi)^{-n}(-\kappa)^n}{n!}\frac{(i+2n)!}{i!}\sum_{j=i+2n}^{k}(2\pi)^{-(j-i-2n)/2}$$
$$\times \sum_{m=0}^{\lfloor\frac{j-i-2n}{2}\rfloor}\frac{C(N-j,j-i-2m-2n)}{m!(j-i-2m-2n)!}$$
$$\times \iint \mathrm{Tr}^{T_t\partial_j M}\left((-\widetilde{R})^m \widetilde{S}_{\nu_{N-j}}^{j-i-2m-2n}\mathbb{1}_{N_t M}(\nu_{N-j})\right)$$
$$\times \mathcal{H}_{N-j-1}(d\nu_{N-j})\mathcal{H}_j(dt)$$
$$= \sum_{n=0}^{\lfloor\frac{N-i}{2}\rfloor}\frac{(4\pi)^{-n}(-\kappa)^n}{n!}\frac{(i+2n)!}{i!}\mathcal{L}_{i+2n}(M,A)$$
$$= \sum_{n=0}^{\infty}\frac{(-4\pi/\kappa)^{-n}}{n!}\frac{(i+2n)!}{i!}\mathcal{L}_{i+2n}(M,A),$$

which is what we require. □

10.6 Volume of Tubes and Gaussian Processes, Continued

With Weyl's tube formula established for submanifolds of $S_\lambda(\mathbb{R}^l)$ we are now able to continue with the volume-of-tubes approximation (10.0.1) to excursion probabilities for Gaussian processes. In (10.2.6), we established that for a Gaussian process f with an orthogonal expansion of order l,

$$\mathbb{P}\left\{\sup_{t\in M} f_t \geq u\right\} = \frac{\Gamma\left(\frac{l}{2}\right)}{2\pi^{l/2}}\mathbb{E}\left\{\mathcal{H}_{l-1}\left(\mathrm{Tube}(\psi(M),\cos^{-1}(u/|\xi|))\right)\mathbb{1}_{\{|\xi|\geq u\}}\right\},$$

where $\xi^2 \sim \chi^2_l$. On the other hand, Theorem 10.5.7 gives an explicit expression for

$$\mathcal{H}_{l-1}\left(\mathrm{Tube}(\psi(M),\cos^{-1}(u/|\xi|))\right)$$

for $\cos^{-1}(u/|\xi|) < \rho_c(M, S(\mathbb{R}^l))$. However, as $|\xi| \to \infty$, the radius of this tube grows and may surpass $\rho_c(M, S(\mathbb{R}^l))$, in which case Weyl's formula is not exact. *Formally ignoring this fact*, we arrive at the following result.

Theorem 10.6.1. *Let f be a finite Karhunen–Loève process such that $\psi(M)$ is a C^2, locally convex, Whitney submanifold of $S(\mathbb{R}^l)$, $l \geq N$. The formal tube approximation is given by*

$$\mathbb{P}\left\{\sup_{t\in M} f_t \geq u\right\} \approx \sum_{j=0}^{N}\rho_j(u)\mathcal{L}_j(\psi(M)), \tag{10.6.1}$$

where

$$\rho_j(u) \triangleq \begin{cases} 1-\Phi(u), & j=0, \\ (2\pi)^{-(j+1)/2} H_{j-1}(u) e^{-u^2/2}, & j \geq 1, \end{cases}$$

and H_j is the jth Hermite polynomial (11.6.9).

Remark 10.6.2. *It is important to understand the (im)precise meaning of* (10.6.1). *The "≈" does* not *indicate approximate equality, asymptotic equality, or any other relationship that at this stage equates the left- and right-hand sides. Rather, it indicates that the right-hand side comes from taking an expression that is eqivalent to the left-hand side, and substituting into it expressions that, while sometimes equivalent, are sometimes definitely wrong. Only in Chapter* 14 *shall we show, via completely different methods, that the purely formal equivalence in* (10.6.1) *can be replaced by asymptotic, in u, equivalence, with specific bounds on the error of doing so.*

Proof. Since the approximation formally ignores the condition $\rho < \rho_c(M)$ in Theorem 10.5.7, we have

$$\begin{aligned} &\mathbb{P}\left\{\sup_{t\in T} f_t \geq u\right\} \\ &\quad \approx \frac{\Gamma\left(\frac{l}{2}\right)}{2\pi^{l/2}} \sum_{j=0}^{N} \mathcal{L}_j(\psi(M)) \\ &\qquad \times \mathbb{E}\left\{ \sum_{n=0}^{\lfloor \frac{j}{2} \rfloor} \frac{(-4\pi)^{-n}}{n!} \frac{j!}{(j-2n)!} G_{j-2n,l-1+2n-j}(\cos^{-1}(u/|\xi|)) \mathbb{1}_{\{|\xi|\geq u\}} \right\}, \end{aligned}$$

and all that remains is to prove that $\rho_j(u)$ is equal to

$$\frac{\Gamma\left(\frac{l}{2}\right)}{2\pi^{l/2}} \sum_{n=0}^{\lfloor \frac{j}{2} \rfloor} \mathbb{E}\left\{ \frac{(-4\pi)^{-n}}{n!} \frac{j!}{(j-2n)!} G_{j-2n,l-1+2n-j}(\cos^{-1}(u/|\xi|)) \mathbb{1}_{\{|\xi|\geq u\}} \right\}.$$

We artificially decompose ξ as

$$|\xi|^2 = X_1 + X_2,$$

where $X_1 \sim \chi^2_{j-2n+1}$, $X_2 \sim \chi^2_{l-1+2n-j}$ are independent, so that

$$\frac{X_1}{X_1+X_2} \sim \text{Beta}\left((j-2n+1)/2, (l-1+2n-j)/2\right),$$

independently of $|\xi|^2$. It follows that

$$\begin{aligned} &G_{j-2n,l-1+2n-j}(\cos^{-1}(u/|\xi|^2)) \\ &\quad = \frac{\Gamma\left(\frac{j-2n+1}{2}\right)}{\Gamma\left(\frac{l}{2}\right)} \pi^{(l-1+2n-j)/2} \mathbb{P}\left\{ \frac{X_1}{X_1+X_2} \geq \frac{u^2}{|\xi|^2} \,\middle|\, |\xi|^2 \right\}. \end{aligned}$$

Continuing with the calculation,

$$\begin{aligned}
&\frac{(-4\pi)^{-n}\Gamma\left(\frac{l}{2}\right)}{2\pi^{l/2}}\mathbb{E}\left\{G_{j-2n,l-1+2n-j}(\cos^{-1}(u/|\xi|))\mathbb{1}_{\{|\xi|\geq u\}}\right\} \\
&\quad= \frac{(-4\pi)^{-n}\Gamma\left(\frac{j-2n+1}{2}\right)}{2}\pi^{-(j-2n+1)/2} \\
&\qquad\times\mathbb{E}\left\{\mathbb{P}\left\{\frac{X_1}{X_1+X_2}\geq\frac{u^2}{|\xi|^2}\bigg||\xi|^2\right\}\mathbb{1}_{\{|\xi|\geq u\}}\right\} \\
&\quad= \frac{(-4\pi)^{-n}\Gamma\left(\frac{j-2n+1}{2}\right)}{2}\pi^{-(j-2n+1)/2}\mathbb{P}\left\{X_1\geq u^2\right\} \\
&\quad= \frac{(-4\pi)^{-n}\Gamma\left(\frac{j-2n+1}{2}\right)}{2}\pi^{-(j-2n+1)/2}\int_{u^2}^{\infty}\frac{x^{(j-2n-1)/2}e^{-x/2}}{2^{(j-2n+1)/2}\Gamma\left(\frac{j-2n+1}{2}\right)}\,dx \\
&\quad= (2\pi)^{-(j+1)/2}\frac{(-1)^n}{2^n}\int_u^{\infty}x^{j-2n}e^{-x^2/2}\,dx.
\end{aligned}$$

Finally,

$$\begin{aligned}
&\frac{\Gamma\left(\frac{l}{2}\right)}{2\pi^{l/2}}\sum_{n=0}^{\lfloor\frac{j}{2}\rfloor}\mathbb{E}\left\{\frac{(-4\pi)^{-n}}{n!}\frac{j!}{(j-2n)!}G_{j-2n,l-1+2n-j}(\cos^{-1}(u/|\xi|))\mathbb{1}_{\{|\xi|\geq u\}}\right\} \\
&\quad= (2\pi)^{-(j+1)/2}\sum_{n=0}^{\lfloor\frac{j}{2}\rfloor}\frac{j!(-1)^n}{n!(j-2n)!2^n}\int_u^{\infty}x^{j-2n}e^{-x^2/2}\,dx \\
&\quad= (2\pi)^{-(j+1)/2}\int_u^{\infty}H_j(x)e^{-x^2/2}\,dx \\
&\quad= \rho_j(u),
\end{aligned}$$

as required, with the last line following from the basic properties of Hermite polynomials (cf. (11.6.12)). □

10.7 Intrinsic Volumes for Whitney Stratified Spaces

So far, we have defined intrinsic volumes for the convex sets of Section 6.3[15] via integral-geometric techniques, and for the Riemannian manifolds of Section 7.6 and this chapter via Lipschitz–Killing curvatures, for spaces that possess a local convexity property. However, for our computations in Part III we shall also need intrinsic volumes of Whitney stratified spaces for which no convexity of any kind need hold.

[15] We also gave a possible, but not very promising, extension to basic complexes via kinematic integrals at (6.3.11).

Thus, in the current section, after describing the basic conditions under which they can be defined, we shall define intrinsic volumes in a more general setting. The defining expression will be a natural generalization of (10.5.4), which covered the locally convex scenario.

However, while most of what we have done so far was devoted to and motivated by the relationship between Weyl's formula and intrinsic volumes, this will no longer be the case. In fact, as we shall soon see in Section 10.8, Weyl's formula need not hold in the non–locally convex setting, and so we need alternative motivation for the new curvatures. You can find this, among other details, in the differential geometry literature (e.g., [24, 33]) or you can wait for Chapter 12, where Lipschitz–Killing curvatures will arise when we compute the expected Euler characteristic of excursion sets. The extended definition follows, after we make one remark about conventions concerning the normal Morse index α of Section 9.2.

Remark 10.7.1. *Referring to the convention adopted in Remark* 10.5.2, *we define*

$$\alpha(0) = 1.$$

Definition 10.7.2. *Let M be an N-dimensional, C^2 Whitney stratified manifold embedded in a manifold $(\widetilde{M}, \widetilde{g})$. Assume that the set* (9.2.6) *of degenerate tangent vectors*[16] *has Hausdorff dimension less than or equal to* $\dim(\widetilde{M}) - 1$, *and that the conventions of Remarks* 10.5.2 *and* 10.7.1 *are adopted. Then the Lipschitz–Killing curvature measures of M are defined by*

$$\begin{aligned} \mathcal{L}_i(M, A) = \sum_{j=i}^{N} (2\pi)^{-(j-i)/2} \sum_{m=0}^{\lfloor \frac{j-i}{2} \rfloor} \frac{(-1)^m C(N-j, j-i-2m)}{m!(j-i-2m)!} \\ \times \int_{\partial_j M \cap A} \int_{S(T_t \partial_j M^{\perp})} \mathrm{Tr}^{T_t \partial_j M} \left(\widetilde{R}^m \widetilde{S}_{\nu_{N-j}}^{j-i-2m} \right) \\ \times \alpha(\nu_{N-j}) \mathcal{H}_{N-j-1}(d\nu_{N-j}) \mathcal{H}_j(dt), \end{aligned} \tag{10.7.1}$$

where $\mathcal{H}_j$ is the volume form on $\partial_j M$ and $\mathcal{H}_{l-j-1}$ the volume form[17] *on $S(T_t \partial_j M^{\perp})$, both determined by $\widetilde{g}$.*

We shall assume that all the signed measures defined here are finite.[18]

As we did before (cf. Lemma 10.5.8), we can again extend these measures to a one-parameter family of Lipschitz–Killing curvature measures, which we do by setting

[16] The assumption on the set of degenerate tangent vectors is prompted by the same issue that arose in Section 9.2, that α may not be well defined in vectors that annihilate a generalized tangent space.

[17] Of course, $\mathcal{H}_{l-j-1}$ really depends on t, but we have more than enough subscripts already.

[18] A priori, there is no reason for the $\mathcal{L}_j(M, \cdot)$ to be finite. However, for all the examples we consider in this section and indeed, all situations of interest in this book, that will in fact be the case. We therefore invoke this as a standing assumption for the remainder of the book.

$$\mathcal{L}_i^{\kappa}(M, A) \triangleq \sum_{n=0}^{\infty} \frac{(-\kappa)^{-n}}{(4\pi)^n} \frac{(i+2n)!}{n!i!} \mathcal{L}_{i+2n}(M, A). \tag{10.7.2}$$

Furthermore, we can define *Lipschitz–Killing curvatures*, or *intrinsic volumes*, of M by

$$\mathcal{L}_j(M) \triangleq \mathcal{L}_j(M, M), \qquad \mathcal{L}_j^{\kappa}(M) \triangleq \mathcal{L}_j^{\kappa}(M, M). \tag{10.7.3}$$

Although it is not at all obvious from the definition (10.7.1), it can be shown that the measures $\mathcal{L}_j^{\kappa}(M; \,)$ are independent of the stratification of M. Furthermore, $\mathcal{L}_j^{\kappa}$ is a finitely additive valuation in its first variable and a (signed) measure in the second one (cf. [24]).

The Lipschitz–Killing curvature $\mathcal{L}_0(M)$ is actually the Euler–Poincaré characteristic of M, as defined via (8.1.1) and a triangulation. The fact that the curvature integral and the Euler characteristic are equivalent is a celebrated result, known as the *Chern–Gauss–Bonnet theorem.*

To get a feel for the fact that (10.7.1) is not quite as forbidding as it seems, you might want to look at the simple example[19] of the unit cube in $\mathbb{R}^N$ now to see how it works for familiar sets.

[19] Consider, for our first example, a rectangle in $\mathbb{R}^N$ equipped with the standard Euclidean metric. Thus, we want to recover (6.3.5), that is,

$$\mathcal{L}_j\left([0, T]^N\right) = \binom{N}{j} T^j. \tag{10.7.4}$$

Since we are in the Euclidean scenario, both the Riemannian curvature and second fundamental form are identically zero, and so the only terms for which the final integral in (10.7.1) is nonzero are those for which $l = k - j = 0$. Thus all of the sums collapse and (10.7.1) simplifies to

$$\pi^{-(N-j)/2}\Gamma\left(\frac{N-j}{2}\right)2^{-1}\int_{\partial_j M}\mathcal{H}_j(dt)\int_{S(T_t^{\perp}\partial_j M)}\alpha(\nu)\mathcal{H}_{N-j-1}(d\nu).$$

Note the following:

(i) Referring to the convex polytope discussion in Section 9.2, $\alpha(\nu) = \mathbb{1}_{N_t \partial_j M}(-\nu)$ is zero except on a $(1/2^{N-j})$th part of the sphere $S(T^{\perp}\partial_j M)$, and the measure $\mathcal{H}_{N-j-1}$ is surface measure on $S(T^{\perp}\partial_j M)$. Consequently,

$$\int_{S(T_t^{\perp}\partial_j M)}\alpha(\nu)\mathcal{H}_{N-j-1}(d\nu) = \frac{2\pi^{(N-j)/2}2^{-(N-j)}}{\Gamma\left(\frac{N-j}{2}\right)}.$$

(ii) There are $2^{N-j}\binom{N}{j}$ disjoint components to $\partial_j M$, each one a j-dimensional cube of edge length T.

(iii) The volume form on $\partial_j M$ is Lebesgue measure, so that each of the cubes in (ii) has volume T^j.

Now combine all of the above, and (10.7.4) follows.

The formulas simplify considerably if M is a C^2 domain of $\mathbb{R}^N$ with the induced Riemannian structure, since then the ambient curvature tensor is identically zero. In that case $\mathcal{L}_N(M, U)$ is the Lebesgue measure of U and

$$\mathcal{L}_j(M, U) = \int_{\partial M \cap U} \frac{1}{s_{N-j}(N-1-j)!} \operatorname{Tr}(S_{\nu_t}^{N-1-j}) \operatorname{Vol}_{\partial M, g}, \tag{10.7.5}$$

for $0 \le j \le N-1$, where s_j is given by (5.7.5) and ν_t is the inward normal at t. Using this formula, you might want to check how to compute the Lipschitz–Killing curvatures for an N-dimensional ball,[20] although it would really be easier to wait for (10.7.6).

In this setting (10.7.5) can also be written in another form that is often more conducive to computation. If we choose an orthonormal frame field $(E_1, \dots, E_{N-1})$ on ∂M, and then extend this to one on M in such a way that E_N is the inward normal, then it follows from (7.2.7) that

$$\mathcal{L}_j(M, U) = \frac{1}{s_{N-j}} \int_{\partial M \cap U} \operatorname{detr}_{N-1-j}(\mathrm{Curv}) \operatorname{Vol}_{\partial M, g}, \tag{10.7.6}$$

where detr_j is given by (7.2.8) and the curvature matrix Curv is given by

$$\mathrm{Curv}(i, j) \stackrel{\Delta}{=} S_{E_N}(E_i, E_j). \tag{10.7.7}$$

It is important to note that while the elements of the curvature matrix may depend on the choice of basis, $\operatorname{detr}_{N-1-j}(\mathrm{Curv})$ is independent of the choice, as will be $\mathcal{L}_j(M, U)$.

Since all of the simplifications (and examples in the footnotes) that we have just looked at are locally convex, in the sense that they have convex support cones, we now describe how to compute the Lipschitz–Killing curvatures for one highly nonconvex example, a many-pointed star.

By a star, we mean the union of a finite collection line segments $L_1, \dots, L_r$ emanating from a point v_0 with endpoints $(v_j)_{1 \le j \le r}$. The ambient space is $\mathbb{R}^2$. The space $M = \bigcup_{j=1} L_r$ has a natural stratification

$$\partial_1 M = \bigcup_{j=1}^{r} L_r^\circ, \qquad \partial_0 M = \bigcup_{j=0}^{r} v_j.$$

[20] For the ball $B^N(T)$ we want to recover (6.3.7). It is easy to check that the second fundamental form of $\partial B^N(T) = S^{N-1}(T)$, with respect to the inner normal, is a constant T^{-1} over the sphere. Thus the term involving its trace can be taken out of the integral in (10.7.5) leaving only the volume of $S^{N-1}(T)$, given by $s_N T^{N-1}$. To compute the trace we use (7.2.11), to see that $\operatorname{Tr} S^{N-1-j} = T^{-(N-j-1)}(N-1)!/j!$, so that

$$\mathcal{L}_j(B^N(T)) = \frac{s_N T^{N-1} T^{-(N-j-1)}(N-1)!}{s_{N-j}(N-1-j)!j!} = T^j \binom{N-1}{j} \frac{s_N}{s_{N-j}} = T^j \binom{N}{j} \frac{\omega_N}{\omega_{N-j}},$$

which is (6.3.7).

First, we note that

$$\mathcal{L}_j(M,\,) = 0, \quad j \geq 2,$$

and

$$\mathcal{L}_1(M, A) = \sum_{j=1}^{r} \mathcal{H}_1(L_j \cap A),$$

so it remains to compute the measure $\mathcal{L}_0(M,\,)$. Clearly, $\mathcal{L}_0(M,\,)$ concentrates on the vertices of M, and considering the case of one segment alone, it follows easily from (10.7.1) that

$$\mathcal{L}_0(M, v_j) = \frac{1}{2}, \quad 1 \leq j \leq r.$$

As for $\mathcal{L}_0\,(M, v_0)$, note that for ν a unit vector emanating from v_0,

$$\alpha(\nu) = 1 - \sum_{j=1}^{r} \mathbb{1}_{\{\langle \nu, L_j\rangle < 0\}},$$

so that it again follows easily from (10.7.1) that

$$\mathcal{L}_0(M, v_0) = 1 - \frac{r}{2}.$$

Putting all this together gives the Euler characteristic $\mathcal{L}_0(M)$ as $1-r/2+r\times(1/2) = 1$, which is as it should be by Chapter 6.

Here is one more example that is simple but informative. You should find it easy to check, using the fact that Lipschitz–Killing curvature measures are additive in M. For $0 \leq \phi \leq 2\pi$, and $(r(x, y), \theta(x, y))$ the polar representation of $(x, y) \in \mathbb{R}^2$, let

$$M_\phi = \left\{(x, y) \in \mathbb{R}^2 : 0 \leq r(x, y) \leq 1, 0 \leq \theta(x, y) \leq \phi\right\}$$

be a solid angle in the plane of angle ϕ. Let $v_0 = (0, 0), v_1 = (1, 0), v_2 = (\cos\phi, \sin\phi)$ be the three vertices of M_ϕ. Then,

$$\begin{aligned}
\mathcal{L}_2(M_\phi, U) &= \mathcal{H}_2(U \cap M_\phi), \\
\mathcal{L}_1(M_\phi, U) &= \frac{1}{2}\mathcal{H}_1(U \cap \partial M_\phi), \\
\mathcal{L}_0(M_\phi, U) &= \frac{1}{2\pi}\mathcal{H}_1(U \cap M_\phi \cap S(\mathbb{R}^2)) \\
&\quad - \frac{\phi - \pi}{2\pi}\mathbb{1}_U(v_0) + \frac{1}{4}\mathbb{1}_U(v_1) + \frac{1}{4}\mathbb{1}_U(v_2).
\end{aligned}$$

10.7.1 Alternative Representation of the Curvature Measures

We conclude this section by noting an alternative representation of the curvature measures of subsets of $\mathbb{R}^l$ and $S(\mathbb{R}^l)$. This representation is sometimes easier to work with, and we shall use it in Chapter 15. Its proof follows from a straightforward application of Lemma 10.5.5.

Lemma 10.7.3. *Let M be a Whitney stratified subspace of $\mathbb{R}^l$. Then*

$$\mathcal{L}_i(M, A) = \sum_{j=i}^{l} (2\pi)^{-(j-i)/2} \frac{1}{(j-i)!} \times \int_{\partial_j M \cap A} \mathbb{E}\left\{ \mathrm{Tr}^{T_t \partial_j M} (S_{X_{l-j,t}})^{j-i} \alpha(X_{l-j,t}) \right\} \mathcal{H}_j(dt), \tag{10.7.8}$$

where for each $t \in \partial_j M$, $X_{l-j,t} \sim \gamma_{T_t \partial_j M^\perp}$ and α is the normal Morse index at $t \in M$ as it sits in $\mathbb{R}^l$.

If M is a Whitney stratified space in $S_{\sqrt{\kappa}}(\mathbb{R}^l)$, then

$$\mathcal{L}_i^\kappa(M, A) = \sum_{j=i}^{l-1} (2\pi)^{-(j-i)/2} \frac{1}{(j-i)!} \int_{\partial_j M \cap A} \times \mathbb{E}\left\{ \mathrm{Tr}^{T_t \partial_j M} (S_{X_{l-1-j,t}})^{j-i} \alpha(X_{l-1-j,t}) \right\} \mathcal{H}_j(dt), \tag{10.7.9}$$

where for each $t \in \partial_j M$, $X_{l-1-j,t} \sim \gamma_{T_t \partial_j M^\perp}$, where $T_t \partial_j M^\perp$ is the orthogonal complement of $T_t \partial_j M$ in $T_t S_{\sqrt{\kappa}} S(\mathbb{R}^l)$, α is the normal Morse index at $t \in M$ as it sits in $S_{\sqrt{\kappa}}(\mathbb{R}^l)$, and S is the second fundamental form of M as it sits in $S_{\sqrt{\kappa}}(\mathbb{R}^l)$.

In general, suppose that $M \subset (\widetilde{M}, g)$ of dimension l is a Whitney stratified space. Then

$$\mathcal{L}_i^\kappa(M, A) = \sum_{j=i}^{l} (2\pi)^{-(j-i)/2} \sum_{m=0}^{\lfloor \frac{j-i}{2} \rfloor} \frac{1}{m!(j-i-2m)!} \int_{\partial_j M \cap A} \times \mathbb{E}\left\{ \mathrm{Tr}^{T_t \partial_j M} ((-\tilde{R} + \kappa I^2/2)^m \tilde{S}_{X_{l-j,t}}^{j-i-2m}) \alpha(X_{l-j,t}) \right\} \mathcal{H}_j(dt), \tag{10.7.10}$$

where for each $t \in \partial_j M$, $X_{l-j,t} \sim \gamma_{T_t \partial_j M^\perp}$, where $T_t \partial_j M^\perp$ is the orthogonal complement of $T_t \partial_j M$ in $T_t \widetilde{M}$, α is the normal Morse index at $t \in M$ as it sits in $\widetilde{M}$, $\tilde{R}$ is the curvature tensor of $(\widetilde{M}, g)$, and $\tilde{S}$ is the shape operator of M as it sits in $\widetilde{M}$.

10.8 Breakdown of Weyl's Tube Formula

While the next three paragraphs may seem to be too brief to demand their own numbered section, the observation that they make is important enough that the honor is deserved. From the definition of the Lipschitz–Killing curvatures for Whitney stratified spaces, one may be tempted to extend Weyl's tube formula to compute the volume of a tube around all Whitney stratified spaces $(M, \mathcal{Z})$ embedded in $\mathbb{R}^l$. Specifically, one may be tempted to conclude that

$$\mathcal{H}_l(\mathrm{Tube}(M,\rho)) = \sum_{j=0}^{l} \rho^{l-j}\omega_{l-j}\mathcal{L}_j(M)$$

even if M does not have convex support cones.

However, this is not true in general, and the same example that appeared in Figure 6.3.2 suffices to provide a counterexample.

It is essentially for this reason that in Chapter 14, we shall have to restrict our attention to locally convex spaces to obtain strong results about the accuracy of the volume-of-tubes approach to computing excursion probabilities, as well as that of the approximation via expected Euler characteristics.

10.9 Generalized Lipschitz–Killing Curvature Measures

As we saw in the earlier sections, the measures $\mathcal{L}_j(M;)$ are signed measures on M that determine the coefficients in a (Lebesgue) volume-of-tubes expansion around M when M is embedded in $\mathbb{R}^l$. These measures are adequate to derive Weyl's tube formula, but they will need to be generalized to continue our general study of the volume-of-tubes problem, that is, computing $\mu(\mathrm{Tube}(M,\rho))$ for measures other than Lebesgue.

Our interest in general measures μ is not purely academic, for the main result in Chapter 15, which treats non-Gaussian random fields, relies heavily on the coefficients in the tube expansion

$$\gamma_{\mathbb{R}^l}(\mathrm{Tube}(M,\rho)).$$

Since this is the main expansion that we shall need, for this section we shall assume that the dimensions of both the underlying manifold and its ambient Euclidean space are l. That is, in our earlier notation,

$$\widetilde{M} = \widehat{M} = \mathbb{R}^l, \qquad \dim(M) = N = l = \dim(\widehat{M}).$$

The generalization of the measures $\mathcal{L}_j(M, \cdot)$ is needed to deal with the term $f_\mu(F_{j,r}(t,s))$ in Theorem 10.4.2, which was, conveniently, identically 1 in Weyl's tube formula. The general strategy is to assume that f_μ is smooth enough so that there is a Taylor series expansion for $f_\mu(F_{j,r}(t,s))$ of the form

$$f_\mu(F_{j,r}(t,s)) = f_\mu\,(t + r\eta(t,s)) = \sum_{m=0}^{\infty} \frac{r^m}{m!}\frac{d^m}{d\eta^m} f_\mu(t), \tag{10.9.1}$$

where, as in (10.4.15), $\eta \equiv \eta(t, s)$ is the unit vector from t to $F_{j,r}(t, s)$, and by $d^m f/d\eta^m$ we mean the mth-order derivative of f in the direction $\eta(t, s)$.

Using this expansion, we proceed to integrate

$$\frac{d^m}{d\eta^m} f_\mu(t)$$

over $S_j(r)$, the hypersurface at distance r from $\partial_j M$ for $0 \leq j \leq l$. After integrating over $S_j(r)$, we integrate r over $[0, \rho]$, much as we did in deriving Weyl's tube formula (10.5.7).

We begin by defining generalized Lipschitz–Killing curvature measures on $\mathrm{SB}(\mathbb{R}^l)$, the sphere bundle of $\mathbb{R}^l$. These measures have finer information than the basic Lipschitz–Killing curvature measures in that, up to a constant, they encode surface measure on the hypersurfaces at distance r from M.

10.9.1 The Generalized Curvature Measures

Before we explicitly define these measures, we note that they *should be* measures on $\mathrm{SB}(\mathbb{R}^l)$, the sphere bundle of $\mathbb{R}^l$, which is canonically isomorphic to the product $S(\mathbb{R}^l) \times \mathbb{R}^l$. For simplicity, however, we choose to define the generalized Lipschitz–Killing measures on $S(\mathbb{R}^l) \times \mathbb{R}^l$ rather than on $\mathrm{SB}(\mathbb{R}^l)$, keeping this canonical isomorphism in mind. We also note that when it is convenient, we shall think of them as measures on $\mathrm{SB}(\mathbb{R}^l)$.

Definition 10.9.1. *Let $(M, \mathcal{Z}) \subset \mathbb{R}^l$ be an l-dimensional Whitney stratified space satisfying the requirements of Definition* 10.7.2. *The generalized Lipschitz–Killing curvature measures of M, defined on Borel sets $A \subset \mathbb{R}^l$, $B \subset S(\mathbb{R}^l)$ and supported on $M \times S(\mathbb{R}^l)$, are defined, for $0 \leq i \leq l-1$, by*

$$\begin{aligned}\widetilde{\mathcal{L}}_i(M, A \times B) \triangleq \sum_{j=i}^{l} (2\pi)^{-(j-i)/2} \sum_{m=0}^{\lfloor \frac{j-i}{2} \rfloor} \frac{(-1)^m C(l-j, j-i-2m)}{m!(j-i-2m)!} \\ \times \int_{\partial_j M \cap A} \int_{S(T^\perp \partial_j M) \cap B} \mathrm{Tr}^{T_t \partial_j M}\left(\widetilde{R}^m \widetilde{S}_{\nu_{l-j}}^{j-i-2m}\right) \\ \times \alpha(\nu_{l-j}) \mathcal{H}_{l-j-1}(d\nu_{l-j}) \mathcal{H}_j(dt),\end{aligned} \tag{10.9.2}$$

where α is the normal Morse index of Section 9.2.

For any Borel function $f : \mathbb{R}^l \times S(\mathbb{R}^l) \to \mathbb{R}$ we use the standard notation

$$\widetilde{\mathcal{L}}_i(M, f)$$

to denote the integral of f with respect to $\widetilde{\mathcal{L}}_i(M, \cdot)$. If $i = l$, we define $\widetilde{\mathcal{L}}_l(M, \cdot)$ only on Borel functions on $\mathbb{R}^l \times S(\mathbb{R}^l)$ that are constant over $S(\mathbb{R}^l)$. Specifically,

$$\widetilde{\mathcal{L}}_l(M, f) \triangleq \begin{cases} \int_M f(t, \nu) \mathcal{H}_l(dt), & f \in \mathcal{B}(\mathbb{R}^l) \times S(\mathbb{R}^l), \\ 0 & \textit{otherwise}, \end{cases} \tag{10.9.3}$$

for some fixed but arbitrary $\nu \in S(\mathbb{R}^l)$, *where*

$$\mathcal{B}(\mathbb{R}^l) \times S(\mathbb{R}^l) = \left\{ A \times S(\mathbb{R}^l) : A \in \mathcal{B}(\mathbb{R}^l) \right\}$$

is the subsigma algebra of $\mathcal{B}(\mathbb{R}^l) \otimes \mathcal{B}(S(\mathbb{R}^l))$ *generated by functions that are constant over* $S(\mathbb{R}^l)$.

The generalized curvature measures $\widetilde{\mathcal{L}}_j(M, \cdot)$ can also be represented nonintrinsically as in (10.5.5) and (10.5.10), though we leave this generalization to the reader. The following two results are immediate corollaries of Definition 10.9.1.

Corollary 10.9.2. *Let* $(M, \mathcal{Z}) \subset \mathbb{R}^l$ *be a Whitney stratified space satisfying the requirements of Definition* 10.7.2. *Then, for any Borel set* $A \subset \mathbb{R}^l$,

$$\mathcal{L}_j(M, A) = \widetilde{\mathcal{L}}_j(M, A \times S(\mathbb{R}^l)).$$

Corollary 10.9.3. *If* $(M, \mathcal{Z})$ *is a locally convex, Whitney stratified space satisfying the requirements of Definition* 10.7.2, *then* (10.9.2) *holds with* $\alpha(\nu_{l-j})$ *replaced by* $\mathbb{1}_{N_t M}(-\nu_{l-j})$.

10.9.2 Surface Measure on the Boundary of a Tube

Having defined the measures $\{\widetilde{\mathcal{L}}_j(M,)\}_{0 \le j \le l}$, we now proceed to describe how they can be used to integrate over $\partial\,\mathrm{Tube}(M, r)$, the hypersurface of distance r from a locally convex space M.

Before we attack this task in earnest, we revisit the *Minkowski functionals*, which we defined, in the context of integral geometry, in Section 6.3. Recall that they were defined as

$$\mathcal{M}_j(M) = \left(j!\omega_j\right) \mathcal{L}_{l-j}(M),$$

although in Chapter 6 we treated only the case $\dim(M) = N = l = \dim(\widetilde{M})$.

Steiner's formula (6.3.10), extended to the case $N \le l$ (or, equivalently, Weyl's tube formula (10.5.7)) gives

$$\mathcal{H}_l(\mathrm{Tube}(M, r)) = \sum_{j=0}^{l} \frac{r^j}{j!} \mathcal{M}_j(M).$$

That is, Weyl's tube formula can be expressed as a (finite) power series with the $\mathcal{M}_j(M)$ as coefficients. This motivates our definition of *Minkowski curvature measures* as

$$\mathcal{M}_j(M, A) \stackrel{\Delta}{=} (j!\omega_j)\mathcal{L}_{l-j}(M, A), \tag{10.9.4}$$

and *generalized Minkowski curvature measures* as

$$\widetilde{\mathcal{M}}_j(M, A \times B) \stackrel{\Delta}{=} (j!\omega_j)\widetilde{\mathcal{L}}_{l-j}(M, A \times B). \tag{10.9.5}$$

The reason we turn to (generalized) Minkowski curvature measures now, instead of (generalized) Lipschitz–Killing curvature measures, is primarily one of notational convenience, and they will mesh in well with the power series expansions of f_μ to come. Ultimately, this will also turn out to be the most convenient approach for the results of Chapter 15.

We now return to describing surface measure on $\partial\,\mathrm{Tube}(M, r)$. In Section 10.4, we invested a substantial amount of effort in computing $F_{j,r}^*(\Omega_{j,r})$, the pullback of the volume form on $S_j(r)$, the hypersurface of distance r from $\partial_j M$. In Section 10.5 we combined this formula with our explicit description of the regions $D_j(\rho)$ to derive (10.5.7), Weyl's formula for locally convex manifolds in $\mathbb{R}^l$. In describing regions, we used the maps $F_{j,r}$ (cf. (10.4.7)), which, when taken together and when r is small enough, form a bijection

$$\bigcup_{j=0}^{l} \bigcup_{t\in\partial_j M} \{t\} \times S(N_t M) \overset{F_r}{\leftrightarrows} \partial\,\mathrm{Tube}(M, r) \simeq M \times S(\mathbb{R}^l),$$

which can explicitly be expressed as

$$\begin{aligned} F_r(t, \nu) &= t + r\nu, \\ F_r^{-1}(s) &= (\xi_M(s), (s - \xi_M(s))/r)\,, \end{aligned}$$

where ξ_M is the metric projection of M, defined in (10.3.2).

By pulling pieces from Sections 10.4 and 10.5, it is now not difficult to prove the following theorem.

Theorem 10.9.4. *Let $(M, \mathcal{Z}) \subset \mathbb{R}^l$ be a C^2, locally convex, Whitney stratified space. For any Borel set $A \subset \mathbb{R}^l$ such that*

$$\mathcal{H}_l(A \cap M) = 0,$$

and for any $r < \rho_c(M, \mathbb{R}^l)$, the surface area of the intersection of A with $\partial\,\mathrm{Tube}(M, r)$ can be written as[21]

$$\mathcal{H}_{l-1}\left(A \cap \partial\,\mathrm{Tube}(M, r)\right) = \sum_{j=0}^{l-1} \frac{r^j}{j!} \widetilde{\mathcal{M}}_{j+1}\left(M, F_{-r}^{-1}(A \cap \partial\,\mathrm{Tube}(M, r))\right). \tag{10.9.6}$$

Proof. We do not provide a detailed proof, since nearly all of the work needed has already appeared in Sections 10.4 and 10.5. In fact, the present case is even simpler, since the integration over r is no longer necessary.

Essentially, all we need to do is to decompose $A \cap \partial\,\mathrm{Tube}(M, r)$ into cross-sections of $\partial\,\mathrm{Tube}(M, r)$. That is, we first find the image of A under the metric projection ξ_M. Next, for each t in $\xi_M(A)$, we find all points in $A \cap \partial\,\mathrm{Tube}(M, r)$ that projected to t. Then, we sum the contribution of all these points, resulting in the integral against the generalized Minkowski curvature measures $\widetilde{\mathcal{M}}_{j+1}(M, \cdot)$. □

[21] The slightly awkward usage of F_{-r} instead of F_r in (10.9.6) is a notational consequence of having used $S_{\nu_{l-j}}$ and $\alpha(\nu_{l-j})$ in Definition 10.9.1 rather than $S_{-\nu_{l-j}}$ and $\alpha_{-\nu_{l-j}}$.

10.9.3 Series Expansions for the Gaussian Measure of Tubes

We are now ready to prove the main result of this section, in which we derive a power series expansion for

$$\mu(\mathrm{Tube}(M,\rho))$$

when μ has a bounded, analytic density with respect to Lebesgue measure. It is possible to replace analytic functions with Schwarz functions, although the power series expansion below is then only formal (cf. [158]).

Theorem 10.9.5. *Let $(M,\mathcal{Z}) \subset \mathbb{R}^l$ be a C^2, locally convex, Whitney stratified manifold, and μ a measure on $\mathbb{R}^l$ that has a bounded density f_μ with respect to Lebesgue measure. Assume that*

$$\mu(\mathrm{Tube}(M,\rho)) = \int_{\mathrm{Tube}(M,\rho)} f_\mu(x)\, d\mathcal{H}_l(x) < \infty.$$

Then, for $\rho < \rho_c(M,\mathbb{R}^l)$,

$$\mu(\mathrm{Tube}(M,\rho)) = \int_0^\rho \sum_{j=0}^{l-1} \frac{r^j}{j!} \widetilde{\mathcal{M}}_{j+1}(M, f_\mu \circ F_{-r})\, dr. \tag{10.9.7}$$

Suppose, in addition, that f_μ is analytic and for every $\varepsilon > 0$ there exists a compact $K(\varepsilon) \subset \mathbb{R}^l$ such that

$$\int_0^\rho \sum_{j=0}^{l-1} \frac{r^j}{j!} |\widetilde{\mathcal{M}}_{j+1}| \left(M, \mathbb{1}_{K(\varepsilon)^c} \circ F_{-r} \times \left| \sum_{m=0}^{n} \frac{(-r)^m}{m!} \frac{d^m}{d\eta^m} f_\mu \right| \right) dr < \varepsilon, \tag{10.9.8}$$

uniformly in n, where $|\widetilde{\mathcal{M}}_j|$ is the (positive) measure corresponding to the signed measure $\widetilde{\mathcal{M}}_j$ and we interpret $\frac{d^m}{d\eta^m}$ as in (10.9.1). Then

$$\begin{aligned} \int_{\mathrm{Tube}(M,\rho)} f_\mu(x)\, d\mathcal{H}_l(x) &= \int_0^\rho \sum_{j=0}^{l-1} \frac{r^j}{j!} \widetilde{\mathcal{M}}_{j+1} \left(M, \sum_{m=0}^{\infty} \frac{(-r)^m}{m!} \frac{d^m}{d\eta^m} f_\mu \right) dr \\ &= \mu(M) + \sum_{j=1}^{\infty} \frac{\rho^j}{j!} \mathcal{M}_j^\mu(M), \end{aligned} \tag{10.9.9}$$

where

$$\mathcal{M}_j^\mu(M) \triangleq \sum_{m=0}^{j-1} \binom{j-1}{m} (-1)^{j-1-m} \widetilde{\mathcal{M}}_{m+1} \left(M, \frac{d^{j-1-m}}{d\eta^{j-1-m}} f_\mu \right). \tag{10.9.10}$$

Proof. The equality in (10.9.7) is a straightforward application of the coarea formula (7.4.13) with the Lipschitz map over whose level sets we integrate being the distance function $d_{\mathbb{R}^l}(M,\cdot)$, along with (10.9.6).

As far as (10.9.9) is concerned, there are two things to show. The first is that the first equality is valid, and the second that the final expression is equal to that on the line above it. The first inequality arises by formally replacing $f_\mu \circ F_{-r}$ in (10.9.7) by its Taylor series expansion as in (10.9.1), that is,

$$f_\mu(F_{-r}(t,s)) = \sum_{m=0}^{\infty} \frac{(-r)^m}{m!} \frac{d^m}{d\eta^m} f_\mu(t).$$

We do, however, need to check that the resulting series expansion in (10.9.9) converges nicely. Fix $\varepsilon > 0$ and choose $K(\varepsilon)$ satisfying (10.9.8). Since we long ago agreed that we consider only tubes of finite measure, we can, without loss of generality, assume that

$$\mu(\mathrm{Tube}(M,\rho) \cap K(\varepsilon)^c) < \varepsilon,$$

so that we need focus our concentration only on what happens on $K(\varepsilon)$. On $F_{-r}^{-1}(K(\varepsilon))$, the convergence

$$\sum_{m=0}^{n} \frac{(-r)^m}{m!} \frac{d^m}{d\eta^m} f_\mu(t) \overset{n\to\infty}{\to} f_\mu(t - r\eta)$$

is uniform in n. Alternatively, on $K(\varepsilon)$ the convergence

$$f_\mu^n(s) = \sum_{m=0}^{n} \frac{r(s)^m}{m!} \frac{d^m}{d\eta(s)^m} f_\mu(t(s)) \overset{n\to\infty}{\to} f_\mu(s)$$

holds uniformly in s. By the boundedness of f_μ and compactness of $K(\varepsilon)$ we can therefore choose $n(\varepsilon)$ large enough so that

$$\begin{aligned}
&\Bigg| \mu(\mathrm{Tube}(M,\rho) \cap K(\varepsilon)) \\
&\quad - \int_0^\rho \sum_{j=0}^{l-1} \frac{r^j}{j!} \widetilde{\mathcal{M}}_{j+1}\left(M, \mathbb{1}_{K(\varepsilon)} \times \sum_{m=0}^{n(\varepsilon)} \frac{(-r)^m}{m!} \frac{d^m}{d\eta^m} f_\mu\right) dr \Bigg| \\
&= \left| \int_{\mathrm{Tube}(M,\rho)\cap K(\varepsilon)} \sum_{m=n(\varepsilon)+1}^{\infty} \frac{(-r(s))^m}{m!} \frac{d^m}{d\eta(s)^m} f_\mu(t(s))\, d\mathcal{H}_l(s) \right| \\
&< \varepsilon.
\end{aligned}$$

Thus there are no problems with the convergence in (10.9.9). As for the equality between the two expressions there, note that

$$
\begin{aligned}
&\int_{\mathrm{Tube}(M,\rho)} f_\mu(x)\, d\mathcal{H}_l(x) \\
&\quad = \mu(M) + \int_0^\rho \sum_{j=0}^{l-1} \frac{r^j}{j!} \widetilde{\mathcal{M}}_{j+1}\left(M, \sum_{m=0}^{\infty} \frac{(-r)^m}{m!} \frac{d^m}{d\eta^m} f_\mu\right) dr \\
&\quad = \mu(M) + \sum_{j=0}^{l-1} \sum_{m=0}^{\infty} (-1)^m \frac{\rho^{j+1+m}}{j!m!(j+1+m)} \widetilde{\mathcal{M}}_{j+1}\left(M, \frac{d^m}{d\eta^m} f_\mu\right) \\
&\quad = \mu(M) + \sum_{j=1}^{l} \sum_{m=j}^{\infty} \frac{\rho^m}{m!} \binom{m-1}{j-1} (-1)^{m-j} \widetilde{\mathcal{M}}_{j+1}\left(M, \frac{d^{m-j}}{d\eta^{m-j}} f_\mu\right) \\
&\quad = \mu(M) + \sum_{m=1}^{\infty} \frac{\rho^m}{m!} \sum_{j=0}^{m-1} \binom{m-1}{j} (-1)^{m-j-1} \widetilde{\mathcal{M}}_j\left(M, \frac{d^{m-j-1}}{d\eta^{m-j-1}} f_\mu\right),
\end{aligned}
$$

which, except for the fact that m and j have switched roles, is precisely what we need. □

We conclude with a corollary to Theorem 10.9.5, which gives an explicit power series representation of

$$\gamma_{\mathbb{R}^l}(\mathrm{Tube}(M,\rho))$$

for suitable locally convex spaces $(M, \mathcal{Z}) \subset \mathbb{R}^l$. This expansion will be crucial for our analysis of non-Gaussian processes in Chapter 15.

Corollary 10.9.6. *Suppose $(M, \mathcal{Z}) \subset \mathbb{R}^l$ is a C^2, locally convex, Whitney stratified manifold, and that for every $\varepsilon > 0$ there exists $K(\varepsilon) \subset \mathbb{R}^l$ compact such that*

$$\int_0^\rho \sum_{j=0}^{l-1} \frac{r^j}{j!} |\widetilde{\mathcal{M}}_{j+1}| \left(M, \mathbb{1}_{K(\varepsilon)^c} \circ F_{-r} \left| \sum_{m=0}^{n} \frac{r^m}{m!} H_m(\langle \eta, t\rangle)\, e^{-|t|^2/2} \right|\right) dr$$

is less than ε, uniformly in n, where H_m is the mth Hermite polynomial (11.6.9). *Then,*

$$\gamma_{\mathbb{R}^l}(\mathrm{Tube}(M,\rho)) = \gamma_{\mathbb{R}^l}(M) + \sum_{j=1}^{\infty} \frac{\rho^j}{j!} \mathcal{M}_j^{\gamma_{\mathbb{R}^l}}(M), \tag{10.9.11}$$

where

$$\mathcal{M}_j^{\gamma_{\mathbb{R}^l}}(M) \triangleq (2\pi)^{-l/2} \sum_{m=0}^{j-1} \binom{j-1}{m} \widetilde{\mathcal{M}}_{m+1}\big(M, H_{j-1-m}(\langle \eta, t\rangle)\, e^{-|t|^2/2}\big). \tag{10.9.12}$$

Proof. Comparing the above with Theorem 10.9.5, it is clear (cf. (10.9.10)) that all we need show is that

$$\frac{d^{j-1-m}}{d\eta^{j-1-m}} f_\mu(t) = (-1)^{j-1-m} H_{j-1-m}\left(\langle \eta, t\rangle\right) \mathrm{e}^{-|t|^2/2}.$$

However, this is easy to check from the generating function definition of Hermite polynomials at (11.6.11). □

To see how this expansion works in the simplest of cases, take $M = [u, \infty) \subset \mathbb{R}$. Since all the $\widetilde{\mathcal{M}}_k$ are then identically zero for $k \geq 2$, there is only one term appearing in the sum defining $\mathcal{M}_j^{\gamma_{\mathbb{R}^l}}(M)$. Furthermore, since $\mathrm{Tube}([u, \infty), \rho) = [u - \rho, \infty))$, we have that $\widetilde{\mathcal{M}}_1$ is supported on $\{u\} \times \{+1\}$ ($+1$ being the right-pointing unit vector at u), and so

$$\mathcal{M}_j^{\gamma_{\mathbb{R}^l}}([u, \infty)) = H_{j-1}(u) \frac{\mathrm{e}^{-u^2/2}}{\sqrt{2\pi}}. \tag{10.9.13}$$

It is elementary calculus to check that this gives the same expansion, when substituted into (10.9.11), that comes from a simple Taylor series expansion of the Gaussian tail probability Ψ of (1.2.1). If you want to see the details, we shall go through them in Section 15.10.1 for expository reasons.

We close with the following simple, but rather important, comment.

Remark 10.9.7. *Unlike the usual Minkowski functionals, the functionals* $\mathcal{M}_j^{\gamma_{\mathbb{R}^l}}$ *are actually normalized independently of the dimension* $\mathbb{R}^l$, *in the sense that for all integers* $k > 0$,

$$\mathcal{M}_j^{\gamma_{\mathbb{R}^l}}(\cdot) = \mathcal{M}_j^{\gamma_{\mathbb{R}^{l+k}}}(\cdot \times \mathbb{R}^k).$$

Hence, from now on, we shall drop the $\mathbb{R}^l$ *in the definition of* $\mathcal{M}_j^{\gamma_{\mathbb{R}^l}}$, *using the simpler notation* $\mathcal{M}_j^{\gamma}$. *Furthermore, although these functionals were derived under the assumption of local convexity, it actually turns out they can also be defined for non–locally convex sets, as long as the integrals with respect to the measures* $\widetilde{\mathcal{M}}_j(M, \cdot)$ *are finite.*

We defer the examples of the applications of the results of this last section to Chapter 15, where you finally will have a chance to see why *we* care about these power series expansions.

Part III

The Geometry of Random Fields

You could not possibly have gotten this far without having read the preface, so you already know that you have finally reached the main part of this book. It is here that the important results lie and it is here that, after over 250 pages of preparation, there will also be new results.

With the general theory of Gaussian processes behind us in Part I, and the geometry of Part II now established, we return to the stochastic setting.

There are three main (classes of) results in this part. The first is an explicit formula for the expected Euler characteristic of the excursion sets of smooth Gaussian random fields. In the same way that we divided the treatment of the geometry into two parts, Chapter 11 will cover the theory of random fields defined over simple Euclidean domains and Chapter 12 will cover fields defined over Whitney stratified manifolds. Unlike the case in Chapter 6, however, even if you are primarily interested in the manifold scenario you will need to read the Euclidean case first, since some of the manifold computations will be lifted from this case via atlas-based arguments.

As an aside, in the final section (Section 12.6) of Chapter 12 we shall return to a purely deterministic setting and use our Gaussian field results to provide a probabilistic proof of the classical Chern–Gauss–Bonnet theorem of differential geometry using nothing[22] but Gaussian processes. This really has nothing to do with anything else in the book, but we like it too much not to include it.

In Chapter 13 we shall see how to lift the results about the mean Euler characteristics of excursion sets to results about mean Lipschitz–Killing curvatures. The argument will rely on a novel extension of the classical Crofton formulas about averaged cross-sections of Euclidean sets to a scenario in which the cross-sections are replaced by intersections of the set with certain random manifolds, and the averaging is against a Gaussian measure. This new "Gaussian Crofton" formula is completely new and would seem to be of significant independent interest.

The second main result appears in Chapter 14, where we shall finally prove the result promised long ago, that not only is the difference

$$\left| \mathbb{P}\left\{ \sup_{t \in M} f(t) \geq u \right\} - \mathbb{E}\{\varphi(A_u(f, M))\} \right|$$

extremely small for large u, but it can even be bounded in a rigorous fashion. Not only will this justify our claims that the mean Euler characteristic of excursion sets yields an excellent approximation to excursion probabilities, but, en passant, we shall also show that the volume-of-tubes approximation of Chapter 10 can be made rigorous as well.

In the closing Chapter 15 we finally leave the Gaussian scenario and develop the third main result, an explicit formula for both the expected Euler characteristics

[22] Of course, this cannot really be true. Establishing the Chern–Gauss–Bonnet theorem without any recourse to algebraic topology would have been a mathematical coup that might even have made probability theory a respectable topic within pure mathematics. What will be hidden in the small print is that everything relies on the Morse theory of Section 9.3, and this, in turn, uses algebraic geometry. However, our approach will save myopic probabilists from having to read the small print.

and Lipschitz–Killing curvatures of excursion sets for a wide range of non-Gaussian random fields, working in the general environment of differential, rather than integral, geometry. The proof and approach of this chapter are quite different from those of Chapters 11–13, and much of it is entirely new.

In fact, the results of Chapter 15 will actually incorporate most of the results of Chapters 11–13, using only a small part of what is developed there. Consequently, you could actually skip these chapters and go straight to the punch line in Chapter 15, although you will, occasionally, have to backtrack to pick up some material from the missed chapters. In doing so, however, you will miss a lot of interesting and useful additional material, and so we advise against this.

11

Random Fields on Euclidean Spaces

The main result of this chapter is Theorem 11.7.2 and its corollaries, which give explicit formulas for the mean Euler characteristic of the excursion sets of smooth Gaussian fields over rectangles in $\mathbb{R}^N$.

The chapter develops in a number of distinct stages. Initially, we shall develop rather general results that give integral formulas for the expected number of points at which a vector-valued random field takes specific values. These are Theorem 11.2.1 and its corollaries. Aside from their basic importance in the current setting, these results will also form the basis of corresponding results for the manifold setting of Chapter 12. In view of the results of Chapter 6, which relate the global topology of excursion sets of a function to its local behavior, it should be clear what this has to do with Euler characteristic computations. However, the results are important beyond this setting and indeed beyond the setting of this book, so we shall develop them slowly and carefully.

Before we tackle all this, however, we shall take a moment to look at the classic Rice formula, which is the simplest special case of our main result and the proof of which incorporates many of the components of the general scenario.

11.1 Rice's Formula

Let f be a real-valued stochastic process on the interval $[0, T]$, and $u \in \mathbb{R}$. Consider the number of *upcrossings* by f of the level u in $[0, T]$, denoted by

$$N_u^+(0, T) \stackrel{\Delta}{=} \#\{t \in [0, T] : f(t) = u, \ f'(t) > 0\}. \tag{11.1.1}$$

In this section, we are going to assume that $N_u^+(0, T)$ is well defined[1] and finite, without worrying about conditions that ensure this. These will appear in the following sections, in horrid detail.

[1] Upcrossings are, of course, an example of a *point process*, with which we imagine most of you will be familiar. For those who are not, a point process defined over a measurable space $(\mathcal{X}, \mathcal{F})$ is a collection of nonnegative, integer-valued random variables $N(B)$, $B \in \mathcal{F}$. The $x \in \mathcal{X}$ for which $N(\{x\}) \geq 1$ are called the *atoms*, or *points* of N, and if $N(\{x\}) \in \{0, 1\}$

Indeed, we shall commit even more serious crimes, including exchanging delicate orders of integration, sometimes involving the Dirac delta function. All will be justified later.

As a first step, we would like to compute $\mathbb{E}\{N_u^+(0,T)\}$. To this end, let δ_x be the Dirac delta function at x, "defined" by the fact that, for any reasonable test function g,

$$\int_{\mathbb{R}} \delta_x(y) g(y)\, dy = g(x).$$

Suppose that the upcrossing points of f, i.e., those $t \in [0, T]$ at which $f(t) = u$ and $f'(t) > 0$, are isolated, so that each one can be covered by a small interval I in which there are no other upcrossings and throughout which $f' > 0$.

Then, treating δ as if it were a smooth function, a simple change of variable argument gives that

$$1 = \int_{\mathbb{R}} \delta_u(y)\, dy = \int_I \delta_u(f(t)) \cdot f'(t)\, dt.$$

Of course, this needs justification, but that is what Theorem 11.2.1 does. Concatenating all such intervals I, and noting that there is no contribution to the following integral from outside of them, we obtain

$$N_u^+(0,T) = \int_0^T \delta_u(f(t)) \cdot \mathbb{1}_{[0,\infty]}(f'(t)) \cdot f'(t)\, dt.$$

Now take expectations, freely exchanging orders of integration and assuming that the pairs of random variables $(f(t), f'(t))$ have joint probability densities p_t, to see that

$$\begin{aligned} \mathbb{E}\left\{N_u^+(0,T)\right\} &= \int_0^T dt \int_{-\infty}^{\infty} dx \int_0^{\infty} dy\, y \delta_u(x) p_t(x,y) \\ &= \int_0^T \int_0^{\infty} y p_t(u,y)\, dy\, dt. \end{aligned} \tag{11.1.2}$$

This is what could be called Rice's formula in its most basic form, and it holds for all processes on $\mathbb{R}$ for which the various operations above are justifiable. Note, however, that it requires no specific distributional assumptions on the process f.

However, it turns out to be remarkably difficult to compute the integral in (11.1.2) unless f is either Gaussian or a function of Gaussian processes. The case that is central to the remainder of this book is that in which f is indeed Gaussian, and has zero mean and constant variance, which for notational convenience we assume is 1. (Otherwise, the variance can be absorbed into u in all the following formulas.) In that case we know from Section 5.6 that $f(t)$ and $f'(t)$ are independent for each t, and so, denoting the variance of $f'(t)$ by λ_t, it immediately follows that in this case,[2]

---

for all x then the point process is called *simple*. All the point processes that will appear in this book will be simple. For further information on point processes, see, for example, [90].

[2] To see an expression in the fully nonstationary case, look at [42, formula (13.2.1)]. Note, however, that this expression is not of closed form, but remains as a rather complicated integral over $[0, T]$.

$$\begin{aligned}\mathbb{E}\{N_u^+(0,T)\} &= \frac{e^{-u^2/2}}{2\pi}\int_0^T\int_0^\infty y\frac{e^{-y^2/2\lambda_t}}{\sqrt{\lambda_t}}\,dy\,dt \qquad (11.1.3)\\ &= \frac{e^{-u^2/2}}{2\pi}\int_0^T \lambda_t^{1/2}\,dt.\end{aligned}$$

If f is actually stationary, so that $\lambda_t \equiv \lambda_2$, where λ_2 is the second spectral moment (cf. (5.5.2)), then (11.1.3) simplifies even further to

$$\mathbb{E}\{N_u^+(0,T)\} = \frac{\lambda_2^{1/2}T}{2\pi}e^{-u^2/2}, \qquad (11.1.4)$$

which is the classic Rice formula.[3]

To see how the Rice formula is related to excursion sets A_u and their Euler characteristics $\varphi(A_u)$, note the trivial fact that

$$\varphi(A_u(f,[0,T])) \equiv \mathbb{1}_{f_0\geq u} + N_u^+(0,T), \qquad (11.1.5)$$

so that

$$\begin{aligned}\mathbb{E}\{\varphi(A_u(f,[0,T]))\} &= \mathbb{P}\{f_0 \geq u\} + \mathbb{E}\{N_u^+(0,T)\} \qquad (11.1.6)\\ &= \Psi(u) + \frac{T\lambda_2^{1/2}}{2\pi}e^{-u^2/2},\end{aligned}$$

the second line holding only in the stationary case.

The role of the second spectral moment λ_2 here is quite easy to understand. It follows from the spectral theory of Section 5.4 that increasing λ_2 increases the "high-frequency components" of f. In other words, the sample paths become locally much rougher. For example, Figure 11.1.1 shows realizations of the very simple Gaussian process with discrete spectrum

$$\frac{a}{\sqrt{2}}\delta_{\pm\sqrt{200}}(\lambda) + \frac{\sqrt{1-a^2}}{\sqrt{2}}\delta_{\pm\sqrt{1,000}}(\lambda), \qquad (11.1.7)$$

for $a = 0$ and $a = \frac{5}{13}$, which gives second spectral moments of $\lambda_2 = 1{,}000$ and $\lambda_2 = 200$, respectively.

From the figure it should be reasonably clear what this means in sample path terms: higher second spectral moments correspond to an increased number of level crossings generated by local fluctuations. Similar phenomena occur also in higher dimensions, as we shall see soon.

[3] Rice's formula is really due to both Rice [131] and Kac [89] in the early 1940s, albeit in slightly different settings. Over the years it underwent significant improvement, being proven under weaker and weaker conditions. In its final form, the only requirement on f (in the current stationary, zero-mean Gaussian scenario) is that its sample paths be, almost surely, absolutely continuous with respect to Lebesgue measure. This is far less than we shall demand in the rigorous proof to follow, since it does not even require a continuous sample path derivative, let alone a continuous second derivative. Furthermore, (11.1.4) turns out to hold whether λ_2 is finite, whereas we have implicitly assumed $\lambda_2 < \infty$. For details see, for example, [97].

Fig. 11.1.1. Realizations of Gaussian sample paths with the spectrum (11.1.7) and second spectral moments $\lambda_2 = 200$ (left) and $\lambda_2 = 1{,}000$ (right).

Most of the current chapter concentrates on extending (11.1.6) from processes on $\mathbb{R}$ to random fields on $\mathbb{R}^N$, still in the stationary case. Along the way, we shall also handle all the technical issues that we ignored above.

The nonstationary case in higher dimensions is qualitatively different[4] from that in one dimension and requires the setting of Riemannian manifolds. We shall turn to this only in Chapter 12, in particular when we look at the nonstationary examples of Section 12.5.

11.2 An Expectation Metatheorem

We start with a metatheorem about the expected number of points at which a vector-valued random field takes values in some set. For the moment, we gain nothing by assuming that our fields are Gaussian and so do not do so. Here is the setting:

For some $N, K \geq 1$ let $f = (f^1, \dots, f^N)$ and $g = (g^1, \dots, g^K)$, respectively, be $\mathbb{R}^N$- and $\mathbb{R}^K$-valued N-parameter random fields. We need two sets, $T \subset \mathbb{R}^N$ and $B \subset \mathbb{R}^K$. As usual, T is a compact parameter set, but now we add the assumption that its boundary ∂T has finite $\mathcal{H}_{N-1}$-measure (cf. footnote 23 in Chapter 7). As for B, we shall assume that it is open and that its boundary $\partial B = \bar{B} \setminus B$ has Hausdorff dimension $K - 1$.

As usual, ∇f denotes the gradient of f. Since f takes values in $\mathbb{R}^N$, this is now an $N \times N$ matrix of first-order partial derivatives of f; i.e.,

$$(\nabla f)(t) \equiv \nabla f(t) \equiv (f^i_j(t))_{i,j=1,\dots,N} \equiv \left(\frac{\partial f^i(t)}{\partial t_j} \right)_{i,j=1,\dots,N}.$$

All the derivatives here are assumed to exist in an almost sure sense.[5]

[4] One might guess that in moving to constant variance, but otherwise nonstationary, random fields on $\mathbb{R}^N$ all that might happen would be that the λ_t of (11.1.3) might be replaced with some other local derivative, or perhaps determinant of derivatives, of the covariance function at the point (t, t). This, however, is not the case, and the actual situation, as we shall soon see, is far more complicated.

[5] This is probably too strong an assumption, since, at least in one dimension, Theorem 11.2.1 is known to hold under the assumption that f is absolutely continuous, and so has only a weak-sense derivative (cf. [109]). However, since we shall need a continuous sample path derivative later for other things, we assume it now.

Theorem 11.2.1. *Let f, g, T and B be as above. Assume that the following conditions are satisfied for some $u \in \mathbb{R}^N$:*

(a) *All components of f, ∇f, and g are a.s. continuous and have finite variances (over T).*
(b) *For all $t \in T$, the marginal densities $p_t(x)$ of $f(t)$ (implicitly assumed to exist) are continuous at $x = u$.*
(c) *The conditional densities*[6] *$p_t(x|\nabla f(t), g(t))$ of $f(t)$ given $g(t)$ and $\nabla f(t)$ (implicitly assumed to exist) are bounded above and continuous at $x = u$, uniformly in $t \in T$.*
(d) *The conditional densities $p_t(z|f(t) = x)$ of* $\det \nabla f(t)$ *given $f(t) = x$, are continuous for z and x in neighborhoods of* 0 *and u, respectively, uniformly in $t \in T$.*
(e) *The conditional densities $p_t(z|f(t) = x)$ of $g(t)$ given $f(t) = x$, are continuous for all z and for x in a neighborhood u, uniformly in $t \in T$.*
(f) *The following moment condition holds:*

$$\sup_{t\in T}\ \max_{1\le i,j\le N}\ \mathbb{E}\left\{\left|f_j^i(t)\right|^N\right\} < \infty. \tag{11.2.1}$$

(g) *The moduli of continuity with respect to the usual Euclidean norm (cf.* (1.3.6)) *of each of the components of f, ∇f, and g satisfy*

$$\mathbb{P}\{\omega(\eta) > \varepsilon\} = o\left(\eta^N\right) \quad \textit{as } \eta \downarrow 0 \tag{11.2.2}$$

for any $\varepsilon > 0$.

Then, if

$$N_u \equiv N_u(T) \equiv N_u(f, g : T, B)$$

denotes the number of points in T for which

$$f(t) = u \in \mathbb{R}^N \quad \textit{and} \quad g(t) \in B \subset \mathbb{R}^K,$$

and $p_t(x, \nabla y, v)$ denotes the joint density of $(f_t, \nabla f_t, g_t)$, we have, with $D = N(N+1)/2 + K$,

$$\mathbb{E}\{N_u\} = \int_T \int_{\mathbb{R}^D} |\det \nabla y|\, \mathbb{1}_B(v) p_t(u, \nabla y, v) d(\nabla y)\, dv\, dt. \tag{11.2.3}$$

It is sometimes more convenient to write this as

$$\mathbb{E}\{N_u\} = \int_T \mathbb{E}\{|\det \nabla f(t)|\, \mathbb{1}_B(g(t))| f(t) = u\} p_t(u)\, dt, \tag{11.2.4}$$

where the p_t here is the density of $f(t)$.

[6] Standard notation would demand that we replace the matrix ∇y by the $N(N+1)/2$-dimensional vector vech(∇y) both here and, even more so, in (11.2.3) following, where the differential $d(\nabla y)$, which we shall generally write in even greater shorthand as $d\nabla y$, is somewhat incongruous. Nevertheless, on the basis that it is clear what we mean, and that it is useful to be able to easily distinguish between reals, vectors, and matrices, we shall always work with this slightly unconventionial notation.

While conditions (a)–(g) arise naturally in the proof of Theorem 11.2.1, they all but disappear in one of the cases of central interest to us, when the random fields f and g are Gaussian. In these cases all the marginal and conditional densities appearing in the conditions of Theorem 11.2.1 are also Gaussian, and so their boundedness and continuity are immediate, as long as all the associated covariance matrices are nondegenerate, which is what we need to assume in this case. This will also imply that all variances are finite and that the moment condition (11.2.1) holds. Thus the only remaining conditions are the a.s. continuity of f, ∇f, and g, and condition (11.2.2) on the moduli of continuity.

Note first, without reference to normality, that if ∇f is continuous, then so must[7] be f. Thus we have only the continuity of ∇f and g to worry about. However, we spent a lot of time in Chapter 1 finding conditions that will guarantee this. For example, we can apply Theorem 1.4.1.

Write $C^i_f = C^i_f(s,t)$ for the covariance function of f^i, so that $C^i_{f_j} = \partial^2 C^i_f/\partial s_j \partial t_j$ is the covariance function of $f^i_j = \partial f^i/\partial t_j$. Similarly, let C^i_g denote the covariance function of g^i. Then, by (1.4.4), ∇f and g will be a.s. continuous if

$$\begin{aligned} \max_{i,j} |C^i_{f_j}(t,t) + C^i_{f_j}(s,s) - 2C^i_{f_j}(s,t)| &\le K |\ln |t-s||^{-(1+\alpha)}, \\ \max_{i} |C^i_{g}(t,t) + C^i_{g}(s,s) - 2C^i_{g}(s,t)| &\le K |\ln |t-s||^{-(1+\alpha)}, \end{aligned} \tag{11.2.5}$$

for some finite $K > 0$, some $\alpha > 0$, and all $|t-s|$ small enough.

All that now remains to check is condition (11.2.2) on the moduli of continuity. Here the Borell–TIS inequality—Theorem 2.1.1—comes into play. Write h for any of the components of ∇f or g, and H for the random field on $T \times T$ defined by $H(s,t) = h(t) - h(s)$. Then, writing

$$\omega(\eta) = \sup_{s,t:|t-s|\le\eta} |H(s,t)|,$$

we need to show, under (11.2.5), that for all $\varepsilon > 0$,

$$\mathbb{P}\{\omega(\eta) > \varepsilon\} \le o(\eta^N). \tag{11.2.6}$$

The Borell–TIS inequality gives us that

$$\mathbb{P}\{\omega(\eta) > \varepsilon\} \le 2\exp\{-(\varepsilon - \mathbb{E}\{\omega(\eta)\})^2/2\sigma^2_\eta\}, \tag{11.2.7}$$

where $\sigma^2_\eta = \sup_{s,t:|t-s|\le\eta} \mathbb{E}\{(H(s,t))^2\}$. But (11.2.5) immediately implies that $\sigma^2_\eta \le K|\ln\eta|^{-(1+\alpha)}$, while together with Theorem 1.4.1 it implies a bound of similar order for $\mathbb{E}\{\omega(\eta)\}$. Substituting this into (11.2.7) gives an upper bound of the form $C_\varepsilon \eta^{|\ln\eta|^\alpha}$, and so (11.2.6) holds with room to spare, in that it holds for any N and not just $N = \dim(T)$.

Putting all of the above together, we have that Theorem 11.2.1 takes the following much more user friendly form in the Gaussian case.

[7] The derivative can hardly exist, let alone be continuous, if f is not continuous! In fact, this condition was vacuous all along and was included only for "symmetry" considerations.

Corollary 11.2.2. *Let f and g be centered Gaussian fields over a T that satisfies the conditions of Theorem* 11.2.1. *If for each $t \in T$, the joint distributions of $(f(t), \nabla f(t), g(t))$ are nondegenerate, and if* (11.2.5) *holds, then so do* (11.2.3) *and* (11.2.4).

We now turn to the proof of Theorem 11.2.1. We shall prove it in a number of stages, firstly by setting up a result that rewrites the random variable N_u in an integral form, more conducive to computing expectations. Indeed, we could virtually end the proof here if we were prepared to work at a level of rigor that would allow us to treat the Dirac delta function with gay abandon and exchange delicate orders of integration without justification. Since it is informative to see how such an argument works, we shall give it, before turning to a fully rigorous proof.

The rigorous proof comes in a number of steps. In the first, we use the integral representation to derive an upper bound to $\mathbb{E}\{N_u\}$, which actually gives the correct result. (The upper bound actually involves little more than replacing the "gay abandon" mentioned above with Fatou's lemma.) The second step involves showing that this upper bound is also a lower bound. The argument here is far more delicate, and involves locally linear approximations of the fields f and g. Both of these steps will involve adding conditions to the already long list of Theorem 11.2.1, and so the third and final step involves showing that these additional conditions can be lifted under the conditions of the theorem.

To state the first result, let $\delta_\varepsilon : \mathbb{R}^N \to \mathbb{R}$ be constant on the N-ball $B(\varepsilon) = \{t \in \mathbb{R}^N : |t| < \varepsilon\}$, zero elsewhere, and normalized so that

$$\int_{B(\varepsilon)} \delta_\varepsilon(t)\, dt = 1. \tag{11.2.8}$$

Theorem 11.2.3. *Let $f : \mathbb{R}^N \to \mathbb{R}^N$, $g : \mathbb{R}^N \to \mathbb{R}^K$ be deterministic, $T \subset \mathbb{R}^N$ closed, and $B \subset \mathbb{R}^K$ open. Suppose, furthermore, that the following conditions are all satisfied for $u \in \mathbb{R}^N$:*

(a) *The components of f, g, and ∇f are all continuous.*
(b) *There are no points $t \in T$ satisfying both $f(t) = u$ and either $g(t) \in \partial B$ or $\det \nabla f(t) = 0$.*
(c) *There are no points $t \in \partial T$ satisfying $f(t) = u$.*

Then

$$N_u(f, g; T, B) = \lim_{\varepsilon \to 0} \int_T \delta_\varepsilon(f(t) - u) \mathbb{1}_B(g(t)) |\det \nabla f(t)|\, dt. \tag{11.2.9}$$

Proof. To save on notation, and without any loss of generality, we take $u = 0$. Consider those $t \in T$ for which $f(t) = 0$. We first note that from (b) there is only a finite number of such points,[8] and none of them lie in ∂T. Consequently, each one can be surrounded by an open ball, of radius η, say, in such a way that the balls

[8] It is a standard result that if there are no points $t \in T$ satisfying both $f(t) = u$ and $\det \nabla f(t) = 0$, then there is only a finite number of points $t \in T$ satisfying $f(t) = u$.

neither overlap nor intersect ∂T. Furthermore, because of (b), we can ensure that η is small enough so that within each ball, $g(t)$ always lies in either B or the interior of its complement, but never both.

Let $\sigma(\varepsilon)$ be the ball $|f| < \varepsilon$ in the image space of f. From what we have just established, we claim that we can now choose ε small enough for the inverse image of $\sigma(\varepsilon)$ in T to be contained within the union of the η spheres. (In fact, if this were not so, we could choose a sequence of points t_n in T not belonging to any η sphere, and a sequence ε_n tending to zero such that $f(t_n)$ would belong to $\sigma(\varepsilon_n)$ for each n. Since T is compact, the sequence t_n would have a limit point t^* in T, for which we would have $f(t^*) = 0$. Since $t^* \notin \partial T$ by (c), we must have $t^* \in T$. Thus t^* is contained in the inverse image of $\sigma(\varepsilon)$ for any ε, as must be infinitely many of the t_n. This contradiction establishes our claim.)

Furthermore, by (b) and the inverse mapping theorem (cf. footnote 6 of Chapter 6) we can choose ε, η so small that for each η sphere in T, $\sigma(\varepsilon)$ is contained in the f image of the η sphere, so that the restriction of f to such a sphere will be one-to-one. Since the Jacobian of the mapping of each η sphere by f is $|\det \nabla f(t)|$, it follows that we can choose ε small enough so that

$$N_0 = \int_T \delta_\varepsilon(f(t)) \mathbb{1}_B(g(t)) |\det \nabla f(t)| \, dt.$$

This follows since each η sphere in T over which $g(t)$ is in B will contribute exactly one unit to the integral, while all points outside the η spheres will not be mapped onto $\sigma(\varepsilon)$. Since the left-hand side of this expression is independent of ε, we can take the limit as $\varepsilon \to 0$ to obtain (11.2.9) and thus the theorem. □

Theorem 11.2.3 does not tell us anything about expectations. Ideally, it would be nice simply to take expectations on both sides of (11.2.9) and then, hopefully, find an easy way to evaluate the right-hand side of the resulting equation. While this requires justification and further assumptions, let us nevertheless proceed in this fashion, just to see what happens. We then have

$$\begin{aligned}
\mathbb{E}\{N_0\} &= \lim_{\varepsilon \to 0} \mathbb{E} \int_T \delta_\varepsilon(f(t)) \mathbb{1}_B(g(t)) |\det \nabla f(t)| \, dt \\
&= \int_T \int_{\mathbb{R}^{N(N+1)/2}} \int_{\mathbb{R}^K} \mathbb{1}_B(v) |\det \nabla y| \\
&\quad \times \left\{ \lim_{\varepsilon \to 0} \int_{\mathbb{R}^N} \delta_\varepsilon(x) p_t(x, \nabla y, v) \, dx \right\} d\nabla y \, dv \, dt,
\end{aligned}$$

To see this, suppose that t is such that $f(t) = u$. (If there are no such t, there is nothing more to prove.) The fact that $f \in C^1$ implies that there is a neighborhood of t in which f is locally linear: that is, it can be approximated by its tangent plane. Furthermore, since $\det \nabla f(t) \neq 0$, not all the partial derivatives can be zero, and so the tangent plane cannot be at a constant "height." Consequently, throughout this neighborhood there is no other point at which $f = u$. Since T is compact, there can therefore be no more than a finite number of t satisfying $f(t) = u$.

where the p_t are the obvious densities. Taking the limit in the innermost integral yields

$$\begin{aligned}\mathbb{E}\{N_0\} &= \int_T \int \int \mathbb{1}_B(v) |\det \nabla y| p_t(0, \nabla y, v)\, d\nabla y\, dv\, dt \\ &= \int_T \mathbb{E}\{|\det \nabla f(t)| \mathbb{1}_B(g(t)) | f(t) = 0\} p_t(0)\, dt.\end{aligned} \tag{11.2.10}$$

Of course, interchanging the order of integration and the limiting procedure requires justification. Nevertheless, at this point we can state the following tongue-in-cheek "corollary" to Theorem 11.2.3.

Corollary 11.2.4. *If the conditions of Theorem* 11.2.3 *hold almost surely, as well as "adequate" regularity conditions, then*

$$\begin{aligned}E\{N_u\} &= \int_T \int_{\mathbb{R}^{N(N+1)/2}} \int_{\mathbb{R}^K} |\det \nabla x| \mathbb{1}_B(v) p_t(u, \nabla x, v)\, d\nabla x\, dv\, dt \\ &= \int_T \mathbb{E}\{|\det \nabla f(t)| \mathbb{1}_B(g(t)) | f(t) = u\} p_t(u)\, dt.\end{aligned} \tag{11.2.11}$$

As will be seen below, "adequate" regularity conditions generally require no more than that the density p_t above be well behaved (i.e., continuous, bounded) and that enough continuous derivatives of f and g exist, with enough finite moments.

At this point we suggest that if you are not the kind of person who cares too much about rigor then you move directly to Section 11.6, where we begin the preparations for using the above results. Indeed, there is probably much to be said in favor of even the mathematically inclined doing the same on the first reading.

We now turn to the rigorous upper bound, and start with a useful lemma. It is easily proven via induction started from Hölder's inequality.

Lemma 11.2.5. *Let $X_1, \dots, X_n$ be any real-valued random variables. Then*

$$\mathbb{E}\{|X_1 \cdots X_n|\} \le \prod_{i=1}^n [\mathbb{E}\{|X_i|^n\}]^{1/n}. \tag{11.2.12}$$

Theorem 11.2.6. *Let f, g, B, and T be as in Theorem* 11.2.3, *but with f and g random and conditions* (a)–(d) *there holding in an almost sure sense. Assume, furthermore, that conditions* (b)–(g) *of Theorem* 11.2.1 *hold, with the notation adopted there. Then*

$$\mathbb{E}\{N_u(f, g; T, B)\} \le \int_T dt \int_B dv \int_{\mathbb{R}^{N(N+1)/2}} |\det \nabla y| p_t(u, \nabla y, v)\, d\nabla y. \tag{11.2.13}$$

Proof. Again assume that $u = 0$. Start by setting, for $\varepsilon > 0$,

$$N^\varepsilon = \int_T \delta_\varepsilon(f(t)) \mathbb{1}_B(g(t)) |\det \nabla f(t)|\, dt,$$

where δ_ε is as in (11.2.8). By expanding the determinant, applying (11.2.12), and recalling the moment assumption (11.2.1), we have that everything is nicely finite, and so Fubini's theorem gives us that

$$\begin{aligned}\mathbb{E}\{N^\varepsilon(T)\} &= \int_T dt \int_{\mathbb{R}^N \times \mathbb{R}^{N(N+1)/2} \times B} \delta_\varepsilon(x) |\det \nabla y| p_t(x, \nabla y, v)\, dx\, d\nabla y\, dv \\ &= \int_T dt \int_{\mathbb{R}^{N(N+1)/2} \times B} |\det \nabla y| p_t(\nabla y, v)\, d\nabla y\, dv \\ &\quad \times \int_{\mathbb{R}^N} \delta_\varepsilon(x) p_t(x \big| \nabla y, v)\, dx.\end{aligned}$$

Since all densities are assumed continuous and bounded, the innermost integral clearly converges to

$$p_t(0 \big| \nabla y, v)$$

as $\varepsilon \to 0$. Furthermore,

$$\begin{aligned}\int_{\mathbb{R}^N} \delta_\varepsilon(x) p_t(x \big| \nabla y, v)\, dx &\le \sup_x p_t(x \big| \nabla y, v) \int_{\mathbb{R}^N} \delta_\varepsilon(x)\, dx \\ &= \sup_x p_t(x | \nabla y, v),\end{aligned}$$

which, by assumption, is bounded.

Again noting the boundedness of $\mathbb{E}\{|\det \nabla f_t|\}$, it follows from (11.2.9), dominated convergence, and Fatou's lemma that

$$\begin{aligned}E\{N_0\} &\le \lim_{\varepsilon \to 0} E\{N^\varepsilon\} \\ &= \int_T dt \int_B dv \int_{\mathbb{R}^{N(N+1)/2}} |\det \nabla y| p_t(0, \nabla y, v)\, d\nabla y.\end{aligned}$$

This, of course, proves the theorem. □

We can now turn to the more difficult part of our problem: showing that the upper bound for $\mathbb{E}\{N(T)\}$ obtained in the preceding theorem also serves as a lower bound under reasonable conditions. We shall derive the following result.

Theorem 11.2.7. *Assume the setup and assumptions of Theorem* 11.2.6, *along with* (11.2.2) *of Theorem* 11.2.1, *that is, that the moduli of continuity of each of the components of f, ∇f, and g satisfy*

$$\mathbb{P}\{\omega(\eta) > \varepsilon\} = o(\eta^N) \quad \text{as } \eta \downarrow 0, \tag{11.2.14}$$

for all $\varepsilon > 0$. Then (11.2.13) *holds with the inequality sign reversed, and so is an equality.*

Since the proof of this theorem is rather involved, we shall start by first describing the principles underlying it. Essentially, the proof is based on constructing a pathwise approximation to the vector-valued process f and then studying the zeros of the approximating process. The approximation is based on partitioning T and replacing f within each cell of the partition by a hyperplane tangential to f at the cell's midpoint. We then argue that if the approximating process has a zero within a certain subset of a given cell, then f has a zero somewhere in the full cell. Thus the number of zeros of the approximating process will give a lower bound to the number of zeros of f.

In one dimension, for example, we replace the real-valued function f on $T = [0, 1]$ by a series of approximations $f^{(n)}$ given by

$$f^{(n)}(t) = f\left(\left(j + \frac{1}{2}\right)2^{-n}\right) + \left[t - \left(j + \frac{1}{2}\right)2^{-n}\right] f'\left(\left(j + \frac{1}{2}\right)2^{-n}\right),$$

for $j2^{-n} \le t < (j+1)2^{-n}$, and study the zeros of $f^{(n)}$ as $n \to \infty$. Although this is perhaps not the most natural approximation to use in one dimension, it generalizes easily to higher dimensions.

Proof of Theorem 11.2.7. As usual, we take the level $u = 0$, and start with some notation. For each $n \ge 1$ let $\mathbb{Z}_n$ denote the lattice of points in $\mathbb{R}^N$ whose components are integer multiples of 2^{-n}, i.e.,

$$\mathbb{Z}_n = \{t \in \mathbb{R}^N : t_j = i2^{-n},\ j = 1, \dots, N,\ i \in \mathbb{Z}\}.$$

Now fix $\varepsilon > 0$, and for each $n \ge 1$ define two half-open hypercubes centered on an arbitrary point $t \in \mathbb{R}^N$ by

$$\begin{aligned}\Delta_n(t) &= \{s \in \mathbb{R}^N : -2^{-(n+1)} \le s_j - t_j < 2^{-(n+1)}\},\\ \Delta_n^{\varepsilon}(t) &= \{s \in \mathbb{R}^N : -(1-\varepsilon)2^{-(n+1)} \le s_j - t_j < (1-\varepsilon)2^{-(n+1)}\}.\end{aligned}$$

Set

$$I_{nt} = \begin{cases} 1 & \text{if } N_0(f, g; \Delta_n(t), B) \ge 1 \text{ and } \Delta_n(t) \subset T,\\ 0 & \text{otherwise,}\end{cases}$$

and define approximations N^n to $N_0(f, g; T, B)$ by

$$N^n = \sum_{t \in \mathbb{Z}_n} I_{nt}.$$

Note (since it will be important later) that only those $\Delta_n(t)$ that lie wholly within T contribute to the approximations. However, since the points being counted are, by assumption, almost surely isolated, and none lie on ∂T, it follows that

$$N^n \overset{\text{a.s.}}{\to} N_0(f, g; T, B) \quad \text{as } n \to \infty.$$

Since the sequence N^n is nondecreasing in n, monotone convergence yields

$$\mathbb{E}\{N_0\} = \lim_{n\to\infty} \mathbb{E}\{N^n\} \geq \lim_{n\to\infty} \sum \mathbb{P}\{N_0(\Delta_n(t)) > 0\}, \tag{11.2.15}$$

where the summation is over all $t \in \mathbb{Z}_n$ for which $\Delta_n(t) \subset T$.

The remainder of the proof involves determining the limiting value of $\mathbb{P}\{N_0(\Delta_n(t)) > 0\}$. Fix $\delta > 0$ small, and $K > 0$ large. For given realizations of f and g define the function $\omega_f^*(n)$ by

$$\omega_f^*(n) = \max\left[\max_j \omega_{f^j}(2^{-n}),\ \max_{ij} \omega_{(\nabla f)_{ij}}(2^{-n}),\ \max_j \omega_{g^j}(2^{-n})\right],$$

where the moduli of continuity are all taken over T. Furthermore, define M_f by

$$M_f = \max\left[\max_{1\le j\le N} \sup_{t\in T} |f^j(t)|,\ \max_{1\le i,j\le N} \sup_{t\in T} \left|\left(\partial f^i/\partial t_j\right)(t)\right|\right].$$

Finally, set

$$\eta = \frac{\delta^2 \varepsilon}{2N!(K+1)^{N-1}}.$$

Then the conditions of the theorem imply that as $n \to \infty$,

$$\mathbb{P}\{\omega^*(n) > \eta\} = o\left(2^{-Nn}\right). \tag{11.2.16}$$

Choose now a fixed n and $t \in \mathbb{Z}_n$ for which $\Delta_n(t) \subset T$. Assume that

$$\omega_f^*(n) < \eta \tag{11.2.17}$$

and that the event $G_{\delta K}$ occurs, where

$$G_{\delta K} = \left\{|\det \nabla f(t)| > \delta,\ M_f < K/N, g(t) \in \left\{x \in B : \inf_{y\in\partial B} \|x-y\|_1 > \delta\right\}\right\},$$

where $\|x\|_1 = \sum_i |x^i|$ is the usual ℓ^1 norm. Finally, define t^* to be the solution of the following equation, when a unique solution in fact exists:

$$f(t) = (t - t^*) \cdot \nabla f(t). \tag{11.2.18}$$

We claim that if both $\omega_f^*(n) < \eta$ and $G_{\delta K}$ occur, then, for n large enough, $t^* \in \Delta_n^\varepsilon(t)$ implies

$$N_0(\Delta_n(t)) > 0 \tag{11.2.19}$$

and

$$\|f(t)\|_1 \leq 2^{-n} K. \tag{11.2.20}$$

These two facts, which we shall establish in a moment, are enough to make the remainder of the proof quite simple. From (11.2.19), it follows that by choosing n large enough for (11.2.16) to be satisfied we have

$$\mathbb{P}\{N_0(\Delta_n(t)) > 0\} \geq \mathbb{P}\{G_{\delta K} \cap [t^* \in \Delta_n^\varepsilon(t)]\} + o(2^{-nN}).$$

Using this, making the transformation $t \to t^*$ given by (11.2.18), and using the notation of the theorem, we obtain

$$\mathbb{P}\{N_0(\Delta_n(t)) > 0\}$$
$$\geq \int_{G_{\delta K} \cap \{t^* \in \Delta_n^\varepsilon(t)\}} |\det \nabla y| p_t((t-t^*)\nabla y, \nabla y, v)\, dt^*\, d\nabla y dv + o(2^{-nN}).$$

Noting (11.2.20), the continuity and boundedness assumptions on p_t, and the boundedness assumptions on the moments of the f^i_j, it follows that, as $n \to \infty$, the last expression, summed as in (11.2.15), converges to

$$(1-\varepsilon)^N \int_T dt \int_{G_{\delta K}} |\det \nabla y| p_t(0, \nabla y, v)\, d\nabla y\, dv.$$

Letting $\varepsilon \to 0$, $\delta \to 0$, $K \to \infty$ and applying monotone convergence to the above expression, we obtain from (11.2.15) that

$$\mathbb{E}\{N_0(T)\} \geq \int_T dt \int_B dv \int_{\mathbb{R}^{N(N+1)/2}} |\det \nabla y| p_t(0, \nabla y, v)\, d\nabla y.$$

This, of course, completes the proof, barring the issue of establishing that (11.2.19) and (11.2.20) hold under the conditions preceding them.

Thus, assume that $\omega^*(n) < \eta$, $G_{\delta K}$ occur, and the t^* defined by (11.2.18) satisfies $t^* \in \Delta_n^\varepsilon(t)$. Then (11.2.20) is immediate. The hard part is to establish (11.2.19). To this end, note that (11.2.18) can be rewritten as

$$t - f(t)[\nabla f(t)]^{-1} \in \Delta_n^\varepsilon(t). \tag{11.2.21}$$

Let τ be any other point in $\Delta_n(t)$. It is easy to check that

$$|\det \nabla f(\tau) - \det \nabla f(t)| < \frac{\varepsilon \delta^2}{2}, \tag{11.2.22}$$

under the conditions we require. Thus, since $\det \nabla f(t) > \delta$, it follows that $\det \nabla f(\tau) \neq 0$ for any $\tau \in \Delta_n(t)$ and so the matrix $\nabla f(\tau)$ is invertible throughout $\Delta_n(t)$. Similarly, one can check that $g(\tau) \in B$ for all $\tau \in \Delta_n(t)$.

Consider first (11.2.19). We need to show that $t^* \in \Delta_n^\varepsilon(t)$ implies the existence of at least one $\tau \in \Delta_n(t)$ at which $f(\tau) = 0$.

The mean value theorem[9] allows us to write

[9] In the form we need it, here is the mean value theorem for Euclidean spaces. A proof can be found, for example, in [13].

Lemma 11.2.8 (mean value theorem for $\mathbb{R}^N$). *Let T be a bounded open set in $\mathbb{R}^N$ and let $f : T \to \mathbb{R}^N$ have first-order partial derivatives at each point in T. Let s and t be two points in T such that the line segment*

$$f(\tau) - f(t) = (\tau - t)\nabla f(t^1, \dots, t^N) \tag{11.2.24}$$

for some points $t^1(\tau), \dots, t^N(\tau)$ lying on the line segment $L(t, \tau)$, for any t and τ, where $\nabla f(t^1, \dots, t^N)$ is the matrix function ∇f with the elements in the kth column evaluated at the point t^k. Using similar arguments to those used to establish (11.2.22) and the invertibility of $\nabla f(\tau)$ throughout $\Delta_n(t)$, invertibility can be shown for $\nabla f(t^1, \dots, t^N)$ as well. Hence we can rewrite (11.2.24) as

$$f(\tau)[\nabla f(t^1, \dots, t^N)]^{-1} = f(t)[\nabla f(t^1, \dots, t^N)]^{-1} + (\tau - t). \tag{11.2.25}$$

Suppose we could show that $f(t)[\nabla f(t^1, \dots, t^N)]^{-1} \in \Delta_n(0)$ if $t^* \in \Delta_n^\varepsilon(t)$. Then by the Brouwer fixed point theorem[10] it would follow that the continuous mapping of $\Delta_n(t)$ into $\Delta_n(t)$ given by

$$\tau \to t - f(t)[\nabla f(t^1, \dots, t^N)]^{-1}, \quad \tau \in \Delta_n(t),$$

has at least one fixed point. Thus, by (11.2.25), there would be at least one $\tau \in \Delta_n(t)$ for which $f(\tau) = 0$. In other words,

$$\{\omega_f^*(n) < \eta,\ G_{\delta K},\ t^* \in \Delta_n^\varepsilon(t),\ n \text{ large}\} \Rightarrow N(T \cap \Delta_n(t)) > 0.$$

But $f(t)[\nabla f(t^1, \dots, t^N)]^{-1} \in \Delta_n(0)$ is easily seen to be a consequence of $t^* \in \Delta_n^\varepsilon(t)$ and $G_{\delta K}$ simply by writing

$$\begin{aligned}
&f(t)[\nabla f(t^1, \dots, t^N)]^{-1} \\
&\quad = f(t)[\nabla f(t)]^{-1}\nabla f(t)[\nabla f(t^1, \dots, t^N)]^{-1} \\
&\quad = f(t)[\nabla f(t)]^{-1}\left(I + \left(\nabla f(t) - \nabla f(t^1, \dots, t^N)\right)[\nabla f(t^1, \dots, t^N)]^{-1}\right),
\end{aligned}$$

noting (11.2.21) and bounding the rightmost expression using basically the same argument employed for (11.2.22). This completes the proof. □

$$L(s, t) = \{u : u = \theta s + (1 - \theta)t, 0 < \theta < 1\}$$

is wholly contained in T. Then there exist points $t^1, \dots, t^N$ on $L(s, t)$ such that

$$f(t) - f(s) = \nabla f(t^1, \dots, t^N) \cdot (t - s), \tag{11.2.23}$$

where by $\nabla f(t^1, \dots, t^N)$ we mean the matrix-valued function ∇f with the elements in the kth column evaluated at the point t^k.

[10] In the form we need it, the Brouwer fixed point theorem is as follows. Proofs of this result are easy to find (e.g., [146]).

Lemma 11.2.9. *Let T be a compact, convex subset of $\mathbb{R}^N$ and $f : T \to T$ a continuous mapping of T into itself. Then f has at least one fixed point; i.e., there exists at least one point $t \in T$ for which $f(t) = t$.*

We now complete the task of this section, i.e., the proof of Theorem 11.2.1, which gave the expression appearing in both Theorems 11.2.6 and 11.2.7 for $\mathbb{E}\{N_u\}$, but under seemingly weaker conditions than those we have assumed. What remains to show is that conditions (b)–(d) of Theorem 11.2.3 are satisfied under the conditions of Theorem 11.2.1. Condition (c) follows immediately from the following rather intuitive result, taking $h = f$, $n = N - 1$, and identifying the T of the lemma with ∂T in the theorems. The claims in (b) and (d) are covered in the three remaining lemmas of the section. Lemma 11.2.10 has roots going back to Bulinskaya [35].

Lemma 11.2.10. *Let T be a compact set of Hausdorff dimension n. Let $h : T \to \mathbb{R}^{n+1}$ be, with probability one, C^1 with bounded partial derivatives. Furthermore, assume that the univariate probability densities of h are bounded in a neighborhood of $u \in \mathbb{R}^{n+1}$, uniformly in t. Then, for such u, there are a.s. no $t \in T$ with $h(t) = u$. That is,*

$$\mathbb{P}\{h^{-1}(u) = \emptyset\} = 1. \tag{11.2.26}$$

Proof. We start with the observation that under the conditions of the lemma, for any $\varepsilon > 0$, there exists a finite C_ε such that

$$\mathbb{P}\left\{ \max_{ij} \sup_{T} \partial h^i(t)/\partial t_j < C_\varepsilon \right\} > 1 - \varepsilon.$$

Writing ω_h for the modulus of continuity of h, it follows from the mean value theorem that

$$\mathbb{P}\{E_\varepsilon\} > 1 - \varepsilon, \tag{11.2.27}$$

where

$$E_\varepsilon \triangleq \left\{ \omega_h(\eta) \le \sqrt{n} C_\varepsilon \eta \text{ for small enough } \eta \right\}.$$

Take now a sequence $\{\delta_m\}$ of positive numbers converging to 0 as $m \to \infty$. The fact that T has Hausdorff dimension n implies that for any any $\delta > 0$, and m large enough, there exists a collection of Euclidean balls $\{B_{mj}\}$ covering T such that

$$\sum_j \left(\operatorname{diam}(B_{mj})\right)^{n+\delta} < \delta_m. \tag{11.2.28}$$

Define the events

$$A = \{\exists t \in T : h(t) = u\}, \qquad A_{mj} = \left\{\exists t \in B_{mj} : h(t) = u\right\}.$$

Fix $\varepsilon > 0$ and note that

$$\mathbb{P}\{A\} \le \sum_j \mathbb{P}\left\{A_{mj} \cap E_\varepsilon\right\} + \mathbb{P}\left\{E_\varepsilon^c\right\}. \tag{11.2.29}$$

In view of (11.2.27) it suffices to show that the sum here can be made arbitrarily small.

To see this, let t_{mj} be the center point of B_{mj}. Then, if both A_{mj} and E_ε occur, it follows that for large enough m,

$$\left|h(t_{mj}) - u\right| \le \sqrt{n}C_\varepsilon \operatorname{diam}(B_{mj}).$$

Since h_t has a bounded density, it thus follows that

$$\mathbb{P}\{A_{mj} \cap E_\varepsilon\} \le M(\sqrt{n}C_\varepsilon \operatorname{diam}(B_{mj}))^{n+1},$$

where M is a bound on the densities. Substituting into (11.2.29) and noting (11.2.28) we are done. □

We now turn to the second part of condition (b) of Theorem 11.2.3, relating to the points $t \in T$ satisfying $f(t) - u = \det \nabla f(t) = 0$. Note firstly that this would follow easily from Lemma 11.2.10 if we were prepared to assume that $f \in C^3(T)$. This, however, is more than we are prepared to assume. That the conditions of Theorem 11.2.1 contain all we need is the content of the following lemma.

Lemma 11.2.11. *Let f and T be as in Theorem* 11.2.1, *with conditions* (a), (b), (d), *and* (g) *of that theorem in force. Then, with probability one, there are no points $t \in T$ satisfying $f(t) - u = \det \nabla f(t) = 0$.*

Proof. As for the proof of the previous lemma, we start with an observation. In particular, for any $\varepsilon > 0$, there exists a continuous function ω_ε for which $\omega_\varepsilon(\eta) \downarrow 0$ as $\eta \downarrow 0$, and a finite positive constant C_ε, such that $\mathbb{P}\{E_\varepsilon\} > 1 - \varepsilon$, where now

$$E_\varepsilon \triangleq \left\{ \max_{ij} \sup_{t\in T} |f^i_j(t)| < C_\varepsilon,\ \max_{ij}\ \omega_{f^i_j}(\eta) \le \omega_\varepsilon(\eta),\ \text{for } 0 < \eta \le 1 \right\}.$$

To see this, the following simple argument[11] suffices: For ease of notation, set

$$\omega^*(\eta) = \max_{ij} \omega_{f^i_j}(\eta).$$

It then follows from (11.2.2) that there are sequences $\{c_n\}$ and $\{\eta_n\}$, both decreasing to zero, such that

$$\mathbb{P}\left\{\omega^*(\eta_n) < c_n\right\} > 1 - 2^{-n}\varepsilon.$$

Defining $\omega_\varepsilon(\eta) = c_n$ for $\eta_{n+1} \le \eta < \eta_n$, Borel–Cantelli then gives, for some $\eta_1 > 0$, that

$$\mathbb{P}\left\{\omega^*(\eta) < \omega_\varepsilon(\eta), 0 < \eta \le \eta_1\right\} > 1 - \varepsilon/2.$$

If $\eta_1 < 1$, set $\omega_\varepsilon = c_0$ for $\eta_1 < \eta \le 1$, where c_0 is large enough so that

[11] Note that this event tells us less about the $\omega_{f^i_j}$ than in the corresponding event in the previous lemma, and so we shall have to work harder to show that $\mathbb{P}\{E_\varepsilon\} > 1 - \varepsilon$. The problem is that this time, we are not assuming that the f^i_j, unlike the h^i of the previous lemma, are C^1.

$$\mathbb{P}\left\{\omega^*(\eta) < c_0,\ \eta_1 < \eta \le 1\right\} > 1 - \varepsilon/2.$$

Now choose C_ε large enough so that

$$\mathbb{P}\left\{\sup_{t\in T} |f(t)| < C_\varepsilon\right\} > 1 - \varepsilon/2,$$

and we are done.

We now continue in a similar vein to that of the previous proof. Let the B_{mj} and t_{mj} be as defined there, although (11.2.28) is now replaced with

$$\sum_j \left(\text{diam } B_{mj}\right)^N \to c_N \lambda_N(T) < \infty \tag{11.2.30}$$

as $m \to \infty$ for some dimension-dependent constant c_N.

Define the events

$$A_{mj} = \left\{\exists t \in B_{mj} : f(t) - u = \det \nabla f(t) = 0\right\}.$$

Fixing $\varepsilon > 0$, as before, we now need only show that $\sum_j \mathbb{P}\left\{A_{mj} \cap E_\varepsilon\right\}$ can be made arbitrarily small.

Allowing C_N to be a dimension-dependent constant that may change from line to line, if A_{mj} and E_ε occur, then, by expanding the determinant, it is easy to see that

$$|\det \nabla f(t_{mk})| \ \le\ C_N \left(\max(1, C_\varepsilon)\right)^{N-1} \omega_\varepsilon(\sqrt{N}\,\text{diam}(B_{mj})),$$

and that, as before,

$$|f(t_{mj}) - u| \ \le\ C_N C_\varepsilon\, \text{diam}(B_{mj}).$$

It therefore follows that

$$\mathbb{P}\{A_{mj} \cap E_\varepsilon\} \le \int \mathbb{P}\Big\{|\det \nabla f(t_{mj})| \le (C_N \left(\max(1, C_\varepsilon)\right)^{N-1} \omega_\varepsilon(\sqrt{N}\,\text{diam}(B_{mj}))$$
$$\Big|\ f(t_{mj}) = x\Big\} p_{t_{mj}}(x)\,dx,$$

where p_t is the density of $f(t)$ and the integral is over a region in $\mathbb{R}^N$ of volume no greater than $C_N\,\text{diam}(B_{mj})^N$.

By assumption (g) of Theorem 11.2.1 the integrand here tends to zero as $m \to \infty$, and so can be made arbitrarily small. Furthermore, p_t is uniformly bounded. Putting all of this together with (11.2.30) proves the result. □

We now turn to the first part of condition (b) of Theorem 11.2.3, relating to the points $t \in T$ satisfying $f(t) = 0$ and $g(t) \in \partial B$. These are covered by the following lemma, whose proof is almost identical to that of Lemma 11.2.11 and so is left to you.

Lemma 11.2.12. *Let f and T be as in Theorem* 11.2.1, *with conditions* (a), (b), (e), *and* (g) *of that theorem in force. Then, with probability one, there are no points $t \in T$ satisfying $f(t) - u = 0$ and $g(t) \in \partial B$.*

11.3 Suitable Regularity and Morse Functions

Back in Chapter 6 we laid down a number of regularity conditions on deterministic functions f for the geometric analyses developed there to be valid. In the integral-geometric setting they were summarized under the "suitable regularity" of Definition 6.2.1. In the manifold setting, we moved to the "Morse functions" of Section 9.3.

We shall soon (in Section 11.7) need to know when a random f is, with probability one, either suitably regular or a Morse function over a finite rectangle $T \subset \mathbb{R}^N$ and later (in Section 12.4) when it is a Morse function over a piecewise smooth manifold.

Despite the fact that this is (logically) not the right place to be addressing these issues, we shall nevertheless do so, while the details of the previous section are still fresh in your mind. In particular, without much further effort, Lemmas 11.2.10 and 11.2.11 will give us what we need.

We consider Morse functions over rectangles first, since here the conditions are tidiest. We start with a reminder (and slight extension) of notation: As usual, $\partial_k T$ is the k-dimensional boundary, or skeleton, of T. Since T is a rectangle, each $\partial_k T$ is made up of $2^{N-k}\binom{N}{k}$ open rectangles, or "faces," of dimension k.

Morse functions can then be defined via the following three characteristics:

(i) f is C^2 on an open neighborhood of T.
(ii) The critical points of $f_{|\partial_k T}$ are nondegenerate for all $k = 0, \dots, N$.
(iii) $f_{|\partial_k T}$ has no critical points on $\bigcup_{j=0}^{k-1} \partial_j T$ for all $k = 1, \dots, N$.

Recall that in the current Euclidean setting, critical points are those t for which $\nabla f(t) = 0$, and "nondegeneracy" means that the Hessian $\nabla^2 f(t)$ has nonzero determinant. Note also that in both conditions (ii) and (iii) there is a strong dependence on dimension. In particular, in (ii) the requirement is that the $\mathbb{R}^{k+1}$-valued function $(\nabla(f_{|\partial_k T}), \det \nabla^2(f_{|\partial_k T}))$ not have zeros over a k-dimensional parameter set. Regarding (iii), the requirement is that the $\mathbb{R}^k$-valued function $\nabla(f_{|\partial_k T})$ defined on a k-dimensional set not have zeros on a subset of dimension $k-1$.

In this light, (ii) and (iii) are clearly related to Lemma 11.2.11, and we leave it to you to check the details that give us the following theorem.

Theorem 11.3.1. *Let f be a real-valued random field over a bounded rectangle T of $\mathbb{R}^N$ with first- and second-order partial derivatives f_i and f_{ij}. Then f is, with probability one, a Morse function over T if the following conditions hold for each face J of T:*

(a) *f is, with probability one, C^2 on an open neighborhood of T, and all second derivatives have finite variance.*
(b) *For all $t \in J$, the marginal densities $p_t(x)$ of $\nabla f_{|J}(t)$ are continuous at 0, uniformly in t.*
(c) *The conditional densities $p_t(z|x)$ of $\det \nabla^2 f_{|J}(t)$ given $\nabla f_{|J}(t) = x$ are continuous for (z, x) in a neighborhood of 0, uniformly in $t \in T$.*

(d) *On J, the moduli of continuity of f and its first- and second-order partial derivatives all satisfy, for any $\varepsilon > 0$,*

$$\mathbb{P}\{\omega(\eta) > \varepsilon\} = o\left(\eta^{\dim(J)}\right) \quad \text{as } \eta \downarrow 0.$$

As usual, we also have a Gaussian corollary (cf. Corollary 11.2.2). It is even simpler than usual, since Gaussianity allows us drop the references to the specific faces J that so cluttered the conditions of the theorem.

Corollary 11.3.2. *Let f be a centered Gaussian field over a finite rectangle T. If for each $t \in T$, the joint distributions of $(f_i(t),\ f_{ij}(t))_{i,j=1,\dots,N}$ are nondegenerate, and if for some finite K and all $s, t \in T$,*

$$\max_{i,j} \left| C_{f_{ij}}(t,t) + C_{f_{ij}}(s,s) - 2C_{f_{ij}}(s,t) \right| \le K \left| \ln |t-s| \right|^{-(1+\alpha)}, \tag{11.3.1}$$

then the sample functions of f are, with probability one, Morse functions over T.

The next issue is to determine when sample functions are, with probability one, suitably regular in the sense of Definition 6.2.1. This is somewhat less elegant because of the asymmetry in the conditions of suitable regularity and their dependence on a particular coordinate system. Nevertheless, the same arguments as above work here as well and it is easy (albeit a little time consuming) to see that the following suffices to do the job.

Theorem 11.3.3. *Under the conditions of Corollary* 11.3.2 *the sample functions of f are, with probability one, suitably regular over bounded rectangles.*

A little thought will show that the assumptions of this theorem would seem to give more than is required. Consider the case $N = 2$. In that case, condition (6.2.6) of suitable regularity requires that there be no $t \in T$ for which $f(t) - u = f_1(t) = f_{11}(t) = 0$. This is clearly implied by Theorem 11.3.3. However, the theorem also implies that $f(t) - u = f_2(t) = f_{22}(t) = 0$, which is not something that we require. Rather, it is a consequence of a desire to write the conditions of the theorem in a compact form.

In fact, Theorem 11.3.3 goes even further, in that it implies that for every *fixed* choice of coordinate system, the sample functions of f are, with probability one, suitably regular over bounded rectangles.[12]

We now turn to what is really the most important case, that of determining sufficient conditions for a random field to be, almost surely, a Morse function over a piecewise C^2 Riemannian manifold (M, g). Writing our manifold as usual as

$$M = \bigcup_{j=0}^{N} \partial_j M \tag{11.3.2}$$

[12] Recall that throughout our discussion of integral geometry there was a fixed coordinate system and that suitable regularity was defined relative to this system.

(cf. (9.3.5)), conditions (i)–(iii) still characterize whether f will be a Morse function. The problem is how to generalize Theorem 11.3.1 to this scenario, since its proof was based on the results of the previous section, all of which were established in a Euclidean setting. The trick, of course, is to recall that each of the three required properties is of a local nature. We can then argue as follows:

Choose a chart (U, φ) from a countable atlas covering M. Let $t^* \in U$ be a critical point of f, and note that this property is independent of the choice of local coordinates. Working therefore with the natural basis for $T_t M$, it is easy to see that $\varphi(t^*)$ is also a critical point of $\varphi(U)$. Furthermore, the covariant Hessian $\nabla^2 f(t^*)$ will be degenerate if and only if the same is true of the regular Hessian of $f \circ \varphi^{-1}$, and since φ is a diffeomorphism, this implies that t^* will be a degenerate critical point of f if and only if $\varphi(t^*)$ is for $f \circ \varphi^{-1}$. It therefore follows that even in the manifold case, we can manage, with purely Euclidean proofs, to establish the following result. The straightforward but sometimes messy details are left to you.

Theorem 11.3.4. *Let f be a real-valued random field over a piecewise C^2, compact Riemannian submanifold (M, g) of a C^3 manifold $\widetilde{M}$. Assume that M has a countable atlas. Then f is, with probability one, a Morse function over M if the following conditions hold for each submanifold $\partial_j M$ in (11.3.2). Throughout ∇ and ∇^2 are to be understood in their Riemannian formulations and implicitly assumed to have the dimension of the $\partial_j M$ where they are being applied.*

(a) *f is, with probability one, C^2 on an open neighborhood of M in $\widetilde{M}$, and $\mathbb{E}\{(XYf)^2\} < \infty$ for $X, Y \in T_t M$, $t \in M$.*
(b) *For each $\partial_j M$, the marginal densities $p_t(x)$ of $\nabla f_{|\partial_j M}(t)$ are continuous at 0, uniformly in t.*
(c) *The densities $p_t(z|x)$ of $\mathrm{Tr}^{\partial_j M}\, \nabla^2 f_{|\partial_j M}(t)$ given $\nabla f_{|\partial_j M}(t) = x$ are continuous for (z, x) in a neighborhood of 0, uniformly in $t \in M$.*
(d) *On $\partial_j M$, the moduli of continuity of f, $\nabla f_{|\partial_j M}$, and $\nabla^2 f_{|\partial_j M}(X, Y)$ all satisfy, for any $\varepsilon > 0$,*

$$\mathbb{P}\{\omega(\eta) > \varepsilon\} = o\left(\eta^{\dim(\partial_j M)}\right) \quad \text{as } \eta \downarrow 0 \tag{11.3.3}$$

for all $X, Y \in S(M) \cap T_t \partial_j M$, where $S(M)$ is the sphere bundle of M and the modulus of continuity is taken with respect to the distance induced by the Riemannian metric g.

As usual, there is a Gaussian corollary to Theorem 11.3.4, which requires only nondegeneracies and condition (d). There are two ways that we could go about attacking condition (d). The more geometric of the two would be to return to the entropy conditions of Section 1.3 and find a natural entropy condition for (11.3.3) to hold. This, however, involves adding a notion of "canonical distance" (under the canonical metric d on M) to the already existing distance corresponding to the Riemannian metric induced by f. The details of carrying this out, while not terribly difficult, would involve some work. Perhaps more importantly, the resulting conditions would not be in a form that would be easy to check in practice.

Thus, we take another route, given conditions that are less elegant but easier to establish and generally far easier to check in practice. As for Theorem 11.3.4 itself, we leave it to you to check the details of the (straightforward) proof of the corollary.

Corollary 11.3.5. *Take the setup of Theorem* 11.3.4 *and let f be a centered Gaussian field over M. Let $\mathcal{A} = (U_\alpha, \varphi_\alpha)_{\alpha \in I}$ be a countable atlas for M such that for every α, the Gaussian field $f_\alpha = f \circ \varphi_\alpha^{-1}$ on $\varphi_\alpha(U_a) \subset \mathbb{R}^N$ satisfies the conditions of Corollary* 11.3.2 *with $T = \varphi_\alpha(U_\alpha)$, $f = f_\alpha$ and some $K_\alpha > 0$. Then the sample functions of f are, with probability one, Morse functions over M.*

11.4 An Alternate Proof of the Metatheorem

The proof of the "metatheorem" Theorem 11.2.1 given in the previous section is but the latest tale in a long history.

The first proof of this kind is probably due to Rice [131] in 1945, who worked in the setting of real-valued processes on the line. His proof was made rigorous by Itô [81] and Ylvisaker [180] in the mid 1960s. Meanwhile, in 1957, Longuet-Higgins [106, 107] was the first to extend it to the setting of random fields and used it to compute the expectations of such variables as the mean number of local maxima of two- and three-dimensional Gaussian fields. Rigorous versions of Theorem 11.2.1, at various levels of generality and with various assumptions, appeared only in the early 1970s as in [1, 7, 16, 108] with the proof of Section 11.2 being essentially that of [1].

Recently, however, Azais and Wschebor [15] have developed a new proof, based on Federer's coarea formula. We bring it up here for two reasons. First of all, it is intrinsically interesting, and secondly, at first glance, it seems to avoid many of the technicalities inherent in the point process approach that we have adopted.

Retaining the notation of the previous section, Federer's coarea formula, in the form (7.4.15), can be rewritten as

$$\int_{\mathbb{R}^N} \left(\sum_{t: f(t)=u} \alpha(t) \right) du = \int_{\mathbb{R}^N} \alpha(t) |\det \nabla f(t)| \, dt,$$

assuming that f and $\alpha : \mathbb{R}^N \to \mathbb{R}^N$ are sufficiently smooth.

Take $\alpha(t) = \varphi(f(t)) \mathbb{1}_T(t)$, where φ is a smooth (test) function. (Of course, α is now no longer smooth, but we shall ignore this for the moment.) The above then becomes

$$\int_{\mathbb{R}^N} \varphi(u) N_u(f : T) \, du = \int_T \varphi(f(t)) |\det \nabla f(t)| \, dt.$$

Now take expectations (assuming that this is allowed) of both sides to obtain

$$\begin{aligned} \int_{\mathbb{R}^N} & \varphi(u) \mathbb{E}\{N_u(f : T)\} \, du \\ &= \int_T \mathbb{E}\left\{\varphi(f(t)) |\det \nabla f(t)|\right\} dt \end{aligned}$$

$$= \int_{\mathbb{R}^N} \varphi(u) \int_T \mathbb{E}\left\{|\det \nabla f(t)| \big| f(t) = u\right\} p_t(u)\, dt du.$$

Since φ was arbitrary, this implies that for (Lebesgue) almost every u,

$$\mathbb{E}\{N_u(f : T)\} = \int_T \mathbb{E}\left\{|\det \nabla f(t)| \big| f(t) = u\right\} p_t(u)\, dt, \tag{11.4.1}$$

which is precisely (11.2.4) of Theorem 11.2.1 with the g there identically equal to 1. Modulo this restriction on g, which is simple to remove, this is the result we have worked so hard to prove. The problem, however, is that since it is true only for almost every u, one cannot be certain that it is true for a specific value of u.

To complete the proof, we need only show that both sides of (11.4.1) are continuous functions of u and that the assumptions of convenience made above are no more than that. This, of course, is not as trivial as it may sound. Going through the arguments actually leads to repeating many of the technical points we went through in the previous section, and eventually Theorem 11.2.1 reappears with the same long list of conditions. However (and this is the big gain), the details have no need of the construction, in the proof of Theorem 11.2.7, of the linear approximation to f.

You can find all the details in [15] and decide for yourself which proof you prefer.

11.5 Higher Moments

While not at all obvious at first sight, hidden away in Theorem 11.2.1 is another result, about higher moments of the random variable $N_u(f, g : T, B)$. To state it, we need, for integral $k \geq 1$, the kth partial factorial of a real x defined by

$$(x)_k \stackrel{\Delta}{=} x(x-1)\cdots(x-k+1).$$

Theorem 11.5.1. *Let f, g, T, and B be as in Theorem* 11.2.1, *and assume that conditions* (a), (f), *and* (g) *there still hold. For $k \geq 1$, write*

$$\begin{aligned}
T^k &= \{\widetilde{t} = (t_1, \dots, t_k) : t_j \in T,\ \forall 1 \leq j \leq k\},\\
\widetilde{f}(\widetilde{t}) &= (f(t_1), \dots, f(t_k)) : T^k \to \mathbb{R}^{Nk},\\
\widetilde{g}(\widetilde{t}) &= (g(t_1), \dots, g(t_k)) : T^k \to \mathbb{R}^{Kk}.
\end{aligned}$$

Replace assumptions (b), (c), *and* (d) *of Theorem* 11.2.1 *by the following, assumed to hold for each $t \in T^k \setminus \{t \in T^k : t_i = t_j$ for some $i \neq j\}$:*

(b′) *The marginal densities $p_{\widetilde{t}}$ of $\widetilde{f}(\widetilde{t})$ are continuous at $\tilde{u} \stackrel{\Delta}{=} (u, \dots, u)$.*
(c′) *The conditional densities of $\widetilde{f}(\widetilde{t})$ given $\widetilde{g}(\widetilde{t})$ and $\nabla \widetilde{f}(\widetilde{t})$ are bounded above and continuous at $\tilde{u}$.*
(d′) *The conditional densities $p_{\widetilde{t}}(z | \widetilde{f}(\widetilde{t}) = x)$ of* $\det \nabla \widetilde{f}(\widetilde{t})$ *given $\widetilde{f}(\widetilde{t}) = x$ are continuous for z and x in neighborhoods of* 0 *and $\tilde{u}$, respectively.*

Then,

$$\mathbb{E}\{(N_u)_k\} = \int_{T^k} \mathbb{E}\left\{\prod_{j=1}^{k} \left|\det \nabla f(t_j)\right| \mathbb{1}_B(g(t_j)) \middle| \widetilde{f}(\widetilde{t}) = \tilde{u}\right\} p_{\widetilde{t}}(\tilde{u})\, d\widetilde{t} \qquad (11.5.1)$$

$$= \int_{T^k} \int_{\mathbb{R}^{kD}} \prod_{j=1}^{k} \left|\det D_j\right| \mathbb{1}_B(v_j) p_{\widetilde{t}}(\tilde{u}, \tilde{D}, \tilde{v})\, d\tilde{D}\, d\tilde{v}\, d\widetilde{t},$$

where

(i) $p_{\widetilde{t}}(\tilde{x})$ *is the density of* $\widetilde{f}(\widetilde{t})$.
(ii) $p_{\widetilde{t}}(\tilde{x}, \tilde{D}, \tilde{v})$ *is the joint density of* $\widetilde{f}(\widetilde{t})$, $\tilde{D}(\widetilde{t})$, *and* $\widetilde{g}(\widetilde{t})$.
(iii) $\tilde{D}(\widetilde{t})$ *represents the* $Nk \times Nk$ *matrix* $\nabla \widetilde{f}(\widetilde{t})$. *Note that* $\tilde{D}(\widetilde{t})$ *is a diagonal block matrix, with the* j*th block* $D_j(\widetilde{t})$ *containing the matrix* $\nabla f(t_j)$, *where* $t_j \in T$ *is the* j*th component of* $\widetilde{t}$.
(iv) $D = N(N+1)/2 + K$.

Proof. For each $\delta > 0$ define the domain

$$T_\delta^k \triangleq \{\widetilde{t} \in T^k : |t_i - t_j| \geq \delta \text{ for all } 1 \leq i < j \leq k\}.$$

Then the field $\widetilde{f}$ satisfies all the assumptions of Theorem 11.2.1, *uniformly* over the parameter set T_δ^k.

It therefore follows from (11.2.3) and (11.2.4) that $\mathbb{E}\{N_{\tilde{u}}(\widetilde{f}, \widetilde{g} : T_\delta^k, B^k)\}$ is given by either of the integrals in (11.5.1), with the outer integrals taken over T_δ^k rather than T^k.

Let $\delta \downarrow 0$. Then, using the fact that $f = u$ only finitely often (cf. footnote 8), it is easy to see that with probability one,

$$N_{\tilde{u}}(\widetilde{f}, \widetilde{g} : T_\delta^k, B^k) \uparrow (N_u(f, g : T, B))_k.$$

The monotonicity of the convergence then implies covergence of the expectations $\mathbb{E}\{N_{\tilde{u}}(\widetilde{f}, \widetilde{g} : T_\delta^k, B^k)\}$.

On the other hand, the integrals in (11.5.1) are trivially the limit of the same expressions with the outer integrals taken over T_δ^k rather than T^k, and so we are done. □

As for the basic expectation result, Theorem 11.5.1 takes a far simpler form if f is Gaussian, and we have the following.

Corollary 11.5.2. *Let* f *and* g *be centered Gaussian fields over a* T *that satisfies the conditions of Theorem* 11.2.1. *If for each* $\widetilde{t} = (t_1, \dots, t_k) \in T^k$, *the joint distributions of* $\{(f(t_j), \nabla f(t_j), g(t_j))\}_{1 \leq j \leq k}$ *are nondegenerate for all choices of distinct* $t_j \in T$, *and if* (11.2.5) *holds, then so does* (11.5.1).

11.6 Preliminary Gaussian Computations

In the following section we shall begin our computations of expectations for the Euler characteristics of excursion sets of Gaussian fields. In preparation, we need to collect a few facts about Gaussian integrals. Some are standard, others particularly tailored to our specific needs. However, since all are crucial components for the proofs of the main results of this chapter, we shall give proofs for all of them.

The first, in particular, is standard fare.

Lemma 11.6.1 (Wick formula). *Let $X_1, X_2, \ldots, X_n$ be a set of real-valued random variables having a joint Gaussian distribution and zero means. Then for any integer m,*

$$\mathbb{E}\{X_1 X_2 \cdots X_{2m+1}\} = 0, \tag{11.6.1}$$

$$\mathbb{E}\{X_1 X_2 \cdots X_{2m}\} = \sum \mathbb{E}\{X_{i_1} X_{i_2}\} \cdots \mathbb{E}\{X_{i_{2m-1}} X_{i_{2m}}\}, \tag{11.6.2}$$

where the sum is taken over the $(2m)!/m!2^m$ different ways of grouping $X_1, \ldots, X_{2m}$ into m pairs.

Note that this result continues to hold even if some of the X_j are identical, so that the lemma also applies to compute joint moments of the form $\mathbb{E}\{X_1^{i_1} \cdots X_k^{i_k}\}$.

Proof. Recall from (1.2.4) that the joint characteristic function of the X_i is

$$\phi(\theta) = \mathbb{E}\left\{e^{i \sum_i \theta_i X_i}\right\} = e^Q, \tag{11.6.3}$$

where

$$Q = Q(\theta) = -\frac{1}{2} \sum_{i=1}^n \sum_{j=1}^n \theta_i \mathbb{E}\{X_i X_j\} \theta_j.$$

Following our usual convention of denoting partial derivatives by subscripting, we have, for all l, k, j,

$$\begin{aligned} Q_j &= -\sum_{k=1}^n \mathbb{E}\{X_j X_k\} \theta_k, \\ Q_{kj} &= -\mathbb{E}\{X_j X_k\}, \\ Q_{lkj} &= 0. \end{aligned} \tag{11.6.4}$$

Successive differentiations of (11.6.3) yield

$$\begin{aligned} \phi_j &= \phi Q_j, \\ \phi_{kj} &= \phi_k Q_j + \phi Q_{kj}, \\ \phi_{lkj} &= \phi_{lk} Q_j + \phi_k Q_{lj} + \phi_l Q_{kj}, \\ &\vdots \\ \phi_{12\ldots n} &= \phi_{12\ldots(j-1)(j+1)\ldots n} Q_j + \sum_{k \neq j} \phi_{r_1 \ldots r_{n-2}} Q_{kj}, \end{aligned} \tag{11.6.5}$$

where in the last equation, the sequence $r_1, \ldots, r_{n-2}$ does not include the two numbers k and j.

The moments of various orders can now be obtained by setting $\theta = 0$ in the equations of (11.6.5). Since from (11.6.4) we have $Q_j(0) = 0$ for all j, the last (and most general) equation in (11.6.5) thus leads to

$$\mathbb{E}\{X_1 \cdots X_n\} = \sum_{k \neq j} \mathbb{E}\{X_{r_1} \cdots X_{r_{n-2}}\} \mathbb{E}\{X_j X_k\}.$$

From this relationship and the fact that the X_j all have zero mean it is easy to deduce the validity of (11.6.1) and (11.6.2). It remains only to determine exactly the number, M say, of terms in the summation (11.6.2). We note first that while there are $(2m)!$ permutations of $X_1, \ldots, X_{2m}$, since the sum does not include identical terms, $M < (2m)!$. Secondly, for each term in the sum, permutations of the m factors result in identical ways of breaking up the $2m$ elements. Thirdly, since $\mathbb{E}\{X_j X_k\} = \mathbb{E}\{X_k X_j\}$, an interchange of the order in such a pair does not yield a new pair. Thus

$$M(m!)(2^m) = (2m)!,$$

implying

$$M = \frac{(2m)!}{m!2^m}$$

as stated in the lemma. □

For the next lemma we need some notation. Let Δ_N be a symmetric $N \times N$ matrix with elements Δ_{ij} such that each Δ_{ij} is a zero-mean normal variable with arbitrary variance but such that the following relationship holds:

$$\mathbb{E}\{\Delta_{ij} \Delta_{kl}\} = \mathcal{E}(i, j, k, l) - \delta_{ij}\delta_{kl}, \tag{11.6.6}$$

where $\mathcal{E}$ is a symmetric function of i, j, k, l, and δ_{ij} is the Kronecker delta. Write $|\Delta_N|$ for the determinant of Δ_N.

Lemma 11.6.2. *Let m be a positive integer. Then, under* (11.6.6),

$$\mathbb{E}\{|\Delta_{2m+1}|\} = 0, \tag{11.6.7}$$

$$\mathbb{E}\{|\Delta_{2m}|\} = \frac{(-1)^m (2m)!}{m!2^m}. \tag{11.6.8}$$

Proof. Relation (11.6.7) is immediate from (11.6.1). Now

$$|\Delta_{2m}| = \sum_P \eta(p) \Delta_{1i_1} \cdots \Delta_{2m,i_{2m}},$$

where $p = (i_1, i_2, \ldots, i_{2m})$ is a permutation of $(1, 2, \ldots, 2m)$, P is the set of the $(2m)!$ such permutations, and $\eta(p)$ equals $+1$ or -1 depending on the order of the permutation p. Thus by (11.6.2) we have

$$\mathbb{E}\{|\Delta_{2m}|\} = \sum_P \eta(p) \sum_Q \mathbb{E}\{\Delta_{1i_1}\Delta_{2i_2}\} \cdots \mathbb{E}\{\Delta_{2m-1,i_{2m-1}}\Delta_{2m,i_{2m}}\},$$

where Q is the set of the $(2m)!/m!2^m$ ways of grouping $(i_1, i_2, \ldots, i_{2m})$ into pairs without regard to order, keeping them paired with the first index. Thus, by (11.6.6),

$$\begin{aligned}\mathbb{E}\{|\Delta_{2m}|\} = &\sum_P \eta(p) \sum_Q \{\mathcal{E}(1, i_1, 2, i_2) - \delta_{1i_1}\delta_{2i_2}\} \times \cdots \\ &\times \{\mathcal{E}(2m-1, i_{2m-1}, 2m, i_{2m}) - \delta_{2m-1,i_{2m-1}}\delta_{2m,i_{2m}}\}.\end{aligned}$$

It is easily seen that all products involving at least one $\mathcal{E}$ term will cancel out because of their symmetry property. Hence

$$\begin{aligned}\mathbb{E}\{|\Delta_{2m}|\} &= \sum_P \eta(p) \sum_Q (-1)^m \left(\delta_{1i_1}\delta_{2i_2}\right) \cdots \left(\delta_{2m-1,i_{2m-1}}\delta_{2m,i_{2m}}\right) \\ &= \frac{(-1)^m(2m)!}{(m!)2^m},\end{aligned}$$

the last line coming from changing the order of summation and then noting that for only one permutation in P is the product of delta functions nonzero. This completes the proof. □

Note that (11.6.8) in no way depends on the specific (co)variances among the elements of Δ_N. These all disappear in the final result due to the symmetry of $\mathcal{E}$.

Before stating the next result we need to introduce the family of *Hermite polynomials*. The nth Hermite polynomial is the function

$$H_n(x) = n! \sum_{j=0}^{\lfloor n/2 \rfloor} \frac{(-1)^j x^{n-2j}}{j!(n-2j)!2^j}, \qquad n \geq 0,\ x \in \mathbb{R}, \tag{11.6.9}$$

where $\lfloor a \rfloor$ is the largest integer less than or equal to a. For convenience later, we also define

$$H_{-1}(x) = \sqrt{2\pi}\,\Psi(x)e^{x^2/2}, \qquad x \in \mathbb{R}, \tag{11.6.10}$$

where Ψ is the tail probability function for a standard Gaussian variable (cf. (1.2.1)). With the normalization inherent in (11.6.9) the Hermite polynomials form an orthogonal (but not orthonormal) system with respect to standard Gaussian measure on $\mathbb{R}$, in that

$$\frac{1}{\sqrt{2\pi}} \int_{\mathbb{R}} H_n(x)H_m(x)e^{-x^2/2}\,dx = \begin{cases} n!, & m = n, \\ 0, & m \neq n. \end{cases}$$

An alternative definition of the Hermite polynomials is via a generating function approach, which gives

$$H_n(x) = (-1)^n e^{x^2/2} \frac{d^n}{dx^n}\left(e^{-x^2/2}\right), \qquad n \geq 0. \tag{11.6.11}$$

From this it immediately follows that

$$\int_u^\infty H_n(x) e^{-x^2/2}\, dx = H_{n-1}(u) e^{-u^2/2}, \qquad n \geq 0. \tag{11.6.12}$$

The centrality of Hermite polynomials for us lies in the following corollary to Lemma 11.6.2.

Corollary 11.6.3. *Let Δ_N be as in Lemma* 11.6.2, *with the same assumptions in force. Let I be the $N \times N$ unit matrix, and $x \in \mathbb{R}$. Then*

$$\mathbb{E}\{\det(\Delta_N - xI)\} = (-1)^N H_N(x).$$

Proof. It follows from the usual Laplace expansion of the determinant that

$$\det(\Delta_N - xI) = (-1)^N [x^N - S_1(\Delta_N) x^{N-1} + S_2(\Delta_N) x^{N-2} + \cdots + (-1)^N S_N(\Delta_N)], \tag{11.6.13}$$

where $S_k(\Delta_N)$ is the sum of the $\binom{N}{k}$ principal minors of order k in $|\Delta_N|$. The result now follows trivially from (11.6.7) and (11.6.8). □

11.7 The Mean Euler Characteristic

We now assume f to be a centered, stationary Gaussian process on a rectangle $T \subset \mathbb{R}^N$ and satisfying the conditions of Corollary 11.3.2. As usual, $C : \mathbb{R}^N \to \mathbb{R}$ denotes the covariance function, and ν the spectral measure. Then f has variance $\sigma^2 = C(0) = \nu(\mathbb{R}^N)$. We change notation a little from Section 5.4, and introduce the *second-order spectral moments*

$$\lambda_{ij} = \int_{\mathbb{R}^N} \lambda_i \lambda_j \nu(d\lambda). \tag{11.7.1}$$

Since we shall use it often, denote the $N \times N$ matrix of these moments by Λ.

Denoting also differentiation via subscripts, so that $f_i = \partial f/\partial t_i$, $f_{ij} = \partial^2 f/\partial t_i \partial t_j$, etc., we have, from (5.5.5), that

$$\mathbb{E}\{f_i(t) f_j(t)\} = \lambda_{ij} = -C_{ij}(0). \tag{11.7.2}$$

Thus Λ is also the variance–covariance matrix of ∇f. We could, of course, define the components of Λ using only derivatives of C, without ever referring to the spectrum.

The covariances among the second-order derivatives can be similarly defined. However, all we shall need is that

$$\mathcal{E}_{ijk\ell} \triangleq \mathbb{E}\{f_{ij}(t) f_{kl}(t)\} = \int_{\mathbb{R}^N} \lambda_i \lambda_j \lambda_k \lambda_\ell \nu(d\lambda) \tag{11.7.3}$$

is a symmetric function of i, j, k, ℓ.

Finally, note, as shown in Section 5.4, that f and its first-order derivatives are independent (at any fixed point t) as are the first- and second-order derivatives (from one another). The field and its second-order derivatives are, however, correlated, and

$$\mathbb{E}\{f(t)f_{ij}(t)\} = -\lambda_{ij}. \tag{11.7.4}$$

Finally, denote the $N \times N$ Hessian matrix (f_{ij}) by $\nabla^2 f$, and recall that the index of a matrix is defined as its number of negative eigenvalues.

Lemma 11.7.1. *Let f and T be as described above, and set*

$$\mu_k = \#\{t \in T : f(t) \ge u,\ \nabla f(t) = 0,\ \text{index}(\nabla^2 f) = k\}. \tag{11.7.5}$$

Then for all $N \ge 1$,

$$\mathbb{E}\left\{\sum_{k=0}^{N}(-1)^k \mu_k\right\} = \frac{(-1)^N |T||\Lambda|^{1/2}}{(2\pi)^{(N+1)/2}\sigma^N} H_{N-1}\left(\frac{u}{\sigma}\right) e^{-u^2/2\sigma^2}. \tag{11.7.6}$$

Before turning to the proof of the lemma, there are some crucial points worth noting. The first is the rather surprising fact that the result depends on the covariance of f only through some of its derivatives at zero, that is, only through the variance and second-order spectral moments. This is particularly surprising in view of the fact that the definition of the μ_k depends quite strongly on the f_{ij}, whose distribution involves fourth-order spectral moments.

As will become clear from the proof, the disappearance of the fourth-order spectral moments has a lot to do with the fact that we compute the mean of the alternating sum in (11.7.6) and do not attempt to evaluate the expectations of the individual μ_k. Doing so would indeed involve fourth-order spectral moments. As we shall see in later chapters, the fact that this is all we need is extremely fortunate, for it is actually *impossible* to obtain closed expressions for any of the $\mathbb{E}\{\mu_k\}$.

Proof. We start with the notationally simplifying assumption that $\mathbb{E}\{f_t^2\} = \sigma^2 = 1$. It is clear that if we succeed in establishing (11.7.6) for this case, then the general case follows by scaling. (Note that scaling f also implies scaling ∇f, which, since Λ contains the variances of the elements of ∇f, gives the factor of σ^{-N} in (11.7.6).)

Our second step is to simplify the covariance structure among the elements of ∇f. Let Q be any square root of Λ^{-1}, so that

$$Q'\Lambda Q = \text{diag}(1, \dots, 1). \tag{11.7.7}$$

Note that $\det Q = (\det \Lambda)^{-1/2}$. Now take the transformation of $\mathbb{R}^N$ given by $t \to tQ^{-1}$, under which $T \to T^Q = \{\tau : \tau = tQ^{-1} \text{ for some } t \in T\}$ and define $f^Q : T^Q \to \mathbb{R}$ by

$$f^Q(t) \triangleq f(tQ).$$

The new process f^Q has covariance function

$$C^Q(s,t) = C(sQ, tQ) = C((t-s)Q)$$

and so is still stationary, with constant, unit variance. Furthermore, simple differentiation shows that $\nabla f^Q = (\nabla f)Q$, from which it follows that

$$\Lambda^Q \triangleq \mathbb{E}\{((\nabla f^Q)(t))'((\nabla f^Q)(t))\} = Q'\Lambda Q = I. \tag{11.7.8}$$

That is, the first-order derivatives of the transformed process are now uncorrelated and of unit variance. We now show that it is sufficient to work with this, much simpler, transformed process.

Firstly, it is crucial to note that the μ_k of (11.7.5) for f over T are *identical* to those for f^Q over T^Q. Clearly, there is a trivial one-to-one correspondence between those points of T at which $f(t) \geq u$ and those of T^Q at which $f^Q(t) \geq u$. We do, however, need to check more carefully what happens with the conditions on ∇f and $\nabla^2 f$.

Since $\nabla f^Q = (\nabla f)Q$, we have that $(\nabla f^Q)(t) = 0$ if and only if $\nabla f(tQ) = 0$. In other words, there is also a simple one-to-one correspondence between critical points. Furthermore, since $\nabla^2 f^Q = Q'(\nabla^2 f)Q$ and Q is a positive definite matrix, $\nabla^2 f^Q(t)$ and $\nabla^2 f(tQ)$ have the same index.

Consequently, we can now work with f^Q rather than f, so that by (11.2.4) the expectation (11.7.6) is given by

$$\begin{aligned}\int_{T^Q} p_t(0)\, dt \sum_{k=0}^{N} (-1)^k \\ \times \mathbb{E}\{|\det \nabla^2 f^Q(t)| \mathbb{1}_{D_k}(\nabla^2 f^Q(t)) \mathbb{1}_{[u,\infty)}(f^Q(t)) | \nabla f^Q(t) = 0\},\end{aligned} \tag{11.7.9}$$

where p_t is the density of ∇f^Q and D_k is set of square matrices of index k. Now note the following:

(i) Since f, and so f^Q, are stationary, the integrand does not depend on t, and we can ignore the t's throughout. The remaining integral then gives the Lebesgue volume of T^Q, which is simply $|\det Q|^{-1}|T| = |\Lambda|^{1/2}|T|$.

(ii) The term $p_t(0)$ is simply $(2\pi)^{-N/2}$, and so can be placed to the side for the moment.

(iii) Most importantly, on the event D_k, the matrix $\nabla^2 f^Q(t)$ has k negative eigenvalues, and so has sign $(-1)^k$. We can combine this with the factor $(-1)^k$ coming immediately after the summation sign, and so remove both it and the absolute value sign on the determinant.

(iv) Recall that from the discussion on stationarity in Section 5.4 (especially (5.5.5)–(5.5.7)), we have the following relationships between the various derivatives of f^Q, for all $i, j, k \in \{1, \ldots, N\}$:

$$\mathbb{E}\{f^Q(t) f_i^Q(t)\} = 0, \qquad \mathbb{E}\{f_i^Q(t) f_{jk}^Q(t)\} = 0, \qquad \mathbb{E}\{f^Q(t) f_{ij}^Q(t)\} = -\delta_{ij},$$

where δ_{ij} is the Kronecker delta. The independence of the first derivatives from all the others means that the conditioning on ∇f^Q in (11.7.9) can be ignored, and so all that remains is to evaluate

$$\int_u^\infty \frac{e^{-x^2/2}}{\sqrt{2\pi}} \mathbb{E}\{\det \Delta_x\}\, dx, \tag{11.7.10}$$

where by (1.2.7) and (1.2.8), Δ_x is a matrix of Gaussian random variables whose elements Δ_{ij} have means $-x\delta_{ij}$ and covariances

$$\mathbb{E}\{\Delta_{ij}\Delta_{k\ell}\} = \mathbb{E}\{f_{ij}^Q(t) f_{k\ell}^Q(t)\} - \delta_{ij}\delta_{k\ell} = \mathcal{E}^Q(i, j, k, \ell) - \delta_{ij}\delta_{k\ell}.$$

By (11.7.3), $\mathcal{E}^Q$ is a symmetric function of its parameters.

This puts us directly into the setting of Corollary 11.6.3, from which it now follows that (11.7.10) is equivalent to

$$\frac{(-1)^N}{\sqrt{2\pi}} \int_u^\infty H_N(x) e^{-x^2/2}\, dx = \frac{(-1)^N}{\sqrt{2\pi}} H_{N-1}(u) e^{-u^2/2},$$

by (11.6.12).

Recalling (i), (ii), and (iv) above, substituting into (11.7.9), and lifting the restriction that $\sigma^2 = 1$ gives us (11.7.6) and we are done. □

With the proof behind us, it should now be clear why it is essentially impossible to evaluate the individual $\mathbb{E}\{\mu_k\}$. In doing so, we would have had to integrate over the various subsets $D_k \subset \mathbb{R}^{N(N+1)/2}$ of (11.7.9), and, with only the rarest exceptions, such integrals do not have explicit forms.

A careful reading of the proof also shows that we never really used the full power of stationarity, but rather only the existence of the matrix Q of (11.7.7) and a number of relationships between f and its first- and second-order derivatives. Actually, given the simplicity of the final result, which depends only on σ^2 and Λ, one is tempted to conjecture that in the nonstationary case one could replace Q by a family Q_t of such matrices. The final result might then be much the same, with the term $|T||\Lambda|^{1/2}$ perhaps being replaced by an integral of the form $\int_T |\Lambda_t|^{1/2} dt$, where Λ_t would be a local version of Λ, with elements $(\Lambda_t)_{ij} = \mathbb{E}\{f_i(t) f_j(t)\}$. That this is *not* the case will be shown later, when we do tackle the (far more complicated) nonstationary scenario as a corollary of results related to fields on manifolds in the following chapter (cf. the discussion of nonstationary fields in Section 12.5).

We can now turn to our first mean Euler characteristic computation, for which we need to set up a little notation, much of it close to that of Section 9.4. Nevertheless, since there are some slight changes, we shall write it all out again. We start with

$$T = \prod_{i=1}^N [0, T_i],$$

a rectangle in $\mathbb{R}^N$. As in Section 9.4, write $\mathcal{J}_k = \mathcal{J}_k(T)$ for the collection of the $2^{N-k}\binom{N}{k}$ faces of dimension k in T. As opposed to our previous conventions, we take these faces as closed. Thus, all faces in $\mathcal{J}_k$ are subsets of some face in $\mathcal{J}_{k'}$ for all $k' > k$. (For example, $\mathcal{J}_N$ contains only T while $\mathcal{J}_0$ contains the 2^N vertices of T.) Let $\mathcal{O}_k$ denote the $\binom{N}{k}$ elements of $\mathcal{J}_k$ which include the origin.

We need one more piece of notation. Take $J \in \mathcal{J}_k$. With the λ_{ij} being the spectral moments of (11.7.1), write Λ_J for the $k \times k$ matrix with elements λ_{ij}, $i, j \in \sigma(J)$.

This is enough to state the following result.

Theorem 11.7.2. *Let f be as described at the beginning of this section and T as above. For real u, let $A_u = A_u(f, T) = \{t \in T : f(t) \geq u\}$ be an excursion set, and let φ be the Euler characteristic. Then*

$$\mathbb{E}\{\varphi(A_u)\} = e^{-u^2/2\sigma^2} \sum_{k=1}^{N} \sum_{J \in \mathcal{O}_k} \frac{|J||\Lambda_J|^{1/2}}{(2\pi)^{(k+1)/2}\sigma^k} H_{k-1}\left(\frac{u}{\sigma}\right) + \Psi\left(\frac{u}{\sigma}\right). \tag{11.7.11}$$

Note that the k-dimensional volume $|J|$ of any $J \in \mathcal{J}_k$ is given by $|J| = \prod_{i \in \sigma(J)} T_i$.

Since (11.7.11) and its extension to manifolds in Section 12.4 is, for our purposes, probably the single most important equation in this book, we shall take a little time to investigate some of its consequences, before turning to a proof. To do so, we first note that it simplifies somewhat if f is isotropic. In that case we have the following corollary.

Corollary 11.7.3. *In addition to the conditions of Theorem 11.7.2, let f be isotropic and T the cube $[0, T]^N$. If λ_2 denotes the variance of f_i (independent of i by isotropy), then*

$$\mathbb{E}\{\varphi(A_u)\} = e^{-u^2/2\sigma^2} \sum_{k=1}^{N} \frac{\binom{N}{k} T^k \lambda_2^{k/2}}{(2\pi)^{(k+1)/2}\sigma^k} H_{k-1}\left(\frac{u}{\sigma}\right) + \Psi\left(\frac{u}{\sigma}\right). \tag{11.7.12}$$

The simplification follows immediately from the spherical symmetry of the spectral measure in this case, which (cf. (5.7.3)) implies that each matrix Λ_J is equal to $\lambda_2 I$. In fact, looking back into the proof of Lemma 11.7.1, which is where most of the calculation occurs, you can see that the transformation to the process f^Q is now rather trivial, since $Q = \lambda^{-1/2} I$ (cf. (11.7.7)). Looked at in this light, it is clear that one of the key points of the proof was a transformation that made the first derivatives of f behave as if they were those of an isotropic process. We shall see this again, but at a far more sophisticated level, when we turn to the manifold setting.

Now consider the case $N = 1$, so that T is simply the interval $[0, T]$. Then, using the definition of the Hermite polynomials given by (11.6.9), it is trivial to check that

$$\mathbb{E}\{\varphi(A_u(f, [0, T]))\} = \Psi(u/\sigma) + \frac{T\lambda_2^{1/2}}{2\pi\sigma} e^{-u^2/2\sigma^2}, \tag{11.7.13}$$

so that we have recovered (11.1.6) and with it, the Rice formula. Figure 11.7.1 gives two examples, with $\sigma^2 = 1$, $\lambda_2 = 200$, and $\lambda_2 = 1{,}000$.

Note from (11.7.13) that as $u \to -\infty$, we have $\mathbb{E}\{\varphi(A_u)\} \to 1$. The excursion set geometry behind this is simple. Once $u < \inf_T f(t)$, we have $A_u \equiv T$, and so $\varphi(A_u) = \varphi(T)$, which in the current case is 1. This is, of course, a general phenomenon, independent of dimension or the topology of T.

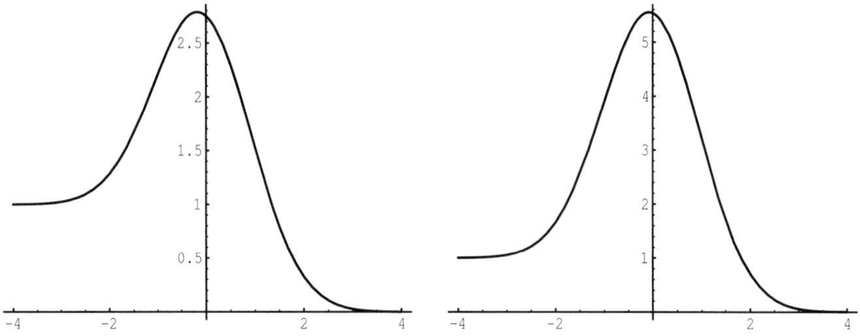

Fig. 11.7.1. $\mathbb{E}\{\varphi(A_u)\} : N = 1$.

To see this analytically, simply look at the expression (11.7.13), or even (11.7.11) for general rectangles. In both cases it is trivial that as $u \to -\infty$ all terms other than $\Psi(u/\sigma)$ disappear, while $\Psi(u/\sigma) \to 1$.

It thus seems not unreasonable to expect that when we turn to a more general theory (i.e., for T a piecewise smooth manifold) the term corresponding to the last term in (11.7.11) might be $\varphi(T)\Psi(u/\sigma)$. That this is in fact the case can be seen from the far more general results of Section 12.4 below.

We now turn to two dimensions, in which case the right-hand side of (11.7.12) becomes, for $\sigma^2 = 1$,

$$\left[\frac{T^2\lambda_2}{(2\pi)^{3/2}}u + \frac{2T\lambda_2^{1/2}}{2\pi}\right]e^{-u^2/2} + \Psi(u). \tag{11.7.14}$$

Figure 11.7.2 gives two examples, again with $\lambda_2 = 200$ and $\lambda_2 = 1{,}000$.

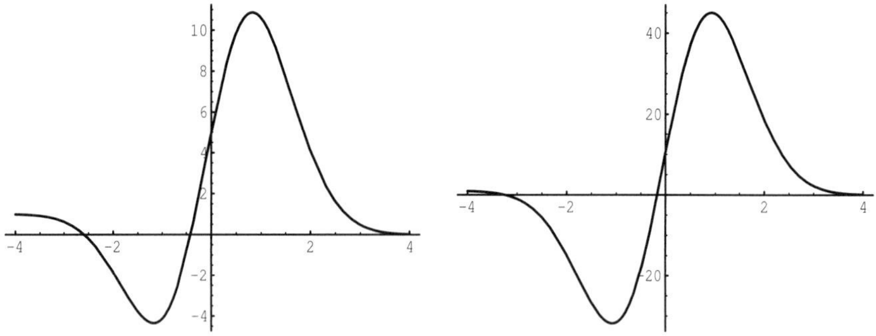

Fig. 11.7.2. $\mathbb{E}\{\varphi(A_u)\} : N = 2$.

Many of the comments that we made for the one-dimensional case have similar analogues here and we leave them to you. Nevertheless, we emphasize three points:

(i) You should note, for later reference, how the expression before the exponential term can be thought of as one of a number of different power series: one in T, one in u, and one in $\sqrt{\lambda_2}$.

(ii) The geometric meaning of the negative values of (11.7.14) are worth understanding. They are due to the excursion sets having, in the mean, more holes than connected components for (most) negative values of u.

(iii) The impact of the spectral moments is not quite as clear in higher dimensions as it is in one. Nevertheless, to get a feel for what is happening, look back at the simulation of a Brownian sheet in Figure 1.4.1. The Brownian sheet is, of course, both nonstationary and nondifferentiable, and so hardly belongs in our current setting. Nevertheless, in a finite simulation, it is impossible to "see" the difference between nondifferentiability and large second spectral moments,[13] so consider the simulation in the latter light. You can then see what is happening. Large spectral moments again lead to local fluctuations generating large numbers of small islands (or lakes, depending on the level at which the excursion set is taken), and this leads to larger variation in the values of $\mathbb{E}\{\varphi(A_u)\}$.

In three dimensions, the last case that we write out, (11.7.12) becomes, for $\sigma^2 = 1$,

$$\left[\frac{T^3\lambda_2^{3/2}}{(2\pi)^2}u^2 + \frac{3T^2\lambda_2}{(2\pi)^{3/2}}u + \frac{3T\lambda_2^{1/2}}{2\pi} - \frac{T^3\lambda_2^{3/2}}{(2\pi)^2}\right]e^{-u^2/2} + \Psi(u).$$

Figure 11.7.3 gives the examples with $\lambda_2 = 200$ and $\lambda_2 = 1{,}000$.

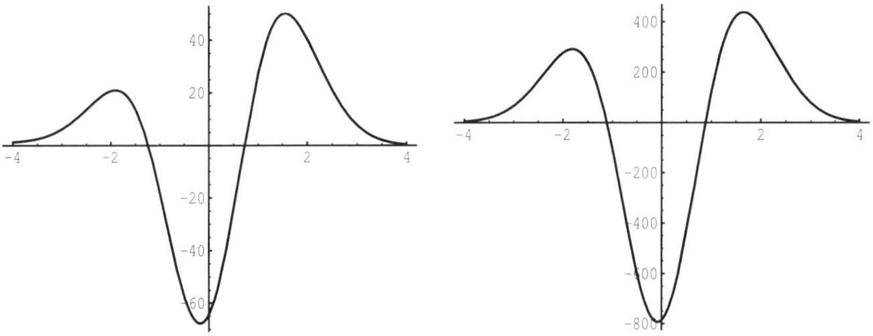

Fig. 11.7.3. $\mathbb{E}\{\varphi(A_u)\} : N = 3$.

Note that once again there are a number of different power series appearing here, although now, as opposed to the two-dimensional case, there is no longer a simple correspondence between the powers of T, $\sqrt{\lambda_2}$, and u.

The two positive peaks of the curve are due to A_u being primarily composed of a number of simply connected components for large u and primarily of simple holes for

[13] Ignore the nonstationarity of the Brownian sheet, since this has no qualitative impact on the discussion.

negative u. (Recall that in the three-dimensional case the Euler characteristic of a set is given by the number of components minus the number of handles plus the number of holes.) The negative values of $\mathbb{E}\{\varphi(A_u)\}$ for u near zero are due to the fact that A_u, at those levels, is composed mainly of a number of interconnected, tubular-like regions, i.e., of handles.

An example[14] is given in Figure 11.7.4 that shows an impression of typical excursion sets of a function on I^3 above high, medium, and low levels.

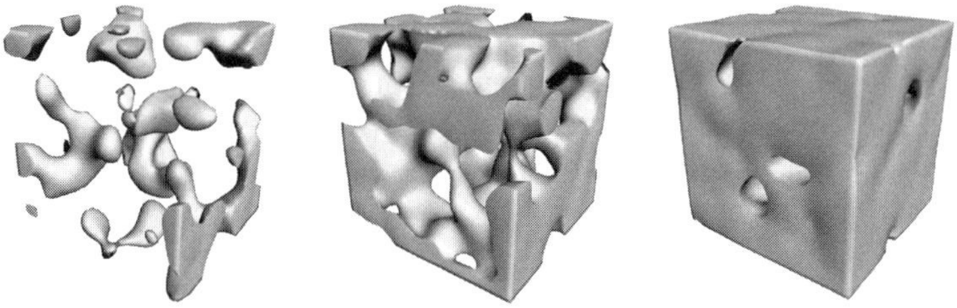

Fig. 11.7.4. Three-dimensional excursion sets above high, medium, and low levels. For obvious reasons, astrophysicists refer to these three cases (from left to right) as "meatball," "sponge," and "bubble" topologies.

We have one more—extremely important—observation to make before we turn to the proof of Theorem 11.7.2. Recall (6.3.5), which gave that the intrinsic volumes, or Lipschitz–Killing curvatures, of $[0, T]^N$ were given by $\binom{N}{j}T^j$. With this in mind, simplifying to the case $\sigma^2 = \lambda_2 = 1$, (11.7.12) can be written far more tidily as

$$\mathbb{E}\{\varphi(A_u(f, T))\} = \sum_{k=0}^{N} \mathcal{L}_k(T)\rho_k(u), \tag{11.7.15}$$

where

$$\rho_k(u) = (2\pi)^{-(k+1)/2} H_{k-1}(u) e^{-\frac{u^2}{2}}, \quad k \geq 0,$$

since $H_{-1}(u) = \sqrt{2\pi}\Psi(u)e^{u^2/2}$ (cf. (11.6.10)).

In fact, the above holds also when T is an N-dimensional, piecewise smooth manifold, and f has constant variance (i.e., there are no assumptions of isotropy, or even stationarity). The Lipschitz–Killing curvatures, however, will be somewhat more complex, depending on a Riemannian metric related to f. This will be the content of Sections 12.2 and 12.4 of the next chapter. Now, however, we (finally) turn to the following proof.

Proof of Theorem 11.7.2. We start by recalling that each k-dimensional face $J \in \mathcal{J}_k$ is determined by a subset $\sigma(J)$ of $\{1, \ldots, N\}$, of size k, and a sequence of $N - k$ zeros and ones, which we write as $\varepsilon(J) = \{\varepsilon_j, j \notin \sigma(J)\}$, so that

[14] Taken, with permission, from Keith Worsley's entertaining and illuminating *Chance* article [175].

$$J = \left\{ t \in \mathbb{R}^N : t_j = \varepsilon_j T_j,\ \text{if } j \notin \sigma(J),\ \ 0 \le t_j \le T_j,\ \text{if } j \in \sigma(J) \right\}.$$

Corresponding to each sequence $\varepsilon(J)$ we define a set $\varepsilon^*(J)$ of ± 1's, according to the rule $\varepsilon_j^* = 2\varepsilon_j - 1$.

Now recall[15] from Section 9.4 (cf. (9.4.1)–(9.4.5)) that the Euler characteristic of A_u is given by

$$\varphi(A_u(f,T)) = \sum_{k=0}^{N} \sum_{i=0}^{k} (-1)^i \sum_{J \in \mathcal{J}_k} \mu_i(J), \tag{11.7.16}$$

where, for $i \le \dim(J)$, $\mu_i(J)$ is the number of $t \in J$ for which

$$f(t) \ge u, \tag{11.7.17}$$

$$f_j(t) = 0, \qquad j \in \sigma(J), \tag{11.7.18}$$

$$\varepsilon_j^* f_j(t) \ge 0, \qquad j \notin \sigma(J), \tag{11.7.19}$$

$$I(t) \stackrel{\Delta}{=} \text{Index}(f_{mn}(t))_{(m,n \in \sigma(J))} = k - i, \tag{11.7.20}$$

where, as usual, the index of a matrix is the number of its negative eigenvalues.

Our first and main step will hinge on stationarity, which we exploit to replace the expectation of the sum over $J \in \mathcal{J}_k$ in (11.7.16) by something simpler.[16] Fix a particular $J \in \mathcal{O}_k$—i.e., a face containing the origin—and let $\mathcal{P}(J)$ denote all faces in $\mathcal{J}_k$ (including J itself) that are affine translates of (i.e., parallel to) J. There are 2^{N-k} faces in each such $\mathcal{P}(J)$. We can then rewrite the right-hand side of (11.7.16) as

$$\sum_{k=0}^{N} \sum_{i=0}^{k} (-1)^i \sum_{J \in \mathcal{O}_k} \sum_{J' \in \mathcal{P}(J)} \mu_i(J'). \tag{11.7.21}$$

Consider the expectation of the innermost sum. By Theorem 11.2.1 (cf. (11.2.4)) we can rewrite this as

$$\sum_{J' \in \mathcal{P}(J)} \int_{J'} \mathbb{E}\{|\det \nabla^2 f_{|J'}(t)| \mathbb{1}_{\{I(t)=k-i\}} \mathbb{1}_{\mathcal{E}_{J'}(t)} | \nabla f_{|J'}(t) = 0\} p_{\nabla f_{|J'}}(0)\, dt,$$

where $\mathcal{E}_J(t)$ denotes the event that (11.7.17) and (11.7.19) hold.

Further simplification requires one more item of notation. For $J \in \mathcal{O}_k$, let $\mathcal{P}^\varepsilon(J)$ denote the collections of all sequences $\{\varepsilon_j,\ j \notin \sigma(J)\}$ of ± 1's. The elements of $\mathcal{P}^\varepsilon(J)$ can be identified with the sequences $\varepsilon^*(J')$ with $J' \in \mathcal{P}(J)$.

---

[15] "Recall" includes extending the results there from cubes to rectangles and changing the order of summation, but both these steps are trivial.

[16] Our approach here is not unlike the first-principles geometry used to prove that

$$\mathcal{L}_j([0,T]^N) = \binom{N}{j} T^j.$$

With this notation, and calling on stationarity, we can replace the last expression by

$$\sum_{\varepsilon^* \in \mathcal{P}^{\varepsilon}(J)} \int_J \mathbb{E}\{|\det \nabla^2 f_{|J}(t)| \mathbb{1}_{\{I(t)=k-i\}} \mathbb{1}_{\mathcal{E}_{\varepsilon^*}(t)} | \nabla f_{|J}(t) = 0\} p_{\nabla f_{|J}}(0)\, dt,$$

where $\mathcal{E}_{\varepsilon^*}(t)$ denotes the event that (11.7.17) and (11.7.19) hold for ε^*.

Now note the trivial fact that for any J,

$$\bigcup_{\varepsilon^* \in \mathcal{P}^{\varepsilon}(J)} \{\varepsilon_j^* f_j(t) \geq 0 \text{ for all } j \notin \sigma(J)\}$$

is the sure event, i.e., has probability one. Applying this to the last sum, we see that it simplifies considerably to

$$\int_J \mathbb{E}\left\{ \left| \det \nabla^2 f_{|J}(t) \right| \mathbb{1}_{\{I(t)=k-i\}} \mathbb{1}_{\{f(t) \geq u\}} \middle| \nabla f_{|J}(t) = 0 \right\} p_{\nabla f_{|J}}(0)\, dt.$$

Going back to Theorem 11.2.1, we have that this is no more that the expected number of points in J for which

$$f(t) \geq u, \qquad \nabla f(t) = 0, \qquad \text{Index}(\nabla^2 f) = k - i.$$

If we call the number of points satisfying these conditions $\mu'_{k-i}(J)$, then putting all of the above together and substituting into (11.7.16) we see that we need the expectation of

$$\sum_{k=0}^{N} \sum_{J \in \mathcal{O}_k} (-1)^k \sum_{i=0}^{k} (-1)^{k-i} \mu'_{k-i}(J).$$

Lemma 11.7.1 gives us a precise expression for the expectation of the innermost sum, at least for $k \geq 1$, namely,

$$\frac{(-1)^k |J| |\Lambda_J|^{1/2}}{(2\pi)^{(k+1)/2} \sigma^k} H_{k-1}\left(\frac{u}{\sigma}\right) e^{-u^2/2\sigma^2}.$$

It is left to you to check that for the case $k = 0$ (i.e., when $\mathcal{J}_k$ contains the 2^N vertices of T), the remaining term is given by $\Psi(u/\sigma)$. Putting all this together immediately gives (11.7.11), and so the proof is complete. □

11.8 Mean Intrinsic Volumes

So far, this entire chapter has concentrated on finding the mean value of the Euler characteristic of Gaussian excursion sets. However, the Euler characteristic is only one of $N + 1$ intrinsic volumes, and it would be useful to know how the remaining N behave. We shall prove, in the setting of manifolds, in Chapter 13, that

$$\mathbb{E}\left\{\mathcal{L}_j(A_u(f,T))\right\} = \sum_{l=0}^{N-j} \begin{bmatrix} j+l \\ l \end{bmatrix} \rho_l(u)\mathcal{L}_{j+l}(T), \qquad (11.8.1)$$

where $\begin{bmatrix} n \\ k \end{bmatrix}$ is the flag coefficient defined by (6.3.12).

As noted above, the proof appears only after we have moved to the manifold setting. Nevertheless, the proof of (11.8.1) in the case of an isotropic field over an N-dimensional rectangle requires none of the manifold material, and so, if you wish, you can read it now in Section 13.2.

11.9 On the Importance of Stationarity

Before leaving the setting of random fields defined over N-dimensional rectangles, it is worthwhile, once again, to recall how crucial stationarity was for our ability to carry out the precise computations that led to Theorem 11.7.2.

In fact, stationarity appeared twice. We first used some consequences of stationarity (although not its full force) in making the crucial transformation to the process f^Q (cf. (11.7.7)) in the proof of Lemma 11.7.1, which is where the really detailed Gaussian calculations were made. The second time, we exploited the full power of stationarity in the proof of Theorem 11.7.2 for handling the expectation of the awkward summation of (11.7.16). At this stage, our argument also relied rather heavily on the assumption that our parameter space was a rectangle.

It should be reasonably clear that handling nonstationary fields and/or nonrectangular domains is going to require new tools. The tools for nonrectangular domains are, not surprisingly, those that will arise in the following chapter, where we turn to the setting of (Riemannian) manifolds. What is a perhaps more than a little surprising, however, are that these will be precisely the tools needed for handling nonstationary fields as well.

12

Random Fields on Manifolds

In essence, this chapter will repeat, for random fields on manifolds, what we have already achieved in the Euclidean setting.

As there, our first step will be to set up a "metatheorem" for computing the mean number of points at which a random field takes a certain value under specific side conditions. This turns out to be rather easy to do, involving little more than taking the Euclidean result and applying it, chart by chart, to the manifold. This is the content of the first section.

Actually computing the resulting expression for special cases—such as finding the mean Euler characteristic of excursion sets over Whitney stratified manifolds—turns out to be somewhat more complicated. We start the computation in Section 12.2, where, for the second time,[1] we shall begin to see why we need our heavy investment in the Riemannian geometry of Chapter 7. The main contribution of this section will be to take the parameter space on which our Gaussian process is defined, endow it with a Riemmanian metric induced by the process, and study this metric a little.

Section 12.3 is devoted to some general Gaussian computations, including the moment formulas for random Gaussian forms that we used when looking at volume-of-tube formulas. The real work is done in Section 12.4, which puts all the preceding material together to devlop mean Euler characteristic formulas.

The strength of the general approach is shown by the examples of Section 12.5, and, as we have already mentioned, we indulge ourselves somewhat with a fun proof of the Chern–Gauss–Bonnet theorem in Section 12.6.

12.1 The Metatheorem on Manifolds

To formulate the metatheorem for manifolds, we need one small piece of notation.

[1] The first was in the volume-of-tubes approximation of Chapter 10, which was something of an aside. It is in the current chapter that we shall see the real need for Riemannian geometry in a general setting, and not just geometry of $\mathbb{R}^N$ or $S(\mathbb{R}^N)$ with the usual Euclidean metrics.

Let (M, g) be an N-dimensional Riemannian manifold, and let $f : M \to \mathbb{R}^N$ be C^1. Fix an orthonormal frame field E. Then ∇f_E denotes the vector field whose coordinates are given by

$$(\nabla f_E)_i \equiv \nabla f_{Ei} \stackrel{\Delta}{=} (\nabla f)(E_i) \equiv E_i f, \tag{12.1.1}$$

where ∇ is the gradient operator defined in (7.3.2). If $f = (f^1, \dots, f^N)$ takes values in $\mathbb{R}^N$, then ∇f_E denotes the $N \times N$ matrix with elements ∇f^j_{Ei}.

Theorem 12.1.1. *Let M be a compact, oriented, N-dimensional C^1 manifold with a C^1 Riemannian metric g. Let $f = (f^1, \dots, f^N) : M \to \mathbb{R}^N$ and $h = (h^1, \dots, h^K) : M \to \mathbb{R}^K$ be random fields on M. For an open set $B \subset \mathbb{R}^K$ for which ∂B has dimension $K - 1$ and a point $u \in \mathbb{R}^N$, let*

$$N_u \equiv N_u(M) \equiv N_u(f, h; M, B)$$

denote the number of points $t \in M$ for which

$$f(t) = u \quad \text{and} \quad h(t) \in B.$$

Assume that the following conditions are satisfied for some orthonormal frame field E:

(a) *All components of f, ∇f_E, and h are a.s. continuous and have finite variances (over M).*
(b) *For all $t \in M$, the marginal densities $p_t(x)$ of $f(t)$ (implicitly assumed to exist) are continuous at $x = u$.*
(c) *The conditional densities $p_t(x|\nabla f_E(t), h(t))$ of $f(t)$ given $h(t)$ and $(\nabla f_E)(t)$ (implicitly assumed to exist) are bounded above and continuous at $x = u$, uniformly in $t \in M$.*
(d) *The conditional densities $p_t(z|f(t) = x)$ of $\det(\nabla f^i_{Ej}(t))$ given $f(t) = x$ are continuous for z and x in neighborhoods of 0 and u, respectively, uniformly in $t \in M$.*
(e) *The conditional densities $p_t(z|f(t) = x)$ of $h(t)$ given $f(t) = x$ are continuous for all z and for x in a neighborhood u, uniformly in $t \in M$.*
(f) *The following moment condition holds:*

$$\sup_{t\in M} \max_{1\le i,j\le N} \mathbb{E}\left\{\left|\nabla f^j_{Ei}(t)\right|^N\right\} < \infty. \tag{12.1.2}$$

(g) *The moduli of continuity with respect to the (canonical) metric induced by g (cf. (7.3.1)) of each component of h, each component of f, and each ∇f^j_{Ei} all satisfy, for any $\varepsilon > 0$,*

$$\mathbb{P}\{\omega(\eta) > \varepsilon\} = o\left(\eta^N\right) \quad \text{as } \eta \downarrow 0. \tag{12.1.3}$$

Then

$$\mathbb{E}\{N_u\} = \int_M \mathbb{E}\left\{|\det(\nabla f_E)|\, \mathbb{1}_B(h) \Big| f = u\right\} p(u)\, \mathrm{Vol}_g, \tag{12.1.4}$$

where p is the density[2] *of f and* Vol_g *the volume element on M induced by the metric g.*

Before turning to the proof of the theorem, there are a few points worth noting. The first is that the conditions of the theorem do not depend on the choice of orthonormal frame field. Indeed, as soon as they hold for one such choice, not only will they hold for all orthonormal frame fields but also for any bounded vector field X. In the latter case the notation will change slightly, and ∇f^j_{Ei} needs to be replaced by $(Xf^j)_i$.

Once this is noted, you should note that the only place that the metric g appears in the conditions is in the definition of the neighborhoods $B_\tau(t, h)$ in the final condition. A quick check of the proof to come will show that any neighborhood system will in fact suffice. Thus the metric does not really play a role in the conditions beyond convenience.

Furthermore, the definition of the random variable N_u is totally unrelated to the metric. From this it follows that the same must be true of its expectation. Consequently, although we require a metric to be able to define the integration in (12.1.4), the final expression must actually yield a result that is independent of the choice of g and so be a function *only* of the "physical" manifold and the distribution of f. However, choosing an appropriate g can greatly simplify the calculation.

Proof. Since M is compact it has a finite atlas. Let (U, φ) be one of its charts and consider the random fields $\bar{f} : \varphi(U) \subset \mathbb{R}^N \to \mathbb{R}^N$ and $\bar{h} : \varphi(U) \subset \mathbb{R}^N \to \mathbb{R}^K$ defined by

$$\bar{f} \triangleq f \circ \varphi^{-1}, \qquad \bar{h} \triangleq h \circ \varphi^{-1}.$$

It is immediate from the definition of N_u that

$$N_u(f, h; U, B) \equiv N_u(\bar{f}, \bar{h}; \varphi(U), B),$$

and so the expectations of both of these random variables are also identical.

Recall the comments made prior to the proof: All conditions in the theorem that involve the orthonormal frame field E also hold for any other bounded vector field on $U \subset M$. In particular, they hold for the natural coordinate vector field $\{\partial/\partial x_i\}_{1 \le i \le N}$ determined by φ.

---

[2] Of course, what is implicit here is that for each $t \in M$, we should really write p as p_t, since it is the density of f_t. There are also a number of additional places in (12.1.4) where we could append a t, but since it has been our habit to drop the subscript when working in the setting of manifolds, we leave it out here as well.

Note that it is implicitly assumed that the integrand in (12.1.4) is a well-defined N-form on M, or, equivalently, that the expectation term is a well-defined Radon–Nikodym derivative. That this is the case will follow from the proof.

Comparing conditions (a)–(g) of the current theorem with those in Theorem 11.2.1, it is clear that $\bar{f}$ and $\bar{h}$ satisfy[3] the conditions of Theorem 11.2.1. Consequently,

$$\mathbb{E}\{N_u(f,h;U,B)\} = \int_{\varphi(U)} \mathbb{E}\left\{\left|\det \nabla \bar{f}(x)\right| \mathbb{1}_B(\bar{h}(x)) \middle| \bar{f}(x) = u\right\} \bar{p}_x(u)\,dx,$$

where $\bar{p}_x$ is the density of $\bar{f}(x)$.

All that remains is to show that this is equivalent to (12.1.4) with the domain of integration restricted to U, and that we can replace U by M throughout. Consider the right-hand side above, and rewrite it in terms of an integral over U. To this end, note that

$$\left(\nabla \bar{f}^j(x)\right)_i = \left(\frac{\partial}{\partial x_i} f^j\right)_{t=\varphi^{-1}(x)},$$

where $\partial/\partial x_i$ is the push-forward under φ^{-1} of the natural basis on $\varphi(U)$. Together with the definition of integration of differential forms in Section 7.4 this gives us that

$$\begin{aligned}&\mathbb{E}\{N_u(f,h;U,B)\}\\ &\quad = \int_U \mathbb{E}\left\{\left|\det\left(\partial/\partial x_i f^j\right)_t\right| \mathbb{1}_B(h(t)) \middle| \bar{f}(t) = u\right\} p_t(u)\partial x_1 \wedge \cdots \wedge \partial x_N.\end{aligned}$$

The next step involves moving from the natural basis on U to the basis given by the orthonormal frame field E. Doing so generates two multiplicative factors, which fortunately cancel. The first comes from the move from the form $\partial x_1 \wedge \cdots \wedge \partial x_N$ to the volume form Vol_g, and generates a factor of $(\det(g_{ij}))^{-1/2}$, where $g_{ij}(t) = g_t(\partial/\partial x_i, \partial/\partial x_j)$ (cf. (7.4.9)).

The second factor comes from noting that

$$\frac{\partial}{\partial x_i} f^j = \sum_k g\left(\frac{\partial}{\partial x_i}, E_k\right) E_k f^j = \sum_k g_{ik}^{1/2} E_k f^j,$$

where $g^{1/2} = \left(g(E_i, \partial x_j)\right)_{1\le i,j\le N}$ is a square root of the matrix $g = (g_{ij})_{1\le i,j\le N}$. Consequently,

$$\det\left(\partial/\partial x_i f^j\right)_t = \sqrt{\det(g_{ij})}\,\det\left(\nabla f_E\right).$$

Putting the pieces together gives us

[3] The only condition that needs any checking is (11.2.2) on the moduli of continuity. It is here that the requirement that g be C^1 over M comes into play. The details are left to you.

$$\mathbb{E}\{N_u(U)\} = \int_U \mathbb{E}\left\{|\det(\nabla f_E)| \, \mathbb{1}_B(h) \middle| f = u\right\} p(u) \, \text{Vol}_g, \tag{12.1.5}$$

for each chart (U, φ).

To finish the proof, note that for each chart (U, φ) the conditions of the theorem imply that there are only a finite number of points in $\varphi(U)$ at which $\bar{f} = u$ (cf. footnote 8 in Chapter 11) and that there are no points of this kind on $\partial\varphi(U)$ (cf. Lemma 11.2.12).

Consequently, the same is true of f over U. In particular, this means that we can refine a given atlas so that each point for which $f = u$ appears in only one chart and no chart contains more than one point of this kind. If this is the case, the integrals in (12.1.5) are either zero or one, and so it is trivial to combine them to obtain a single integral over M and so the theorem. □

As usual, we have the following corollary for the Gaussian case.

Corollary 12.1.2. *Let (M, g) be a Riemannian manifold satisfying the conditions of Theorem* 12.1.1. *Let f and h be centered Gaussian fields over M. Then if f, h, and ∇f_E are a.s. continuous over M, and if for each $t \in M$, the joint distributions of $(f(t), \nabla f_E(t), h(t))$ are nondegenerate, then* (12.1.4) *holds.*

Ultimately, we shall apply the above corollary to obtain, among other things, an expression for the expected Euler characteristic of Gaussian excursion sets over manifolds. Firstly, however, we need to set up some machinery.

12.2 Riemannian Structure Induced by Gaussian Fields

Up until now, all our work with Riemannian manifolds has involved a general Riemannian metric g. Using this, back in Section 7.5 we developed a number of concepts, starting with connections and leading up to curvature tensors and shape operators, in corresponding generality.

For our purposes, however, it will turn out that for each random field f on a piecewise C^2 manifold M, there is only one Riemannian metric that we shall need. It is induced by the random field f, which we shall assume has zero mean and, with probability 1, is C^2 over M. It is defined by

$$g_t(X_t, Y_t) \triangleq \mathbb{E}\{(X_t f) \cdot (Y_t f)\}, \tag{12.2.1}$$

where $X_t, Y_t \in T_t M$, the tangent manifold to M at t.

Since the notation of (12.2.1) is rather heavy, we shall in what follows generally drop the dependence on t. Thus (12.2.1) becomes

$$g(X, Y) = \mathbb{E}\{XfYf\}. \tag{12.2.2}$$

We shall call g the *metric induced by the random field*[4] f. The fact that this definition actually gives a Riemannian metric follows immediately from the positive semidefiniteness of covariance functions.

Note that at this stage, there is nothing in the definition of the induced metric that relies on f being Gaussian.[5] The definition holds for any C^2 random field. Furthermore, there are no demands related to stationarity, isotropy, etc.

One way to develop some intuition for this metric is via the geodesic metric τ that it induces on M. Since τ is given by

$$\tau(s,t) = \inf_{c \in D^1([0,1];M)_{(s,t)}} \int_{[0,1]} \sqrt{g_t(c',c')(t)}\,dt \tag{12.2.3}$$

(cf. (7.3.1)), it follows that the geodesic between two points on M is the curve along which the expected variance of the derivative of f is minimized.

It is obvious that g is closely related to the covariance function $C(s,t) = \mathbb{E}(f_s f_t)$ of f. In particular, it follows from (12.2.1) that

$$g_t(X_t, Y_t) = X_s X_t C(s,t)|_{s=t}. \tag{12.2.4}$$

Consequently, it is also obvious that the tools of Riemannian manifolds—connections, curvatures, etc.—can be expressed in terms of covariances. In particular, in the Gaussian case, to which we shall soon restrict ourselves, all of these tools also have interpretations in terms of conditional means and variances. Since these interpretations will play a crucial role in the extension of the results of Sections 11.6 and 11.7 to Gaussian fields over manifolds, we shall now spend some time developing them.

12.2.1 Connections and Curvatures

Our first step is to describe the Levi-Civitá connection ∇ determined by the induced metric g. Recall from Chapter 6 that the connection is uniquely determined by Koszul's formula,

$$\begin{aligned} 2g(\nabla_X Y, Z) = {} & Xg(Y,Z) + Yg(X,Z) - Zg(X,Y) \\ & + g(Z,[X,Y]) + g(Y,[Z,X]) + g(X,[Z,Y]), \end{aligned} \tag{12.2.5}$$

[4] A note for the theoretician: Recall that a Gaussian process has associated with it a natural L^2 space, which we denoted by $\mathcal{H}$ in Section 3.1. The inner product between two random variables in $\mathcal{H}$ is given by their covariance. There is also a natural geometric structure on $\mathcal{H}$, perhaps seen most clearly through orthogonal expansions of the form (3.1.7). In our current scenario, in which f is indexed by a manifold M, it is easy to see that the Riemannian structure induced on M by f (i.e., via the associated metric (12.2.2)) is no more than the pullback of the natural structure on $\mathcal{H}$.

[5] Or even on f being C^2. The induced metric is well defined for f, which is merely C^1. However, it is not possible to go much further—e.g., to a treatment of curvature—without more derivatives.

where X, Y, Z are C^1 vector fields (cf. (7.3.12)).

Since $g(X, Y) = \mathbb{E}\{Xf \cdot Yf\}$, it follows that

$$Zg(X, Y) = Z\mathbb{E}\{Xf \cdot Yf\} = \mathbb{E}\{ZXf \cdot Yf + Xf \cdot ZYf\}$$
$$= g(ZX, Y) + g(X, ZY).$$

Substituting this into (12.2.5) yields

$$g(\nabla_X Y, Z) = \mathbb{E}\{(\nabla_X Yf)(Zf)\} = \mathbb{E}\{(XYf)(Zf)\}, \tag{12.2.6}$$

and so we have a characterization of the connection in terms of covariances. We shall see how to exploit this important relationship to obtain more explicit representations of ∇ when we turn to specific examples in a moment.

We now turn to the curvature tensor R of (7.5.2), given by

$$R(X, Y, Z, W) = g\left(\nabla_X \nabla_Y Z - \nabla_Y \nabla_X Z - \nabla_{[X,Y]} Z,\ W\right).$$

In order also to write R in terms of covariances, we recall (cf. (7.3.22) the covariant Hessian of a C^2 function f:

$$\nabla^2 f(X, Y) = XYf - \nabla_X Yf. \tag{12.2.7}$$

It follows from the fact that ∇ is torsion-free (cf. (7.3.10)) that $\nabla^2 f(X, Y) = \nabla^2 f(Y, X)$, and so ∇^2 is a symmetric form.

With this definition, we now prove the following useful result, which relates the curvature tensor R to covariances[6] and is crucial for later computations.

Lemma 12.2.1. *If f is a zero-mean, C^2 random field on a C^3 Riemannian manifold equipped with the metric induced by f, then the curvature tensor R on M is given by*

$$-2R = \mathbb{E}\left\{\left(\nabla^2 f\right)^2\right\}, \tag{12.2.8}$$

where the square of the Hessian is to be understood in terms of the dot product of tensors developed at (7.2.4).

Proof. Note that for C^1 vector fields it follows from the definition[7] (12.2.7) that

$$\begin{aligned}
&\left(\nabla^2 f\right)^2 ((X, Y), (Z, W)) \\
&\quad = 2\left[\nabla^2 f(X, Z)\nabla^2 f(Y, W) - \nabla^2 f(X, W)\nabla^2 f(Y, Z)\right] \\
&\quad = 2\left[(XZf - \nabla_X Zf)(YWf - \nabla_Y Wf)\right. \\
&\qquad \left. - (XWf - \nabla_X Wf)(YZf - \nabla_Y Zf)\right].
\end{aligned}$$

---

[6] Keep in mind, however, that while (12.2.8) looks as if it has only geometry on the left-hand side and covariances on the right, the truth is a little more complicated, since ∇^2 involves the connection, which depends on the metric, which depends on covariances.

[7] Alternatively, apply (7.2.5), treating ∇^2 as a $(1, 1)$ rather than $(2, 0)$ tensor.

Take expectations of this expression and exploit (12.2.6) to check (after a little algebra) that

$$\mathbb{E}\left\{\left(\nabla^2 f\right)^2((X,Y),(Z,W))\right\} = 2\left(\mathbb{E}[XZf \cdot YWf] - g(\nabla_X Z, \nabla_Y W)\right.$$
$$\left. - \mathbb{E}[XWf \cdot YZf] - g(\nabla_X W, \nabla_Y Z)\right).$$

Now apply (7.3.11) along with (12.2.6) to see that the last expression is equal to

$$\begin{aligned}
&2\big(X\mathbb{E}[Zf \cdot YWf] - \mathbb{E}[Zf \cdot XYWf] - g(\nabla_X Z, \nabla_Y W) \\
&\quad - Y\mathbb{E}\big[XWf \cdot Zf\big] + \mathbb{E}[Zf \cdot YXWf] + g(\nabla_X W, \nabla_Y Z)\big) \\
&\quad = 2\big(Xg(Z, \nabla_Y W) - g(\nabla_X Z, \nabla_Y W) - g(Z, \nabla_{[X,Y]} W) \\
&\qquad - Yg(\nabla_X W, Z) - g(\nabla_X W, \nabla_Y Z)\big) \\
&\quad = 2\big(g(Z, \nabla_X \nabla_Y W) - g(\nabla_Y \nabla_X W, Z) - g(Z, \nabla_{[X,Y]} W)\big) \\
&\quad = 2R\big((X,Y),(W,Z)\big) \\
&\quad = -2R\big((X,Y),(Z,W)\big),
\end{aligned}$$

the first equality following from the definition of the Lie bracket, the second from (7.3.11), the third from the definition of the curvature tensor R, and the last is trivial.

This establishes[8] (12.2.8), which is what we were after. □

12.2.2 Some Covariances

Many of the Euclidean computations of Section 11.7 were made possible as a result of convenient independence relationships between f and its first- and second-order derivatives. The independence of f and ∇f followed from the fact that f had constant variance, while that of ∇f and the matrix $\nabla^2 f$ followed from stationarity. Computations were further simplified by a global transformation (cf. (11.7.7)) that transformed f to being isotropic.

While we shall continue to assume that f has constant variance, we no longer can assume stationarity nor find easy transformations to isotropy. However, we have invested considerable effort in setting up the geometry of our parameter space with the metric induced by f, and now we are about to start profiting from this. We start with some general computations, which require no specific assumptions.

We start with the variance function

$$\sigma_t^2 = \mathbb{E}\left\{f_t^2\right\}.$$

[8] If you are a stickler for detail, you may have noticed that since our assumptions require only that f be C^2, it is not at all clear that the terms $XYWf$ and $YXWf$ appearing in the derivation make sense. However, their difference, $[X,Y]Wf$, is well defined, and that is all we have really used.

Assuming, as usual, that $f \in C^2(M)$, we also have that $\sigma^2 \in C^2(M)$, in which case there are no problems in changing the order of differentiation and expectation to see that, for C^1 vector fields X and Y,

$$X\sigma^2 = X\mathbb{E}\left\{f^2\right\} = 2\mathbb{E}\left\{f \cdot Xf\right\}. \tag{12.2.9}$$

Note that an important consequence of this is that if σ^2 is constant, so that $X\sigma^2 = 0$ for all X, then $f(t)$ and $Xf(t)$ are independent for all t.

Continuing in this vein, but for general σ^2, we have

$$XY\sigma^2 = 2\left(\mathbb{E}\left\{XYf \cdot f\right\} + \mathbb{E}\left\{Xf \cdot Yf\right\}\right)$$

and

$$XY\sigma^2 - \nabla_X Y\sigma^2 = 2\left(\mathbb{E}\left\{XYf \cdot f\right\} + \mathbb{E}\left\{Xf \cdot Yf\right\} - \mathbb{E}\left\{\nabla_X Yf \cdot f\right\}\right).$$

Rearranging the last line yields

$$\begin{aligned} \mathbb{E}\left\{\nabla^2 f(X,Y) \cdot f\right\} &= -\mathbb{E}\left\{Xf \cdot Yf\right\} + \frac{1}{2}\nabla^2\sigma^2(X,Y) \\ &= -g(X,Y) + \frac{1}{2}\nabla^2\sigma^2(X,Y). \end{aligned} \tag{12.2.10}$$

Now note that Xf and $\nabla^2 f(Y,Z)$ are uncorrelated (and so independent in our Gaussian scenario), since

$$\mathbb{E}\left\{Xf \cdot \nabla^2 f(Y,Z)\right\} = \mathbb{E}\left\{Xf \cdot (YZf - \nabla_Y Zf)\right\} = 0 \tag{12.2.11}$$

by (12.2.6). You should note that this result requires no assumptions whatsoever. It is an immediate consequence of the geometry that f induces on M via the induced metric and the fact that the covariant Hessian ∇^2 incorporates this metric in its definition.

Putting all the above together gives that

$$\mathbb{E}\left\{\nabla^2 f_t \,\middle|\, f_t = x, \nabla f_t = v\right\} = -\frac{x}{\sigma_t^2} I + \frac{x}{2\sigma_t^2}\nabla^2\sigma_t^2,$$

where I is the identity double form determined by g, defined in (7.2.10).

Assume now that f has constant variance, which we take to be 1. Then $X\sigma^2 \equiv 0$, and the last equality simplifies to give

$$\mathbb{E}\left\{\nabla^2 f_t \,\middle|\, \nabla f_t = v, f_t = x\right\} = -xI. \tag{12.2.12}$$

The conditional variance of $\nabla^2 f$ is also easy to compute, since combining (12.2.8) and (12.2.10) gives what is perhaps the most crucial equality for the detailed computations that we shall carry out in Section 12.4:

$$\mathbb{E}\left\{\left(\nabla^2 f - \mathbb{E}\left\{\left(\nabla^2 f\right) \middle| \nabla f = v, f = x\right\}\right)^2 \middle| \nabla f = v, f = x\right\} = -(2R + I^2). \tag{12.2.13}$$

The above correlations change somewhat if instead of concentrating on the covariant Hessian $\nabla^2 f$, we look at simple second derivatives. For example, it follows from (12.2.11) that

$$\mathbb{E}\{XYf \cdot Zf\} = \mathbb{E}\{\nabla_X Yf \cdot Zf\} = g(\nabla_X Y, Z). \tag{12.2.14}$$

Continuing to assume that f has unit variance, let E be an orthonormal frame field for the induced metric g. It then immediately follows from the above and the fact that $g(E_i, E_j) = \delta_{ij}$ that

$$\begin{aligned} \mathbb{E}&\left\{XYf_t \,\middle|\, E_k f_t = v_k, k = 1, \ldots, N, f_t = x\right\} \\ &= -xI + \sum_{k=1}^{N} v_k g(\nabla_X Y, E_k) \\ &= -xI + g(\nabla_X Y, v). \end{aligned} \tag{12.2.15}$$

Now might be a good time to take some time off to look at a few examples.

12.2.3 Gaussian Fields on $\mathbb{R}^N$

An extremely important example, which can be treated in detail without too much pain, is given by the differential structure induced on a compact domain M in $\mathbb{R}^N$ by a zero-mean, C^2 Gaussian field. For the moment we shall assume that M has a C^2 boundary, although at the end of the discussion we shall also treat the piecewise C^2 case.

We shall show how to explicitly compute both the curvature tensor R and the shape operator S in terms of the covariance function C, as well as traces of their powers. We shall also discuss the volume form Vol_g.

We shall not, in general, assume that f is either stationary or isotropic. In fact, one of the strengths of the manifold approach is that it handles the nonstationary case almost as easily as the stationary one.

The basis for our computations is Section 7.7, where we saw how to compute what we need after starting with a convenient basis. Not surprisingly, we start with $\{E_i\}_{1\le i\le N}$, the standard coordinate vector fields on $\mathbb{R}^N$. This also gives the natural basis in the global chart $(\mathbb{R}^N, i)$, where i is the inclusion map. We give $\mathbb{R}^N$ the metric g induced by f.

From Section 7.7 we know that, as far as the curvature operator is concerned, everything depends on two sets of functions, the covariances

$$g_{ij}(t) = g(E_{ti}, E_{tj}) = \left.\frac{\partial^2 C(r,s)}{\partial r_i \partial r_j}\right|_{(t,t)} \tag{12.2.16}$$

and the Christoffel symbols of the first kind,

$$\Gamma_{ijk} \stackrel{\Delta}{=} g(\nabla_{E_i} E_j, E_k) = \left.\frac{\partial^3 C(r,s)}{\partial r_i \partial r_j \partial s_k}\right|_{(t,t)}, \tag{12.2.17}$$

where the last equality is a trivial consequence of (12.2.6) and (5.5.5).

All other terms appearing in Section 7.7 are derivable from these two either via simple algebraic operations or by taking inverses and square roots of matrices. In other words, there is nothing that cannot be computed directly from the first three derivatives of the covariance function. Of course, for each example, considerable perseverence or, even better, computer algebra might come in handy to actually carry out the computations.

Nevertheless, there is one rather important case in which the expressions simplify considerably. If f is stationary, then the $g_{ij}(t)$ are actually independent of t. Consequently, it follows from (5.5.5) and the symmetry of the spectral measure that

$$\Gamma_{ijk} \equiv 0 \quad \text{for all } i, j, k, \tag{12.2.18}$$

and so the curvature tensor, its powers, and their traces are identically zero. As a consequence, most of the complicated formulas of Section 7.7 simply disappear. The isotropic situation is, of course, simpler still, since then

$$g_{ij}(t) \equiv \lambda_2 \delta_{ij}, \tag{12.2.19}$$

where λ_2 is the variance of any first-order directional derivative of f (cf. (5.7.3)).

The computation of the shape operator S also follows from the considerations of Section 7.7. For a specific example, assume that ∂M is a C^2 manifold of codimension 1 in M. Thus the normal space to ∂M in M consists of a one-dimensional vector field, which we take to be generated by an inward-pointing unit normal vector field ν on ∂M.

As we saw in Section 7.7, the natural choice of basis in this setting is an orthonormal frame field $E^* = \{E_i^*\}_{1 \le i \le N}$ chosen so that on ∂M, $E_N^* = \nu$. In this case, we only need to know how to compute the Γ_{jiN}^* of (7.7.13). While this is not always easy, if f is stationary and isotropic, then as for the curvature tensor things do simplify somewhat. In particular, if it is possible to explicitly determine functions a_{ij} such that

$$E_{it}^* = \sum_{k=1}^{N} a_{ik}(t) E_{kt},$$

then, as in footnote 33 of Chapter 6, we have that

$$\Gamma_{jiN}^*(t) = \lambda_2 \sum_{k,l=1}^{N} a_{jk}(t) \frac{\partial}{\partial t_k} \big(a_{Nl}(t) a_{il}(t) \big), \tag{12.2.20}$$

so that the summation has dropped one dimension. Far more significant, however, are the facts that the information about the Riemannian structure of (M, g) is now summarized in the single parameter λ_2 and that this information has been isolated from the "physical" structure of ∂M inherent in the functions a_{ik}.

In fact, this can also be seen directly from the definition of the shape operator. From (12.2.19) it is also easy to check that for any vectors X, Y,

$$g(X, Y) = \lambda_2 \langle X, Y \rangle,$$

where the right-hand side denotes the Euclidean inner product of X and Y. Consequently, writing S^g for the shape operator under the induced Gaussian metric and S^E for the standard Euclidean one, we have

$$S^g(X, Y) = \lambda_2 S^E(X, Y), \tag{12.2.21}$$

a result that will be useful for us later.

There is another scenario in which significant simplification occurs. Suppose that $A \subset \partial M$ is locally flat, in the sense that A is a subset of an $(N-1)$-dimensional hyperplane. In that case the $a_{jk}(t)$ are constant over A, and so it follows from (12.2.20) that $\Gamma^*_{jiN}(t) = 0$ for all $t \in A$.

The last issue that we need to look at for this class of examples is that of the form of the volume element Vol_g corresponding to the metric induced by the Gaussian field. Since our parameter space is a compact domain $M \subset \mathbb{R}^N$, we can take an atlas consisting of the single chart (M, i), where i is the identity mapping on $\mathbb{R}^N$. This being the case, it is immediate from (7.4.9) that for any $A \subset \mathbb{R}^N$,

$$\int_A \mathrm{Vol}_g = \int_A |\det \Lambda_t|^{1/2}\, dt, \tag{12.2.22}$$

where Λ_t is the matrix with entries

$$\lambda_t(i, j) = \mathbb{E}\left\{ \frac{\partial f}{\partial t_i} \frac{\partial f}{\partial t_j} \right\} = \left. \frac{\partial^2 C(r, s)}{\partial r_i \partial s_j} \right|_{(r,s)=(t,t)}.$$

If f is stationary, then Λ_t is independent of t and is simply the matrix of second-order spectral moments that we met at (11.7.1) and (11.7.2).

If f is also isotropic, then $\Lambda_t = \lambda_2 I$, where I is the identity matrix.

12.3 Another Gaussian Computation

At the core of the calculation of the expected Euler characteristic in the Euclidean case were the results of Lemma 11.6.2 and Corollary 11.6.3 about mean values of determinants of Gaussian matrices. In the manifold case we shall need a somewhat more general result.

To start, recall the discussion following (7.2.6). If we view an $N \times N$ matrix Δ as representing a linear mapping T_Δ from $\mathbb{R}^N$ to $\mathbb{R}^N$, with $\Delta_{ij} = \langle e_i, T_\Delta e_j \rangle$, then Δ can also be represented as a double form $\gamma_\Delta \in \Lambda^{1,1}(\mathbb{R}^N)$, and from the discussion in Section 7.1,

$$\det \Delta = \frac{1}{N!} \mathrm{Tr}\left((\gamma_\Delta)^N \right). \tag{12.3.1}$$

Thus it should not be surprising that we now turn our attention to the expectations of traces of random double forms, for which we need a little notation.

Let V be a vector space and $\mu \in \Lambda^{1,1}(V)$ a double form on V. Let Cov : $(V \otimes V) \times (V \otimes V) \to \mathbb{R}$ be bilinear, symmetric, and nonnegative definite. We think of μ as a mean function and Cov as a covariance function, and call W a random, Gaussian 2-form on $V \otimes V$ with mean function μ and covariance function Cov if for all finite collections of pairs $(v_{i_1}, v_{i_2}) \in V \otimes V$, the $W(v_{i_1}, v_{i_2})$ have a joint Gaussian distribution with means

$$\mathbb{E}\left\{W(v_{i_1}, v_{i_2})\right\} = \mu(v_{i_1}, v_{i_2})$$

and covariances

$$\begin{aligned}\mathbb{E}\left\{\left(W(v_{i_1}, v_{i_2}) - \mu(v_{i_1}, v_{i_2})\right) \cdot \left(W(v_{j_1}, v_{j_2}) - \mu(v_{j_1}, v_{j_2})\right)\right\} \\ = \mathrm{Cov}\left((v_{i_1}, v_{i_2}), (v_{j_1}, v_{j_2})\right).\end{aligned}$$

Lemma 12.3.1. *With the notation and conditions described above, and understanding all powers and products of double forms as being with respect to the double wedge product of* (7.2.4)*,*

$$\mathbb{E}\left\{W^k\right\} = \sum_{j=0}^{\lfloor \frac{k}{2} \rfloor} \frac{k!}{(k-2j)!\,j!\,2^j} \mu^{k-2j} C^j, \tag{12.3.2}$$

in the sense that for all $v_1, \dots, v_k, v'_1, \dots, v'_k \in V$,

$$\begin{aligned}&\mathbb{E}\left\{W^k\left((v_1, \dots, v_k), (v'_1, \dots, v'_k)\right)\right\} \\ &= \sum_{j=0}^{\lfloor \frac{k}{2} \rfloor} \frac{k!}{(k-2j)!\,j!\,2^j} \left(\mu^{k-2j} C^j\right)\left((v_1, \dots, v_k), (v'_1, \dots, v'_k)\right),\end{aligned}$$

where $C \in \Lambda^{2,2}(V)$ *is defined by*

$$\begin{aligned}&C\left((v_1, v_2), (v'_1, v'_2)\right) \\ &= \mathbb{E}\left\{\left(W - \mathbb{E}\{W\}\right)^2\left((v_1, v_2), (v'_1, v'_2)\right)\right\} \\ &= 2\left(\mathrm{Cov}\left((v_1, v'_1), (v_2, v'_2)\right) - \mathrm{Cov}\left((v_1, v'_2), (v_2, v'_1)\right)\right).\end{aligned}$$

Proof. Since it is easy to check that the standard binomial expansion works also for dot products, the general form of (12.3.2) will follow from the special case $\mu = 0$, once we show that for this case,

$$\mathbb{E}\left\{W^k\right\} = \begin{cases} 0, & k \text{ odd}, \\ \frac{(2j)!}{j!\,2^j} C^j, & k = 2j. \end{cases} \tag{12.3.3}$$

Thus, assume now that $\mu = 0$. The case of odd k in (12.3.3) follows immediately from (11.6.1), and so we have only the even case to consider.

Recalling the definition (7.2.4) of the dot product of double forms, we have that

$$W^{2j}\big((v_1,\dots,v_{2j}),(v'_1,\dots,v'_{2j})\big) = \sum_{\pi,\sigma\in S(2j)} \varepsilon_\pi\varepsilon_\sigma \prod_{k=1}^{2j} W(v_{\pi(k)}, v'_{\sigma(k)}),$$

where, as usual, $S(2j)$ is the symmetric group of permutations of $2j$ letters and ε_σ is the sign of the permutation σ. It then follows immediately from (11.6.2) that the expectation on the right-hand side is given by

$$\frac{(2j)!}{j!} \sum_{\pi,\sigma\in S(2j)} \varepsilon_\pi\varepsilon_\sigma \prod_{k=1}^{j} \mathbb{E}\left\{W(v_{\pi(2k-1)}, v'_{\sigma(2k-1)})W(v_{\pi(2k)}, v'_{\sigma(2k)})\right\},$$

where the combinatorial factor comes from the different ways of grouping the vectors $(v_{\pi(k)}, v'_{\sigma(k)})$, $1\le k\le 2j$, into ordered pairs.[9]

The last expression can be rewritten as

$$\begin{aligned}
&\frac{(2j)!}{j!2^j} \sum_{\pi,\sigma\in S(2j)} \varepsilon_\pi\varepsilon_\sigma \prod_{k=1}^{j} \mathbb{E}\Big\{W(v_{\pi(2k-1)}, v'_{\sigma(2k-1)})W(v_{\pi(2k)}, v'_{\sigma(2k)}) \\
&\qquad\qquad - W(v_{\pi(2k)}, v'_{\sigma(2k-1)})W(v_{\pi(2k-1)}, v'_{\sigma(2k)})\Big\} \\
&\quad = \frac{(2j)!}{j!2^j} \sum_{\pi,\sigma\in S(2j)} \varepsilon_\pi\varepsilon_\sigma \prod_{k=1}^{j} C\big((v_{\pi(2k-1)}, v_{\pi(2k)}),(v'_{\sigma(2k-1)}, v'_{\sigma(2k)})\big) \\
&\quad = \frac{(2j)!}{j!2^j} C^j\big((v_1,\dots,v_{2j}),(v'_1,\dots,v'_{2j})\big),
\end{aligned}$$

which completes the proof. □

The following corollary is immediate from Lemma 12.3.1 and the definition (7.2.6) of the trace operator.

Corollary 12.3.2. *With the notation and conditions of Lemma* 12.3.1,

$$\mathbb{E}\left\{\mathrm{Tr}(W^k)\right\} = \sum_{j=0}^{\lfloor \frac{k}{2}\rfloor} \frac{k!}{(k-2j)!j!2^j} \mathrm{Tr}\left(\mu^{k-2j}C^j\right). \tag{12.3.4}$$

---

9 In comparing this with (11.6.2), note that there we had an extra summation over the groupings into *unordered* pairs rather than a simple multiplicative factor. We already have each possible grouping due to the summation over π and σ in $S(2j)$, and since we are keeping pairs ordered we also lose the factor of 2^{-j} there.

12.4 The Mean Euler Characteristic

We now have everything we need to undertake the task of developing an explicit formula for $\mathbb{E}\{\varphi(A_u(f, M))\}$ for a smooth, Gaussian random field f over a manifold M. We shall treat the cases in which M does and does not have a boundary separately, even though the first scenario is a special case of the second. Nevertheless, it is best to see the calculations, for the first time, in the simpler scenario. When we get around to adding in all the structure of piecewise smooth manifolds, the notation will become somewhat heavier, although the main result, Theorem 12.4.2, will not look very different from the nonboundary case of Theorem 12.4.1.

12.4.1 Manifolds without Boundary

Throughout this section, we shall assume that M is a C^2 submanifold of a C^3 manifold. Here is the main result:

Theorem 12.4.1. *Let f be a centered, unit-variance Gaussian field on an N-dimensional, C^2 manifold M, and satisfying the conditions of Corollary* 11.3.5. *Then*

$$\mathbb{E}\{\varphi(A_u)\} = \sum_{j=0}^{N} \mathcal{L}_j(M)\rho_j(u), \tag{12.4.1}$$

where the ρ_j are given by

$$\rho_j(u) = (2\pi)^{-(j+1)/2} H_{j-1}(u) e^{-\frac{u^2}{2}}, \quad j \geq 0, \tag{12.4.2}$$

H_j is the jth Hermite polynomial, given by (11.6.9) *and* (11.6.10)*, and the $\mathcal{L}_j(M)$ are the Lipschitz–Killing curvatures* (7.6.1) *of M, calculated with respect to the metric* (12.2.2) *induced by f, i.e.,*

$$\mathcal{L}_j(M, U) = \begin{cases} \frac{(-2\pi)^{-(N-j)/2}}{(N-j)!} \int_U \mathrm{Tr}^M \left(R^{(N-j)/2}\right) \mathrm{Vol}_g & \text{if } N-j \text{ is even,} \\ 0 & \text{if } N-j \text{ is odd.} \end{cases} \tag{12.4.3}$$

Proof. The first consequence of the assumptions on f is that the sample functions of f are almost surely Morse functions over M. Thus Corollary 9.3.5 gives us that

$$\begin{aligned} \varphi(A_u) &= \sum_{\{t \in M : \nabla f_t = 0\}} (-1)^{\iota_{-f,M}(t)} \\ &= \sum_{k=0}^{N} (-1)^k \#\{t \in M : f_t \geq u, \nabla f_t = 0, \mathrm{index}(-\nabla^2 f_t) = k\}. \end{aligned}$$

To compute the expectation here first choose an orthonormal frame field $E = (E_1, \ldots, E_N)$ for M and apply Theorem 12.1.1 $N + 1$ times. If we write the f

there as f^*, then we set $f^* = \nabla f_E$ and h to be $(f, -(E_iE_j)f)$, where, to save on subscripts, we shall write (E_iE_jf) to denote the matrix $(E_iE_jf)_{i,j=1,\dots,N}$.

We read off the components of (E_iE_jf) (and later of $\nabla^2 f_E$) in lexicographic order to get a vector. To avoid confusion, we write everything out once more in full:

$$\nabla f_{E,k} = E_k f, \qquad (E_iE_jf)_{kl} = E_kE_lf, \qquad \nabla^2 f_{E,kl} = \nabla^2 f(E_k, E_l).$$

Finally, for the kth application of Theorem 12.1.1, we set

$$B = (u,\infty) \times A_k \stackrel{\Delta}{=} (u,\infty) \times \{H : \mathrm{index}(-H) = k\} \subset \mathbb{R} \times \mathrm{Sym}_{N\times N},$$

where $\mathrm{Sym}_{N\times N}$ is the space of symmetric $N \times N$ matrices. Then the second consequence of our assumptions on f is that Theorem 12.1.1 is applicable in this setting. Applying it gives

$$\begin{aligned}\mathbb{E}\{\varphi(A_u)\} = \sum_{k=0}^{N} \int_M (-1)^k \mathbb{E}\big\{|\det(E_iE_jf)| \\ \times \mathbb{1}_{A_k}(\nabla^2 f_E)\mathbb{1}_{(u,\infty)}(f) | \nabla f = 0\big\} p_{\nabla f}(0)\,\mathrm{Vol}_g\,.\end{aligned} \tag{12.4.4}$$

Recall now that $\nabla^2 f(E_i, E_j) = E_iE_jf - \nabla_{E_i}E_jf$. However, conditioning on $\nabla f_E = 0$ gives $\nabla^2 f(E_i, E_j) \equiv E_iE_jf$, so that we can replace (E_iE_jf) by $\nabla^2 f_E$ in the last equation above to obtain

$$\sum_{k=0}^{N} \int_M (-1)^k \mathbb{E}\{|\det \nabla^2 f_E| \mathbb{1}_{A_k}(\nabla^2 f_E)\mathbb{1}_{(u,\infty)}(f) | \nabla f = 0\} p_{\nabla f}(0)\,\mathrm{Vol}_g\,.$$

If we now interchange summation and integration and bracket the factor of $(-1)^k$ together with $|\det \nabla^2 f_E|$, then we can drop the absolute value sign on the latter, although there is a remaining factor of -1. This allows us to exchange expectation with summation, and the factor of $\mathbb{1}_{A_k}(\nabla^2 f_E)$ and the sum over k disappear completely.

Now recall that f has constant variance and note that since E is an orthonormal frame field (with respect to the induced metric g), the components of ∇f_E at any $t \in M$ are all independent standard Gaussians. Furthermore, as we saw in Section 12.2.2, the constant variance of f implies that they are also independent of $f(t)$. Consequently, the joint probability density of $(f, \nabla f_E)$ at the point $(x, 0)$ is simply

$$\frac{e^{-x^2/2}}{(2\pi)^{(N+1)/2}}.$$

Thus not only is it known, but it is constant over M.

Noting all this, conditioning on f and integrating out the conditioning allows us to rewrite the above in the much simpler format

$$\begin{aligned}\mathbb{E}\{\varphi(A_u)\} = (2\pi)^{-(N+1)/2} \int_u^\infty e^{-x^2/2}\,dx \\ \times \int_M \mathbb{E}\{\det(-\nabla^2 f_E) | \nabla f_E = 0, f = x\}\,\mathrm{Vol}_g\,.\end{aligned} \tag{12.4.5}$$

Recalling the definition of the trace (cf. (12.3.1)), the innermost integrand can be written as

$$\frac{1}{N!}\mathbb{E}\{\mathrm{Tr}((-\nabla^2 f)^N)|\nabla f_E = 0,\ f = x\}.$$

Since $\nabla^2 f$ is a Gaussian double form, we can use Corollary 12.3.2 to compute the above expectation, once we recall (12.2.12) and (12.2.13) to give us the conditional mean and covariance of $\nabla^2 f$. These give

$$\begin{aligned}
&\frac{(-1)^N}{N!}\mathbb{E}\{\mathrm{Tr}^M(\nabla^2 f)^N|\nabla f_E = 0,\ f = x\}\\
&\quad = \sum_{j=0}^{\lfloor \frac{N}{2}\rfloor} \frac{(-1)^j}{(N-2j)!j!2^j}\,\mathrm{Tr}^M((xI)^{N-2j}(I^2+2R)^j)\\
&\quad = \sum_{j=0}^{\lfloor \frac{N}{2}\rfloor}\sum_{l=0}^{j} \frac{(-1)^j}{j!2^j(N-2j)!}x^{N-2j}\,\mathrm{Tr}^M\left(I^{N-2l}\binom{j}{l}(2R)^l\right)\\
&\quad = \sum_{l=0}^{\lfloor \frac{N}{2}\rfloor} \frac{(-1)^l}{l!}\,\mathrm{Tr}^M\left(R^l\sum_{k=0}^{\lfloor \frac{N-2l}{2}\rfloor}\frac{(-1)^k}{2^k(N-2k-2l)!k!}x^{N-2k-2l}I^{N-2l}\right)\\
&\quad = \sum_{l=0}^{\lfloor \frac{N}{2}\rfloor} \frac{(-1)^l}{l!}\,\mathrm{Tr}^M(R^l)H_{N-2l}(x),
\end{aligned}$$

where in the last line, we have used (7.2.11) and the definition (11.6.9) of the Hermite polynomials.

Substituting back into (12.4.5) we conclude that $\mathbb{E}\{\varphi(A_u)\}$ is given by

$$\begin{aligned}
&\sum_{l=0}^{\lfloor \frac{N}{2}\rfloor}\left[\int_u^\infty (2\pi)^{-(N+1)/2}H_{N-2l}(x)e^{-x^2/2}\,dx\right]\frac{(-1)^l}{l!}\int_M \mathrm{Tr}^M(R^l)\,\mathrm{Vol}_g\\
&\quad = \sum_{l=0}^{\lfloor \frac{N}{2}\rfloor}(2\pi)^{-(N+1)/2}H_{N-2l-1}(x)e^{-u^2/2}\frac{(-1)^l}{l!}\int_M \mathrm{Tr}^M(R^l)\,\mathrm{Vol}_g\\
&\quad = \sum_{j=0}^{N}\mathcal{L}_j(M)\rho_j(u),
\end{aligned}$$

where the first equality follows from (11.6.12) and the second from the definitions (12.4.2) of the ρ_j and (12.4.3) for the $\mathcal{L}_j$, along with a little algebra.

That is, we have (12.4.1) and so the theorem. □

12.4.2 Manifolds with Boundary

We now turn to the case of manifolds with boundaries and shall incorporate, in one main result, both the results of the previous section in which the manifold had no boundary, and the results of Section 11.7. There, as you will recall, the parameter space was an N-dimensional rectangle, and the Gaussian process was required to be stationary.

Thus, we return to the setting of Sections 8.1 and 9.3, and take M to be an N-dimensional regular stratified manifold.

Theorem 12.4.2. *Let M be an N-dimensional regular stratified manifold, and f, as in Theorem 12.4.1, a centered, unit-variance Gaussian field on M satisfying the conditions of Corollary 11.3.5.*

Then $\mathbb{E}\{\varphi(A_u)\}$ is again given by (12.4.1), so that

$$\mathbb{E}\left\{\varphi\left(A_u\right)\right\} = \sum_{i=0}^{N} \mathcal{L}_i(M)\rho_i(u), \tag{12.4.6}$$

with the single change that the Lipschitz–Killing curvatures $\mathcal{L}_j$ are now defined by (10.7.1), i.e.,

$$\begin{aligned}\mathcal{L}_k(M) = \sum_{j=k}^{N}(2\pi)^{-(j-k)/2} \sum_{l=0}^{\lfloor \frac{j-k}{2} \rfloor} C(N-j, j-k-2l)\frac{(-1)^l}{l!(j-k-2l)!} \\ \times \int_{\partial_j M}\int_{S(T_t\partial_j M^{\perp})} \mathrm{Tr}^{T_t\partial_j M}\left(R^l S_{\nu_{N-j}}^{j-k-2l}\right) \\ \times \alpha(\nu_{N-j})\mathcal{H}_{N-j-1}(d\nu_{N-j})\mathcal{H}_j(dt).\end{aligned} \tag{12.4.7}$$

Remark 12.4.3. *If we also require that M be locally convex, then $\alpha(\nu) \equiv \mathbb{1}_{N_t M}(-\nu)$, and so α disappears from (12.4.7), although the integral over $S(T_t\partial_j M^{\perp})$ then becomes an integral over $S(N_t M)$.*

Before turning to the proof of Theorem 12.4.2, we need to establish a slightly extended version of the metatheorem Theorem 12.1.1 and its Gaussian corollary, Corollary 12.1.2. The extension is needed due to the fact that for general Whitney stratified manifolds, Morse theory "counts" critical points with multiplicities, given by their normal Morse index α. Since $\alpha \equiv 1$ for locally convex spaces, this issue does not arise for these spaces. Consequently, if these are the cases that you care about, you can skip directly to the proof of Theorem 12.4.2 without worrying about the following result. However, if you want to see a more complete proof of Theorem 12.4.2 in its full generality, then you will have to bear with a little notation and some technicalities.

We start by adopting all the notation of Theorem 12.1.1 and adding to it a new random field $F : M \to \mathbb{R}^L$ for some finite L. Next, we take a (deterministic) function

$$\alpha : M \times \mathbb{R}^L \to \mathbb{Z}$$

and define, for $u \in \mathbb{R}^N$,

$$N_u^{\alpha}(f, h, F; M, B) \equiv N_u^{\alpha}(M) \equiv N_u^{\alpha}$$

by

$$N_u^\alpha(f, h, F; M, B) \triangleq \sum_{t \in M : f(t)=u \text{ and } h(t) \in B} \alpha\,(F(t)),$$

assuming, of course, that the sum is over a countable number of points.

We can now state our next result.

Theorem 12.4.4. *Retaining the notation and all conditions of Theorem* 12.1.1 *and adopting the notation above, assume also the following:*

(i) *F is almost surely continuous over M.*
(ii) *For each $t \in M$, the marginal densities $p_t(x)$ of $F(t)$ (implicitly assumed to exist) are continuous in x.*
(iii) *The conditional densities $p_t(x|\nabla f_E(t), h(t), F(t))$ of $f(t)$ given $h(t)$, $(\nabla f_E)(t)$, and $F(t)$ (implicitly assumed to exist) are bounded above and continuous at $x = u$, uniformly in $t \in M$.*
(iv) *The conditional densities $p_t(z|f(t) = x)$ of $F(t)$ given $f(t) = x$ are continuous for all z and for x in a neighborhood of u, uniformly in $t \in M$.*
(v) *The modulus of continuity of F satisfies* (12.1.3).
(vi) *$|\alpha|$ is bounded and α is piecewise constant on $M \times \mathbb{R}^L$. Furthermore, for each $t \in M$, the set of discontinuities of $\alpha(t, \cdot)$ has dimension no greater than $L - 1$ in $\mathbb{R}^L$, while for each $x \in \mathbb{R}^L$ the set of discontinuities of $\alpha(\cdot, x)$ has dimension no greater than $N - 1$ in $\mathbb{R}^N$.*

Then

$$\mathbb{E}\{N_u^\alpha\} = \int_M \mathbb{E}\left\{|\det(\nabla f_E)|\, \mathbb{1}_B(h)\alpha(F)\middle| f = u\right\} p(u)\,\mathrm{Vol}_g\,. \tag{12.4.8}$$

Proof. The proof of the theorem is really no different from that of Theorem 12.1.1, but we are missing an analogue of its Euclidean precursor, Theorem 11.2.1, with the extra complication of the α term. However, if you had the patience to follow the details of the proof of Theorem 11.2.1, you will agree that it *really* is easy to see that under the conditions we have added, the proof that we had of that theorem requires only minor changes to cover the current scenario.

Further (unnecessary) details are left to you. □

The Gaussian corollary to Theorem 12.4.4 is the following.

Corollary 12.4.5. *Let (M, g) be a Riemannian manifold satisfying the conditions of Theorem* 12.4.4. *Let f, h, and F be centered Gaussian fields over M and α as in the theorem. Then if f, h, ∇f_E, and F are a.s. continuous over M, and if for each $t \in M$, the joint distributions of $(f(t), \nabla f_E(t), h(t), F(t))$ are nondegenerate, then* (12.4.8) *holds.*

Proof of Theorem 12.4.2. In essence, although Theorem 12.4.2 subsumes many others that have gone before, it will not be hard to prove. We have already done most of the hard work in proving some of the earlier cases, and so we now face a proof that is more concerned with keeping track of notation than needing any serious new computations. Nevertheless, there is something new here, and that is the way the Morse index and integration over normal cones are handled. These two points actually take up most of the proof.

We start by recalling the setup, which implies that M has the unique decomposition

$$M = \bigcup_{j=0}^{\dim M} \partial_j M,$$

as in (9.3.5). Morse's theorem, as given in Corollary 9.3.5 and slightly rewritten, states that

$$\varphi(A_u) = \sum_{j=0}^{\dim M} \sum_{k=-K}^{K} \sum_{n=0}^{j} (-1)^n k \mu_{jkn}, \tag{12.4.9}$$

where

$$\mu_{jkn} \stackrel{\Delta}{=} \#\Big\{ t \in \partial_j M : f(t) \geq u,\ \nabla f(t) \in N_t M,$$
$$\alpha(-P^{\perp}_{T_t \partial_j M} \nabla f(t)) = k,\ \iota_{-f, \partial_j M}(t) = n \Big\}.$$

If $t \in M^{\circ}$ then the normal cone is $\{0\}$, and so the expectation of the $j = N$ term in (12.4.9) has already been computed, since this is the computation of the previous proof. Nevertheless, we shall rederive it, en passant, in what follows. For this case, however, it will be important that you recall the conventions set out in Remark 10.5.2, which we shall use freely.

Fix a j, $0 \leq j \leq N$, and choose an orthonormal frame field E such that $E_{j+1}, \ldots, E_N$ are normal to $\partial_j M$, from which it follows that $E_1 f(t), \ldots, E_N f(t)$ are independent standard Gaussians for each t.

To compute the expectation of (12.4.9) we argue much as we did in the proof of Theorem 12.4.1, applying Theorem 12.4.4 rather than Theorem 12.1.1 to allow for the new factor of α (cf. the argument leading up to (12.4.5)).[10]

This gives us the following, in which ν is an $(N - j)$-dimensional vector, which, for convenience, we write as $(\nu_{j+1}, \ldots, \nu_N)$:

$$\mathbb{E}\left\{ \sum_{j=0}^{\dim M} \sum_{k=-K}^{K} \sum_{n=0}^{j} (-1)^n k \mu_{jkn} \right\}$$
$$= (2\pi)^{-(N+1)/2} \int_u^{\infty} e^{-x^2/2}\, dx \int_{\partial_j M} \mathcal{H}_j(dt)$$
$$\times \int_{T_t \partial_j M^{\perp}} e^{-|\nu|^2/2} \theta_t^j(\nu, x) \alpha(-\nu) \mathcal{H}_{N-j}(d\nu),$$

[10] The fact that α is piecewise constant, as required by Theorem 12.4.4, is a consequence of the fact that we are working with Whitney stratified manifolds. See [72] for details.

where $\theta_t^j(\nu, x)$ is given by

$$\mathbb{E}\left\{\det\left(-\left(\nabla^2 f_{|\partial_j M}(E_m, E_n)\right)_{1\leq n,m\leq j}\right)\right.$$
$$\left.\Big| f = x,\ E_m f = 0, 1 \leq m \leq j,\ E_m f = \nu_m, j+1 \leq m \leq N\right\}.$$

As usual, $\mathcal{H}_j$ denotes the volume (Hausdorff) measures that g induces on the $\partial_j M$ and $\mathcal{H}_{N-j}$ denotes the corresponding measure on $T_t \partial_j M^{\perp}$.

We need to compute a few basic expectations before we can continue. The conditional covariances of the Hessian remain the same as in the previous section, since we are still conditioning on the vector $(f, \nabla f_E)$. Specifically,

$$\mathrm{Var}\left(\nabla^2 f_{|\partial_k M}\Big| f = x, \nabla f_E = (0, \nu)\right) = -(2R + I^2),$$

where the zero vector here is of length j.

Conditional means, however, do change, and from (12.2.15) we have, for $X, Y \in C^2(T(\partial_j M))$,

$$\begin{aligned}
\mathbb{E}&\left\{XYf\Big| f = x, \nabla f_E = (0, \nu)\right\}\\
&= \mathbb{E}\left\{\nabla^2 f_{|\partial_j M}(X, Y)\Big| f = x, \nabla f_E = (0, \nu)\right\}\\
&= -xg(X, Y) + g\left(\nabla_X Y, (0, \nu)\right)\\
&= -xg(X, Y) + S_\nu(X, Y),
\end{aligned}$$

where S is the scalar second fundamental form, given by (7.5.8) and (7.5.12). Equivalently,

$$\mathbb{E}\left\{\nabla^2 f_{|\partial_j M}\Big| f = x, \nabla f_E = (0, \nu)\right\} = -xI + S_\nu = -\left(xI + S_{-\nu}\right).$$

We now have all we need to evaluate $\theta_t^j(\nu, x)$, which, following the argument of the preceding section and using the above conditional expectations and variance, is equal to

$$\begin{aligned}
\frac{(-1)^j}{j!}&\mathbb{E}\left\{\mathrm{Tr}^{\partial_j M}\left(\nabla^2 f_{|\partial_j M}\right)^j \Big| f = x, \nabla f_E = (0, \nu)\right\}\\
&= \sum_{k=0}^{\lfloor \frac{j}{2} \rfloor} \frac{(-1)^k}{(j-2k)!k!2^k}\,\mathrm{Tr}^{\partial_j M}\left((xI + S_{-\nu})^{j-2k}\left(I^2 + 2R\right)^k\right)
\end{aligned}$$

by Lemma 12.3.1. Rearranging somewhat, this is equal to

$$\sum_{k=0}^{\lfloor \frac{j}{2} \rfloor} \sum_{l=0}^{k} \frac{(-1)^k}{(j-2k)!} \frac{1}{2^{k-l} l!(k-l)!} \operatorname{Tr}^{\partial_j M} \left((xI + S_{-\nu})^{j-2k} I^{2k-2l} R^l \right)$$

$$= \sum_{k=0}^{\lfloor \frac{j}{2} \rfloor} \sum_{m=0}^{j-2k} \sum_{l=0}^{k} \frac{(-1)^k (j-2l-m)!}{(j-2k-m)!m!} \frac{1}{2^{k-l} l!(j-l)!} \times x^{j-2k-m} \operatorname{Tr}^{\partial_j M} \left(S_{-\nu}^m R^l \right)$$

by (7.2.11). Further rearrangement gives that this is the same as

$$\sum_{m=0}^{j} \sum_{k=0}^{\lfloor \frac{j-m}{2} \rfloor} \sum_{l=0}^{k} \frac{(-1)^k (j-2l-m)!}{(j-2k-m)!m!} \frac{1}{2^{k-l} l!(k-l)!} \times x^{j-2k-m} \operatorname{Tr}^{\partial_j M} \left(S_{-\nu}^m R^l \right)$$

$$= \sum_{m=0}^{j} \sum_{l=0}^{\lfloor \frac{j-m}{2} \rfloor} \sum_{k=l}^{\lfloor \frac{j-m}{2} \rfloor} \frac{(-1)^k (j-2l-m)!}{(j-2k-m)!m!} \frac{1}{2^{k-l} l!(k-l)!} \times x^{j-2k-m} \operatorname{Tr}^{\partial_j M} \left(S_{-\nu}^m R^l \right)$$

$$= \sum_{m=0}^{j} \sum_{l=0}^{\lfloor \frac{j-m}{2} \rfloor} \frac{(-1)^l \operatorname{Tr}^{\partial_j M} \left(S_{-\nu}^m R^l \right)}{m!l!} \times \sum_{i=0}^{\lfloor \frac{j-2l-m}{2} \rfloor} \frac{(-1)^i (j-2l-m)!}{(j-2l-m-2i)!} \frac{1}{2^i i!} x^{j-2l-m-2i}$$

$$= \sum_{m=0}^{j} \sum_{l=0}^{\lfloor \frac{j-m}{2} \rfloor} \frac{(-1)^l}{l!m!} \operatorname{Tr}^{\partial_j M} (S_{-\nu}^m R^l) H_{j-2l-m}(x)$$

$$= \sum_{m=0}^{j} \sum_{l=0}^{\lfloor \frac{m}{2} \rfloor} \frac{(-1)^l}{l!(m-2l)!} \operatorname{Tr}^{\partial_j M} (S_{-\nu}^{m-2l} R^l) H_{j-m}(x)$$

$$= \sum_{k=0}^{j} H_k(x) \sum_{l=0}^{\lfloor \frac{j-k}{2} \rfloor} \frac{(-1)^l}{l!(j-k-2l)!} \operatorname{Tr}^{\partial_j M} (S_{-\nu}^{j-k-2l} R^l).$$

We now fix a $t \in \partial_j M$ and concentrate on computing the expectation

$$(2\pi)^{-N/2} \int_{T_t \partial_j M^\perp} e^{-|\nu|^2/2} \operatorname{Tr}^{\partial_j M} (S_{-\nu}^{j-k-2l} R^l) \alpha(-\nu) \mathcal{H}_{N-j}(d\nu)$$

$$= (2\pi)^{-N/2} \int_{T_t \partial_j M^\perp} e^{-|\nu|^2/2} \operatorname{Tr}^{\partial_j M} (S_{\nu}^{j-k-2l} R^l) \alpha(\nu) \mathcal{H}_{N-j}(d\nu).$$

We essentially computed this integral in proving Lemmas 10.5.1 and 10.5.3. Specifically, by a simple generalization of Lemma 10.5.5, using the fact that α is constant along rays, i.e., that for $c > 0$,

$$\alpha(c\nu) = \alpha(\nu),$$

the above integral can be evaluated as

$$\begin{aligned}(2\pi)^{-N/2} &\int_{T_t\partial_j M^\perp} e^{-|\nu|^2/2}\, \mathrm{Tr}^{\partial_j M}(S_\nu^{j-k-2l} R^l)\alpha(\nu)\mathcal{H}_{N-j}(d\nu) \\ &= C(N-j, j-k-2l)\int_{S(T_t\partial_j M^\perp)} \mathrm{Tr}^{\partial_j M}(S_\nu^{j-k-2l} R^l)\alpha(\nu)\mathcal{H}_{N-j-1}(d\nu).\end{aligned}$$

Collecting all the above, we finally have that

$$\begin{aligned}\mathbb{E}\{\varphi(A_u)\} &= \sum_{j=0}^{\dim M} (2\pi)^{-(N+1)/2} \int_u^\infty e^{-x^2/2} \sum_{k=0}^{j} H_k(x) \\ &\quad\times \sum_{l=0}^{\lfloor \frac{j-k}{2} \rfloor} \frac{(-1)^l}{l!(j-k-2l)!} C(N-j, j-k-2l) \\ &\quad\times \int_{S(T_t\partial_j M^\perp)} \mathrm{Tr}^{\partial_j M}(S_\nu^{j-k-2l} R^l)\alpha(\nu)\mathcal{H}_{N-j-1}(d\nu) \\ &= \sum_{k=0}^{\dim M} \rho_k(u) \times \left(\sum_{j=k}^{N} \sum_{l=0}^{\lfloor \frac{j-k}{2} \rfloor} \frac{(-1)^l}{l!(j-k-2l)!} C(N-j, j-k-2l) \right. \\ &\quad\left. \times \int_{S(T_t\partial_j M^\perp)} \mathrm{Tr}^{\partial_j M}(S_\nu^{j-k-2l} R^l)\alpha(\nu)\mathcal{H}_{N-j-1}(d\nu) \right)\end{aligned}$$

after integrating out x (via (11.6.12)) and changing the order of summation.

Comparing the last expression with the definitions (12.4.2) of the ρ_k and (12.4.7) of the $\mathcal{L}_k$, the proof is complete. □

12.5 Examples

With the hard work behind us, we can now look at some applications of Theorems 12.4.1, and 12.4.2. One of the most powerful implications of the formula

$$\mathbb{E}\{\varphi(A_u)\} = \sum_{j=0}^{\dim M} \mathcal{L}_j(M)\rho_j(u), \tag{12.5.1}$$

is that for any example, all that needs to be computed are the Lipschitz–Killing curvatures $\mathcal{L}_j(M)$, since the ρ_j are well defined by (12.4.2) and dependent neither on the geometry of M nor the covariance structure of f.

Nevertheless, this is not always easy, and there is no guarantee that explicit forms for the $\mathcal{L}_j(M)$ exist. In fact, more often than not, this will unfortunately be the case, and one needs to turn to a computer for assistance, performing either (or often both) symbolic or numeric evaluations.

However, there are some cases that are not too hard, and so we shall look at these.

Stationary Fields over Rectangles

This is the example that we treated in Theorem 11.7.2 via the techniques of integral rather than differential geometry.

Nevertheless, since N-dimensional rectangles are definitely piecewise C^2 manifolds, we should be able to recover Theorem 11.7.2 from Theorem 12.4.2. Doing so is in fact quite easy, so we set $M = \prod_{i=1}^N [0, T_i]$ and, to make life even easier, assume that f has unit variance and is isotropic with the variance of its derivatives given by λ_2. Thus, what we are trying to recover is (11.7.12) with $\Lambda_J = \lambda_2 I$, where I is the identity matrix.

The first point to note is that induced Riemannian g is given by

$$g(X, Y) = \mathbb{E}\{XfYf\} = \lambda_2 \langle X, Y \rangle,$$

where the last term is the simple Euclidean inner product. Thus g changes the usual flat Euclidean geometry of $\mathbb{R}^N$ only by scaling, and so the geometry remains flat.

This being the case, (6.3.6) gives us the necessary Lipschitz–Killing curvatures, although each $\mathcal{L}_j$ of (6.3.6) needs to be multiplied by a factor of $\lambda_2^{j/2}$. Substituting this into (12.5.1) gives the required result, (11.7.12).

The few lines of algebra needed along the way are left to you.

Also left to you is the nonisotropic case, which is not much harder, although you will need a slightly adapted version of (6.3.6) that allows for a metric that is a constant times the Euclidean metric on hyperplanes in $\mathbb{R}^N$, but for which the constant depends on the hyperplane. This will give Theorem 11.7.2 in its full generality.

Isotropic Fields over Smooth Domains

In the previous example, we assumed isotropy only for convenience, and leaving the argument for the general stationary case to you was simply to save us having to do more complicated algebra.

However, once one leaves the setting of rectangles it is almost impossible to obtain simple closed-form expressions for the $\mathcal{L}_j$ if f is not isotropic. To see how isotropy helps, we now take f isotropic over a compact, piecewise C^2 domain in $\mathbb{R}^N$, and also assume that $\mathbb{E}\{f^2\} = \mathbb{E}\{(\partial f/\partial t_k)^2\} = 1$, so as to save carrying through an awkward constant.

What makes the Lipschitz–Killing curvatures simpler in this case is not that their defining formula (12.4.7) changes in any appreciable way, but that the symbols appearing in it have much simpler meanings. In particular, $\mathcal{H}_k$ is no more than standard Hausdorff measure over $\partial_k M$, while $\mathcal{H}_{N-k-1}$ becomes surface measure on the unit sphere S^{N-k-1}. In other words, the $\mathcal{L}_k$ no longer carry any information related to f.

The Riemannian curvature R is now zero, and the second fundamental form S simplifies considerably. To see how this works in an example, assume that M is a C^2

domain, so that there is only one C^2 component to its boundary, of dimension $N-1$. Then $\mathcal{L}_N(M) = \mathcal{H}_N(M)$, and from (10.7.6) we have that for $0 \le j \le N-1$,

$$\mathcal{L}_j(M) = \frac{1}{s_{N-j}} \int_{\partial M} \operatorname{detr}_{N-1-j}(\mathrm{Curv})\mathcal{H}_{N-1}, \tag{12.5.2}$$

where detr_j is given by (7.2.8) and the curvature matrix Curv is given by (10.7.7)

A simple example was given in Figure 6.2.3, for which we discussed finding, via integral-geometric techniques, the Euler characteristic of $A_u(f, M)$, where M was a C^2 domain in $\mathbb{R}^2$. Although we found a point process representation for the $\varphi(A_u)$, we never actually managed to use integral-geometric techniques to find its expectation. The reason is that differential-geometric techniques work much better, since applying (12.5.2) with $N = 2$ immediately gives us the very simple result that

$$\mathcal{L}_2(M) = \mathrm{Area}(M), \qquad \mathcal{L}_1(M) = \frac{\mathrm{length}(\partial M)}{2}, \qquad \mathcal{L}_0(M) = 1,$$

which, when substituted into (12.5.1), gives the required expectation.

A historical note is appropriate here: It was the example of isotropic fields on C^2 domains that, in Keith Worsley's paper [173], was really the genesis of the manifold approach to Gaussian geometry that has been the central point of this chapter and the reason for writing this book.

Under isotropy, you should now be able to handle other examples yourself. All reduce to calculations of Lipschitz–Killing curvatures under a constant multiple of the standard Euclidean metric, and for many simple cases, such as balls and spheres, these have already been computed in Chapter 6. What is somewhat harder is the nonisotropic case.

Stationary and Nonstationary Fields on $\mathbb{R}^N$

It would be nice to be able to find nice formulas for nonstationary Gaussian fields over smooth domains, and even for stationary but nonisotropic fields, as we have just done for isotropic fields over smooth domains and stationary processes over rectangles. Unfortunately, although Theorems 12.4.1 and 12.4.2 allow us to do this in principle, it is not so simple to do in practice.

The basic reason for this can already be seen in the stationary scenario of Theorem 11.7.2, from which one can see that the Lipschitz–Killing curvatures for a stationary process over a rectangle $T = (\prod_{i=1}^N [0, T_i])$ are given by

$$\mathcal{L}_j(T) = \sum_{J \in \mathcal{O}_j} |J||\Lambda_J|^{1/2}, \tag{12.5.3}$$

the sum being over faces of dimension j in T containing the origin and the rest of the notation as in Theorem 11.7.2. What (12.5.3) shows is that there is no simple averaging over the boundary of T as there is in the isotropic case. Each piece of the boundary has its own contribution to make, with its own curvature and second fundamental form. In the case of a rectangle this is not too difficult to work with. In the case of a general domain it is not so simple.

In Section 7.7 we saw how to compute curvatures and second fundamental forms on Euclidean surfaces in terms of Christoffel symbols. In Section 12.2.3 we saw how to compute Christoffel symbols for the metric induced by f in terms of its covariance function. For any given example, these two computations need to be coordinated and then fed into definitions such as (12.4.3) and (12.4.7) of Lipschitz–Killing curvatures. From here to a final answer is a long path, often leading through computer algebra packages.

There is, however, one negative result that is worth mentioning here, since it is sufficiently counterintuitive that it has led many to an incorrect conjecture. As we mentioned following the proof of Lemma 11.7.1, a first guess at extending (12.5.3) to the nonstationary case would be to replace the terms $|J||\Lambda_J|^{1/2}$ there by integrals of the form $\int_J |\Lambda_t|^{1/2}\, dt$, where the elements of Λ_t are the covariances $\mathbb{E}\{f_i(t) f_j(t)\}$. Without quoting references, it is a fact that this has been done more than once in the past. However, it is clear from the computations of the Christoffel symbols in Section 12.2.3 that this does not work, and additional terms involving third-order derivatives of the covariance function enter into the computation. The fact that these terms are all identically zero in the stationary case is probably what led to the errors.[11]

Stationary Fields over Lie Groups

Lie groups provide an example that, while perhaps a little abstract, still yields a simple form for the expected Euler characteristic of excursion sets. We first met Lie groups back in Section 5.3, where we discussed them in the framework of stationarity.

Recall that a Lie group G is a group that is also a C^∞ manifold such that the map taking g to g^{-1} is C^∞ and the map taking (g_1, g_2) to $g_1 g_2$ is also C^∞. We need a little more notation than we did in Section 5.3. We denote the identity element of G by e, the left and right multiplication maps by L_g and R_g, and the inner automorphism of G induced by g by $I_g = L_g \circ R_g^{-1}$.

Recall that a vector field X on G is said to be *left invariant* if for all $g, g' \in G$, $(L_g)_* X_{g'} = X_{gg'}$. Similarly, a covariant tensor field ω is said to be left invariant (right invariant) if for every g_0, g in G, $L^*_{g_0}\omega_{g_0 g} = \omega_g$ ($R^*_{g_0}\omega_{g g_0} = \Phi_g$). As usual, ω is said to be *bi-invariant* if it is both left and right invariant. If h is a (left-, right-, bi-)invariant Riemannian metric on G, then it is clear that for every g, the map (L_g, R_g, I_g) is an isometry[12] of (G, h). In particular, the curvature tensor R of h, is (left, right, bi-)invariant. This means that for Gaussian random fields that

[11] On the other hand, if one thinks of the expression for $\mathbb{E}\{\varphi(A_u)\}$ as $e^{-u^2/2\sigma^2}$ multiplied by a power series in the level u, then it *is* correct to say that this conjecture gives the correct coefficient for the leading term of the power series. This, of course, follows from the fact that the coefficient of the leading term is $\mathcal{L}_N(T)$, which is the "volume" of T measured under the volume form derived from the induced metric. The corresponding integrals involve no curvatures.

[12] An *isometry* between two C^k Riemannian manifolds (M, g) and $(\hat{M}, \hat{g})$ is a C^{k+1} diffeomorphism $F : M \to \hat{M}$ for which $F^*\hat{g} = g$.

induce such Riemannian metrics, the integrals needed to evaluate $\mathbb{E}\{\varphi(A_u(f, M))\}$ are significantly easier to calculate.

Theorem 12.5.1. *Let G be a compact N-dimensional Lie group and f a centered, unit-variance Gaussian field on G satisfying the conditions of Corollary 11.3.5 for G. Suppose that the Riemannian metric g induced by f is (left, right, bi-)invariant. Then*

$$\mathbb{E}\left\{\varphi(A_u(f, M))\right\} = \mathrm{Vol}_g(G) \sum_{k=0}^{\lfloor \frac{N}{2} \rfloor} \frac{(-1)^k \rho_{N-2k}(u)}{(2\pi)^k k!} \mathrm{Tr}^{T_eG}(R_e^k), \tag{12.5.4}$$

where T_eG is the tangent space to G at e.

Proof. This is really a corollary of Theorem 12.4.1, since G has no boundary. Applying that result, and comparing (12.5.4) with (12.4.1), it is clear that all we need to show is that

$$\int_G \mathrm{Tr}^G\left(R^l\right)_{g'} \mathrm{Vol}_g(dg') = \mathrm{Tr}^{T_eG}\left(R_e^l\right) \mathrm{Vol}_g(G).$$

Suppose X, Y, Z, W are left-invariant vector fields. Since $g' \mapsto L_g g'$ is an isometry for every g, we have

$$\begin{aligned} R_g & \left((X_g, Y_g), (Z_g, W_g)\right) \\ &= R_e\left((L_{g^{-1}*} X_g, L_{g^{-1}*} Y_g), (L_{g^{-1}*} Z_g, L_{g^{-1}*} W_g)\right) \\ &= R_e\left((L_{g*})^{-1} X_g, (L_{g*})^{-1} Y_g, (L_{g*})^{-1} Z_g, (L_{g*})^{-1} W_g\right) \\ &= R_e\left(X_e, Y_e, Z_e, W_e\right). \end{aligned}$$

Therefore, if $(X_i)_{1 \le i \le n}$ is an orthonormal set of left-invariant vector fields, then

$$\begin{aligned} (R_g)^l & \left((X_{i_1 g}, \ldots, X_{i_l g}), (X_{j_1 g}, \ldots, X_{j_l g})\right) \\ &= (R_e)^l \left((X_{i_1 e}, \ldots, X_{i_l e}), (X_{j_1 e}, \ldots, X_{j_l e})\right), \end{aligned}$$

from which it follows that

$$\mathrm{Tr}^{T_gG}\left((R_g)^l\right) = \mathrm{Tr}^{T_eG}\left((R_e)^l\right),$$

which completes the proof. □

12.6 Chern–Gauss–Bonnet Theorem

As promised at the beginning of the chapter, we now give a purely probabilistic proof of the classical Chern–Gauss–Bonnet theorem. Of course, "purely" is somewhat of an overstatement, since the results on which our proof is based were themselves based on Morse's critical point theory.

The Chern–Gauss–Bonnet theorem is one of the most fundamental and important results of differential geometry, and it gives a representation of the Euler characteristic of a deterministic manifold in terms of curvature integrals. It has a long and impressive history, starting in the early nineteenth century with simple Euclidean domains. Names were added to the result as the setting became more and more general. While it has nothing to do with probability, it can be obtained as a simple corollary to Theorems 12.4.1 and 12.4.2. Here is a version very close to that originally proven in 1943 by Allendoerfer and Weil [11], for what they called "Riemannian polyhedra."

Theorem 12.6.1 (Chern–Gauss–Bonnet theorem). *Let (M, g) be a C^2 compact, orientable, N-dimensional Riemannian manifold, or a C^2 regular stratified manifold, in either case isometrically embedded in some C^3 Riemannian manifold $(\widetilde{M}, \widetilde{g})$ of dimension N. Then, if M has no boundary,*

$$\varphi(M) \equiv \mathcal{L}_0(M) = \frac{(-1)^{N/2}}{(2\pi)^{N/2} N!} \int_M \mathrm{Tr}^M (R^{N/2}) \, \mathrm{Vol}_g \tag{12.6.1}$$

if N is even, and 0 if N is odd. In the piecewise smooth case,

$$\begin{aligned} \varphi(M) = & \sum_{j=0}^{\dim M} (2\pi)^{-j/2} \sum_{m=0}^{\lfloor \frac{j}{2} \rfloor} C(N-j, j-2m) \frac{(-1)^m}{m!(j-2m)!} \\ & \times \int_{\partial_j M} \int_{S(T_t \partial_j M^{\perp})} \mathrm{Tr}^{T_t \partial_j M} (R^m S_{\nu_{N-j}}^{j-2m}) \\ & \times \alpha(\nu_{N-j}) \mathcal{H}_{N-j-1}(d\nu_{N-j}) \mathcal{H}_j(dt), \end{aligned}$$

where we adopt the notation of (10.7.1) *and the convention of Remark* 10.5.2.

Proof. Suppose f is a Gaussian random field on M such that f induces[13] the metric g. Suppose, furthermore, that f satisfies all the side conditions of either Theorem 12.4.1 or Theorem 12.4.2, depending on whether M does, or does not have, a boundary.

To save on notation, assume now that M does not have a boundary. Recall that in computing $\mathbb{E}\{\varphi(A_u(f, M))\}$ we first wrote $\varphi(A_u(f, M)))$ as an alternating sum of N different terms, each one of the form

$$\mu_k(u) \stackrel{\Delta}{=} \#\{t \in M : f_t > u, \ \nabla f_t = 0, \ \mathrm{index}(-\nabla^2 f_t) = k\}.$$

[13] Note that this is the opposite situation to that which we have faced until now. We have always started with the field f and used it to define the metric g. Now, however, g is given and we are assuming that we can find an appropriate f.

Clearly, as $u \to -\infty$, these numbers increase to

$$\mu_k \triangleq \#\{t \in M : \nabla f_t = 0, \ \text{index}(-\nabla^2 f_t) = k\},$$

and $\varphi(M)$ is given by an alternating sum of the μ_k. Since $\mu_k(u)$ is bounded by the total number of critical points of f, which is an integrable random variable, dominated convergence gives us that

$$\varphi(M) = \lim_{u \to -\infty} \mathbb{E}\{\varphi(A_u(f, M))\},$$

and the statement of the theorem then follows by first using Theorem 12.4.1 to evaluate $\mathbb{E}\{\varphi(A_u(f, M))\}$ and then checking that the right-hand side of (12.6.1) is, in fact, the above limit.

If we do not know that g is induced by an appropriate Gaussian field, then we need to adopt a nonintrinsic approach via Nash's embedding theorem in order to construct one.

Nash's embedding theorem [118] states that for any C^3, N-dimensional Riemannian manifold (M, g), there is an isometry $i_g : M \to i_g(M) \subset \mathbb{R}^{N'}$ for some finite N' depending only on N. More importantly, $\mathbb{R}^{N'}$ is to be taken with the usual Euclidean metric.

The importance of this embedding is that it is trivial to find an appropriate f when the space is Euclidean with the standard metric. Any unit zero-mean unit-variance isotropic Gaussian random field, f' say, on $\mathbb{R}^{N'}$ whose first partial derivatives have unit variance and that satisfies the nondegeneracy conditions of Theorem 12.4.1 will do. If we now define

$$f = f'_{|i_g(M)} \circ i_g^{-1}$$

on M, then it is easy to see that f induces the metric g on M, and so our construction is complete.

Finally, we note that the case of piecewise smooth M follows exactly the same argument, simply appealing to Theorem 12.4.2 rather than Theorem 12.4.1 and to a version of the embedding theorem for stratified spaces (cf. [71, 119, 120]). □

Another way to prove the above Chern–Gauss–Bonnet theorem would be to repeat the arguments of Theorem 12.4.2, simply removing the condition $\{f_t \geq u\}$ when applying Corollary 12.4.5. This leads to the following corollary, which can be thought of as a random field version of the Chern–Gauss–Bonnet theorem. This corollary will be useful for us in Chapters 13 and 15 as a computational trick to compute the Euler characteristic.

Corollary 12.6.2. *Let (M, g) be as in Theorem 12.6.1, and f a centered, unit-variance Gaussian field on M satisfying the conditions of Corollary 11.3.5 and inducing the metric g on M. Then*

$$\varphi(M) = \sum_{j=0}^{\dim M} \frac{(2\pi)^{-j/2}}{j!} \times \int_{\partial_j M} \mathbb{E}\left\{\alpha(P^{\perp}_{T_t\partial_j M}\nabla\widetilde{y}; M) \times \mathrm{Tr}^{T_t\partial_j M}\left((-\nabla^2\widetilde{y}_{|\partial_j M})^j\right)\right\} \mathcal{H}_j(dt).$$

More generally,

$$\mathcal{L}_0(M, A) = \sum_{j=0}^{\dim M} \frac{(2\pi)^{-j/2}}{j!} \times \int_{\partial_j M\cap A} \mathbb{E}\left\{\alpha(P^{\perp}_{T_t\partial_j M}\nabla\widetilde{y}; M) \times \mathrm{Tr}^{T_t\partial_j M}\left((-\nabla^2\widetilde{y}_{|\partial_j M})^j\right)\right\}\mathcal{H}_j(dt).$$

13

Mean Intrinsic Volumes

In the preceding two chapters we devoted a considerable amount of energy to computing the mean Euler characteristics of the excursion sets of smooth Gaussian fields. However, we know from both Chapters 6 and 7 that the Euler characteristic is but one of the family of geometric quantifiers known as the Lipschitz–Killing curvatures.

Even when we were dealing with the Euler characteristic alone, the Lipschitz–Killing curvatures of the parameter space arose in the expression for the mean Euler characteristics of the excursion sets. In this chapter we want to compute the mean of the Lipschitz–Killing curvatures of the excursion sets themselves. The main result will be given in Theorem 13.4.1, which states that, under appropriate conditions,

$$\mathbb{E}\left\{\mathcal{L}_j(A_u(f,M))\right\} = \sum_{l=0}^{\dim M-j} \begin{bmatrix} j+l \\ l \end{bmatrix} \rho_l(u)\mathcal{L}_{j+l}(M), \tag{13.0.1}$$

where the $\mathcal{L}_j$ on both sides of this equation are computed with respect to the metric induced on the parameter space M by the zero-mean, unit-variance Gaussian field f (cf. (12.2.1)) and the $A_u(f,M)$ are, as always, the excursion sets of f above the level u. The functions ρ_l remain as in the preceding two chapters (cf. (12.4.2)).

In all, we shall give three proofs of (13.0.1). The first two, which make up the content of the current chapter, exploit the results of Chapters 11 and 12 and use some nonstochastic results of geometry to move from Euler characteristics to general Lipschitz–Killing curvatures.

The first proof, in Section 13.2, will be for isotropic fields with unit variance defined over subsets of $\mathbb{R}^N$. In this case the $\mathcal{L}_k$ on both sides of (13.0.1) are the basic intrinsic volumes (volume, surface area, etc.) that we met first in Chapter 6 and the proof is quite straightforward. Indeed, all we shall need for the proof are two basic results of integral geometry—the formulas of Crofton and Hadwiger—which we shall give in the following section.

The second proof of (13.0.1) will be for the case in which f is more general and M a Whitney stratified space. For this we shall need a non-Euclidian analogue of Crofton's formula, which we give in Section 13.3. This is a new result and should be of independent interest outside the particular application that we have for it.

The third proof will come in Chapter 15 when, via a completely new approach, we shall re-prove the main results of this and the preceding two chapters.

13.1 Crofton's Formula

Before formulating Crofton's formula we need to review the idea of the *affine Grassmannian* manifold $\text{Graff}(N,k)$, the set of k-dimensional subspaces of $\mathbb{R}^N$. The affine Grassmannian $\text{Graff}(N,k)$ is the set of all k-dimensional flats in $\mathbb{R}^N$, i.e, all linear spaces in $\mathbb{R}^N$ that do not necessarily pass through the origin. The affine Grassmannian is diffeomorphic to $\text{Gr}(N,k)\times\mathbb{R}^N$. To see this, for $V\in\text{Graff}(N,k)$ let $p(V)$ be the unique point in V closest to the origin and take the mapping

$$\text{Graff}(N,k)\ni V\leftrightarrow (V-p(V),p(V))\in\text{Gr}(N,k)\times\mathbb{R}^N.$$

The space $\text{Graff}(N,k)$ is a homogeneous space, being a quotient space of G_N, the set of rigid motions on $\mathbb{R}^N$. Hence it inherits a measure λ_k^N invariant on G_N. Under the above diffeomorphism, this measure factors as the invariant measure, say ν_k^N, on $\text{Gr}(N,k)$ and Lebesgue measure on $\mathbb{R}^N$.

There are of course many ways to normalize ν_k^N, and hence λ_k^N. One useful normalization (useful because of Crofton's formula below) is to require that

$$\nu_k^N(\text{Gr}(N,k))=\begin{bmatrix}N\\k\end{bmatrix},$$

the flag coefficients of (6.3.12).

With our constants set, we can now formulate Crofton's formula. In its original form, it was derived for M in the convex ring.[1] We, however, need it in the setting of Whitney stratified spaces, and in this setting it is due to Bröckner and Kuppe [33].

Theorem 13.1.1 (Crofton's formula). *Suppose $M\subset\mathbb{R}^N$ is a compact, tame, Whitney stratified space. Then*

$$\int_{\text{Graff}(N,N-k)}\mathcal{L}_j(M\cap V)\,d\lambda_{N-k}^N(V)=\begin{bmatrix}k+j\\j\end{bmatrix}\mathcal{L}_{k+j}(M).\tag{13.1.1}$$

The $\mathcal{L}_j$ in (13.1.1) are all computed with respect to the standard (Euclidean) Riemannian metric on $\mathbb{R}^N$. This is important and somewhat limiting. To avoid this limitation we shall develop a different, but analogous, result in Section 13.3 below, after first seeing, in the following section, what we can do with results of this kind.

The special case $k=0$ of Crofton's formula is generally known as *Hadwiger's formula*,[2] and is given by

[1] See [92, 139] for a full treatment in the setting of the convex ring and also an extension to basic complexes due originally to Hadwiger [75]. There is also a nice survey of the more modern theory, which goes considerably beyond what we shall need, in [80].

[2] Note that (13.1.2) differs a little from the version of Hadwiger's formula that we met earlier in (6.3.11). The difference, however, is superficial, and results only from where one places the various normalizing constants.

$$\mathcal{L}_k(M) = \int_{\mathrm{Graff}(N,N-k)} \mathcal{L}_0(M \cap V)\, d\lambda_{N-k}^N(V). \tag{13.1.2}$$

13.2 Mean Intrinsic Volumes: The Isotropic Case

The existence of Hadwiger's formula linking, as it does, Lipschitz–Killing curvatures and Euler characteristics in Euclidean spaces, allows a very simple derivation of the mean Lipschitz–Killing curvatures of excursion sets of isotropic processes from the formula for their mean Euler characteristic.

The key point to the argument will rely on taking an isotropic f with unit second spectral moment. In that case, as we have noted a number of times, the Riemannian metric that f induces on the parameter space is the standard Euclidean one, and the Lipschitz–Killing curvatures appearing in Theorems 12.4.1 and 12.4.2 are the simple Euclidean ones of Section 13.1.

Theorem 13.2.1. *Suppose that $M \subset \mathbb{R}^N$ satisfies the conditions of Theorem 13.1.1 with the additional assumption that both M and its cross-sections $M \cap V$ are (λ_K^N almost surely) C^2, tame, Whitney stratified manifolds. Suppose also that f is a centered, isotropic Gaussian process on M satisfying the conditions of Corollary 11.3.2 and that*

$$\mathbb{E}\left\{f^2(t)\right\} = 1 \quad \text{and} \quad \mathbb{E}\left\{f_i^2(t)\right\} = \lambda,$$

for all $1 \le i \le N$, where the f_i are the usual partial derivatives of f. Then, for every $0 \le j \le N$,

$$\mathbb{E}\left\{\mathcal{L}_j(A_u(f,M))\right\} = \sum_{l=0}^{N-j} \begin{bmatrix} j+l \\ l \end{bmatrix} \lambda^{j+l} \rho_l(u) \mathcal{L}_{j+l}(M), \tag{13.2.1}$$

where the $\mathcal{L}_j$, on both sides of the equality, are computed with respect to the standard Euclidean metric on $\mathbb{R}^N$.

Proof. We first note that it suffices to establish (13.2.1) for the case $\lambda = 1$. The general case then follows from the scaling properties (6.3.1) of the $\mathcal{L}_j$.

Thus, assuming that $\lambda = 1$, Hadwiger's formula and Theorem 12.4.2 immediately yield

$$\begin{aligned}
\mathbb{E}\left\{\mathcal{L}_j(A_u(f,M))\right\} &= \int_{\mathrm{Graff}(N,N-j)} \mathbb{E}\left\{\mathcal{L}_0\left(A_u(f,M) \cap V\right)\right\} d\lambda_{N-j}^N(V) \\
&= \sum_{l=0}^{N-j} \rho_l(u) \int_{\mathrm{Graff}(N,N-j)} \mathcal{L}_l(M \cap V)\, d\lambda_{N-j}^N(V) \\
&= \sum_{l=0}^{N-j} \begin{bmatrix} j+l \\ l \end{bmatrix} \rho_l(u) \mathcal{L}_{j+l}(M),
\end{aligned}$$

and we are done. □

For an alternative approach to computing the mean Lipschitz–Killing curvatures in the setting of this section see [176].

13.3 A Gaussian Crofton Formula

Given the simplicity of the above approach for bootstrapping from the mean Euler characteristics of excursion sets to their mean Lipschitz–Killing curvatures, it is tempting to try to adopt a similar approach in the general case. The problem in doing so, however, lies in the fact that Crofton's formulas are rather specific to manifolds in $\mathbb{R}^N$ with its standard Riemannian metric and there does not seem to be an analogous result for more general manifolds. Thus, we shall now spend a little time developing a tool that will work in the general setting.

The basic idea will be to replace the cross-sections $M \cap V$ by something more appropriate for the manifold setting. In particular, we shall introduce an auxiliary set of *random* manifolds D for which the union of the $M \cap D$ over all D again gives M and, more importantly, for which $\mathcal{L}_j(M)$ can be computed by averaging the Euler characteristics $\mathcal{L}_0(M \cap D)$ over D.

The main result, which would seem to be of significant interest beyond the application that we have in mind, is as follows.

Theorem 13.3.1. *Let $\widetilde{M}$ be a C^2, n-dimensional manifold (without boundary) embedded in a C^3 manifold. Let $y^1, \dots, y^k$ be centered, unit-variance, independent, identically distributed Gaussian fields on $\widetilde{M}$ satisfying the conditions of Corollary* 11.3.5. *For $u \in \mathbb{R}^k$ define the (random) submanifold*

$$D_u = \left\{ t \in \widetilde{M} : y_t^1 = u_1, \dots, y_t^k = u_k \right\}$$

of $\widetilde{M}$, and suppose $Z_k = (Z_k^1, \dots, Z_k^k) \sim \gamma_{\mathbb{R}^k}$ independently of the field $y = (y^1, \dots, y^k)$. Then, for $0 \le j \le \dim(\widetilde{M}) - k$,

$$\mathbb{E}\left\{ \mathcal{L}_j(\widetilde{M} \cap D_{Z_k}) \right\} = (2\pi)^{-k/2} \frac{[k+j]!}{[j]!} \mathcal{L}_{k+j}(\widetilde{M}), \tag{13.3.1}$$

where the $\mathcal{L}_j$ on both sides of the equation are computed with respect to the Riemannian metric induced on $\widetilde{M}$ by the y^i.

The reason we call this a "Gaussian Crofton formula' should be clear by analogy with the original, Euclidean, Crofton formula (13.1.1). The formulas are similar, but we have replaced the kinematic averaging under the measure λ_{N-k}^N with a Gaussian average. Note, also, that the averaging in (13.3.1) is over both y and Z_k.

Proof. We start by noting that it suffices to prove (13.3.1) for the Euler characteristic $\varphi = \mathcal{L}_0$. That is, we need only prove that

$$\mathbb{E}\left\{ \varphi(\widetilde{M} \cap D_{Z_k}) \right\} = (2\pi)^{-k/2} [k]! \mathcal{L}_k(\widetilde{M}). \tag{13.3.2}$$

This follows from the following observation: For $j > 0$, take $Z_j \sim \gamma_{\mathbb{R}^j}$ independent of everything else, and write Z_{k+j} for the concatenation of Z_k and Z_j. Note that, in distribution,

$$\widetilde{M} \cap D_{Z_{k+j}} = \left(\widetilde{M} \cap D_{Z_j}\right) \cap D_{Z_k}.$$

Conditioning first on Z_k, apply (13.3.2) to the manifold $\widetilde{M} \cap D_{Z_k}$ to see that

$$\begin{aligned}\mathbb{E}\left\{\mathcal{L}_j\left(\widetilde{M} \cap D_{Z_k}\right)\right\} &= \frac{(2\pi)^{j/2}}{[j]!}\mathbb{E}\left\{\varphi\left(\widetilde{M} \cap D_{Z_{k+j}}\right)\right\} \\ &= (2\pi)^{-k/2}\frac{[k+j]!}{[j]!}\mathcal{L}_{k+j}(\widetilde{M}),\end{aligned}$$

which establishes the general case.

We now turn to establishing (13.3.2). To this end, note that for fixed $u \in \mathbb{R}^k$, and for almost all realizations of y, it follows from the regularity properties of y that D_u is a regular stratified manifold, embedded in $\widetilde{M}$.

Consequently, an attractive route to try to take to prove the theorem would be to apply our version of the Chern–Gauss–Bonnet theorem (Corollary 12.6.2) to the (random) manifold $\widetilde{M} \cap D_{Z_k}$ and so write the mean Euler characteristic of $\widetilde{M} \cap D_{Z_k}$ as

$$\begin{aligned}\frac{(2\pi)^{-(\dim(\widetilde{M})-k)/2}}{(\dim(\widetilde{M})-k)!}\int_{D_{Z_k}} \mathbb{E}\Big\{&\alpha(P^{\perp}_{T_t D_{Z_k}}\nabla\widetilde{y};\widetilde{M}) \\ &\times \mathrm{Tr}^{T_t D_{Z_k}}\left(\left(-\nabla^2\widetilde{y}_{|D_{Z_k}}\right)^{(\dim(\widetilde{M})-k)}\right)\Big\}\mathcal{H}_j(dt),\end{aligned}$$

for some suitably regular random field $\widetilde{y}$, which we can take to be a copy of y^1, independent of $y^1, \dots, y^k$. All that would then remain would be to compute the expectation here, over both $\widetilde{y}$ and Z_k.

Unfortunately, the technicalities involved in applying this appealingly simple argument are not trivial, and so we shall take a more basic route, via Morse theory.

We shall break the proof into in two parts, firstly fixing Z_k and applying Morse theory, which will eventually involve an averaging over y and $\widetilde{y}$. At the second stage we shall also average over Z_k.

Thus, with $Z_k = u$, we look for a way to express the Euler characteristic of $\widetilde{M} \cap D_u = D_u$ in terms of points $\{t : y_t = u\}$ and $\nabla\widetilde{y}_t$, which is in the span of $\nabla y_t^1, \dots, \nabla y_t^k$. Denote the latter space by

$$L_t \stackrel{\Delta}{=} \mathrm{span}(\nabla y_t^i,\ 1 \le i \le k).$$

With the manifold D_u fixed, straightforward calculations show that the Hessian of $\widetilde{y}$ on D_u can be written as

$$\nabla^2 \widetilde{y}_{|D,t} = \nabla^2 \widetilde{y}_t - \sum_{i,j=1}^{k} \langle \nabla \widetilde{y}_t, \nabla y_t^i \rangle g_t^{ij} \nabla^2 y_t^j \tag{13.3.3}$$
$$= \nabla^2 \widetilde{y}_t - \sum_{i,j=1}^{k} \langle P_{L_t} \nabla \widetilde{y}_t, \nabla y_t^i \rangle g_t^{ij} \nabla^2 y_t^j,$$

where G_t is the $k \times k$ matrix with elements

$$g_{ij,t} \stackrel{\Delta}{=} \langle \nabla y_t^i, \nabla y_t^j \rangle,$$

and g_t^{ij} are the elements of G_t^{-1}. We note for later use that the matrices G_t have a Wishart$(n, I_{k\times k})$ distribution.[3]

We can now apply Morse's theorem (Theorem 9.3.2) in an almost sure fashion, for each realization of the pair y and $\widetilde{y}$, to characterize the Euler characteristic of $\widetilde{M} \cap D_u$. This gives a point set characterization of $\varphi(\widetilde{M} \cap D_u)$ in the usual way. A representation such as (11.2.9) turns this into an integral over $\widetilde{M}$, and an argument such as that used to prove Corollary 12.4.5 allows[4] us to conclude, under the conditions of the theorem, that

$$\begin{aligned} &\mathbb{E}\left\{\varphi\left(\widetilde{M} \cap D_u\right)\right\} &(13.3.4)\\ &\quad = \frac{1}{(n-k)!} \int_{\widetilde{M}} \mathbb{E}\left\{ \mathrm{Tr}^{L_t^\perp}\left((-\nabla^2 \widetilde{y}_{|D})^{n-k} \right) J_t \,\Big|\, y = u,\, P_{L_t}^\perp \nabla \widetilde{y} = 0 \right\} \\ &\quad\quad \times p_{y, P_{L_t}^\perp \nabla \widetilde{y}}(u, 0) \mathcal{H}_n(dt), \end{aligned}$$

where $p_{y, P_{L_t}^\perp \nabla \widetilde{y}}$ is the joint density of y and $P_{L_t}^\perp \nabla \widetilde{y}$, and

$$J_t \stackrel{\Delta}{=} \sqrt{\det(G_t)}. \tag{13.3.5}$$

While this formula is similar to many that appear in Chapter 12, there is a new Jacobian term, J_t, here that deserves a few words. It arises due the fact that whereas previously we counted points that were critical with respect to the underlying manifold $\widetilde{M}$, we are now counting points of the form

$$\left\{ t : y_t = u,\, P_{L_t}^\perp \nabla \widetilde{y}_t = 0 \right\} = \left\{ t : y_t^1 = u_1, \ldots, y_t^k = u_k,\, P_{L_t}^\perp \nabla \widetilde{y}_t = 0 \right\}.$$

That is, we now have points critical with respect to a random submanifold. By choosing a convenient basis of $T_t\widetilde{M}$ it is fairly straightforward to see that the Jacobian

[3] Recall that the Wishart(n, Σ) distribution is defined as the distribution of a matrix W with elements of the form $W_{ij} = \sum_{m=1}^n X_{mi} X_{mj}$, where the n vectors $X_m = (X_{m1}, \ldots, X_{mk})$ are independent, each distributed as $N(0, \Sigma)$.

[4] Obviously, we have allowed ourselves considerable latitude in this argument. Filling in all the details would take a few pages, but would involve little more than taking the references we have made and rewriting them in the notation of the present problem. We leave the details to the masochists and/or pedants among you.

(13.3.5) is precisely the one that arises in an application of Corollary 12.4.5 needed to establish (13.3.4).

Furthermore, noting that since y_t and ∇y_t are independent the same is true of y_t and L_t, and y_t and J_t, we can integrate out u in the above to obtain

$$\begin{aligned}
&\mathbb{E}\left\{\varphi\left(\widetilde{M} \cap D_{Z_k}\right)\right\} \\
&\quad = \frac{1}{(n-k)!} \int_{\widetilde{M}} \mathbb{E}\left\{ \mathrm{Tr}^{L_t^\perp}\left((-\nabla^2 \widetilde{y}_{|D})^{n-k} \right) J_t \,\middle|\, P_{L_t}^\perp \nabla \widetilde{y} = 0 \right\} \\
&\qquad \times p_{P_{L_t}^\perp \nabla \widetilde{y}}(0) \mathcal{H}_n(dt).
\end{aligned}$$

Consider the integrand.

$$\begin{aligned}
&\mathbb{E}\left\{ \mathrm{Tr}^{L_t^\perp}\left((-\nabla^2 \widetilde{y}_{|D})^{n-k} \right) J_t \,\middle|\, P_{L_t}^\perp \nabla \widetilde{y} = 0 \right\} p_{P_{L_t}^\perp \nabla \widetilde{y}}(0) \\
&\quad = \lim_{\varepsilon \to 0} \mathbb{E}\left\{ \mathrm{Tr}^{L_t^\perp}\left((-\nabla^2 \widetilde{y}_{|D})^{n-k} \right) J_t \mathbb{1}_{\{P_{L_t}^\perp \nabla \widetilde{y}_t}(B(0,\varepsilon))\} \right\} \\
&\quad = (2\pi)^{-(n-k)/2} \mathbb{E}\left\{ \mathrm{Tr}^{L_t^\perp}\left((-\nabla^2 \widetilde{y}_{|D})^{n-k} \right) J_t \right\},
\end{aligned}$$

where we have used the fact that, given the subspace L_t, the pair $(\nabla^2 \widetilde{y}_{|D,t}, J_t)$ is conditionally independent of $P_{L_t}^\perp \nabla \widetilde{y}$, which then has a standard Gaussian distribution on $L_t^\perp$.

We now condition on $(\nabla^2 \widetilde{y}_{|D,t}, J_t)$, so that the next step is to compute the conditional expectation

$$\mathbb{E}\left\{ \mathrm{Tr}^{L_t^\perp}\left((-\nabla^2 \widetilde{y}_{|D})^{n-k} \right) J_t \,\middle|\, \nabla^2 \widetilde{y}_{|D,t}, J_t \right\}.$$

Because of the conditioning, this is just the expected value of the trace of a fixed form $\alpha \in \bigwedge^{n-k,n-k}(T_t\widetilde{M})$, restricted to a random subspace of dimension $n-k$ of $T_t\widetilde{M}$. Lemma 13.5.1 below shows how to evaluate such expectations, and in our case it follows from the lemma that

$$\mathbb{E}\left\{ \mathrm{Tr}^{L_t^\perp}\left((-\nabla^2 \widetilde{y}_{|D})^{n-k} \right) J_t \,\middle|\, \nabla^2 \widetilde{y}_{|D,t}, J_t \right\} = \binom{n}{k}^{-1} \mathrm{Tr}^{T_t\widetilde{M}}\left((-\nabla^2 \widetilde{y}_{|D})^{n-k} \right) J_t. \tag{13.3.6}$$

To complete the computation, we now need to evaluate

$$\mathbb{E}\left\{ \mathrm{Tr}^{T_t\widetilde{M}}\left((-\nabla^2 \widetilde{y}_{|D})^{n-k} \right) J_t \right\}.$$

Recall (13.3.3), which gives a different way of writing $\nabla^2 \widetilde{y}_{|D}$. In particular, consider the second term in the last expression there, which involves the term

$$\sum_{i,j=1}^{k} \langle P_{L_t} \nabla \widetilde{y}_t, \nabla y_t^i \rangle g_t^{ij} \nabla^2 y_t^j.$$

We want a more user-friendly version of this. To obtain it, for the moment we drop the dependence on t and define

$$V_i \triangleq \sum_{j=1}^{k} g_{ij}^{-1/2} \langle P_L \nabla \widetilde{y}, \nabla y^i \rangle,$$

where the $g_{ij}^{-1/2}$ are the elements of the matrix $G^{-1/2}$ $(= G_t^{-1/2})$. Then, conditional on $\nabla y^1, \dots, \nabla y^k, \nabla^2 y^1, \dots, \nabla^2 y^k$, we have $V \sim N(0, I_{k\times k})$. Since the conditional distribution does not depend on the conditioning variables, the V_i are actually independent of them. Furthermore, since the ∇y^i and $\nabla^2 y^j$ are all independent of one another, we have that for each t,

$$\sum_{i,j=1}^{k} \langle P_{L_t} \nabla \widetilde{y}_t, \nabla y_t^i \rangle g_t^{ij} \nabla^2 y_t^j \overset{\mathcal{L}}{=} \sum_{i,j=1}^{k} V_i W_{ij}^{-1/2} \nabla^2 y^j,$$

where $W \sim \text{Wishart}(n, I_{k\times k})$, independently of everything else.

Consequently, by (13.3.3), we can write

$$\begin{aligned} &\text{Tr}^{T_t \widetilde{M}} \left((-\nabla^2 \widetilde{y}_{|D})^{n-k} \right) J_t \qquad (13.3.7) \\ &\quad \overset{\mathcal{L}}{=} \text{Tr}^{T_t \widetilde{M}} \left(\left(\nabla^2 \widetilde{y} - \sum_{i,j=1}^{k} V_i W_{ij}^{-1/2} \nabla^2 y^i \right)^{n-k} \right) \sqrt{\det(W)}, \end{aligned}$$

where, as above, $W \sim \text{Wishart}(n, I_{k\times k})$ and the random matrices $\nabla^2 \widetilde{y}_t, \nabla^2 y_t^1, \dots, \nabla^2 y_t^k$ are independent and identically distributed.

From Lemma 13.5.2 below it then follows that

$$\mathbb{E}\left\{ \text{Tr}^{T_t \widetilde{M}} \left((-\nabla^2 \widetilde{y}_{|D})^{n-k} \right) J_t \right\} = (2\pi)^{-k/2} \frac{n!}{(n-k)!} \omega_k \mathbb{E}\left\{ \text{Tr}^{T_t \widetilde{M}} \left((-\nabla^2 \widetilde{y})^{n-k} \right) \right\}. \qquad (13.3.8)$$

This last expectation, however, is not new, for we have already computed almost identical terms in Chapter 12. In particular, it follows from Corollary 12.3.2 and the computations at the end of the proof of Theorem 12.4.2 that

$$\int_{\widetilde{M}} \mathbb{E}\left\{ \text{Tr}^{T_t \widetilde{M}} ((-\nabla^2 \widetilde{y}_t)^{n-k}) \right\} \mathcal{H}_n(dt) = (2\pi)^{(n-k)/2} (n-k)! \mathcal{L}_k(\widetilde{M}).$$

Putting all the pieces together, we have enough to prove (13.3.1), and so we are done, modulo proving Lemmas 13.5.1 and 13.5.2, which we defer to Section 13.5. □

Our next step will be to extend Theorem 13.3.1 from manifolds to Whitney stratified manifolds. In fact, one can go marginally further than this. Define a family of additive functionals Ψ_k^j on C^2 Whitney stratified submanifolds M of a manifold $\widetilde{M}$ by setting, in the notation of Theorem 13.3.1,

$$\Psi_k^j(M) \stackrel{\Delta}{=} \mathbb{E}\left\{\mathcal{L}_j(M \cap D_{Z_k})\right\}.$$

Then the following result shows that the Ψ_k^j and the $\mathcal{L}_l$ are very closely related.

Theorem 13.3.2. *Let $\widetilde{M}$, $y^1, \ldots, y^k$, Z_k and D_u be as in Theorem 13.3.1, and let M be a regular stratified submanifold of $\widetilde{M}$. Then, for all $j \le \dim(M) - k$,*

$$\mathbb{E}\left\{\mathcal{L}_j(M \cap D_{Z_k})\right\} = (2\pi)^{-k/2} \frac{[k+j]!}{[j]!} \mathcal{L}_{k+j}(M).$$

In the notation of the linear functionals described above, this can be written as

$$\Psi_k^j = (2\pi)^{-k/2} \frac{[k+j]!}{[j]!} \mathcal{L}_{k+j}.$$

Proof. Most of the hard work has been done in the proof of Theorem 13.3.1, and what remains is some bookkeeping. As was the case for the proof of Theorem 13.3.1, the strategy of the proof is to derive an expression to which we can apply Lemma 13.5.2. Similar arguments to those appearing in the proof of Theorem 13.3.1, but taking into account that we now have boundary terms to deal with, show that

$$\begin{aligned}
&\mathbb{E}\left\{\varphi\left(M \cap D_{Z_k}\right)\right\} \qquad (13.3.9)\\
&= \sum_{l=k}^{n} \frac{1}{(l-k)!} \int_{\partial_l M} \mathbb{E}\Big\{\alpha\left(\eta_t; M \cap D_{Z_k}\right) \mathrm{Tr}^{L_t^\perp}((-\nabla^2 \widetilde{y}_{|D})^{l-k}) J_t \Big| \\
&\qquad\qquad P_{L_t}\nabla\widetilde{y} = 0\Big\} p_{P_{L_t}\nabla\widetilde{y}}(0) \mathcal{H}_l(dt),
\end{aligned}$$

where $\widetilde{y}$ is again an independent copy of the y^j being used as a Morse function and α is the normal Morse index of Section 9.2. As for the other terms,

$$J_t = \sqrt{\det(G_t)},$$

where G_t is the $k \times k$ matrix with elements

$$g_{ij,t} \stackrel{\Delta}{=} \langle P_{T_t \partial_l M} \nabla y_t^i, P_{T_t \partial_l M} \nabla y_t^j \rangle,$$

while

$$L_t = \mathrm{span}\left\{P_{T_t \partial_l M} \nabla y_t^i, 1 \le i \le k\right\} \qquad (13.3.10)$$

and we write $L_t^\perp$ to denote the normal space to L_t in $T_t \partial_l M$. Finally, we have

$$\begin{aligned}
\eta_t &= \nabla\widetilde{y}_t - P_{L_t}\nabla\widetilde{y}_t \qquad (13.3.11)\\
&= P^\perp_{T_t \partial_l M} \nabla\widetilde{y}_t + \sum_{i,j=1}^{l} \langle \nabla\widetilde{y}_t, P_{T_t \partial_l M} \nabla y_t^i \rangle g_t^{ij} P_{T_t \partial_l M} \nabla y_t^j,
\end{aligned}$$

and

$$\nabla^2 \widetilde{y}_{|D,t} = \nabla^2 \widetilde{y}_t - \sum_{i,j=1}^{l} \langle \nabla \widetilde{y}_t, P_{T_t \partial_l M} \nabla y_t^i \rangle g_t^{ij} \nabla^2 y_t^j + S_{\nu_t}. \tag{13.3.12}$$

Here the g_t^{ij} are the elements of G_t^{-1}, S is the second fundamental form of $\partial_l M$ as it sits in $\widetilde{M}$, and

$$\nu_t = P^{\perp}_{T_t \partial_l M} \nabla \widetilde{y}_t - \sum_{i,j=1}^{l} \langle \nabla \widetilde{y}_t, P_{T_t \partial_l M} \nabla y_t^i \rangle g_t^{ij} P^{\perp}_{T_t \partial_l M} \nabla y_t^j. \tag{13.3.13}$$

The term $\nabla^2 \widetilde{y}_{|D,t}$ in (13.3.12) deserves some explanation. It is the Hessian of the restriction of $\widetilde{y}$ to $\partial_l M \cap D_{Z_k}$, and seeing that it has the above form takes some work. Consider two vector fields X, W on $\partial_l M \cap D_{Z_k}$ and let $\widehat{\nabla}$ denote the Levi-Civitá connection of $\partial_l M \cap D_{Z_k}$. Then, from the definition (7.3.22) of the Hessian and the compatability property (7.3.11) of $\widehat{\nabla}$, we have

$$\begin{aligned}
\nabla^2 \widetilde{y}_{|D}(W, X) &= W \langle \nabla \widetilde{y}, X \rangle - \langle \nabla \widetilde{y}, \widehat{\nabla}_W X \rangle \\
&= W \langle \nabla \widetilde{y}, X \rangle - \langle \nabla \widetilde{y}, \nabla_W X \rangle + \langle P^{\perp}_{T_t \partial_l M} \nabla_W X, \nabla \widetilde{y} \rangle \\
&\quad + \sum_{r,s=1}^{l} \langle \nabla_W X, P_{T_t \partial_l M} \nabla y^r \rangle g_t^{rs} \langle P_{T_t \partial_l M} \nabla y^s, \nabla \widetilde{y} \rangle,
\end{aligned}$$

with the second line coming from the form of the projection of vectors in $T_t \widetilde{M}$ to $T_t(\partial_l M \cap D_{Z_k})$.

Now note that

$$\langle X, P_{T_t \partial_l M} \nabla y^r \rangle = \langle W, P_{T_t \partial_l M} \nabla y^r \rangle = 0,$$

for all $1 \le r \le k$, which implies, again by (7.3.11) and (7.3.22), that

$$\langle \nabla_W X, P_{T_t \partial_l M} \nabla y^r \rangle = -\nabla^2 y^r_{|\partial_l M}(W, X).$$

Substituting this in the above gives

$$\begin{aligned}
\nabla^2 \widetilde{y}_{|D}(W, X) &= \nabla^2 \widetilde{y}(W, X) + \langle \nabla_W X, P^{\perp}_{T_t \partial_l M} \nabla \widetilde{y} \rangle \\
&\quad - \sum_{r,s=1}^{l} \langle \nabla \widetilde{y}, P_{T_t \partial_l M} \nabla y^s \rangle g_t^{rs} \nabla^2 y^r_{|\partial_l M}(W, X) \\
&= \nabla^2 \widetilde{y}(W, X) + S_{P^{\perp}_{T_t \partial_l M} \nabla \widetilde{y}}(W, X) \\
&\quad - \sum_{r,s=1}^{l} \langle \nabla \widetilde{y}, P_{T_t \partial_l M} \nabla y^s \rangle g_t^{rs} \nabla^2 y^r_{|\partial_l M}(W, X),
\end{aligned}$$

on applying the Weingarten equation (7.5.12). Furthermore, by the definition (7.5.8) of the second fundamental form and the Weingarten equation, the rightmost term in the last summand can be rewritten as

$$\nabla^2 y^r_{|\partial_l M}(W, X) = \nabla^2 y^r(W, X) + \langle \nabla_W X, P^{\perp}_{T_t \partial_l M} \nabla y^r \rangle.$$

Substituting this into the previous equation gives

$$\begin{aligned}
&\nabla^2 \widetilde{y}_{|D}(W, X) \\
&\quad = \nabla^2 \widetilde{y}(W, X) - \sum_{r,s=1}^{l} \langle \nabla \widetilde{y}, P_{T_t \partial_l M} \nabla y^s \rangle g_t^{rs} \nabla^2 y^r(W, X) + S_{\nu_t}(W, X),
\end{aligned}$$

which is (13.3.12). Thus, all the terms in (13.3.9) are well defined and we can turn to evaluating the right-hand side there.

As in the proof of Theorem 13.3.1, this evaluation relies on getting everything into just the right form for applying Lemmas 13.5.1 and 13.5.2.

Applying Lemma 13.5.1 to (13.3.9) and using the independence of $\nabla \widetilde{y}$ and $\nabla^2 \widetilde{y}$, we immediately find that

$$\begin{aligned}
&\mathbb{E}\left\{\varphi\left(M \cap D_{Z_k}\right)\right\} \\
&= \sum_{l=k}^{n} \frac{1}{(l-k)!} \\
&\quad \times \int_{\partial_l M} \mathbb{E}\{\alpha(\eta_t; M \cap D_{Z_k}) \operatorname{Tr}^{L_t^{\perp}}((-\nabla^2 \widetilde{y}_{|D})^{l-k}) J_t \,|\, P_{L_t} \nabla \widetilde{y} = 0\} p_{P_{L_t} \nabla \widetilde{y}}(0) \mathcal{H}_l(dt) \\
&= \sum_{l=k}^{n} (2\pi)^{-(l-k)/2} \frac{k!}{l!} \\
&\quad \times \int_{\partial_l M} \mathbb{E}\{\alpha(\eta_t; M \cap D_{Z_k}) \operatorname{Tr}^{T_t \partial_l M}((-\nabla^2 \widetilde{y}_{|D})^{l-k}) J_t\} \mathcal{H}_l(dt).
\end{aligned}$$

We now want to apply Lemma 13.5.2 to the expectation here. The expression above for $\nabla^2 \widetilde{y}_{|D}$ is of the right form, but this is not so for the term involving the normal Morse index. However, an application of Theorem 9.2.6 shows that in fact,

$$\alpha(\eta_t; M \cap D_{Z_k}) = \alpha(\nu_t; M),$$

which *is* of the right form. (Recall that η_t and ν_t are defined, respectively, by (13.3.11) and (13.3.13).)

Consequently, using this equivalence and Lemma 13.5.2, we find that

$$\begin{aligned}
&\mathbb{E}\left\{\varphi\left(M \cap D_{Z_k}\right)\right\} \\
&\quad = \sum_{l=k}^{n} (2\pi)^{-(l-k)/2} \frac{k!}{l!} \int_{\partial_l M} \mathbb{E}\Big\{\alpha\left(\nu_t; M\right) \operatorname{Tr}^{T_t \partial_l M}((-\nabla^2 \widetilde{y}_{|D})^{l-k}) J_t\Big\} \mathcal{H}_l(dt) \\
&\quad = \sum_{l=k}^{n} (2\pi)^{-l/2} \frac{[k]!}{(l-k)!} \\
&\qquad \times \int_{\partial_l M} \mathbb{E}\Big\{\alpha\left(P_{T_t \partial_l M} \nabla \widetilde{y}_t; M\right) \operatorname{Tr}^{T_t \partial_l M}((-\nabla^2 \widetilde{y})^{l-k})\Big\} \mathcal{H}_l(dt).
\end{aligned}$$

Once again, similar computations to those appearing at the end of the proof of Theorem 12.4.2 show that the final expression above is equal to

$$(2\pi)^{-k/2} [k]! \mathcal{L}_k(M),$$

and so we are done. □

13.4 Mean Intrinsic Volumes: The General Case

We can now formulate and prove the main theorem of this chapter, which generalizes to general Whitney stratified manifolds and to nonisotropic processes the Euclidean result of Theorem 13.2.1.

Theorem 13.4.1. *Suppose that M is a regular stratified manifold and that f is a centered, unit-variance, Gaussian process on M satisfying the conditions of Corollary* 11.3.2.

Then, for every $0 \le j \le \dim(M)$,

$$\mathbb{E}\left\{\mathcal{L}_j(A_u(f, M))\right\} = \sum_{l=0}^{\dim M - j} \begin{bmatrix} j+l \\ l \end{bmatrix} \rho_l(u) \mathcal{L}_{j+l}(M), \tag{13.4.1}$$

where the $\mathcal{L}_k$ on both sides of this equation are computed with respect to the metric induced by f as at (12.2.1), *the combinatorial coefficients are defined in* (6.3.12), *and the ρ_l remain as in* (12.4.2).

Proof. The proof mimics that of the Euclidean, isotropic version of the result in Theorem 13.2.1, using the Gaussian Crofton formula of Theorem 13.3.2 rather than the standard Crofton formula.

We start by fixing $0 \le j \le \dim(M)$ and again introducing an auxiliary set of Gaussian processes $y^1, \dots, y^j$ and a Gaussian random variable Z_j as in Theorem 13.3.1, satisfying the conditions required there. As before, write

$$D_u = \left\{t \in M : y_t^1 = u_1, \dots, y_t^j = u_j\right\}.$$

Then, by Theorem 13.3.2,

$$\mathbb{E}\left\{\mathcal{L}_j(A_u(f,M))\right\} = \frac{(2\pi)^{j/2}}{[j]!}\mathbb{E}\left\{\mathcal{L}_0(A_u(f,M)\cap D_{Z_j})\right\}$$
$$= \frac{(2\pi)^{j/2}}{[j]!}\mathbb{E}\left\{\mathbb{E}\left\{\mathcal{L}_0(A_u(f,M)\cap D_{Z_j})\middle|D_{Z_j}\right\}\right\}.$$

We now note that with probability one, the sets $M \cap D_{Z_j}$ are Whitney stratified manifolds,[5] and so we can apply Theorem 12.4.2 to compute the inner expectation above, giving

$$\mathbb{E}\left\{\mathcal{L}_j(A_u(f,M))\right\} = \frac{(2\pi)^{j/2}}{[j]!}\sum_{l=0}^{\dim M-j}\rho_l(u)\mathbb{E}\left\{\mathcal{L}_l(M\cap D_{Z_k})\right\}$$
$$= \frac{(2\pi)^{j/2}}{[j]!}\sum_{l=0}^{\dim M-j}\rho_l(u)(2\pi)^{-j/2}\frac{[l+j]!}{[l]!}\mathcal{L}_{j+l}(M),$$

again by Theorem 13.3.2. However, this is (13.4.1) and so we are done. □

13.5 Two Gaussian Lemmas

We conclude this chapter with the two lemmas we promised during the proof of Theorem 13.3.1. They are stated in somewhat more generality than what is needed there. It turns out to be not much harder to prove the more general results, and as currently formulated they may also turn out to have broader application.

Both lemmas turn out to be exercises in (statistical) multivariate analysis, for which the standard reference is Anderson [12]. In particular, both proofs exploit some basic facts of *Wishart distributions*, which we now summarize and proofs of which can be found in [12, Section 13.3].

As noted earlier, the Wishart (n, Σ) distribution is the distribution of a matrix W with elements of the form

$$W_{ij} = \sum_{m=1}^{n} X_{mi}X_{mj},$$

where the k n-dimensional vectors $X_m = (X_{m1},\dots,X_{mk})$ are independent, each distributed as $N(0,\Sigma)$. Alternatively, writing X for the $n\times k$ matrix with X_m being the mth row, we have $W \stackrel{\mathcal{L}}{=} X'X$.

[5] This follows from arguments similar to those used in Chapter 12 to establish the fact that $A_u(f,M)$ is a Whitney stratified manifold. Specifically, note first that

$$M\cap D_{Z_j} = \left\{t\in M : y_t^i - Z_j = 0,\ 1\le i\le k\right\}.$$

Since the components y_t^i are assumed suitably regular and Z_i is just a random shift of each component, the regularity of $M\cap D_{Z_i}$ is essentially the same as the zero sets of the vector-valued field y.

The probability density of (the elements of) W is of the form

$$\frac{|W|^{(n-k-1)/2} e^{-(\mathrm{Tr}\, W)/2}}{2^{nk/2}\Gamma_k(\frac{n}{2})}, \tag{13.5.1}$$

where

$$\Gamma_k(n) \triangleq \pi^{k(k-1)/4} \prod_{j=1}^{k} \Gamma\left(n - \frac{1}{2}(j-1)\right).$$

We shall need two properties of the Wishart$(n, I_{k\times k})$ distribution. The first is the rather immediate fact that for any $W \sim$ Wishart$(n, I_{k\times k})$ and any orthogonal matrix A,

$$AWA' \stackrel{\mathcal{L}}{=} W. \tag{13.5.2}$$

The second follows immediately from the fact that the density (13.5.1) can be rewritten as

$$\frac{\prod_{j=1}^{p} \lambda_j^{(n-k-1)/2} e^{-\sum_{j=1}^{k} \lambda_j/2}}{2^{nk/2}\Gamma_k(\frac{n}{2})},$$

where $\lambda_1 \leq \cdots \leq \lambda_k$ are the (real) eigenvalues of W. It states that for $W \sim$ Wishart$(n, I_{k\times k})$,

$$\det(W) \stackrel{\mathcal{L}}{=} \prod_{j=1}^{k} \chi^2_{n+1-j}, \tag{13.5.3}$$

where we read the right-hand side as "the product of k independent χ^2 variables, with degrees of freedom running from $n+1-k$ to n." An immediate consequence of this is that

$$\mathbb{E}\{\det(W)\} = \prod_{j=1}^{k}(n+1-j) = \frac{n!}{(n-k)!}. \tag{13.5.4}$$

Here is the first of our two lemmas.

Lemma 13.5.1. *Let $(V, \langle , \rangle)$ be an inner product space with* $\dim(V) = n$. *Suppose that $L_k \subset V$ is a uniformly distributed random subspace of V of dimension k, in the sense that $gL_k \stackrel{\mathcal{L}}{=} L_k$ for all orthonormal transformations g of V. Then, for all double forms $\alpha \in \bigwedge^{j,j}(V)$,*

$$\mathbb{E}\left\{\mathrm{Tr}^{L_k}(\alpha)\right\} = \begin{cases} \frac{\binom{k}{j}}{\binom{n}{j}} \mathrm{Tr}^V(\alpha), & j \leq k, \\ 0 & \textit{otherwise.} \end{cases}$$

Proof. Firstly, we note that the case $j > k$ is trivial, since then $\alpha \equiv 0$ by definition. Secondly, by the linearity of double forms we note that it suffices to consider the case

$$\alpha = \left(\omega_1 \wedge \cdots \wedge \omega_j\right) \cdot \left(\omega_1 \wedge \cdots \wedge \omega_j\right) \tag{13.5.5}$$

for some orthonormal set $\{\omega_1, \ldots, \omega_j\}$. Note that from the definition (7.2.6) of the trace operator,

$$\mathrm{Tr}^V(\alpha) = \mathrm{Tr}^V\left(\left(\omega_1 \wedge \cdots \wedge \omega_j\right) \cdot \left(\omega_1 \wedge \cdots \wedge \omega_j\right)\right) = 1. \tag{13.5.6}$$

Finally, we claim that it also suffices to consider only the case $j = k$. To see this, note that for an α of the above type,

$$\mathrm{Tr}^{L_k}(\alpha) = \sum_{\{i_1,\ldots,i_j\}\subset\{1,\ldots,k\}} \alpha\left(\left(\theta_{i_1}, \ldots, \theta_{i_j}\right), \left(\theta_{i_1}, \ldots, \theta_{i_j}\right)\right),$$

where $\{\theta_1, \ldots, \theta_k\}$ is a (random) orthonormal basis for L_k. Note that the individual terms in the sum all have an identical expectation, due to the assumed distributional invariance of L_k (and its subspaces) under orthonormal transformation. Hence

$$\begin{aligned} \mathbb{E}\left\{\mathrm{Tr}^{L_k}(\alpha)\right\} &= \binom{k}{j} \mathbb{E}\left\{\alpha\left(\left(\theta_1, \ldots, \theta_j\right), \left(\theta_1, \ldots, \theta_j\right)\right)\right\} \\ &= \binom{k}{j} \mathbb{E}\left\{\mathrm{Tr}^{L_j}(\alpha)\right\}. \end{aligned} \tag{13.5.7}$$

Thus, to compute the expected value of the trace of an $\alpha \in \bigwedge^{j,j}(V)$ on a uniform subspace of dimension k, it suffices to compute it on a uniform subspace of dimension j. Consequently, we shall now concentrate on the form (13.5.5) for the case $j = k$.

To this end we introduce an auxiliary set of random vectors, $X_1, \ldots, X_k$, taken to be independent, identicially distributed with distribution γ_V.[6] Then, for α of the form (13.5.5), it follows from the definition of the trace operator that

$$\alpha\left((X_1, \ldots, X_k), (X_1, \ldots, X_k)\right) = \det(g)\, \mathrm{Tr}^{L_k}(\alpha),$$

where

$$g_{ij} = \langle X_i, X_j \rangle \sim \mathrm{Wishart}(n, I_{k\times k})$$

is independent of L_k.

Let $\widetilde{V}^* = \mathrm{span}\{\omega_1, \ldots, \omega_k\}$ and, using the usual identification of V and its dual V^*, let $\widetilde{V}$ be the corresponding subspace of V. Define

$$Y_i = P_{\widetilde{V}} X_i \sim \gamma_{\widetilde{V}}.$$

Note that

[6] Recall that this means that for any $x \in V^*$, the dual of V, the real-valued random variable $x(X)$ is distributed as $N(0, \|x\|)$.

$$\alpha\left((X_1,\ldots,X_k),(X_1,\ldots,X_k)\right)=\alpha\left((Y_1,\ldots,Y_k),(Y_1,\ldots,Y_k)\right).$$

However, the right-hand side here has a particularly simple distribution, since

$$\alpha\left((Y_1,\ldots,Y_k),(Y_1,\ldots,Y_k)\right)=\det\left(\widetilde{g}\right),$$

where

$$\widetilde{g}_{ij}=\langle Y_i,Y_j\rangle\sim \mathrm{Wishart}(k,I_{k\times k}).$$

Collecting equivalences, we now find that

$$\det(g)\,\mathrm{Tr}^{L_k}(\alpha)=\det(\widetilde{g}).$$

Taking expectations over L_k, g, and $\widetilde{g}$, recalling the independence of L_k and g, and applying (13.5.4), we find that

$$\mathbb{E}\left\{\mathrm{Tr}^{L_k}(\alpha)\right\}=\binom{n}{k}^{-1}=\binom{n}{k}^{-1}\mathrm{Tr}^{V}(\alpha),$$

the last equality following from (13.5.6). Combining this with (13.5.7) completes the proof. □

The second, unrelated, lemma is the following.

Lemma 13.5.2. *We consider three sets of random variables: a vector Z, a matrix W, and a sequence $U_1,\ldots,U_{k+1}$ of vectors. All are independent, with distributions*

$$Z\sim N(0,I_{k\times k}),\qquad W\sim \mathrm{Wishart}(n,I_{k\times k}),\qquad U_j\sim N(0,\Sigma),$$

where Σ is an arbitrary covariance matrix.

Then, for any homogeneous functions G_l of degree l and $\widetilde{G}_r$ of degree r,

$$\mathbb{E}\left\{G_l\left(U_{k+1}+\sum_{i,j=1}^{k}Z_iW_{ij}^{-1/2}U_j\right)\widetilde{G}_r\left(\sum_{i,j=1}^{k}Z_iW_{ij}^{-1/2}\right)\sqrt{\det(W)}\right\}$$
$$=(2\pi)^{-(k+r)/2}\frac{n!}{(n-k)!}\frac{\omega_{n-l}}{\omega_{n-l-k-r}}\mathbb{E}\left\{G_l(U_1)\right\}\mathbb{E}\{\widetilde{G}_r(Z)\}.$$

In particular,

$$\mathbb{E}\left\{\left(U_{k+1}+\sum_{i,j=1}^{k}Z_iW_{ij}^{-1/2}U_j\right)^{2l}\sqrt{\det(W)}\right\}$$
$$=\frac{(2l)!}{l!2^l}(2\pi)^{-k/2}\frac{n!}{(n-k)!}\frac{\omega_{n-2l}}{\omega_{n-2l-k}}\mathbb{E}\left\{U_1^{2l}\right\}.$$

Proof. For ease of notation, we first consider the case $\widetilde{G}_r \equiv 1$, since there is really little change until the final step.

Let $\mathbb{P}_W$ denote the distribution of W on $\mathrm{Sym}_{k\times k}$, the space of $k \times k$ symmetric matrices. Then

$$\begin{aligned}
&\mathbb{E}\left\{G_l\left(U_{k+1}+\sum_{i,j=1}^{k} Z_i W_{ij}^{-1/2} U_j\right)\sqrt{\det(W)}\right\} \qquad (13.5.8)\\
&= \int_{\mathrm{Sym}_{k\times k}}\int_{\mathbb{R}^k}\sqrt{\det(w)}\\
&\quad\times \mathbb{E}\left\{G_l\left(U_{k+1}+\sum_{i,j=1}^{k} z_i w_{ij}^{-1/2} U_j\right)\right\}\frac{e^{-|z|^2/2}}{(2\pi)^{k/2}}\,dz\mathbb{P}_W(dw)\\
&= \int_{\mathrm{Sym}_{k\times k}}\int_{\mathbb{R}^k}\det(w)\\
&\quad\times \mathbb{E}\left\{G_l\left(U_{k+1}+\sum_{j=1}^{k} z_j U_j\right)\right\}\frac{e^{-zwz'/2}}{(2\pi)^{k/2}}\,dz\mathbb{P}_W(dw).
\end{aligned}$$

In order to simplify this expression, fix Z and consider the expectation over W. Take an orthornormal matrix, say O_Z, with first row equal to $Z/|Z|$ and the remaining rows chosen arbitrarily (but consistently with the first). Then, as we noted at (13.5.2), $O_Z W O_Z' \stackrel{\mathcal{L}}{=} W$, and we can write $\det(O_Z W O_Z')$ as the product of χ^2 random variables (cf. (13.5.3)) the first of which is actually $ZWZ'/|Z|^2$.

Taking the expectation over these terms, we have $k-1$ expectations of χ^2 variables, along with

$$\mathbb{E}\left\{\frac{ZWZ^t}{|Z|^2}e^{-ZWZ'/2}\,\middle|\, Z\right\} = \mathbb{E}\left\{X_n e^{-|Z|^2 X_n/2}\,\middle|\, Z\right\} = \frac{d}{d\lambda}M_{X_n}(\lambda)\bigg|_{\lambda=-|Z|^2/2},$$

where $X_n \sim \chi_n^2$ and M_{X_n} is its moment-generating function. However, this is equal to

$$\frac{d}{d\lambda}\left(\frac{1}{1-2\lambda}\right)^{n/2}\bigg|_{\lambda=-|Z|^2/2} = n\left(\frac{1}{1+|Z|^2}\right)^{n/2-1}.$$

Substituting this into the last line of (13.5.8) gives that the expectation there is equivalent to

$$\begin{aligned}
&(2\pi)^{-k/2}\prod_{j=1}^{k-1}\mathbb{E}\left\{\chi_{n-j}^2\right\}\int_{\mathbb{R}^k} n(1+|z|^2)^{l/2-n/2-1} \qquad (13.5.9)\\
&\quad\times \mathbb{E}\left\{G_l\left(\frac{U_{k+1}+\sum_{i=1}^k z_i U_i}{\sqrt{1+|z|^2}}\right)\right\}dz\\
&= (2\pi)^{-k/2}\frac{n!}{(n-k)!}\mathbb{E}\left\{G_l\left(U_1\right)\right\}\int_{\mathbb{R}^k}(1+|z|^2)^{l/2-n/2-1}\,dz.
\end{aligned}$$

Finally, since the integral here is, effectively, that of a multivariate t density with $n-l-k+2$ degrees of freedom and covariance parameter $(n-l-k+2)^{-1}I_{k\times k}$ (cf. [12]), we have

$$\int_{\mathbb{R}^k} (1+|z|^2)^{l/2-n/2-1}\,dz = \pi^{k/2}\frac{\Gamma\left(\frac{n-l-k}{2}+1\right)}{\Gamma\left(\frac{n-l}{2}+1\right)} = \frac{\omega_{n-l}}{\omega_{n-l-k}},$$

and after putting all the pieces together, the proof is done.

Now suppose that $\widetilde{G}_r$ is not identically 1. Then, the proof proceeds exactly as above up to the point (13.5.9), keeping in mind that we had made the substitution

$$z_i \mapsto \widetilde{z}_i = \sum_{i,j=1}^{k} W_{ij}^{-1/2} z_j,$$

and the integral

$$\int_{\mathbb{R}^k} (1+|z|^2)^{l/2-n/2-1}\,dz$$

is replaced with

$$\int_{\mathbb{R}^k} \widetilde{G}_r(z)(1+|z|^2)^{l/2-n/2-1}\,dz.$$

Finally, exploit the homogeneity of $\widetilde{G}_r$ to see that

$$\begin{aligned}
&\mathbb{E}\left\{\widetilde{G}_r(T_{n-l-k+2})\right\} \\
&\quad = \mathbb{E}\left\{\left(\chi^2_{n-l-k+2}\right)^{-r/2}\right\}\mathbb{E}\left\{\widetilde{G}_r(Z)\right\} \\
&\quad = 2^{-r/2}\cdot\pi^{-(n-l-k)/2}\omega_{n-l-k}\frac{\Gamma\left(\frac{n-l-k+2-r}{2}\right)}{\Gamma\left(\frac{n-l-k+2}{2}\right)}\mathbb{E}\left\{\widetilde{G}_r(Z)\right\}. \qquad \square
\end{aligned}$$

14

Excursion Probabilities for Smooth Fields

As we have mentioned more than once before, one of the oldest and most important problems in the study of stochastic processes of any kind is to evaluate the *excursion probabilities*

$$\mathbb{P}\left\{\sup_{t\in T} f(t) \geq u\right\}, \tag{14.0.1}$$

where f is a random process over some parameter set T. As usual, we shall restrict ourselves to the case in which f is centered and Gaussian and T is compact for the canonical metric of (1.3.1).

In this chapter we shall restrict ourselves to the case of parameter sets that are locally convex manifolds, and so shall write M rather than T for the parameter space. The reason why we require local convexity will become clear from a counterexample in Section 14.4.4.

In particular, we shall show that for such M and for smooth Gaussian random fields f,

$$\left|\mathbb{P}\left\{\sup_{t\in M} f(t) \geq u\right\} - \mathbb{E}\left\{\varphi\left(A_u(f, M)\right)\right\}\right| < O\left(e^{-\alpha u^2/(2\sigma^2)}\right), \tag{14.0.2}$$

where φ is the Euler characteristic, $A_u(f, M)$ is the excursion set of f over M, σ^2 is the variance of f (assumed constant), and $\alpha > 1$ is a constant that we shall be able to identify. Since we already have explicit expressions for $\mathbb{E}\{\varphi(A_u(f, M))\}$ in Chapters 11 and 12, we can rewrite (14.0.2) as

$$\begin{aligned}\mathbb{P}\left\{\sup_{t\in M} f(t) \geq u\right\} = C_0\Psi\left(\frac{u}{\sigma}\right) + u^N e^{-u^2/2\sigma^2} \sum_{j=1}^{N} C_j u^{-j} \\ + o\left(e^{-\alpha u^2/2\sigma^2}\right),\end{aligned} \tag{14.0.3}$$

where the C_j are constants depending on σ^2 and the Lipschitz–Killing curvatures of M with the Riemannian structure induced by f, and $N = \dim(M)$.

A little thought shows that (14.0.3) is really a remarkable result. If we think of the right-hand side as an expansion of the form

$$C_0\Psi\left(\frac{u}{\sigma}\right) + e^{-u^2/2\sigma^2}\sum_{j=1}^{N} C_j u^{N-j} + \text{error},$$

it would be natural to expect that the error term here would be the "next" term of what seems like the beginning of an infinite expansion for the excursion probability, and so of order $u^{-1}e^{-u^2/2\sigma^2}$. However, (14.0.3) indicates that this is *not* the case. Since $\alpha > 1$, the error is actually *exponentially smaller* than this.

Our main result, which formalizes the above, will be given in Theorem 14.3.3. It has a number of precursors, the closest being a recent paper [159] with Akimichi Takemura, from which most of this chapter is taken.[1] Takemura and Kuriki [148] had a version of Theorem 14.3.3 for Gaussian fields with finite Karhunen–Loève expansions. Their approach was somewhat different from the one that is adopted here, in that they used the tube formulas of Chapter 10 to compute the excursion probability, and showed, via a kinematic formula on the sphere, that this is equivalent (up to a superexponentially small error) to computing the expected Euler characteristic. In both [148] and [159] it was recognized that the expansion in (14.0.3) is actually the expected Euler characteristic of an excursion set.

An important precursor to all of these results was an approximation of Piterbarg [126, 127], who obtained a result like (14.0.3), over a class of subsets $\mathbb{R}^N$, although without realizing that the expected Euler characteristic of an excursion set was involved. We shall return to Piterbarg's seminal results at the end of the chapter, when we can appreciate them fully with the benefits of hindsight.

There is also a rich literature of results, dating back over half a century to Rice's formula (11.1.4) on $\mathbb{R}^1$ and thirty years to [2] for the first calculation of an expected Euler characteristic, which uses these formulas to approximate excursion probabilities. Beyond the theory, these are approximations that have been heavily used by practicioners in a wide variety of areas (cf. [8], our planned next volume). The literature of applications, naturally enough, had little, if anything, to say about either the validity or the accuracy of the approximation. The theoretical papers (which generally treated only stationary processes over smooth Euclidean domains) were able to give various justifications for the validity of the approximation, but were never able to identify its precise level of accuracy. Doing so, particularly for the wide class of parameter spaces with which we are working, is the central aim of this chapter.

The main result of this chapter is contained in Theorem 14.3.3, which gives a precise bound on the difference between excursion probabilities and expected Euler characteristics of excursion sets for Gaussian fields.

The path toward proving Theorem 14.3.3 is surprisingly technical, although the idea behind it is actually quite simple. In principle, since we do not know much

---

[1] In [159], piecewise smooth manifolds were defined in a somewhat different fashion, but this leads to no significant differences in either the results or their proofs.

about how to compute excursion probabilities, but we do know a lot about computing expectations, we turn the probability problem into an expectation problem by introducing the random variable

$$M_u \stackrel{\Delta}{=} \text{the number of global suprema above the level } u.$$

Suppose that f has, almost surely, a unique global supremum, a situation that will generally hold (as we shall show) for smooth random fields. Then

$$M_u = \begin{cases} 1, & \sup_M f(t) \geq u, \\ 0 & \text{otherwise.} \end{cases}$$

It is therefore immediate that

$$\mathbb{P}\Big\{ \sup_{t \in M} f(t) \geq u \Big\} = \mathbb{E}\{M_u\}.$$

In order to compute $\mathbb{E}\{M_u\}$ we need to find a point process representation for M_u itself, much as we did for the Euler characteristic, in which case the representation came ready made from Morse theory. We can then use the theory of Chapter 12 to develop an integral expression for $\mathbb{E}\{M_u\}$, analogous to those we developed for computing $\mathbb{E}\{\varphi(A_u)\}$. Although we were actually able to evaluate the integral expression for $\mathbb{E}\{\varphi(A_u)\}$, it turns out to be impossible to evaluate it for $\mathbb{E}\{M_u\}$. Consequently, we shall take the difference between the two expressions, and work on this in order to bound $|\mathbb{E}\{M_u\} - \mathbb{E}\{\varphi(A_u)\}|$ or equivalently, $|\mathbb{P}\{\sup_{t \in M} f(t) \geq u\} - \mathbb{E}\{\varphi(A_u)\}|$. All of this is done in Sections 14.1 and 14.2 for quite general smooth random fields. In other words, Gaussianity is not assumed, and so these arguments lay the basis for quite a general theory. In Section 14.3 we specialize to the Gaussian case and derive bounds that begin to look computable. This is the central Theorem 14.3.3 we mentioned above. Finally, in Section 14.4, we shall see how to apply Theorem 14.3.3 in specific Gaussian scenarios, for which, after some work, the bounds become simple and explicit.

The first step of this procedure, that of finding a useful point process representation for M_u, turns out to be surprisingly difficult and extremely technical. We therefore start with it.

14.1 On Global Suprema

Our setting will be that of Chapter 12, so that (M, g) will be an N-dimensional, locally convex, regular stratified manifold ($M = \bigcup_{j=0}^{N} \partial_j M$) embedded in a C^3 manifold $(\widetilde{M}, \widetilde{g})$.

Furthermore, we take a random process $\widetilde{f} \in C^2(\widetilde{M})$, write f for the restriction of $\widetilde{f}$ to M, and assume that f is a Morse function on M.

Of course, since f is random, each of the above requirements on $\widetilde{f}$ and f is assumed to hold with probability one.

However, our argument starts in a purely deterministic setting, searching for a way to characterize global suprema of smooth functions. Thus, so as to make it clear which parts of the argument are deterministic and which stochastic, we start with (a realization of $\widetilde{f}$, denoted by) $\widetilde{h}$, which satisfies

$$\widetilde{h} \in C^2(\widetilde{M}), \quad h \stackrel{\Delta}{=} \widetilde{h}_{|M} \text{ is a Morse function on } M.$$

14.1.1 A First Representation

Our first representation of global suprema lies in the following lemma, whose main contribution lies in its third set of conditions.

Lemma 14.1.1. *With M, $\widetilde{h}$, and h as above, fix $t \in \partial_j M$, for some $0 \le j \le N$. Choose $\alpha^t \in C^2(M, (-\infty, 1])$ such that*

$$\alpha^t(s) = 1 \Rightarrow h(t) = h(s), \quad \alpha^t(t) = 1.$$

Then t is a maximizer of h above the level u if and only if the following three conditions hold:

(i) $h(t) \ge u$.
(ii) $\nabla\widetilde{h}(t) \in N_t M$, *where $N_t M$ is the normal cone to M at t. Thus t is an extended outward critical point of h (cf. Definition* 9.3.4).
(iii) *For all $t \in M$,*

$$h(t) \ge \sup_{s \in M \setminus \{t\}} h^t(s),$$

where for all $s \in M$,

$$h^t(s) \stackrel{\Delta}{=} \begin{cases} \frac{h(s) - \alpha^t(s) h(t)}{1 - \alpha^t(s)}, & \alpha^t(s) \ne 1, \\ h(s), & \alpha^t(s) = 1. \end{cases}$$

Furthermore, if $\nabla^2 \alpha^t(t)$ is nondegenerate, and t is a critical point of $h_{|\partial_j M}$, then for any C^2 curve $c : (-\delta, \delta) \to \partial_j M$ with $c(0) = t$, $\dot{c}(0) = X_t \in T_t \partial_j M$,

$$\lim_{u \to 0} h^t(c(u)) = \frac{\nabla^2 h_{|\partial_j M}(t)(X_t, X_t) - \nabla^2 \alpha^t_{|\partial_j M}(t)(X_t, X_t) h(t)}{-\nabla^2 \alpha^t_{|\partial_j M}(t)(X_t, X_t)}. \tag{14.1.1}$$

Note that the condition that $\alpha^t(t) = 1$ ensures that for each $t \in M$, t is a critical point of α^t.

Proof. The condition $h(t) \ge u$ is self-evident. Suppose, then, that $t \in \partial_i M$, $0 \le i \le k$, is a maximizer of h. Then, $\nabla\widetilde{h}(t) \in N_t M$. Otherwise, there exists a direction $X_t \in S_t M$ such that $\widetilde{g}_t(X_t, \nabla\widetilde{h}(t)) > 0$ and t cannot be a maximizer.

Because t is a maximizer, for all s such that $\alpha^t(s) < 1$ it follows that

$$\frac{h(s) - \alpha^t(s)h(t)}{1 - \alpha^t(s)} < h(t).$$

On the other hand, if $\alpha^t(s) = 1$, then, by choice of α^t, $h(s) = h(t)$, which proves that

$$h(t) \geq \sup_{s \in M \setminus \{t\}} h^t(s).$$

To prove the reverse implication, assume that t is an extended outward critical point of $h_{|\partial_i M}$ and

$$h(t) \geq \sup_{s \in M \setminus \{t\}} h^t(s).$$

Now suppose that t is not a maximizer of h. Then there exists $s \in M \setminus \{t\}$ such that

$$h(t) < h(s).$$

In particular, for such an s, our choice of α^t implies that $\alpha^t(s) < 1$. It follows that

$$h(t) < \frac{h(s) - \alpha^t(s)h(t)}{1 - \alpha^t(s)},$$

which yields a contradiction.

The limit (14.1.1) follows from two applications of l'Hôpital's rule. Specifically, we note that t is a critical point of $h_{|\partial_i M}$ by assumption and the properties of α^t imply that t must also be a nondegenerate critical point of α^t. Therefore,

$$\begin{aligned}
\lim_{u \to 0} \frac{h(c(u)) - \alpha^t(c(u))h(t)}{1 - \alpha^t(c(u))} &= \lim_{u \to 0} \frac{\frac{d}{du}\left(h(c(u)) - \alpha^t(c(u))h(t)\right)}{\frac{d}{du}\left(1 - \alpha^t(c(u))\right)} \\
&= \lim_{u \to 0} \frac{\frac{d^2}{du^2}\left(h(c(u)) - \alpha^t(c(u))h(t)\right)}{\frac{d^2}{du^2}\left(1 - \alpha^t(c(u))\right)}.
\end{aligned}$$

To compute the numerator and denominator here note the following: Fix a $t \in \partial_i M$ and take $\beta \in C^2(\partial_i M)$ for which t is a critical point of β. Let $c : (-\delta, \delta) \to \partial_i M$ be a C^2 curve with $c(0) = t$ and $\dot{c}(0) = X_t$. Then

$$\lim_{u \to 0} \frac{d^2}{du^2} \beta(c(u)) = \nabla^2 \beta(t)(X_t, X_t).$$

Apply this fact to each term in the last ratio above and use the assumed properties of α^t to complete the proof. □

Although we are currently in a deterministic setting, it is nevertheless worthwhile to "think stochastically" for a moment, to see what Lemma 14.1.1 has gained us. A much simpler and, indeed, trivial result would be to say that t is a maximizer of h if and only if conditions (i) and and (ii) of the lemma hold, and if (iii) is replaced by

$$h(t) \geq \sup_{s \in M \setminus \{t\}} h(s).$$

(Of course, this replacement for (iii) means that (i) and (ii) must automatically hold.) This seems far simpler than the original (iii) and does not require the introduction of the seemingly unnecessary function α^t.

However, now think of $h \equiv f$ as a mean zero, unit-variance Gaussian field, and let $\alpha^t(s) = \rho(t, s)$, where ρ is the correlation function of f. It is trivial to check that for each $t \neq s$ in M,

$$\mathbb{E}\left\{f(t) f^t(s)\right\} \equiv 0,$$

so that $f(t)$ and $\sup_{s \in M \setminus \{t\}} f^t(s)$ are independent, something that is not true of $f(t)$ and $\sup_{s \in M \setminus \{t\}} f(s)$. It is to gain this independence that we have introduced α^t and h^t. In fact, if life were good, we would stop here and count ourselves lucky at having already found a useful point process representation for global suprema, as required in the outline of our plan of attack. However, life is not always good, and indeed, we shall soon see that the function $(s, t) \mapsto h^t(s)$ can have some unpleasant properties.

14.1.2 The Problem with the First Representation

Despite all that we have said so far about the functions h^t, we have not looked too closely at their properties. From (14.1.1) we know that $h^t(s)$ is continuous as $s \to t$ in $\partial_j M$, but we have not looked at other directions. In fact, in other directions h^t is badly behaved. It is straightforward to check, again by two applications of l'Hôpital's rule, that

$$\nabla \widetilde{h}(t) \in N_t M \iff \sup_{s \in M \setminus \{t\}} h^t(s) < \infty, \tag{14.1.2}$$

$$\nabla \widetilde{h}(t) \in N_t M \quad \Rightarrow \quad \inf_{s \in M \setminus \{t\}} h^t(s) = -\infty.$$

In other words, there are paths of approach to t, in M but not in $\partial_j M$, along which the behavior of h^t is highly singular. We now begin attempting to resolve this singularity, with the solution finally appearing in Corollary 14.1.5, which gives a useable point process representation for global maximizers. The stochastic version of the solution appears in Corollary 14.1.6.

14.1.3 A Second Representation

Our aim in this section is to replace the function h^t we studied above by something smoother, although, unfortunately, it will be somewhat more involved. In order to profit from the characterization of global maxima as certain types of extended outward critical points (cf. Lemma 14.1.1), we shall ensure that the replacement function will be identical to our original h^t at critical points. Furthermore, we shall show that for this new function, which we shall continue to denote by h^t, the function

$$W(t) \triangleq \sup_{s \in M \setminus \{t\}} h^t(s) \tag{14.1.3}$$

is continuous at critical points of h if additional smoothness conditions are placed on the functions α^t. This will be the main result of this section.

Our strategy will be to augment M around a neighborhood of $t \in \partial_i M$ with additional information that allows us to examine functions on M in finer scales. This will lead to a "correction term" to h^t that will compensate for the singular behavior inherent in (14.1.2).

To start, we build the augmented space. Recall that we have assumed that M is a piecewise smooth space, and so at any $t \in M$ it is locally homeomorphic to the support cone $\mathcal{S}_t M$. Furthermore, from Section 8.2, we have that if $t \in \partial_i M$ then we can decompose the support cone by writing

$$\mathcal{S}_t M = T_t \partial_i M \times K_t,$$

where K_t is a convex cone that contains no subspace.

For ease of exposition, for the moment we move to a simple Euclidean setting and replace $\mathcal{S}_t M$ by

$$\widehat{K} = L \times K \subset \mathbb{R}^q = \widetilde{M},$$

with L a subspace and K a convex cone containing no subspaces. We describe the construction in this scenario now and return to piecewise smooth spaces later.

In this scenario, the augmented space $B(\widehat{K})$ of $\widehat{K}$ along L is taken to be the disjoint union of the spaces

$$\mathcal{X} = L \times (\widehat{K} \setminus \{0\}),$$
$$\partial \mathcal{X} = L \times (\widehat{K} \cap S(\mathbb{R}^q)).$$

You should think of $\mathcal{X}$ as the image of

$$(L \times \widehat{K}) \setminus \{(t, s) \in L \times \widehat{K} : t = s\}$$

under the map Ψ given by

$$\Psi(t, s) = (t, s - t), \tag{14.1.4}$$

and the second space, $\partial \mathcal{X}$, as the "boundary" of $\mathcal{X}$. The boundary is attached as follows: A sequence of points $(t_n, s_n) \in \mathcal{X}$ converges to $(t_0, s_0) \in \partial \mathcal{X}$ if and only if

$$t_n \to t_0, \qquad |s_n| \to 0, \qquad \frac{s_n}{|s_n|} \to s_0. \tag{14.1.5}$$

This notion of convergence corresponds to a sequence $\Psi^{-1}(t_n, s_n) = (t_n, t_n + s_n)$ converging to the diagonal

$$\left\{(t, s) \in L \times \widehat{K} : s = t\right\}$$

along a well-defined direction.

Note that by identifying the tangent bundle $T(\mathbb{R}^q)$ with $\mathbb{R}^q \times \mathbb{R}^q$ we can think of $B(\widehat{K})$ as a subset of $T(\mathbb{R}^q)$ in which the t's above are replaced with $X_t \in \mathcal{S}_t M \subset T_t\mathbb{R}^q$. When it is convenient, we will consider $B(\widehat{K})$ as either a subset of $\mathbb{R}^{2q}$ or a subset of $T(\mathbb{R}^q)$.

With an augmented space at hand, we now take our first steps toward a new definition of h^t. Note firstly, however, that the $\{\alpha^t\}_{t\in M}$ of Lemma 14.1.1 were a family of functions that did not necessarily arise as the partial map of a function $\alpha : \widetilde{M} \times \widetilde{M} \to (-\infty, 1]$. In some cases, particularly in the stochastic setting below, this is a natural assumption to make. Doing so allows us to prove the following.

Lemma 14.1.2. *Suppose $\alpha \in C^2(\mathbb{R}^q \times \mathbb{R}^q)$ is such that the partial map*

$$\alpha^t(s) = \alpha(t, s)$$

satisfies the conditions of Lemma 14.1.1 at every $t \in L$ and such that the Hessian of the partial map α^t is nondegenerate at every $t \in L$. Then, any $\widetilde{h} \in C^2(\mathbb{R}^q)$ maps to a continuous function $\widetilde{h}^{\alpha,\widehat{K}}$ on $B(\widehat{K})$ as follows:

$$\widetilde{h}^{\alpha,\widehat{K}}(t, s) \triangleq \frac{\widetilde{h}(t+s) - \alpha(t, t+s)\widetilde{h}(t) - \langle \nabla\widetilde{h}_t, s\rangle_{\mathbb{R}^q}}{1 - \alpha(t, t+s)}$$

if either t or s is not in $\partial\mathcal{X}$, and

$$\lim_{(u,v)\to(t,s)} \widetilde{h}^{\alpha,\widehat{K}}(u, v) \triangleq \widetilde{h}(t) - \frac{\nabla^2\widetilde{h}(t)(s, s)}{\nabla^2\alpha(t)(s, s)}$$

for $(t, s) \in \partial\mathcal{X}$.

Proof. All we need to prove is the continuity of $\widetilde{h}^{\alpha,\widehat{K}}$. This, however, follows from two applications of l'Hôpital's rule. □

The term

$$\frac{\langle \nabla\widetilde{h}_t, s\rangle_{\mathbb{R}^q}}{1 - \alpha(t, t+s)}$$

above "resolves" the singularity along the diagonal in some sense. In effect, it forces every $t \in L$ to be a critical point of the map

$$s \mapsto \frac{\widetilde{h}(t+s) - \alpha(t, t+s)\widetilde{h}(t) - \langle \nabla\widetilde{h}(t), s\rangle_{\mathbb{R}^q}}{1 - \alpha(t, t+s)}.$$

To see how to exploit this fact, recall that our motivation for introducing $\widetilde{h}^{\alpha,\widehat{K}}$ was to describe the singularities in the function $h^t(s)$ at critical points t of $h_{|L}$. (Recall also that L takes the place of $T_t\partial_i M$ in our general cone $\widehat{K}$). We are therefore interested in critical points t of $h_{|L} = \widetilde{h}_{|L}$. Note that if t is a critical point of $h_{|L}$, then, for all $s \in \mathbb{R}^q$,

$$h^t(s) = \frac{h(s) - \alpha(t,s)h(t) - \langle \nabla h(t), s-t\rangle_{\mathbb{R}^q}}{1-\alpha(t,s)} + \frac{\langle P_L^\perp \nabla h(t), s-t\rangle_{\mathbb{R}^q}}{1-\alpha(t,s)} \tag{14.1.6}$$
$$= \widetilde{h}^{\alpha,\widehat{K}}(t, s-t) + \frac{\langle P_L^\perp \nabla h(t), s-t\rangle_{\mathbb{R}^q}}{1-\alpha(t,s)},$$

where $P_L^\perp$ represents orthogonal projection onto $L^\perp$, the orthogonal complement of L in $\mathbb{R}^q$. The expression (14.1.6) indicates that at critical points t of $h_{|L}$, $h^t(s)$ is the sum of a well-behaved term, $\widetilde{h}^{\alpha,\widehat{K}}$, and a singular term.

This representation holds for all critical points of $h_{|L}$. However, for later arguments we would like to have something of this form for all $t \in L$ and therefore redefine h^t by setting

$$h^t(s) \triangleq \begin{cases} \frac{h(s)-\alpha^t(s)h(t)-\langle P_L \nabla h(t), s\rangle_{\mathbb{R}^q}}{1-\alpha^t(s)}, & \alpha^t(s) \neq 1, \\ h(s), & \alpha^t(s) = 1. \end{cases} \tag{14.1.7}$$

With this final (re)definition of h, (14.1.6) holds for all $t \in L$, and furthermore, for each critical point t of $h_{|L}$ the two definitions of h^t coincide. For the remainder of this section, as long as we remain in the Euclidean settting, we use the definition (14.1.7). We shall, however, need to adjust it somewhat when we return to the piecewise smooth scenario.

We now derive some properties of our new h^t.

Lemma 14.1.3. *$P_L^\perp \nabla h(t) \in K^*$, the dual of $K \subset L^\perp$, if and only if for any bounded neighborhood O_t of t,*

$$\sup_{s \in O_t \setminus \{t\}} h^t(s) < +\infty.$$

Proof. If $P_L^\perp \nabla h(t) \in K^*$, then (14.1.6) implies that for all s,

$$h^t(s) \leq \widetilde{h}^{\alpha,\widehat{K}}(t, s-t).$$

However, by Lemma 14.1.2, $\widetilde{h}^{\alpha,\widehat{K}}$ is continuous in s, and therefore bounded on bounded sets. This proves one direction of the lemma.

For the other, if $P_L^\perp \nabla h(t) \notin K^*$, then there exists a unit vector $v \in \widehat{K}$ such that $\langle P_L^\perp \nabla h(t), v\rangle_{\mathbb{R}^q} > 0$ and $t + \delta v \in O_t$ for sufficiently small δ. Equation (14.1.6) then implies that

$$\lim_{\delta \downarrow 0} h^t(t+\delta v) = +\infty,$$

since the numerator on the right of (14.1.6) is strictly positive of order $O(\delta)$ for δ small, while the denominator is of order $O(\delta^2)$. □

Lemma 14.1.4. *If $P_L^\perp \nabla h(t) \in (K^*)^\circ$, then, for any bounded neighborhood O of the origin in $\mathbb{R}^q$,*

$$
\begin{aligned}
W_O(t) &\stackrel{\Delta}{=} \sup_{s\in\widehat{K}\cap(\{t\}\oplus O)\setminus\{t\}} h^t(s) \\
&= \sup_{v\in\widehat{K}\cap(O\setminus\{0\})} \left(\widetilde{h}^{\alpha,\widehat{K}}(x,v) + \frac{\langle P_L^{\perp}\nabla h(t), v\rangle_{\mathbb{R}^q}}{1-\alpha(t,t+v)} \right)
\end{aligned}
$$

is continuous in t.

Proof. We first note that by (14.1.6) the two suprema above are equal, and so it suffices to consider the second one.

Consider a convergent sequence

$$
\{(t, v_n(t)\}_{n\geq 0} \to (x, v^*(t))
$$

in $B(\widehat{K})$ along which the supremum $W_O(t)$ is approached. Then either $\|v_n(t)\|_{\mathbb{R}^q} > 0$ for all n sufficiently large or $\|v_n(t)\|_{\mathbb{R}^q} \to 0$. In the first case it is immediate that $(t, v^*(t))$ is in $\mathcal{X}$, and as we will show in a moment, in the second case $(t, v^*(t)) \in \partial\mathcal{X} \cap L \times S(L)$, where $S(L)$ is the unit sphere in L. In other words, the limiting direction $v^*(t)$ is in $S(L)$. Furthermore, if $\|v_n(t)\|_{\mathbb{R}^q} \to 0$ then it is easy to see that the sequence $(t, P_L v_n(t))$ also achieves the supremum $W_O(t)$.

To see why $v^*(t)$ must be in $S(L)$ in the second case, suppose that $(t, v_n(t))$ converges to $(x, v^*(t))$ with $v^*(t) \in \widehat{K} \setminus L$. Since $P_L^{\perp}\nabla h(t) \in (K^*)^{\circ}$ it follows that

$$
\langle P_L^{\perp}\nabla h(t), v^*(t)\rangle_{\mathbb{R}^q} < 0.
$$

Therefore, applying the same argument as in (14.1.6),

$$
\lim_{n\to\infty} \frac{\langle P_L^{\perp}\nabla h(t), v_n(t)\rangle_{\mathbb{R}^q}}{1-\alpha(t,t+v_n(t))} = -\infty,
$$

which contradicts the assumption that

$$
\lim_{n\to\infty} h^t(t+v_n(t)) = W_O(t).
$$

The above implies that for any $t \in L$ and any convergent sequence $(x, v_n(t))$ achieving $W_O(t)$ we can either assume that $|v_n(t)|$ is bounded uniformly below, or that $v_n(t) \in L$ for all n. Continuity of $W_O(t)$ now follows, since for such sequences both

$$
\sup_{s\in B(t,\delta(\varepsilon))} \left| \widetilde{h}^{\alpha,\widehat{K}}(t, v_n(t)) - \widetilde{h}^{\alpha,\widehat{K}}(s, v_n(t)) \right|
$$

and

$$
\sup_{s\in B(t,\delta(\varepsilon))} \left| \frac{\langle P_L^{\perp}\nabla h(t), v_n(t)\rangle_{\mathbb{R}^q}}{1-\alpha(t,t+v_n(t))} - \frac{\langle P_L^{\perp}\nabla h(s), v_n(t)\rangle_{\mathbb{R}^q}}{1-\alpha(s,s+v_n(t))} \right|
$$

go to 0 as $n \to \infty$. This completes the proof. □

Lemmas 14.1.3 and 14.1.4 give us what we need for the Euclidean scenario, that is, a candidate for h^t that is well behaved and that we can plug into our characterization for global suprema. The same arguments also work for piecewise smooth spaces, after some slight modifications.

Specifically, the map Ψ, defined in (14.1.4), has no natural replacement for a piecewise smooth space. However, we can think of $B(\widehat{K})$ as a subset of the tangent bundle $T(\mathbb{R}^q)$, so that we can think of Ψ as a map from $\mathbb{R}^q \times \mathbb{R}^q$ to $T(\mathbb{R}^q)$. Translating this to the piecewise smooth setting, we should therefore look for a replacement map, say H, from $\widetilde{M} \times \widetilde{M}$ to $T(\widetilde{M})$ such that for each $t \in \widetilde{M}$,

$$H\left(\{t\} \times \widetilde{M}\right) \subset T_t\widetilde{M}. \tag{14.1.8}$$

One of the key properties of Ψ was that for sequences $(t_n, s_n) \in \mathcal{X}$ converging to $(t, 0) \in \partial\mathcal{X}$ for which the unit vector

$$\frac{t_n - s_n}{\|t_n - s_n\|_{\mathbb{R}^q}}$$

converged in $S(\mathbb{R}^q)$ (cf. (14.1.5)), $\Psi(t_n, s_n)$ converged to a point in $\mathcal{X}$. We can replace this property of Ψ by asking the following of H. For any C^1 curve $c : (-\delta, \delta) \to \widetilde{M} \times \widetilde{M}$ with $c(0) = (c_1(0), c_2(0)) = (t, t)$ and $\dot{c}_2(0) = X_t$ we require that

$$\lim_{u \to 0} \frac{H(t, c_2(u)) - H(t, t)}{u} = X_t. \tag{14.1.9}$$

Given such an H, we can, once again, redefine h^t for a C^2 function on a piecewise smooth space M. For $t \in \partial_i M$, we redefine h^t to be

$$h^t(s) \triangleq \begin{cases} \frac{h(s) - \alpha^t(s)h(t) - \widetilde{g}(\nabla h_{|\partial_i M}(t), H(t,s) - \alpha^t(s)H(t,t))}{1 - \alpha^t(s)}, & \alpha^t(s) \neq 1, \\ h(s), & \alpha^t(s) = 1. \end{cases} \tag{14.1.10}$$

Note that at critical points of $h_{|\partial_i M}(t)$ we always have that $\nabla h_{|\partial_i M} = 0$, so that the term in $\widetilde{g}$ in (14.1.10) disappears, as it does for (i) in Lemma 14.1.5 below. Having made this observation, by working in suitably chosen charts, it is not difficult to prove the following lemma.

Lemma 14.1.5. *With M, $\widetilde{h}$, and h as above, suppose that $\alpha \in C^2(\widetilde{M} \times \widetilde{M}, (-\infty, 1])$ is such that the partial maps $\alpha^t(s) = \alpha(t, s)$ satisfy*

$$\alpha^t(s) = 1 \Rightarrow h(t) = h(s), \quad \alpha^t(t) = 1,$$

for every $t \in M$ and such that the Hessian of the partial map α^t is nondegenerate at every $t \in M$. Furthermore, suppose $H : \widetilde{M} \times \widetilde{M} \to T(M)$ satisfies (14.1.8) *and* (14.1.9), *and let h^t be defined as in* (14.1.10).

Then t is a maximizer of h above the level u if and only if the following three conditions hold:

(i) $h(t) - \widetilde{g}(H(t,t), \nabla h_{|\partial_i M}(t)) \geq u$.
(ii) $\nabla \widetilde{h}(t) \in N_t M$.
(iii) *For all* $t \in M$,

$$h(t) \geq W(t) \stackrel{\Delta}{=} \sup_{s \in M \setminus \{t\}} h^t(s).$$

Furthermore, for every $t \in \partial_i M$ *and any* C^2 *curve* $c : (-\delta, \delta) \to \partial_i M$ *with* $c(0) = t$ *and* $\dot{c}(0) = X_t$,

$$\lim_{t \to 0} h^t(c(t)) = \frac{\nabla^2 \left(h(\cdot) - \widetilde{g}\left(H^t(\cdot), \nabla h_{|\partial_i M}(t)\right)\right)(X_t, X_t)}{-\nabla^2 \alpha^t_{|\partial_i M}(t)(X_t, X_t)} - \left(h(t) - \widetilde{g}(H^t(t), \nabla h_{|\partial_i M}(t))\right), \tag{14.1.11}$$

where H^t *is the partial map* $H^t(s) = H(t,s)$.

In addition, the function

$$W(t) \stackrel{\Delta}{=} \sup_{s \in M \setminus \{t\}} h^t(s)$$

is, for each $0 \leq i \leq k$, *continuous on the set*

$$\left\{ t \in \partial_i M : P^{\perp}_{T_t \partial_i M} \nabla h(t) \in (P^{\perp}_{T_t \partial_i M} N_t M)^{\circ} \right\},$$

where $P^{\perp}_{T_t \partial_i M}$ *represents projection onto the orthogonal complement of* $T_t \partial_i M$ *in* $T_t \widetilde{M}$.

To appreciate why all of this hard work has been necessary to characterize something as simple as a supremum, we need first to translate it all to a stochastic setting.

14.1.4 Random Fields

We now turn to a stochastic version of the previous section, with M a C^2, locally convex, regular, stratified manifold embedded in a C^3 manifold $\widetilde{M}$ and $\widetilde{f}$ a C^2 process on $\widetilde{M}$ such that $f = \widetilde{f}_{|M}$ is a Morse function on M.

We assume that f is L^2, with correlation function

$$\rho(s,t) = \frac{\mathbb{E}\{f(s)f(t)\}}{\sigma(t)\sigma(s)},$$

where $\sigma^2(t) = \mathbb{E}\{f^2(t)\}$.

Although we are not, at this stage, assuming that $\widetilde{f}$ is Gaussian, we nevertheless endow $\widetilde{M}$ with the usual Riemannian metric induced by $\widetilde{f}$, so that for $X_t, Y_t \in T_t \widetilde{M}$,

$$\widetilde{g}(X_t, Y_t) \stackrel{\Delta}{=} \mathbb{E}\left\{X_t \widetilde{f} \cdot Y_t \widetilde{f}\right\}.$$

As usual, M is given the metric induced by $\widetilde{g}$, and we also assume that both metrics are C^2.

We also place a new condition on ρ that we have not met before. It ensures that the map $t \mapsto \widetilde{f}(t)$ is an embedding of $\widetilde{M}$ into $L^2(\Omega, \mathcal{F}, \mathbb{P})$ and rules out "global" singularities in the process, a term whose meaning will become clearer later. The condition is

$$\rho(t, s) = 1 \iff t = s. \tag{14.1.12}$$

This condition rules out, for example, periodic processes on $\mathbb{R}$. However, as we shall see in Section 14.4.4, there is a good reason why this should be the case.

With the above definitions, we can now choose candidates for our functions α and H of the previous subsection. Specifically, we take an orthonormal (with respect to $\widetilde{g}$) frame field $(X_{1,t}, \dots, X_{\dim \widetilde{M},t})$ on $T_t\widetilde{M}$ and set

$$\alpha(t, s) \stackrel{\Delta}{=} \rho(t, s), \tag{14.1.13}$$

$$H(t, s) = F(t, s) \stackrel{\Delta}{=} \sum_{j=1}^{\dim \widetilde{M}} \mathrm{Cov}(\widetilde{f}(s), X_j \widetilde{f}(t)) X_{j,t}. \tag{14.1.14}$$

It is easy to check that the definition of F is independent of the choice of frame field.

We can then state the stochastic analogy of Lemma 14.1.5, i.e., a point process representation for the maximizers of the random field f.

Corollary 14.1.6. *In addition to the above assumptions, assume that the maximizers of $\widetilde{f}$ are almost surely isolated and that there are no critical points of $f_{|\partial_i M}$ such that $P^{\perp}_{T_t\partial_i M} \nabla \widetilde{f}(t) \in \partial N_t M \subset T_t \partial_i M^{\perp}$. Then the maximizers of f are the points $t \in \partial_i M$, $0 \le i \le N$, satisfying the following four conditions:*

(i) $f(t) - \widetilde{g}(F(t,t), \nabla f_{|\partial_i M}(t)) \ge u$.
(ii) $\nabla f_{|\partial_i M}(t) = 0$.
(iii) $P^{\perp}_{T_t\partial_i M} \nabla \widetilde{f}(t) \in (P^{\perp}_{T_t\partial_i M} N_t M)^{\circ}$.
(iv) $f(t) \ge \sup_{s \in M \setminus \{t\}} f^t(s)$,

where

$$f^t(s) \stackrel{\Delta}{=} \begin{cases} \frac{f(s) - \rho(t,s) f(t) - \widetilde{g}(F(t,s) - \rho(t,s)F(t,t), \nabla f_{|\partial_i M}(t))}{1 - \rho(t,s)}, & \rho(t,s) \neq 1, \\ f(s), & \rho(t,s) = 1. \end{cases} \tag{14.1.15}$$

Note that for later convenience, we have split the condition that t be an extended outward critical point into two parts—(ii) and (iii).

Proof. Note firstly that since there are no critical points of $f_{|\partial_i M}$ such that $P^{\perp}_{T_t\partial_i M} \nabla \widehat{f}(t) \in \partial N_t M \subset T_t \partial_i M^{\perp}$, all global maximizers will be such that $P^{\perp}_{T_t\partial_i M} \nabla \widehat{f}(t)$ is in the relative interior of $N_t M$ in $T_t \partial_i M^{\perp}$. Thus (ii) and (iii) are equivalent to the requirement that t be an extended outward critical point.

Next, note that while α and ρ have been assumed to have almost identical properties, they differ in one important aspect: While we assumed that $\alpha^t(s) = 1$ implied

$h(t) = h(s)$, the same cannot be said to hold when $\rho(t,s) = 1$ unless we also assume that $\mathbb{E}\{f(t)\}$ is constant and $\mathbb{E}\{f^2(t)\} = 1$ for all t. Nevertheless, a quick check through the previous proofs will confirm the fact that given the definition of f^t, this difference affects none of the arguments.

Having noted these two facts, we can apply Lemma 14.1.5 on an almost sure basis, with h^t replaced by f^t, α by ρ, and H by F. □

14.1.5 Suprema and Euler Characteristics

We now have enough to finally see what the link is between $\varphi(A_u)$, the Euler characteristic of the excursion set $\{t \in M : f(t) \geq u\}$, and the global supremum of f.

To do this, we need one more assumption, that the global maximizer of the process f is almost surely unique. In this case, it follows from Corollary 14.1.6, under the assumptions made there along with uniqueness of the global supremum, that[2]

$$\mathbb{1}_{\{\sup_{t\in M} f(t)\geq u\}} = \#\Big\{\text{extended outward critical points } t \text{ of } f \text{ for which} \\ f(t) \geq u,\ \sup_{s\in M\setminus\{t\}} f^t(s) \leq f(t)\Big\}. \tag{14.1.16}$$

On the other hand, if we return to Corollary 9.3.4, which gives a point set characterization for the Euler characteristic for excursion sets over piecewise smooth spaces, and compare this to the above, we immediately obtain that

$$\begin{aligned}&\left|\mathbb{1}_{\left\{\sup_{t\in M} f(t)\geq u\right\}} - \varphi(A_u(f,M))\right| \\ &\quad \leq \#\Big\{\text{extended outward critical points } t \text{ of } f \text{ for which} \\ &\qquad f(t) \geq u,\ \sup_{s\in M\setminus\{t\}} f^t(s) > f(t)\Big\}.\end{aligned}$$

While correct, this relationship is not terribly useful, since we have no way of computing the right-hand side of the inequality. What turns out to be a lot more accessible is to begin evaluating the expectations of each term on the left, and then compute with the difference. To do this, we need additional regularity conditions on f, and adopt those of Theorem 11.3.4 along with the assumption[3] that for all $t \in M$,

$$\mathbb{P}\{W(t) = f(t)\} = \mathbb{P}\{W(t) = u\} = 0, \tag{14.1.17}$$

---

[2] Recall, once again, that at extended outward critical points $\nabla f_{|\partial_i M} = 0$, so that all the expressions in Corollary 14.1.6 involving $\nabla f_{|\partial_i M}$ disappear.

[3] Sufficient conditions to guarantee that (14.1.17) holds are that for all pairs $1 \leq i, j \leq N$ and all pairs $\{(t,s) : t \in \partial_i M, s \in \partial_j M\}$, the random vector

$$V(t,s) \stackrel{\Delta}{=} (f(t) - f(s), \nabla f_{|\partial_i M, E_i}(t), \nabla f_{|\partial_j M, E_j}(s))$$

have a density, bounded by some constant $K(t,s)$. Then it is an easy consequence of Lemma 11.2.10 that

where

$$W(t) \stackrel{\Delta}{=} \sup_{s \in M \setminus \{t\}} f^t(s),$$

and f^t is as defined in (14.1.15). Note that none of these additional assumptions require that f be Gaussian.

Noting (14.1.16), it is then a straightforward[4] application of Theorem 12.1.1, to see, in the notation of Chapter 12, that

$$\begin{aligned} \mathbb{P}\Big\{ \sup_{t \in M} f(t) \geq u \Big\} &= \mathbb{E}\{M_u\} \\ &= \sum_{i=0}^{N} \int_{\partial_i M} \mathbb{E}\{\det(-\nabla^2 f_{|\partial_i M, E_i}(t)) \mathbb{1}_{A_t^{\text{SUP}}} | \nabla f_{|\partial_i M, E_i}(t) = 0\} \\ &\quad \times p_{\nabla f_{|\partial_i M, E_i}(t)}(0)\, d\mathcal{H}_i(t), \end{aligned}$$

where

$$A_t^{\text{SUP}} \stackrel{\Delta}{=} \{f(t) \geq u \wedge W(t),\ \nabla \widetilde{f}(t) \in N_t M\},$$

and where $p_{\nabla f_{|\partial_i M, E_i}(t)}$ is the density of $\nabla f_{|\partial_i M, E_i}(t)$ and $\mathcal{H}_i$ is the volume measure induced by the Riemannian metric g.

Arguing as in the proof of Theorem 12.4.2 (cf. also (12.4.4)) it is not hard to see that

$$\begin{aligned} \mathbb{E}\{\varphi(A_u(f, M))\} &= \sum_{i=0}^{N} \int_{\partial_i M} \mathbb{E}\{\det(-\nabla^2 f_{|\partial_i M, E_i}(t)) \mathbb{1}_{A_t^{EC}} | \nabla f_{|\partial_i M, E_i}(t) = 0\} \\ &\quad \times p_{\nabla f_{|\partial_i M, E_i}(t)}(0)\, d\mathcal{H}_i(t), \end{aligned}$$

where

$$A_t^{EC} \stackrel{\Delta}{=} \{f(t) \geq u,\ \nabla \widetilde{f}(t) \in N_t M\}.$$

$$\mathbb{P}\left\{\exists (t, s) : t \in \partial_i M,\ y \in \partial_j M,\ V(t, s) = 0 \in \mathbb{R}^{i+j+1}\right\} = 0,$$

which implies (14.1.17).

[4] Well... almost straightforward. A trivial application of Theorem 12.1.1 would require that W be continuous, which is not the case. However, (an almost sure version of) Corollary 14.1.5 does give us that W is continuous on the open set

$$\left\{ t \in \partial_i M : P^{\perp}_{T_t \partial_i M} \nabla \widehat{f}(t) \in (N_t M)^{\circ} \right\},$$

and, if you check through the proofs leading to Theorem 12.1.1—which is still "straightforward" but which would take some time—you will see that this is all we actually need.

The discrepancy between the mean Euler characteristic of an excursion set and the supremum distribution is therefore

$$\begin{aligned}
&\operatorname{Diff}_{f,M}(u) \qquad (14.1.18)\\
&\triangleq \mathbb{E}\{\varphi(A_u(f,M))\} - \mathbb{P}\Big\{\sup_{t\in M} f(t) \geq u\Big\}\\
&= \sum_{i=0}^{N} \int_{\partial_i M} \mathbb{E}\{\det(-\nabla^2 f_{|\partial_i M, E_i}(t)) \mathbb{1}_{A_t^{\mathrm{ERR}}} | \nabla f_{|\partial_i M, E_i}(t) = 0\}\\
&\quad \times p_{\nabla f_{|\partial_i M, E_i}(t)}(0)\, d\mathcal{H}_i(t),
\end{aligned}$$

where

$$A_t^{\mathrm{ERR}} \triangleq \{u \leq f(t) \leq W(t),\ \nabla \widetilde{f}(t) \in N_t M\}.$$

There are three remarkable facts to note about (14.1.18). The first is that it is an identity, and not an approximation or a bound. The second is that it holds for a wide class of smooth random fields over a large class of piecewise smooth parameter spaces. The last is that these smooth random fields were never assumed to be Gaussian.

Given (14.1.18), it is not difficult to guess how the rest of the proof should proceed. The key point is that the expectation is taken only over the event A_t^{ERR}, which is included in the event $\{u \leq f(t) \leq W(t)\}$. Now, if you remember our brief stochastic diversion at the end of Section 14.1.1, you will remember that if f is Gaussian with constant mean and unit variance, $f(t)$ and $W(t)$ are independent. Thus

$$\begin{aligned}
\mathbb{P}\left\{A_t^{\mathrm{ERR}}\right\} &\leq \mathbb{P}\{u \leq f(t) \leq W(t)\} \qquad (14.1.19)\\
&\leq \mathbb{P}\{u \leq f(t)\} \times \mathbb{P}\{u \leq W(t)\}.
\end{aligned}$$

We know from the very general theory of Chapter 4 that for the smooth random fields that we are considering it is likely that

$$\mathbb{P}\Big\{\sup_{t\in M} f(t) \geq u\Big\} \approx u^{\alpha} \sup_{t\in M} \mathbb{P}\{u \leq f(t)\},$$

for some α, and where, since the current argument is heuristic, we make no attempt to give a precise mathematical meaning to the $\approx$ here. Furthermore, we should also be able to bound the rightmost probability in (14.1.19) by the general theory of Chapter 4, since it is just a Gaussian excursion probability. If so, it too is going to be exponentially small. Putting these two facts into (14.1.19) makes $\mathbb{P}\left\{A_t^{\mathrm{ERR}}\right\}$ *super*exponentially small (when compared to $\mathbb{P}\left\{\sup_{t\in M} f(t) \geq u\right\}$) and, modulo technicalities, we basically have the result promised at the beginning of the chapter. However, since that result also promised explicit bounds on the rate at which $\operatorname{Diff}_{f,M}(u) \to 0$ as $u \to \infty$, we are also going to have to do some more detailed computation.

Of course, what is clear from the above description is that to truly exploit (14.1.18) we are going to have to make distributional assumptions. Above, we have assumed

Gaussianity. Perhaps not surprisingly, despite the universality of (14.1.18), we do not know how to handle non-Gaussian processes.

This is a shame, since in Chapter 15 we shall obtain explicit formulas for $\mathbb{E}\{\varphi(A_u(f, M))\}$ for a wide class of non-Gaussian random fields f, and it would be nice to be able to exploit (14.1.18) to also approximate their excursion probabilities. We do hope/believe that results akin to those that we are about to encounter for Gaussian processes will also one day be proven in non-Gaussian scenarios, but for the moment, we are stumped by the details.

14.2 Some Fine Tuning

While (14.1.18) is an explicit formula for the error in the expected Euler characteristic approximation, it is still a little hard to compute with. The problem lies in the fact that while we finally managed to find a version of $W(t) = \sup_{s \in M \setminus \{t\}} f^t(s)$ that is continuous in t (at least where it counts), f^t itself is still rather badly behaved. In particular, it is generally singular near t, with infinite variance.[5] This, of course, means that the approach outlined at the end of the last subsection for the Gaussian case—of using the general theory of Chapter 4 to handle the distribution of W—is doomed to failure, unless something is done.

---

[5] To see why this should be so, it suffices to look at one example. Take f to be the restriction of an isotropic field to $[0, 1]^2$. Fix a point $t = (x, 0)$. We compute the variance of f^t along the curve

$$c(u) = (x, 0) - u(1, 0).$$

Straightforward calculations show that

$$F(t, c(u)) = 0, \qquad F(t, t) = 0,$$

so that

$$\widetilde{f}^t(c(u)) = \frac{f(c(u)) - \rho(t, (c(u))f(t)}{1 - \rho(t, c(u))}.$$

The variance of $f^t(c(u))$ is

$$\mathrm{Var}\left(f^t(c(u))\right) = \frac{1 + \rho(t, c(u))}{1 - \rho(t, c(u))},$$

and

$$\lim_{u \to 0} \mathrm{Var}\left(f^t(c(u))\right) = \lim_{u \to 0} \frac{1 + \rho(t, c(u))}{1 - \rho(t, c(u))} = +\infty.$$

In general, if $t \in \partial_i M$, then along a curve $c : (-\delta, 0] \to M$ with $\dot{c}(0) = -X_t \in S_t M \setminus T_t \partial_i M$, we have

$$\lim_{u \uparrow 0} \mathrm{Var}\left(f^t(c(u))\right) = +\infty.$$

The key to circumventing this problem relies on two facts. Firstly, we are really concerned only with large *positive* values of f^t, and, secondly, we care about the behavior of f^t only on the set

$$\left\{(t,\omega) : P^{\perp}_{T_t\partial_i M}\nabla\widetilde{f}(t)(\omega) \in (P^{\perp}_{T_t\partial_i M}N_tM)^{\circ}\right\}. \tag{14.2.1}$$

We exploit these facts and introduce a process $\widehat{f^t}$ in this section that has, under appropriate conditions, finite variance while dominating[6] f^t on the set (14.2.1). The good news is that this is the last time we shall have to "adjust" f^t. The final adjustment is important, for not only does it make the proofs work, but the variance of $\widehat{f^t}$ will appear as an important parameter in our final result, related to the α in (14.0.3).

The process $\widehat{f^t}$ is defined by

$$\widehat{f^t}(s) \stackrel{\Delta}{=} \begin{cases} \frac{f(s)-\rho(t,s)f(t)-\widetilde{g}\left(\widehat{F^t}(s)-P_{N_tM}\widehat{F^t}(s),\nabla\widetilde{f}(t)\right)}{1-\rho(t,s)}, & \rho(t,s)\neq 1, \\ f(s)-\widetilde{g}\left(\widehat{F^t}(s)-P_{N_tM}\widehat{F^t}(s),\nabla\widetilde{f}(t)\right), & \rho(t,s)=1, \end{cases} \tag{14.2.2}$$

where $P_{N_tM} : T_t\widetilde{M} \to N_tM$ represents orthogonal projection onto N_tM and $\widehat{F} : \widetilde{M}\times\widetilde{M} \to T(\widetilde{M})$ is given by

$$\widehat{F^t}(s) = \begin{cases} F(t,s)-\rho(t,s)F(t,t), & \rho(t,s)\neq 1, \\ F(t,s), & \rho(t,s)=1, \end{cases} \tag{14.2.3}$$

where F is defined in (14.1.14).

Lemma 14.2.1. *We adopt the setting of Section* 14.1.4 *and the assumptions of Corollary* 14.1.6. *Then, for every t in the set* (14.2.1) *of extended outward critical points and for every $s \in M$,*

$$\widehat{f^t}(s) \geq f^t(s).$$

If $t \in \partial_N M = M^{\circ}$, then equality holds above.

Proof. First, we note that

$$(1-\rho(t,s))\cdot(f^t(s)-\widehat{f^t}(s)) = \widetilde{g}\left(\widehat{F^t}(s)-P_{N_tM}\widehat{F^t}(s), P^{\perp}_{T_t\partial_i M}\nabla\widetilde{f}(t)\right).$$

Since N_tM is a convex cone, it follows that for any $Y_t \in T_t\widetilde{M}$,

$$Y_t - P_{N_tM}Y_t \in N_tM^*,$$

where N_tM^* is the dual cone of N_tM, which is just the convex hull of $\mathcal{S}_tM$. By duality,

$$\widetilde{g}\left(Y_t - P_{N_tM}Y_t, V_t\right) \leq 0$$

[6] Obviously, $\widehat{f^t}$ does not dominate the *absolute* value of f^t. Were this true, $\widehat{f^t}$ would also have infinite variance and we would have gained nothing.

for every $V_t \in N_t M$. Consequently, on the set (14.2.1),

$$\widetilde{g}\left(Y_t - P_{N_t M} Y_t, P^{\perp}_{T_t \partial_i M} \nabla \widetilde{f}(t)\right) \le 0$$

for every $Y_t \in T_t \widetilde{M}$. Since $\widehat{F}^t(s) \in T_t \widetilde{M}$ for each s, the first claim holds.

As for the second, if $t \in \partial_N M$, then $N_t M = T_t \partial_N M^{\perp}$ and $P_{N_t M} V_t = 0$ for all $V_t \in T_t \partial_N M$. Similarly,

$$\widetilde{g}\left(V_t, P^{\perp}_{T_t \partial_i M} \nabla \widetilde{f}(t)\right) = 0.$$

Therefore, on this set

$$\widetilde{g}\left(\widehat{F}^t(s) - P_{N_t M} \widehat{F}^t(s), \nabla \widetilde{f}(t)\right) = 0. \qquad \square$$

As far as the continuity (in t) of

$$\widehat{W}(t) \stackrel{\Delta}{=} \sup_{s \in M \setminus \{t\}} \widehat{f}^t(s) \tag{14.2.4}$$

is concerned, it is not difficult to show that, almost surely, Lemma 14.1.5 holds with f^t replaced by $\widehat{f}^t$, i.e., that $\widehat{W}(t)$ is continuous on the set

$$\{t \in \partial_i M : P^{\perp}_{T_t \partial_i M} \nabla \widetilde{f}(t) \in (P^{\perp}_{T_t \partial_i M} N_t M)^{\circ}\}.$$

As we will see in a moment in the proof of Theorem 14.2.2, Lemma 14.2.1 provides the basic bounds for $\mathrm{Diff}_{f,M}(u)$.

Lemma 14.2.1 now easily implies the following theorem, for which, since it is the main result of this section, we collect all the conditions and notation.

Theorem 14.2.2. *Suppose M is a C^2 compact, orientable, piecewise smooth manifold, embedded in a C^3 manifold $\widetilde{M}$. Let $\widetilde{f}$ be a C^2 random field on $\widetilde{M}$, and f its restriction to M. Let ρ be the correlation function of f, and assume that the induced metric is C^2 and that* (14.1.12) *holds. Furthermore, assume that for each $t \in \widetilde{M}$ the partial map $\rho(t, \cdot)$ has a nondegenerate Hessian.*

Assume that the maximizer of $\widetilde{f}$ is unique and that there are no critical points of $f_{|\partial_i M}$ such that $P^{\perp}_{T_t \partial_i M} \nabla \widetilde{f}(t) \in \partial N_t M \subset T_t \partial_i M^{\perp}$.

Finally, assume that all the conditions of Theorem 11.3.4 *are in force and that for every $t \in M$,*

$$\mathbb{P}\{\widehat{W}(t) = f(t)\} = \mathbb{P}\{\widehat{W}(t) = u\} = 0,$$

where $\widehat{W}$ is defined by (14.2.4), *where $\widehat{f}^t$ there is defined by* (14.2.2).

Then, with $\mathrm{Diff}_{f,M}(u)$ *denoting the difference* (14.1.18),

$$\begin{aligned} &|\mathrm{Diff}_{f,M}(u)| \\ &\quad \le \sum_{i=0}^{N} \int_{\partial_i M} \mathbb{E}\{|\det(-\nabla^2 f_{|\partial_i M, E_i}(t))| \mathbb{1}_{B_t^{\mathrm{ERR}}} | \nabla f_{|\partial_i M, E_i}(t) = 0\} \\ &\qquad \times p_{\nabla f_{|\partial_i M, E_i}(t)}(0)\, d\mathcal{H}_i(t), \end{aligned} \tag{14.2.5}$$

where

$$B_t^{\mathrm{ERR}} = \{u \le f(t) \le \widehat{W}(t)\}.$$

Of course, there is the usual simple Gaussian corollary to the theorem.

Corollary 14.2.3. *With M as in Theorem* 14.2.2, *assume that $\widetilde{f}$ is a zero-mean Gaussian field satisfying the conditions of Corollary* 11.3.5 *and that* (14.1.12) *holds. Then* (14.2.5) *holds.*

14.3 Gaussian Fields with Constant Variance

We now come to what is the main section of this chapter, and the last part of the program we described at its beginning. That is, we shall now actually obtain an explicit bound for the difference between excursion probabilities and $\mathbb{E}\{\varphi(A_u)\}$.

Our first step will be to show that for each $t \in M$, the process $\widehat{f}^t$ of (14.2.2) is uncorrelated with the random vector $\nabla f_{|\partial_i M}(t)$. Hence, in the Gaussian case, $\widehat{f}^t(s)$ is independent of $\nabla f_{|\partial_i M}(t)$. The following lemma requires no side conditions beyond the basic ones on M and the requirement that f and ρ be smooth enough for all terms to be well defined. Since this is much weaker than the conditions of Theorem 14.2.2, we will leave more details for later. It also does not require that f be Gaussian, nor that it have constant variance.

Lemma 14.3.1. *For every $t \in \partial_i M$, $0 \le i \le k$, and every $s \in M \setminus \{t\}$,*

$$\mathrm{Cov}\left(\widehat{f}^t(s), X_t f\right) = 0$$

for every $X_t \in T_t \partial_i M$.

Proof. Noting the definition (14.2.2) of $\widehat{f}^t$, if $\rho(t,s) \neq 1$, then

$$\begin{aligned}
(1-\rho(t,s))\,\mathrm{Cov}\left(\widehat{f}^t(s), X_t f\right)
&= \mathrm{Cov}\left(f(s) - \rho(t,s) f(t), X_t f\right) \\
&\quad - \mathrm{Cov}\left(\widetilde{g}\left(\widehat{F}^t(s) - P_{N_t M}\widehat{F}^t(s), \nabla \widetilde{f}(t)\right), X_t f\right) \\
&= \mathrm{Cov}\left(f(s) - \rho(t,s) f(t), X_t f\right) - \widetilde{g}\left(\widehat{F}^t(s) - P_{N_t M}\widehat{F}^t(s), X_t\right),
\end{aligned}$$

where probably the best way to check the last line is to expand all the vectors in the middle expression and then compute.

If, on the other hand, $\rho(t,s) = 1$, then a similar computation shows that

$$\begin{aligned}
\mathrm{Cov}\left(\widehat{f}^t(s), X_t f\right) &= \mathrm{Cov}\left(f(s), X_t f\right) \\
&\quad - \mathrm{Cov}\left(\widetilde{g}\left(\widehat{F}^t(s) - P_{N_t M}\widehat{F}^t(s), \nabla \widetilde{f}(t)\right), X_t f\right) \\
&= \mathrm{Cov}\left(f(s), X_t f\right) - \widetilde{g}\left(\widehat{F}^t(s) - P_{N_t M}\widehat{F}^t(s), X_t\right).
\end{aligned}$$

The conclusion will therefore follow once we prove that for every $t \in M$,

$$\begin{aligned} \operatorname{Cov}\left(f(s), X_t f\right) &= \widetilde{g}\left(\widehat{F}^t(s) - P_{N_t M}\widehat{F}^t(s), X_t\right), \\ \operatorname{Cov}\left(f(s) - \rho(t,s) f(t), X_t f\right) &= \widetilde{g}\left(\widehat{F}^t(s) - P_{N_t M}\widehat{F}^t(s), X_t\right). \end{aligned}$$

Since the two arguments are similar, we prove only the first equality. The map $\widehat{F}^t$ can be decomposed as

$$\widehat{F}^t(s) = P_{T_t \partial_i M}\widehat{F}^t(s) + P^{\perp}_{T_t \partial_i M}\widehat{F}^t(s),$$

where

$$\begin{aligned} P_{T_t \partial_i M}\widehat{F}^t(s) &= \sum_{j=1}^{i} \operatorname{Cov}\left(f(s), X_j f(t)\right) X_{j,t}, \\ P^{\perp}_{T_t \partial_i M}\widehat{F}^t(s) &= \sum_{j=i+1}^{q} \operatorname{Cov}\left(f(s), X_j f(t)\right) X_{j,t}, \end{aligned}$$

and $(X_{1,t}, \dots, X_{q,t})$ is chosen such that $(X_{1,t}, \dots, X_{i,t})$ forms an orthonormal basis for $T_t \partial_i M$, while $(X_{i+1,t}, \dots, X_{q,t})$ forms an orthonormal basis for $T_t \partial_i M^{\perp}$, the orthogonal complement of $T_t \partial_i M$ in $T_t \widetilde{M}$.

Furthermore, since $\widetilde{g}(X_t, V_t) = 0$ for every $X_t \in T_t \partial_i M$ and $V_t \in N_t M$, it follows that

$$P_{N_t M}\widehat{F}^t(s) = P_{N_t M} P^{\perp}_{T_t \partial_i M}\widehat{F}^t(s),$$

and, for every $X_t \in T_t \partial_i M$,

$$\begin{aligned} \widetilde{g}(\widehat{F}^t(s) - P_{N_t M}\widehat{F}^t(s), X_t) &= \widetilde{g}(P_{T_t \partial_i M}\widehat{F}^t(s), X_t) \\ &\quad + \widetilde{g}(P^{\perp}_{T_t \partial_i M}\widehat{F}^t(s) - P_{N_t M} P^{\perp}_{T_t \partial_i M}\widehat{F}^t(s), X_t) \\ &= \widetilde{g}(P_{T_t \partial_i M}\widehat{F}^t(s), X_t) \\ &= \sum_{j=1}^{i} \operatorname{Cov}(f(s), X_j f(t)) \widetilde{g}(X_{j,t}, X_t) \\ &= \operatorname{Cov}(f(s), X_t f), \end{aligned}$$

and we are done. □

If we now assume that f is Gaussian, then the lack of correlation of Lemma 14.3.1 becomes independence, and the independence between the process f^t and $\nabla f_{|\partial_i M}(t)$ allows us to remove the conditioning on $\nabla f_{|\partial_i M}(t)$ in the expression for $\operatorname{Diff}_{f,M}(u)$, regardless of whether f has constant variance.

Corollary 14.3.2. *Suppose f is a Gaussian process satisfying the conditions Corollary* 14.2.3. *Then, with the notation of Theorem* 14.2.2,

$$\begin{aligned} |\operatorname{Diff}_{f,M}(u)| \le \sum_{i=0}^{N} \int_{\partial_i M} & \mathbb{E}\{|\det(-\nabla^2 f_{|\partial_i M, E_i}(t))| \mathbb{1}_{C_t^{\mathrm{ERR}}}\} \\ & \times p_{\nabla f_{|\partial_i M, E_i}(t)}(0)\, d\mathcal{H}_i(t), \end{aligned} \tag{14.3.1}$$

where

$$C_t^{\mathrm{ERR}} = \{u \le f(t) - \widetilde{g}(P_{T_t\partial_i M}\widehat{F}^t(t), \nabla \widetilde{f}(t)) \le \widehat{W}(t)\}.$$

If f has constant variance, then $\widehat{F}^t(t) = 0 \in T_tM$ and

$$C_t^{\mathrm{ERR}} = \{u \le f(t) \le \widehat{W}(t)\}.$$

Proof. Comparing this result with Theorem 14.2.2, it is clear that the only thing we need to prove is that if the event C_t^{ERR} occurs, then the condition $\{u \le f(t) \le W(t)\}$ can be replaced with

$$\{u \le f(t) - \widetilde{g}(P_{T_t\partial_i M}\widehat{F}^t(t), \nabla \widetilde{f}(t)) \le \widehat{W}(t)\}$$

and the conditioning can be removed.

The replacement is justified by the fact that on the event $\{\nabla f_{|\partial_i M}(t) = 0\}$, we trivially have

$$f(t) = f(t) - \widetilde{g}(P_{T_t\partial_i M}\widehat{F}^t(t), \nabla \widetilde{f}(t)).$$

Furthermore, $f(t) - \widetilde{g}\left(P_{T_t\partial_i M}\widehat{F}^t(t), \nabla \widetilde{f}(t)\right)$ is independent of $\nabla f_{|\partial_i M}(t)$, and Lemma 14.3.1 implies that $\widehat{W}(t)$ is also independent of $\nabla f_{|\partial_i M}(t)$. Therefore, the conditioning on $\nabla f_{\partial_i M}(t)$ can be removed. □

We are now ready to prove the main result of this chapter, and one of the main ones of the book. The remainder of the chapter is about special cases, examples, and counterexamples.

Theorem 14.3.3. *Let $\widetilde{f}$ be a Gaussian process on a locally convex, regular, stratified manifold M, with the assumptions of Corollary* 14.2.3 *in force. Assume, furthermore, that $\widetilde{f}$ has constant variance* 1. *Then,*

$$\liminf_{u\to\infty} -u^{-2}\log|\mathrm{Diff}_{f,M}(u)| \ge \frac{1}{2}\left(1 + \frac{1}{\sigma_c^2(f)}\right), \tag{14.3.2}$$

where

$$\sigma_c^2(f,t) \stackrel{\Delta}{=} \sup_{s\in M\setminus\{t\}} \mathrm{Var}\left(\widehat{f}^t(s)\right) \tag{14.3.3}$$

and

$$\sigma_c^2(f) \stackrel{\Delta}{=} \sup_{t\in M} \sigma_c^2(f,t). \tag{14.3.4}$$

Proof. We must find an upper bound for (14.3.1). Write

$$\begin{aligned}
\nabla^2 f_{|\partial_i M, E_i}(t) &= \nabla^2 f_{|\partial_i M, E_i}(t) - \mathbb{E}\{\nabla^2 f_{|\partial_i M, E_i}(t) | f(t)\} \\
&\quad + \mathbb{E}\{\nabla^2 f_{|\partial_i M, E_i}(t) | f(t)\} \\
&= \nabla^2 f_{|\partial_i M, E_i}(t) + f(t)I - f(t)I
\end{aligned}$$

(cf. (12.2.12)). Now expand the determinant in (14.3.1) via a standard Laplace expansion as we did in (11.6.13) and apply Hölder's inequality for conjugate exponents p and q to see that

$$\begin{aligned}|\operatorname{Diff}_{f,M}(u)| \le \sum_{i=0}^{N} \int_{\partial_i M} \sum_{j=0}^{i} & \mathbb{E}\{f(t)^j \mathbb{1}_{\{f(t)\ge u\}}\} \\ & \times (\mathbb{E}\{|\operatorname{detr}_{i-j}(-\nabla^2 f_{|\partial_i M, E_i}(t) - f(t)I)|^p\})^{1/p} \\ & \times (\mathbb{P}\{\widehat{W}(t) \ge u\})^{1/q}\, d\mathcal{H}_i(t),\end{aligned}$$

where detr_k is the "det-trace" defined in (7.2.8). Define

$$\mu(t) \triangleq \mathbb{E}\Big\{\sup_{s\in M\setminus t} \widehat{f}^t(s)\Big\}, \qquad \mu^+ \triangleq \sup_{t\in M} \mu(t)\mathbb{1}_{\{\mu(t)\ge 0\}}.$$

For

$$u \ge \sup_{t\in M} \mathbb{E}\Big\{\sup_{s\in M\setminus\{t\}} (\widehat{f}^t(s) - \mathbb{E}(\widehat{f}^t(s)))\Big\} + \mu^+$$

the Borell–TIS inequality (Theorem 2.1.1) then implies that

$$\mathbb{P}\{\widehat{W}(t) \ge u\} \le 2\mathrm{e}^{-(u-\mu^+)^2/2\sigma_c^2(f,x)}.$$

Recalling that $\widehat{f}^t(t) = f(t)$, for such u it also follows that

$$\mathbb{E}\Big\{f(t)^j \mathbb{1}_{\{f(t)\ge u\}}\Big\} \le C_j u^{j-1} \mathrm{e}^{-(u-\mu^+)^2/2}.$$

Putting these facts together, for any conjugate exponents p, q,

$$\begin{aligned}&|\operatorname{Diff}_{f,M}(u)| \\ &\quad \le C u^{N-1} \mathrm{e}^{-\frac{(u-\mu^+)^2}{2}\left(1+\frac{1}{q\sigma_c^2(f)}\right)} \\ &\qquad \times \sum_{i=0}^{N} \int_{\partial_i M} \sum_{j=0}^{i} (\mathbb{E}\{|\operatorname{detr}_{i-j}(-\nabla^2 f_{|\partial_i M, E_i}(t) - f(t)I)|^p\})^{1/p}\, d\mathcal{H}_i(t).\end{aligned}$$

The result now follows after noting that we can choose q arbitrarily close to 1, and $u(q)$ such that for $u \ge u(q)$, the remaining terms are arbitrarily small logarithmically when compared to u^2. □

There is, of course, a glaring weakness in Theorem 14.3.3, in that (14.3.2) contains a liminf rather a lim, and it contains an inequality rather than an equality.

Fortunately, the theorem provides a lower bound on the exponential decay of $\operatorname{Diff}_{f,M}(u)$, and this is almost always the important one in any practical situation. We believe that this lower bound is generally tight when a maximizer of $\sigma_c^2(f)$ occurs in the interior of M, in the sense that then the term corresponding to $\partial_N M$ in the sum defining $\text{Diff}_{f,M}(u)$ in (14.1.18) is exponentially of the same order as the upper bound. However, we have not been able to round off the desired proof, since it seems difficult to determine the sign of the lower-order terms in the error. This opens the (unlikely) possibility that some terms in the sum (14.1.18) cancel each other out, leading to a faster rate of exponential decay.

Thus, in the piecewise smooth setting, it is therefore still open as to whether (14.3.2) can be improved, and if so, under what conditions.

14.4 Examples

We now look at a number of applications of Theorem 14.3.3. The theorem contains two "unknowns." One is the expected Euler characteristic and the other is the parameter $\sigma_c^2(f)$. The expected Euler characteristic is the material of Chapters 11 and 12, so we have no need to treat it again here. Consequently, looking at examples means computing, or trying to understand, $\sigma_c^2(f)$, which, for reasons that will become clear below, we now refer to as the *critical variance*. We shall start with the simplest possible example—stationary processes on $\mathbb{R}$—and then jump to a rather general scenario. We shall then close the chapter with a number of rather instructive special cases.

It is the general scenario that is actually the most instructive, for the simpler ones are so special that they give little or no insight into what is really going on. Hopefully, after reading the general case, with its elegant geometric interpretation of $\sigma_c^2(f)$, it will be clearer to you from where the constructions of the previous sections of this chapter, particularly $\widehat{f^t}$, came.

The general case also allows us to investigate some very special cases in which the approximation of Theorem 14.3.3 is either perfect ($\sigma_c^2(f) = 0$) or meaningless ($\sigma_c^2(f) = \infty$). Along the way, we shall also spend a little time with the "cosine random field" (which actually provides the only example for which the approximation is perfect) and see how it provides both examples and counterexamples for various approximation techniques.

14.4.1 Stationary Processes on $[0, T]$

A simple but illustrative example is given by a stationary Gaussian $f(t)$, $t \in [0, T]$ satisfying the conditions of Theorem 14.3.3. The ambient manifold will be $\mathbb{R}$. Thus f has variance 1, and we write ρ for its correlation (and covariance) function. To save on notation, we shall also assume that the time scale has been chosen such that $\text{Var}(\dot{f}(t)) = -\ddot{\rho}(0) = 1$. Under these assumptions the Riemannian metric $\widetilde{g}$ reduces to the usual Euclidean metric on $\mathbb{R}$.

We now need to find the supremum of $\sigma_c^2(f, t)$, which we shall do by dividing the argument into two cases, t in the open set $(0, T)$ and t one of the boundary points $\{0, T\}$. This division into separate cases will be a theme in virtually all of the (more complicated) examples to come. For this we introduce the notation

$$\sigma_c^2(f, A) \triangleq \sup_{t \in A} \sigma_c^2(f, t). \tag{14.4.1}$$

For $t \in (0, T)$ the process $\widehat{f}^t$ is given by

$$\widehat{f}^t(s) = \frac{f(s) - \rho(t-s)f(t) - \dot{\rho}(t-s)\dot{f}(t)}{1 - \rho(t-s)},$$

where we apply (5.5.5) and the fact that the normal cone at interior points is empty, so that the projection $P_{N_t M}$ in the definition (14.2.2) disappears.

To compute the critical variance at such t we use the relationships (5.5.5)–(5.5.7). These easily give

$$\begin{aligned} \sigma_c^2(f,t) &= \sup_{\substack{-t \le s \le T-t \\ s \ne 0}} \frac{1 - \rho(s)^2 - \dot{\rho}(s)^2}{(1-\rho(s))^2} \qquad (14.4.2) \\ &= \sup_{s \in [0,T] \setminus \{t\}} \frac{\mathrm{Var}(f(s) \mid f(t), \dot{f}(t))}{(1 - \rho(t-s))^2}, \end{aligned}$$

where the second line also uses (1.2.8).

By (14.4.2) the critical variance in the interior of $[0, T]$ is

$$\sigma_c^2(f, (0,T)) = \sup_{0<t<T} \frac{1 - \rho(t)^2 - \dot{\rho}(t)^2}{(1-\rho(t))^2}. \qquad (14.4.3)$$

The critical variance at the endpoints $\{0, T\}$ needs a little more care. We consider only the point $t = 0$, since by stationarity the critical variance at T is the same. In the notation of Section 7.1, the normal cone at $t = 0$ is

$$N_0[0,T] = \bigcup_{c \le 0} c \frac{\partial}{\partial t} \subset T_0\mathbb{R}.$$

For ease of notation, we write N_0 for $N_0[0, T]$. The projection onto N_0 is

$$P_{N_0}\left(a \frac{\partial}{\partial t}\right) = \mathbb{1}_{\{a \le 0\}} \left(a \frac{\partial}{\partial t}\right),$$

and so the process $\widehat{f}^0(s)$ is given by

$$\widehat{f}^0(s) = \frac{f(s) - \rho(s)f(0) - \mathbb{1}_{\{\dot{\rho}(s) \le 0\}}\dot{\rho}(s)\dot{f}(0)}{(1-\rho(s))^2}.$$

The critical variance at $t = 0$ is therefore given by

$$\begin{aligned} \sigma_c^2(f,0) &= \sup_{0 \le t \le T} \frac{1 - \rho(t)^2 - \dot{\rho}(t)^2 + \max(\dot{\rho}(t), 0)^2}{(1-\rho(t))^2} \qquad (14.4.4) \\ &\ge \sup_{0 \le t \le T} \frac{1 - \rho(t)^2 - \dot{\rho}(t)^2}{(1-\rho(t))^2}. \end{aligned}$$

This implies that $\sigma_c^2(f, 0) \ge \sigma_c^2(f, t),\ 0 \le t \le T$. Therefore the critical variance $\sigma_c^2(f)$ is attained at the endpoints $t = 0, T$.

We can summarize the above discussion as follows.

Example 14.4.1. *Suppose f is a centered, unit-variance C^2 stationary Gaussian process on $\mathbb{R}$ satisfying the conditions of Theorem* 14.3.3 *and such that* $\mathrm{Var}(\dot{f}(t)) = -\ddot{\rho}(0) = 1$. *Then, for all $T > 0$,*

$$\sigma_c^2(f, [0, T]) = \sup_{0 \le t \le T} \frac{1 - \rho(t)^2 - \dot{\rho}(t)^2 + \max(\dot{\rho}(t), 0)^2}{(1 - \rho(t))^2}.$$

While this is a reasonably simple expression to compute, it is actually quite trivial if ρ is monotone over $[0, T]$, since then $\dot{\rho}(t) \le 0$ for all $t > 0$ and the max term disappears. The remaining supremum then turns out to be easy to compute, as we shall see in the next section.

There are a number of results akin to Example 14.4.1 to be found in the literature, some even slightly more general (cf. [14, 95, 126] and the references therein). All, however, have proofs that rely on the fact that the random process under consideration is defined over $\mathbb{R}$, proofs that do not generalize to more general scenarios. As we are about to see, the approach we take generalizes easily.

14.4.2 Isotropic Fields with Monotone Covariance

Carrying on in the spirit of the previous example, we now consider isotropic fields over convex sets M in $\mathbb{R}^N$. We add a new condition related to a monotonicity property for the covariance function. Thus, while this example is more general than Example 14.4.1 in that it treats more general parameter spaces, it is more specific in that it requires more of the covariance function. Nevertheless, it is quite general, and gives a very simple and computable formula for the critical variance σ_c^2.

Example 14.4.2. *Suppose f is a centered, unit-variance, C^2 isotropic Gaussian process on $\mathbb{R}^N$ satisfying the conditions of Theorem* 14.3.3 *and such that*

$$\mathrm{Var}\left(\frac{\partial f(t)}{\partial t_j}\right) = -\left.\frac{\partial^2 \rho(t)}{\partial t_j^2}\right|_{t=0} = 1,$$

for all j. Suppose that M is compact, piecewise smooth, and convex, and set $T = \mathrm{diam}(M)$, where the diameter is computed in the standard Euclidean sense.

Suppose also that

$$\frac{\partial \rho(t)}{\partial t_j} \le 0, \tag{14.4.5}$$

for all $t \in [0, T]^N$. Then,

$$\sigma_c^2(f, M) = \mathrm{Var}\left(\frac{\partial^2 f(t)}{\partial t_1^2} \,\middle|\, f(t)\right) = \left.\frac{\partial^4 \rho(t)}{\partial t_1^4}\right|_{t=0} - 1. \tag{14.4.6}$$

Proof. We start by noting that since f is isotropic with unit second spectral moment, the metric induced by f is the standard Euclidean one. Furthermore, we can take $\widetilde{M} = \mathbb{R}^N$.

We need to compute the maximal variance of the random field $\widehat{f}^t(s)$ given by (14.2.2). Consider the term $P_{N_tM}\widehat{F}^t(s)$ that appears there, where

$$\widehat{F}^t(s) = \sum_{i=1}^{N} \mathrm{Cov}(f(s), X_i f(t))X_i \tag{14.4.7}$$

and the X_i are the usual orthonormal basis elements of $\mathbb{R}^N$.

We claim that under the above assumptions $P_{N_tM}\widehat{F}^t(s)$ is identically zero, in which case $\widehat{f}^t(s)$ trivially simplifies to

$$\begin{aligned} \widehat{f}^t(s) &= \frac{f(s) - \rho(t,s)f(t) - \widetilde{g}(\widehat{F}^t(s), \nabla\widetilde{f}(t))}{1-\rho(t,s)} \\ &= \frac{f(s) - \rho(t,s)f(t) - \sum_{i=1}^N \mathrm{Cov}(f(s), X_i f(t))X_i f(t)}{1-\rho(t,s)}. \end{aligned} \tag{14.4.8}$$

To prove this claim, we start with $t \in \partial_N M$, the highest-order stratum of M. In this case $N_tM = \{0\}$, so that the projection P_{N_tM} is trivially zero and there is nothing to prove.

Thus, take $t \in \partial_k M$ with $k < N$ and $s \in M$, $s \neq t$. Let X_t^s be the unit vector in $T_t\mathbb{R}^N$ in the direction $s - t$. Since M is convex, $X_t^s \in \mathcal{S}_t M$, and, as we shall see in a moment, X_t^s is parallel to $\widehat{F}^t(s)$. Consequently, $X_t^s \in \mathcal{S}_t M$ implies that $P_{N_tM}\widehat{F}^t(s) = 0$, as required

To see that X_t^s is parallel to $\widehat{F}^t(s)$ we use isotropy twice. To make the calculations easier, we use it first to note that isotropy implies that there is no loss of generality involved by taking $t = 0$ and $s = (\tau, 0, \dots, 0)$, where $\tau = |s-t|$. In this case X_t^s is simply X_1.

Secondly, it is easy to check from (5.5.5) that, again because of isotropy,

$$\mathrm{Cov}(f(\tau, 0, \dots, 0), X_i f(0)) = \begin{cases} -\dot{\rho}(t) \stackrel{\Delta}{=} -\frac{\partial\rho(t)}{\partial t_1}\big|_{(\tau,0,\dots,0)}, & i = 1, \\ 0, & i \neq 1. \end{cases} \tag{14.4.9}$$

Substituting into (14.4.7) and using the fact that $\dot{\rho}(s) \le 0$ gives us that $\widehat{F}^t(s)$ is proportional to X_1, so that X_t^s is parallel to $\widehat{F}^t(s)$.

We turn now to computing the variance of (14.4.8) for given s, t. Once again, we can choose $t = 0$ and $s = (\tau, 0, \dots, 0)$, for $\tau = |s-t|$. If we then substitute (14.4.9) into (14.4.8), a little algebra easily leads to the fact that

$$\mathrm{Var}(\widehat{f}^t(s)) = \frac{1 - \rho^2(|s-t|) - \dot{\rho}^2(|s-t|)}{(1-\rho(|s-t|))^2},$$

and so

$$\sigma_c^2(f, M) = \sup_{s \neq t \in M} \frac{1 - \rho^2(|s-t|) - \dot{\rho}^2(|s-t|)}{(1 - \rho(|s-t|))^2}. \tag{14.4.10}$$

By isotropy, we need not take the supremum over t here, and so fix t, taking $t = 0$. We now claim is that the supremum over s is achieved as $s \to 0$. If this is true, then isotropy implies that we can take $s \to 0$ along any Euclidean axis. This being the case, two applications of l'Hôpital's rule show that

$$\sigma_c^2(f, M) = \left. \frac{\partial^4 \rho(t)}{\partial t_1^4} \right|_{t=0} - 1,$$

while (5.5.5) and the basic properties of Gaussian distributions give that this, in turn, is the same as $\mathrm{Var}(\frac{\partial^2 f(t)}{\partial t_1^2} | f(t))$.

Consequently, modulo the issue of where the supremum in (14.4.10) is achieved, we have completed the proof. We shall return to this issue at the end of the next section. □

14.4.3 A Geometric Approach

In this section we want to look a little more carefully at the critical variance

$$\sigma_c^2(f) = \sup_{t \in M} \sigma_c^2(f, t), \tag{14.4.11}$$

where

$$\sigma_c^2(f, t) \stackrel{\Delta}{=} \sup_{s \in M \setminus \{t\}} \mathrm{Var}\left(\widehat{f}^t(s)\right) \tag{14.4.12}$$

and $\widehat{f}^t$ is given by (14.2.2) (cf. (14.3.3) and (14.3.4)). Our aim is to give a geometrical interpretation of $\sigma_c^2(f)$ in a quite general setting. With this, we shall return to complete the proof of Example 14.4.2.

We shall start by looking closely at (14.4.12), for points[7] $t \in \partial_N M = M^\circ$. In later examples we shall see that these are often the most important points to consider. They are also the points for which a geometric intepretation of (14.4.12) is most accessible.

The geometry that comes into play here is that of H, the reproducing kernel Hilbert space (RKHS) of Section 3.1. In particular, since we shall retain the assumption that f has constant unit variance, we shall be concerned with the spherical geometry of the unit sphere $S(H)$ in H.

Recall, however, that there is a canonical isometry between H and $\mathcal{H}$, the span of $f_t,\ t \in M$. Consequently, we can, and shall, work in $\mathcal{H}$ rather than H. Recall also

[7] Actually, the arguments will work in slightly greater generality. For example, if M is composed of a number of disjoint piecewise smooth manifolds, then they will also hold for all t in the top-dimensional stratum of the separate piecewise smooth manifolds. For example, if $M = [0, 1]^2 \cup (\{3\} \times [0, 1])$ then it will hold for all $t \in (0, 1)^2 \cup (\{3\} \times (0, 1))$.

that the unit-variance assumption also implies that the $\mathcal{H}$ inner product between $f(t)$ and $f(s)$ in $S(\mathcal{H})$ is given by

$$\langle f(t), f(s)\rangle_{\mathcal{H}} = \rho(s,t),$$

where ρ is the correlation function of f.

As a first step, we define a map $\Psi : M \to S(\mathcal{H})$ by

$$\Psi(t) = f(t).$$

Under the assumptions of Theorem 14.3.3 it is immediate that Ψ is a piecewise C^2 embedding of M in $S(\mathcal{H})$. Furthermore, it is not hard to see that

$$\Psi_*(X_t) = X_t f,$$

so that the tangent space $T_{f(t)}\Psi(M)$ is spanned by $(X_{1,t}f, \dots, X_{N,t}f)$ for any basis $\{X_{1,t}, \dots, X_{N,t}\}$ of T_tM. We denote the orthogonal complement of $T_{f(t)}\Psi(M)$ in $T_{f(t)}\mathcal{H}$ by $T^{\perp}_{f(t)}\Psi(M)$, and the orthogonal complement of $T_{f(t)}\Psi(M)$ in $T_{f(t)}S(\mathcal{H})$ by $N_{f(t)}\Psi(M)$.

Given a point $f(t)$ and a unit normal vector $v_{f(t)} \in N_{f(t)}\Psi(M)$ we denote the geodesic, in $S(\mathcal{H})$, originating at $f(t)$ in the direction $v_{f(t)}$ by $c_{f(t),v_{f(t)}}$. It is easy to check that this is given by

$$c_{f(t),v_{f(t)}}(r) = \cos r \cdot f(t) + \sin r \cdot v_{f(t)}, \quad 0 \le r < \pi. \tag{14.4.13}$$

As discussed in Section 10.3.1, when we looked at tube formulas, up to a certain point along $c_{f(t),v_{f(t)}}$ the points on this geodesic project uniquely to t. The largest r for which this is true is the local critical radius at $f(t)$ in the direction $v_{f(t)}$, and we denote it by $\theta(f(t), v_{f(t)})$.[8] The local critical radius at $f(t)$ is then

$$\theta_c(f(t)) = \inf_{v_{f(t)} \in S(N_{f(t)}\Psi(M))} \theta(f(t), v_{f(t)}), \tag{14.4.14}$$

and the global critical radius is

$$\theta_c = \theta_c(\Psi(M)) = \inf_{t\in M} \theta_c(f(t)). \tag{14.4.15}$$

The following lemma links these critical radii of $\Psi(M)$ to the critical variances (14.4.11) and (14.4.12).

Lemma 14.4.3. *Under the setup and assumptions of Theorem* 14.3.3,

$$\sigma_c^2(f,t) = \begin{cases} \cot^2(\theta_c(f(t))), & 0 \le \theta_c < \frac{\pi}{2}, \\ 0, & \frac{\pi}{2} \le \theta_c < \pi, \end{cases} \tag{14.4.16}$$

for all $t \in \partial_N M$.

[8] It is important to note at this point that although $\mathcal{H}$ is made up of random variables, the quantity $\theta(f(t), v_{f(t)}$ is deterministic. Furthermore, by the isometry between the RKHS and $\mathcal{H}$ it has essentially the same definition in the RKHS that it has in $\mathcal{H}$.

Proof. Much as in the proof of Lemma 14.2.1 and Example 14.4.2, we start by noting that if $t \in \partial_N M$, then $N_t M = T_t \partial_N M^\perp$ and $P_{N_t M} V_t = 0$ for all $V_t \in T_t \partial_N M$. Now choose an orthonormal frame field for $T(\widetilde{M})$ such that at every $t \in \partial_N M$, the first N vectors in the frame field generate $T_t M$, while the remaining vectors generate $N_t M$ (as a subspace of $T_t \widetilde{M}$). Then $\widehat{f^t}$ simplifies somewhat, and for $s \neq t$, $t \in \partial_N M$, it is easy to check that

$$\begin{aligned}\widehat{f^t}(s) &= \frac{f(s) - \rho(t,s) f(t) - \widetilde{g}(\widehat{F}^t(s), \nabla \widetilde{f}(t))}{1-\rho(t,s)} \\ &= \frac{f(s) - \rho(t,s) f(t) - \sum_{i=1}^N \mathrm{Cov}(f(s), X_i f(t)) X_i f(t)}{1-\rho(t,s)}.\end{aligned}$$

We now compute the variance of $\widehat{f^t}(s)$. Note that

$$\begin{aligned}&\rho(t,s) f(t) + \sum_{i=1}^N \mathrm{Cov}(f(s), X_i f(t)) X_i f(t) \\ &\qquad = \langle f(t), f(s)\rangle_{\mathcal{H}} f(t) + \sum_{i=1}^N \langle f(s), X_i f(t)\rangle_{\mathcal{H}} X_i f(t) \\ &\qquad = P_{L_t} f(s),\end{aligned}$$

where $L_t = \mathrm{span}\{f(t), X_1 f(t), \ldots, X_N f(t)\}$.

However, $N_{f(t)}\Psi(M)$ is the orthogonal complement (in $T_{f(t)} H$) of the subspace L_t. Consequently,

$$\begin{aligned}\mathrm{Var}\left(\widehat{f^t}(s)\right) &= \mathrm{Var}\left(f(s) - \rho(t,s) f(t) - \sum_{i=1}^N \mathrm{Cov}(f(s), X_i f(t)) X_i f(t)\right) \\ &= \| P_{N_{f(t)}\Psi(M)} f(s)\|_{\mathcal{H}}^2,\end{aligned}$$

where $P_{N_{f(t)}\Psi(M)}$ represents orthogonal projection onto $N_{f(t)}\Psi(M)$.

Finally, we have that

$$\sigma_c^2(f,t) = \sup_{s \in M \setminus \{t\}} \frac{\| P_{N_{f(t)}\Psi(M)} f(s)\|_{\mathcal{H}}^2}{(1-\rho(t,s))^2}. \tag{14.4.17}$$

We now need to show that the same expression holds for $\cot^2(\theta_c(f(t)))$. Turning to the picture in terms of geodesics, fix $f(t)$ and $v_{f(t)}$. Suppose that for a certain r, the point $c_{f(t),v_{f(t)}}(r)$ does not metrically project to $f(t)$. This and a little basic Hilbert sphere geometry implies there is a point $f(s) \in \Psi(M)$ such that

$$\begin{aligned}\cos^{-1}\left(\langle c_{f(t),v_{f(t)}}(r), f(s)\rangle_{\mathcal{H}}\right) &= \tau\left(c_{f(t),v_{f(t)}}(r), f(s)\right) \\ &< \tau\left(c_{f(t),v_{f(t)}}(r), f(t)\right) \\ &= \cos^{-1}\left(\langle c_{f(t),v_{f(t)}}(r), f(t)\rangle_{\mathcal{H}}\right),\end{aligned} \tag{14.4.18}$$

where τ is geodesic distance on $S(H)$.

Alternatively, applying (14.4.13),

$$\begin{aligned}\left\langle c_{f(t),v_{f(t)}}(r),\, f(s)\right\rangle_{\mathcal{H}} &= \cos r \cdot \langle f(t), f(s)\rangle_{\mathcal{H}} + \sin r \cdot \left\langle v_{f(t)}, f(s)\right\rangle_{\mathcal{H}} \\ &= \cos r \cdot \rho(t,s) + \sin r \cdot \left\langle v_{f(t)}, f(s)\right\rangle_{\mathcal{H}} \\ &> \left\langle c_{f(t),v_{f(t)}}(r),\, f(t)\right\rangle_{\mathcal{H}} \\ &= \cos r.\end{aligned}$$

A little rearranging shows that this is true if and only if

$$\cot t < \frac{\left\langle v_{f(t)}, f(s)\right\rangle_{\mathcal{H}}}{1-\rho(t,s)}.$$

Therefore,

$$\cot\left(\theta(f(t), v_{f(t)})\right) = \sup_{s\in M\setminus\{t\}} \frac{\left\langle v_{f(t)}, f(s)\right\rangle_{\mathcal{H}}}{1-\rho(t,s)}.$$

Taking the supremum over all $v_{f(t)} \in S(N_{f(t)}\Psi(M))$ we obtain

$$\begin{aligned}\cot\left(\theta_c(f(t))\right) &= \sup_{v_{f(t)}\in S(N_{f(t)}\Psi(M))} \sup_{s\in M\setminus\{t\}} \frac{\left\langle v_{f(t)}, f(s)\right\rangle_{\mathcal{H}}}{1-\rho(t,s)} \\ &= \sup_{s\in M\setminus\{t\}} \frac{\|P_{N_{f(t)}\Psi(M)} f(s)\|_{\mathcal{H}}}{1-\rho(t,s)}.\end{aligned}$$

Comparing this to (14.4.17) completes the proof. □

Lemma 14.4.3 gives a very succinct representation for the critical variance $\sigma_c^2(f)$ that requires nothing about the rather complicated process $\widehat{f^t}$ that led to its original definition. Here is another result, part of which relates to the geometric structure of H and part of which has a rather down-to-earth implication for the correlation function ρ.

Lemma 14.4.4. *Retain the assumptions of Lemma* 14.4.3, *and suppose that the pair* $(t^*, s^*) \in \partial_N M \times \partial_N M$ *achieves the supremum in*

$$\sup_{t\in M}\sup_{s\in M\setminus\{t\}} \frac{\|P_{N_{f(t)}\Psi(M)} f(s)\|^2_{\mathcal{H}}}{(1-\rho(t,s))^2}. \tag{14.4.19}$$

Then,

$$f(s^*) = \cos(2\theta_c)\cdot f(t^*) + \sin(2\theta_c)\cdot v^*_{f(t^*)}, \tag{14.4.20}$$

where

$$v^*_{f(t^*)} = \frac{P_{N_{f(t)}\Psi(M)} f(s^*)}{\|P_{N_{f(t)}\Psi(M)} f(s^*)\|_{\mathcal{H}}}$$

is the direction of the geodesic between $f(t^*)$ and $f(s^*)$. For any $X_{t^*} \in T_{t^*}M$ and any $X_{s^*} \in T_{s^*}M$,

$$\mathrm{Cov}(X_{t^*}f, f(s^*)) = \langle \Psi_*(X_{t^*}), f(s^*)\rangle_{\mathcal{H}} = 0, \tag{14.4.21}$$

$$\mathrm{Cov}(X_{s^*}f, f(t^*)) = \langle \Psi_*(X_{s^*}), f(t^*)\rangle_{\mathcal{H}} = 0. \tag{14.4.22}$$

Furthermore, the partial map

$$\rho^{t^*}(s) \triangleq \rho(t^*, s)$$

has a local maximum at s^*, and the partial map

$$\rho^{s^*}(t) \triangleq \rho(t, s^*)$$

has a local maximum at t^*.

In order to appreciate the geometry behind (14.4.20) you might think as follows: Given a pair of points in $\partial_N M$ that maximize (14.4.19), the three equations (14.4.20), (14.4.21), and (14.4.22) imply that the tube of radius θ_c around $\Psi(M)$ should self-intersect along a geodesic from t^* to s^* in such a way that the tube, viewed locally from the point t^*, shares a hyperplane with the tube viewed locally from the point s^*. Alternatively, at the point of self-intersection the outward-pointing unit normal vectors should be pointing in opposite directions.

Proof. Our first requirement is to establish (14.4.20). Looking back through the proof of Lemma 14.4.3 (cf. (14.4.18)), it is immediate that if a pair (t^*, s^*) achieves the supremum in (14.4.19) then there exists a point $z \in \mathcal{H}$ equidistant from $f(t^*)$ and $f(s^*)$ and unit vectors $v_{f(t^*)} \in N_{f(t^*)}\Psi(M)$ and $w_{f(s^*)} \in N_{f(s^*)}\Psi(M)$ such that

$$z = \cos\theta_c \cdot f(t^*) + \sin\theta_c \cdot v_{f(t^*)} = \cos\theta_c \cdot f(s^*) + \sin\theta_c \cdot w_{f(s^*)}.$$

It is not a priori obvious that $w_{f(s^*)} \in N_{f(s^*)}\Psi(M)$. However, if this were not the case, then there would exist $t \in M$, $t \neq s$, such that $d(t, z) < \theta_c$. This would contradict the assumption that (t^*, s^*) achieve the supremum in (14.4.19) and the critical radius is θ_c. This proves (14.4.20).

To prove (14.4.21) and (14.4.22) it is enough to show that the unit tangent vectors $\dot{c}_{f(t^*),v_{f(t^*)}}(\theta_c)$ and $\dot{c}_{f(s^*),w_{f(s^*)}}(\theta_c)$ at z satisfy

$$u = \dot{c}_{f(t^*),w_{f(t^*)}}(\theta_c) + \dot{c}_{f(s^*),w_{f(s^*)}}(\theta_c) = 0 \in T_z\mathcal{H}. \tag{14.4.23}$$

Suppose that $u \neq 0$. A simple calculation shows that

$$\langle u, f(t^*)\rangle_{\mathcal{H}} = \langle u, f(s^*)\rangle_{\mathcal{H}} = \frac{\cos(2\theta_c) - \langle f(t^*), f(s^*)\rangle_{\mathcal{H}}}{\sin\theta_c}.$$

If $u \neq 0$, then $f(t^*)$, $f(s^*)$, and z are not on the same geodesic, and the triangle inequality implies that

$$\cos(2\theta_c) < \left\langle f(t^*), f(s^*) \right\rangle_{\mathcal{H}},$$

which implies that

$$\left\langle u, f(t^*) \right\rangle_{\mathcal{H}} = \left\langle u, f(s^*) \right\rangle_{\mathcal{H}} < 0.$$

Consider the geodesic originating at z in the direction $u^* = u/\|u\|_H$ and given by

$$c_{z,u^*}(t) = \cos t \cdot z + \sin t \cdot u^*.$$

Assuming that $u^* \neq 0$, we have

$$\left\langle \dot{c}_{z,u^*}(0), f(s^*) \right\rangle_{\mathcal{H}} = \left\langle \dot{c}_{z,u^*}(0), f(t^*) \right\rangle_{\mathcal{H}} = \left\langle u^*, f(t^*) \right\rangle_{\mathcal{H}} < 0.$$

This implies that for $r < 0$ and sufficiently small $|r|$,

$$\langle f(t^*), c_{z,u^*}(r) \rangle_{\mathcal{H}} = \langle f(s^*), c_{z,u^*}(r) \rangle_{\mathcal{H}} > \cos\theta_c.$$

For such an r, there exist distinct points $\widehat{t}(r)$ and $\widehat{s}(r)$ such that

$$d(c_{z,u^*}(r), \widehat{t}(r)) < \theta_c, \qquad d(c_{z,u^*}(r), \widehat{s}(r)) < \theta_c,$$

such that the geodesic connecting $\widehat{t}(r)$ and $c_{z,u^*}(r)$ is normal to M at $\widehat{t}(r)$, and the geodesic connecting $\widehat{s}(r)$ and $c_{z,u^*}(r)$ is normal to M at $\widehat{s}(r)$.

Without loss of generality we assume that

$$d(c_{z,u^*}(r), \widehat{t}(r)) \leq d(c_{z,u^*}(r), \widehat{s}(r)).$$

In this case, the geodesic connecting $\widehat{s}$ to $c_{z,u^*}(r)$ is no longer a minimizer of distance once it passes the point $c_{z,u^*}(r)$, which is of distance strictly less than θ_c from $\widehat{s}(r)$. That is, for some point $\widehat{z}(r)$ along the geodesic connecting $\widehat{s}(r)$ and $c_{z,u^*}(r)$, beyond $c_{z,u^*}(r)$ but of distance strictly less than θ_c from $\widehat{s}(r)$, there exist points in M strictly closer to $\widehat{z}(r)$ than $\widehat{s}(r)$. We therefore have a contradiction, since the critical radius of $\Psi(M)$ is θ_c.

Consequently, we have that the u of (14.4.23) is indeed zero, and so (14.4.21) and (14.4.22) are proven.

To prove claims about the partial maps $\rho^{t^*}(s)$ and $\rho^{s^*}(t)$ note first that $f(s^*)$ is a linear combination of $f(t^*)$ and $v_{f(t^*)}$, which are perpendicular to every vector in $T_{f(t^*)}\Psi(M)$. Similarly, $f(t^*)$ is a linear combination of $f(s^*)$ and $w_{f(s^*)}$, which are perpendicular to every vector in $T_{f(t^*)}\Psi(M)$. The fact that the partial maps $\rho^{t^*}(s)$ and $\rho^{s^*}(t)$ have local maxima then follows from the same contradiction argument used above. $\square$

Conclusion of the proof of Example 14.4.2. Recall that what remained was to show the equivalence of the two suprema in

$$\sup_{s \neq t \in M} \frac{1-\rho^2(|s-t|)-\dot{\rho}^2(|s-t|)}{(1-\rho(|s-t|))^2} = \sup_{t \to 0} \frac{1-\rho^2(t)-\dot{\rho}^2(t)}{(1-\rho^2(t))} \tag{14.4.24}$$

(cf. (14.4.10)).

It is trivial that the right-hand side of (14.4.24) is no larger than the left-hand side. It is also trivial that left-hand side is less than

$$\sup_{t \in [0, \mathrm{diam}(M)]^N} \frac{1-\rho^2(t)-\dot{\rho}^2(t)}{(1-\rho(t))^2}. \tag{14.4.25}$$

Thus if we can show that this is equal to the right-hand side of (14.4.24) we shall be done.

Let t^* be one of the points achieving the supremum in (14.4.25) with $N = 1$. Then Lemma 14.4.4 implies that $\dot{\rho}(t^*) = 0$, and that t^* is a local maximum of $\rho(t)$. But under the monotonicity assumption of the example, there is only one such point, given by $t^* = 0$. □

14.4.4 The Cosine Field

We now turn to what is perhaps the grandfather of all Gaussian random fields, the so-called *cosine random field* on $\mathbb{R}^N$. It is defined as

$$f(t) \stackrel{\Delta}{=} N^{-1/2} \sum_{k=1}^{N} f_k(\omega_k t_k), \tag{14.4.26}$$

where each f_k is the process on $\mathbb{R}$ given by

$$f_k(t) \stackrel{\Delta}{=} \xi_k \cos t + \xi_k' \sin t.$$

The ξ_k and ξ_k' are independent, standard Gaussians and the ω_k are positive constants.

It is a trivial computation that f is a stationary, centered Gaussian process on $\mathbb{R}^N$ with covariance and correlation functions

$$C(t) = \rho(t) = N^{-1} \sum_{k=1}^{N} \cos(\omega_k t_k).$$

We shall concentrate on the case in which $\omega_k = 1$ for all k, in which case it is easy to check that

$$\mathbb{E}\{f^2(t)\} = \mathbb{E}\{(X_j f(t))^2\} = \mathbb{E}\{(X_j X_j f(t))^2\} = 1,$$

for all j, where $X_1, \ldots, X_j$ is the usual orthonormal basis of $\mathbb{R}^N$. Similarly,

$$\mathbb{E}\{X_i f(t) X_j f(t)\} = 0$$

if $i \neq j$.

The cosine field is important for a number of reasons, perhaps the most significant of which is the fact that it is simple enough to enable explicit computations. For example, set $N = 1$ and consider the excursion probability

$$\mathbb{P}\left\{ \sup_{0 \le t \le T} f(t) \ge u \right\} \tag{14.4.27}$$

for $u > 0$. Note that

$$\xi \cos t + \xi' \sin t = R \cos(t - \theta), \tag{14.4.28}$$

where $R^2 = \xi^2 + (\xi')^2$ has an exponential distribution with mean $\frac{1}{2}$, $\theta = \arctan(\xi/\xi')$ has a uniform distribution on $[0, 2\pi]$, and R and θ are independent. You can use this information to compute the excursion probability (14.4.27) directly, but here is an easier way:

Assume that $T < \pi$ and write N_u for the number of upcrossings of u in $(0, T]$. Then

$$\mathbb{P}\left\{ \sup_{0 \le t \le T} f(t) \ge u \right\} = \mathbb{P}\{f(0) \ge u\} + \mathbb{P}\{f(0) < u, N_u \ge 1\}.$$

But since $T < \pi$, the event $\{f(0) \ge u, N_u \ge 1\}$ is empty, implying that

$$\mathbb{P}\{f(0) < u, N_u \ge 1\} = \mathbb{P}\{N_u \ge 1\}.$$

Again using the fact that $T < \pi$, we note that N_u is either 0 or 1, so that

$$\mathbb{P}\left\{ \sup_{0 \le t \le T} f(t) \ge u \right\} = \Psi(u) + \mathbb{E}\{N_u\}.$$

Looking back at Chapter 11, in particular (11.7.13), what we have just proven is that

$$\mathbb{P}\left\{ \sup_{0 \le t \le T} f(t) \ge u \right\} \equiv \mathbb{E}\{\varphi(A_u(f, [0, T])\}. \tag{14.4.29}$$

In other words, the approximation that has been at the center of this chapter is *exact* for the one-dimensional cosine process on an interval of length less than π.

On the other hand, if $\varphi < T \le 2\pi$, then much of the above argument breaks down and (14.4.29) no longer holds. If $T > 2\pi$, then, since the cosine process is periodic, increasing T has no effect whatsoever on $\sup_{[0,T]} f(t)$, and the left-hand side of (14.4.29) is independent of T. The right-hand side, however, contains the term $E\{N_u\}$, which grows linearly with T. Hence we see that the assumption in the central Theorem 14.3.3 that $\rho(t, s) = 1 \iff t = s$, which does not hold for periodic processes, was necessary.

But more than this can be seen from this very simple example. For example, if we take an interval of length greater than 2π, so that we get at least a full cycle of the process, the excursion probability is simply

$$\mathbb{P}\{R \geq u\} = e^{-u^2/2},$$

where R is defined in (14.4.28). Note that this does not depend on T. On the other hand, $\mathbb{E}\{\varphi(A_u(f, [0, T]))\}$ contains a term growing linearly in T, and so here the approximation breaks down. To relate these phenomena to the rest of this chapter, we need to look at the critical variance σ_c^2 for this example.

Unfortunately, we do not have among the results we have already proven one that immediately yields σ_c^2. However, we do have a technique. The technique is that of the proof of Example 14.4.2. Although the cosine process over $\mathbb{R}^N$ is not isotropic, as that example requires, if we restrict f to a unit cube of the form $[0, T]^N$ with $T \leq \pi/2$ then the same argument as given there also works now. The conclusion is again that

$$\sigma_c^2(f, [0, T]^N) = \text{Var}\left(\frac{\partial^2 f(t)}{\partial t_1^2}\Big| f(t)\right) = N - 1, \tag{14.4.30}$$

the rightmost expression resulting from a straightforward computation.

Taking $N = 1$ in (14.4.30) we immediately see why (14.4.29) is true. In this case the critical variance is zero, and so Theorem 14.3.3 tells us that the mean Euler characteristic approximation to the excursion probability is, in fact, exact. In higher dimensions, however, this is no longer the case.

The cosine process can also be exploited to show why the basic approximation of Theorem 14.3.3 works only for piecewise smooth manifolds, and not for Whitney stratified manifolds.[9] Consider a two-dimensional cosine process restricted to the set L_α depicted in Figure 14.4.1. We take the angle α (measured from the horizontal line) in $[0, \pi]$ and we take each edge to be of length $T < \pi/4$. (Our convention is that if $\alpha = \pi/2$ then the resulting vertical edge is *above* the horizontal one.)

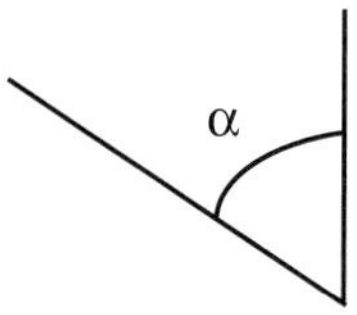

Fig. 14.4.1. The stratified manifold L_α.

It is easy to check that regardless of the value of α, when restricted to an edge, the cosine field is stationary with variance one and second spectral moment $\lambda_2 = \frac{1}{2}$. Furthermore, the expected Euler characteristic of the excursion set $A_u(f, L_\alpha)$ is, with one exception, independent of α and is given by

[9] Recall that the main difference between these two classes of objects is that piecewise smooth manifolds have convex support cones. Back in Chapter 12 we were able to obtain expected Euler characteristics of excursion sets in both settings. In the current chapter, however, we have been restricted to the locally convex scenario.

$$\mathbb{E}\{\varphi(A_u(f, L_\alpha))\} = \begin{cases} \Psi(u) + \frac{T}{\sqrt{2\pi}} e^{-u^2/2}, & \alpha > 0, \\ \Psi(u) + \frac{T}{2\sqrt{2\pi}} e^{-u^2/2}, & \alpha = 0. \end{cases}$$

When $\alpha = \pi$ or $\alpha = 0$, so that L_α is simply a straight line of length $2T$ or T, then the argument above for the one-dimensional cosine process gives that the excursion probability is identical to $\mathbb{E}\{\varphi(A_u(f, L_\alpha))\}$. On the other other hand, while $\mathbb{E}\{\varphi(A_u(f, L_\alpha))\}$ is constant on $(0, \pi]$ and discontinuous at $\alpha = 0$, it seems reasonable to claim that the excursion probability must be continuous in α. Consequently, it is unreasonable to expect that it is exponentially accurate over all $\alpha \in [0, \pi]$, as we might have hoped for from Theorem 14.3.3. In fact, in the notation of that theorem we have, for $\alpha \notin \{0, \pi\}$, that the critical variance σ_c is infinite, or, in the notation of Lemma 14.4.3, that the critical radius θ_c is zero.[10]

All this comes from the non(local) convexity of L_α.

Before leaving the cosine field, we note that it actually deserves far more credit than we have given it so far. To this point, it has been a curiosity useful for providing examples and counterexamples. In truth, however, it was the cosine process that gave birth to this entire chapter and also had much to do with the Lipschitz–Killing curvatures that appeared in Chapter 12. The history, which we describe only briefly, is as follows.

Firstly, we restrict our attention to the cosine process on a rectangle of the form $T = \prod_{k=1}^N [0, T_k]$, where $T_k \in (0, \pi/2]$. Then, given the explicit form of the cosine process (we are assuming still that $\omega_k = 1$ for all k), note that

$$M_N = \sup_{t \in T} \sum_{k=1}^N \frac{1}{\sqrt{N}} \sup_{0 \le t_k \le T_k} \{\xi_k \cos t_k + \xi_k' \sin t_k\} \triangleq \frac{1}{\sqrt{N}} \sum_{k=1}^N m_k,$$

where the random variables m_k are independent.

However, we know the precise distribution of each of the m_k by (14.4.29), so that in principle at least, we should now be able to find the distribution of M_N.

It turns out that this computation is a little too involved to be done exactly, and some approximations need to be made along the way. This is what Piterbarg did in [126], and what he derived was basically Theorem 14.3.3 for this case. You can find all the details of the calculations in [126].

Theorem 14.3.3 is much richer, however, since it treats quite general processes over piecewise smooth spaces. To go from the cosine process to a wider class of processes, Piterbarg used Slepian's inequality (Theorem 2.2.1). While Slepian's inequality allowed an extension beyond the cosine process, it did restrict the result to random fields for which the matrix of second-order spectral moments is diagonal. After all we have seen, we now know this to be a form of local isotropy and that the way to remove this assumption is via the Riemannian metric induced by the field.

To go from rectangles to a wider class of parameter sets made up of locally convex cells, not unlike the way we built up the basic complexes of Chapter 6, Piterbarg used

[10] For more sophisticated examples of what can, and usually does, go wrong when the critical radius is zero, see [149].

a result not unlike our Theorem 6.3.1. As a consequence, he obtained an expression for the excursion probability involving Minkowski functionals, which we now know to be a Euclidean version of Lipschitz–Killing curvatures.

What Piterbarg did not realize was that the approximation that he obtained for the excursion probability, involving second-order spectral moments and Minkowski functionals, was in fact the expected Euler characteristic of excursion sets. This was realized by one of us in [4]. In the meantime, Keith Worsley had worked out some expected Euler characteristics for isotropic random fields on two-dimensional surfaces (e.g., [173]), and Lipschitz–Killing curvatures (although not yet recognized as such) began to appear.

All this came together in the thesis [157], where the induced Riemannian metric appeared for the first time. As we have seen, this allowed the development of a theory that was no longer restricted to stationary (and often also isotropic) random fields.

The rest is history, and the (currently) final result is what you have been reading.

15

Non-Gaussian Geometry

This final chapter is, for two reasons, somewhat of an outlier as far as this book is concerned.

The first reason is that it was initially primarily motivated by applications that required a non-Gaussian theory, while most of the book has been about purely Gaussian random fields.

The second reason is that in large part, it stands independent of the preceding four chapters of Part III of the book and is able to recoup, with completely different proofs, the main results of Chapters 11–13, although not those of Chapter 14.

Consider the first issue first. If you, like us, have applications in mind, it will take no effort whatsoever to convince you that not all random fields occurring in the "real world" are Gaussian.[1] The term "non-Gaussian" is, however, not well defined and covers too wide a class of generalizations. Consequently, throughout this chapter we shall limit ourselves to random fields of the form

$$f(t) = F(y(t)) = F(y_1(t), \ldots, y_k(t)), \tag{15.0.1}$$

where the $y_j(t)$ are a collection of independent, identically distributed *Gaussian* random fields of the kind considered in the previous four chapters, all defined over a nice parameter space M, and $F : \mathbb{R}^k \to \mathbb{R}$ is a smooth function.[2] Such non-Gaussian random fields appear naturally in many statistical applications of smooth random fields (e.g., [172] or wait for [8]).

Choosing $k = 1$ and $F(x) = x$ takes us back to the Gaussian case, but other simple choices take us to a wide range of interesting random fields. For example,

[1] If you are a theoretician who has been reading this material for pure intellectual enjoyment, then you will just have to take our word for the fact that such is the case. In any case, you will be rewarded by some elegant theory in the coming pages, so you need concern yourself no further with the "real world."

[2] In the previous chapters we assumed that the Gaussian fields $f = \tilde{f}_{|M}$ were the restriction of processes defined on some larger ambient space $\widetilde{M}$, itself necessary to describe M. The random fields y_j in (15.0.1) should therefore be thought of as restrictions of such fields and so the left-hand side of (15.0.1) should actually read $f = (F \circ y)_{|M}$.

suppose that the y_j are centered and of unit variance and consider the following three choices for F, where in the third we set $k = n + m$:

$$\sum_{i=1}^{k} x_i^2, \qquad \frac{x_1\sqrt{k-1}}{(\sum_{i=2}^{k} x_i^2)^{1/2}}, \qquad \frac{m\sum_{i=1}^{n} x_i^2}{n\sum_{i=n+1}^{n+m} x_i^2}. \tag{15.0.2}$$

The corresponding random fields are known as χ^2 fields with k degrees of freedom, Student's t field with $k-1$ degrees of freedom, and the F field with n and m degrees of freedom. If you have any familiarity with basic statistics you will know that the corresponding distributions are almost as fundamental to statistical theory as is the Gaussian distribution. If you are not statistically literate you can consider these merely as specific examples of general non-Gaussian random fields of the form (15.0.1). Although we shall return to two of these examples in Section 15.10, the rest of this chapter will treat the general case.

As usual, we shall concentrate on the excursion sets of these random fields, a problem that can be tackled in at least two quite distinct ways. The standard approach, which held sway until JET's thesis [157] appeared in 2001 (cf. also [158]), was to treat each particular F as a special case and to handle it accordingly. This invariably involved detailed computations of the kind we met first in Chapter 11 and that were always related to the underlying Gaussian structure of y and to the specific transformation F. This approach led to a large number of papers, which provided not only a nice theory but also a goodly number of doctoral theses, promotions, and tenure successes.

A more unified and far more efficient approach is based, in essence, on the observation that

$$\begin{aligned} A_u(f, M) &= A_u(F(y), M) = \{t \in M : (F \circ y)(t) \geq u\} \\ &= \{t \in M : y(t) \in F^{-1}[u, \infty)\} = M \cap y^{-1}(F^{-1}[u, +\infty)). \end{aligned} \tag{15.0.3}$$

Thus, the excursion set of a *real-valued* non-Gaussian $f = F \circ y$ above a level u is equivalent to the excursion set for a *vector-valued* Gaussian y in $F^{-1}[u, \infty)$, which, under appropriate assumptions on F, is a manifold with piecewise smooth boundary.

This chapter is all about exploiting this observation in order to compute explicit expressions for all the mean Lipschitz–Killing curvatures

$$\mathbb{E}\{\mathcal{L}_i(A_u(f, M))\} = \mathbb{E}\left\{\mathcal{L}_i(M \cap y^{-1}\left(F^{-1}([u, +\infty))\right)\right\}. \tag{15.0.4}$$

In fact, since the function F in (15.0.4) is quite general, so is the set $F^{-1}([u, +\infty))$, which will generally be a stratified manifold in $\mathbb{R}^k$. Consequently, rather than concentrating on the usual excursion sets, there is no reason not to take a general subset D of $\mathbb{R}^k$ and study the Lipschitz–Killing curvatures of $y^{-1}(D)$. Indeed, this is the path that we shall take, and we shall show that for suitable stratified manifolds M and suitable $D \subset \mathbb{R}^k$, we have

$$\mathbb{E}\left\{\mathcal{L}_i(M \cap y^{-1}(D)\right\} = \sum_{j=0}^{\dim M - i} \begin{bmatrix} i+j \\ j \end{bmatrix} (2\pi)^{-j/2} \mathcal{L}_{i+j}(M) \mathcal{M}_j^{\gamma}(D), \tag{15.0.5}$$

where the combinatorial flag coefficients $\left[{n \atop m} \right]$ are those we met at (6.3.12) and the $\mathcal{M}_j^{\gamma} = \mathcal{M}_j^{\gamma_{\mathbb{R}^k}}$ are the generalized (Gaussian) Minkowski functionals that we met in the Gaussian tube formula of Corollary 10.9.6.

The Lipschitz–Killing curvatures in both (15.0.4) and (15.0.5) are computed with respect to the Riemannian metric[3] induced on M by the individual Gaussian fields y_j. This is good news in the latter case, since the fact that they are therefore identical to those appearing in the purely Gaussian results of Chapters 11 and 12 means that they do not need to be recomputed for the non-Gaussian scenario. Consequently, obtaining a useful final formula for a specific case—such as those of (15.0.2)—involves only knowing how to compute the $\mathcal{M}_j^{\gamma}(F^{-1}[u, +\infty))$. This usually turns out to be an exercise in multivariate calculus, and in Section 15.10 we shall look at some examples.

Before describing how we plan to do all the above, we note two important points. The first you might have already noticed yourself: The parameter spaces for the random fields of this chapter are denoted by M and not T. Thus we are in the manifold setting of Chapter 12 rather than the Euclidean setting of Chapter 11. There is no way around this. Even if we wanted to treat only simple Euclidean parameter spaces, the structure of $F^{-1}([u, \infty))$ is that of a manifold, and it is the structure of this set that is crucial to the central computations of the chapter. There is no way that we know to write this chapter for the differential-geometrically challenged.

The other point is that we shall not be able to develop a geometric theory of random excursion sets when dealing with fields that do not have the underlying Gaussian structure of (15.0.1). We have nothing satisfactory to say here. Indeed, we know of no "purely" non-Gaussian random field for which excursion sets have been successfully studied.[4] While this is undesirable, it is probably to be expected, since the sad fact is that even in the case of processes on $\mathbb{R}$ very little is known in this generality. We therefore leave it as a (probably very hard) challenge for future generations.

15.1 A Plan of Action

The derivation that we shall give of (15.0.5) will be somewhat different from what we developed earlier in treating its Gaussian analogue. In particular, the Morse theory and the point process characterization of the Euler characteristic that were so important then will now play only a subsidiary role.[5] Nor, in the main parts of the argument,

[3] Recall that as far as $\mathcal{L}_0$ is concerned, the Riemannian metric used to compute it is irrelevant, since $\mathcal{L}_0$ always gives the Euler characteristic. By the Chern–Gauss–Bonnet theorem (cf. Section 12.6) this is independent of the metric.

[4] There is a partial exception to this statement, in that smoothed Poisson fields have been studied in [130]. The theory there, however, is not as rich as in the Gaussian case.

[5] One negative aspect of this approach is that we have not been able to develop an analogue of the results of Chapter 14 for the non-Gaussian case; i.e., we are unable to show that the mean Euler characteristic provides a good asymptotic approximation to excursion probabilities. Whether this is true, and under what conditions, is still an open question.

will the Euler characteristic be singled out as a special Lipschitz–Killing curvature from which the others may be derived, as was the case in Chapter 13.

Our first step, in Section 15.2, will be to compute a general form for expectation in (15.0.5), showing that it must be of the form

$$\mathbb{E}\left\{\mathcal{L}_i(M \cap y^{-1}(D))\right\} \sum_{j=0}^{\dim M - i} \mathcal{L}_{i+j}(M)\widetilde{\rho}(i, j, D), \tag{15.1.1}$$

where $\widetilde{\rho}$ depends on all the parameters given, but not on the distribution of the underlying Gaussian fields y_j.

With (15.1.1) established it is clear that we need only find the form of the function $\widetilde{\rho}$. Moreover, we are free to choose simple examples of the manifold M and the Gaussian fields y_j in order to do this, since the result cannot depend on these choices. The choice we make is that of a specific rotationally invariant field restricted to subsets of a sphere.

However, even with this choice the computations are nontrivial, and so we approach them in an indirect fashion. In particular, we shall start by looking at non-Gaussian fields with finite orthogonal expansions and with coefficients coming from random variables distributed uniformly over high-dimensional spheres. This will enable us to take expectations using an appropriate version of the kinematic fundamental formula on spheres, Euclidean (flat) relatives of which we already met in Chapter 13 under the titles of Hadwiger's formula (13.1.2) and Crofton's formula (13.1.1). These preliminary expectations are computed in Section 15.6, after we establish the requisite kinematic fundamental formulas in Section 15.5.

The passage from this scenario to the Gaussian one will be via a limit theorem for projections of uniform random variables on S^n as $n \to \infty$, historically associated with the name of Poincaré, although we shall need a slight extension of some more recent versions due to Diaconis and Freedman [46] and their generalizations to matrices in [47]. Poincaré's result is given in Section 15.4 and applied to our setting in Section 15.9, after we first see how it works in the case of the Euler characteristic, $\mathcal{L}_0$, in Section 15.7. This section effectively rederives the main results of Chapters 11 and 12.

It is in Section 15.9 that we shall finally exploit the non-Lebesgue tube-volume results of Section 10.9 to see how the $\mathcal{M}_j^\gamma$ arise in (15.0.5).

The argument throughout will involve a continual and delicate intertwining of probability and geometry to obtain a result that, since it involves both expectations and Lipschitz–Killing curvatures, lies in both areas. As such, we see it as being at the core of what this book was supposed to be about, and we hope that you will enjoy reading it as much as we did finding it.[6]

---

[6] Actually, finding the proofs for this chapter was not always enjoyable. As you will soon see, the basic idea behind everything is not too difficult, although it is often well camouflaged by heavy notation and subject to heavy combinatorial manipulation. If you find following the details occasionally oppressive you can imagine what it was like going through it all for the first time. However, the final result, Theorem 15.9.5 (i.e., the formal version of (15.0.5)) came out to be so simple and elegant that "enjoy" is still the most appropriate verb.

15.2 A Representation for Mean Intrinsic Volumes

In this section, we give a formal statement of (15.1.1). The proof appears only in the following section, so as to encourage you not to read it immediately and so spoil what we feel is otherwise a rather elegant flow of ideas.

Theorem 15.2.1. *Let M be an N-dimensional, regular stratified manifold embedded in $\widetilde{M}$, also of dimension N, and let D be a regular, stratified manifold in $\mathbb{R}^k$, $k \geq 1$.*

Let $y = (y_1, \dots, y_k) : M \to \mathbb{R}^k$ be a vector-valued Gaussian field, the components y_i of which are independent, identically distributed, zero-mean, unit-variance Gaussian fields satisfying the conditions of Corollary 11.3.5.

Let $\mathcal{L}_j$, $j = 0, \dots, N$, be the Lipschitz–Killing measures on M with respect to the metric induced by the y_i, as defined in (10.7.1). *Then there exist functions $\widetilde{\rho}(i, j, D)$ dependent on all the parameters displayed, but not on the distribution of the underlying Gaussian fields y_i, such that*

$$\mathbb{E}\left\{\mathcal{L}_i(M \cap y^{-1}(D))\right\} = \sum_{j=0}^{N-i} \mathcal{L}_{i+j}(M)\widetilde{\rho}(i, j, D). \tag{15.2.1}$$

The proof is not short, and it proceeds via a series of smaller calculations. The basic idea is to write $M \cap y^{-1}(D)$ as

$$M \cap y^{-1}(D) = \bigcup_{j=0}^{N} \bigcup_{l=0}^{j} \partial_j M \cap y^{-1}\left(\partial_{k-l} D\right). \tag{15.2.2}$$

Since both M and D are regular stratified manifolds and y is, with probability one, C^2, it follows from the basic properties of these manifolds (cf. Section 8.1) that such a partition exists. If we also assume, without loss of generality,[7] that $k \geq \dim(\widetilde{M})$, then it follows from the properties of y that, with probability one, each $\partial_j M \cap y^{-1}(\partial_{k-l} D)$ will be a (random) manifold of dimension $j - l$ in $\widetilde{M}$.

To simplify notation below, we write the strata of $M \cap y^{-1} D$ as

$$M_{jl} \stackrel{\Delta}{=} \partial_j M \cap y^{-1}(\partial_{k-l} D). \tag{15.2.3}$$

Using the alternative definition (10.7.10) of the Lipschitz–Killing curvatures and applying additivity, it follows that the Lipschitz–Killing curvatures of $M \cap y^{-1} D$ can be expressed as

$$\mathcal{L}_i(M \cap y^{-1} D) = \sum_{j=i}^{N} \sum_{l=0}^{j-i} \mathcal{L}_i\left(M, M_{jl}\right). \tag{15.2.4}$$

What we shall prove in the following section is that for $i \leq j$,

[7] If $k < N$, then simply add $N - k$ superfluous copies of the coordinate processes to y_j, yielding a new vector-valued process $\tilde{y}$. Now set $\tilde{D} = \pi_k^{-1} D$, where $\pi_k : \mathbb{R}^N \to \mathbb{R}^k$ is projection onto the first k coordinates of $\mathbb{R}^N$, and work with $\widetilde{y}$ and $\tilde{D}$ rather than y and D.

$$\mathbb{E}\left\{\mathcal{L}_i\left(M, M_{jl}\right)\right\} = \sum_{m=0}^{j-i} \mathcal{L}_{i+m}(M, \partial_j M) \cdot \widehat{\rho}(i, m, \partial_l D). \tag{15.2.5}$$

By summing the above over strata $\partial_j M$ of M, we reach the desired conclusion, in which case the functionals $\widetilde{\rho}$ will eventually be expressed as some linear combination of the $\widehat{\rho}$, summed over strata $\partial_l D$ of D. To actually identify the precise form of $\widetilde{\rho}$ will take quite some time and effort, and it will not be until Section 15.9 that we shall see how the $\mathcal{M}_j^{\gamma}$ enter the picture.

Nevertheless, if this is your first reading, we recommend you start on this path now, turning to Poincaré's limit (Section 15.4) and the kinematic fundamental formula on the sphere (Section 15.5), and continuing on from there, returning only later to the following section for a proof of Theorem 15.2.1.

15.3 Proof of the Representation

In this section we prove Theorem 15.2.1, noting, however, that you will not need the techniques of the proof for anything else in this chapter, and that on first reading it should probably be skipped.

In the notation of the previous section, we shall prove that under the conditions of Theorem 15.2.1,

$$\mathbb{E}\left\{\mathcal{L}_0\left(M, M_{jl}\right)\right\} = \sum_{i=0}^{j} \mathcal{L}_i^1(M, \partial_j M) \cdot \widehat{\rho}(i, \partial_{k-l} D), \tag{15.3.1}$$

for some functions $\widehat{\rho}$ depending only on the parameters displayed.

Once (15.3.1) is established, we can apply the linear relationships between the $\mathcal{L}_i^1$ and the $\mathcal{L}_i$ (cf. Lemma 10.5.8) to move from a representation involving the $\mathcal{L}_i$ to one based on the $\mathcal{L}_i^1$. While the functionals $\widehat{\rho}$ will then change,[8] this is not important for us. The proof of Theorem 15.2.1 is then completed using the addivity of the curvature measures $\mathcal{L}_i^1(M, \cdot)$ and the Crofton formula of Theorem 13.3.2 as in the proof of Theorem 13.4.1.

The calculations that follow are not particularly enlightening. However, if the difficulties in this section do have one saving grace, it is to demonstrate the power of the approach that we shall soon adopt via Poincaré's limit. If we did not adopt that approach, then we would need to compute an explicit formula for the functionals $\widehat{\rho}$ above and relate these to the Gaussian Minkowski functionals appearing in (10.9.12). This was the approach taken in [158] when M and D were C^2 manifolds with smooth boundary. It was difficult then, and would be even more so in the current scenario of stratified spaces.

[8] To see how they might change, look at the combinatorial manipulations in Section 15.6.2, where a similar exchange occurs. As you will then see, we are indeed fortunate that we do not need to repeat these manipulations here.

Nevertheless, despite the technical nature of what follows, it is not completely foreign, for the arguments have a lot in common with those in Chapters 12 and 13. In particular, in the proof of Theorem 13.3.2 we defined an independent field $\widetilde{y}$ with the same distribution as each of the components of y to act as a (random) Morse function for computing the Euler characteristic of $M \cap D_{Z_k}$. We shall do the same again, although now we need to compute only $\mathcal{L}_0(M; M_{jl})$, and not the full Euler characteristic $\mathcal{L}_0(M, y^{-1}D)$.

As in Chapter 13 we shall compute $\mathbb{E}\{\mathcal{L}_0(M; M_{jl})\}$ by appealing to an expectation metatheorem, in this case Corollary 12.4.5. However, in order to use this line of argument, we shall require a little notation and shall need to write the strata of D in a form that will simplify computations. In particular, we shall assume that each $\partial_{k-l}D$ is a relatively open subset of the zero set of a nice function.

More formally, suppose that $F_l : \mathbb{R}^k \to \mathbb{R}^l$ is a C^2 function with $F_l^{-1}(0)$ a $(k-l)$-dimensional C^2 submanifold containing $\partial_{k-l}D$. The stratum $\partial_{k-l}D$ will generally have to be broken up into smaller pieces to achieve this, in which case we can think of $\partial_{k-l}D$ as a generic stratum of dimension $k-l$ rather than the union of all $(k-l)$-dimensional strata.[9] With this in mind, for the remainder of the proof we shall assume that

$$\partial_{k-l}D \subset \{x \in \mathbb{R}^k : F_{l,r}(x) = 0, 1 \le r \le l\}.$$

We now define $f_t = F_l(y_t)$, an $\mathbb{R}^l$-valued random field on M with components $f_r(t) = F_{l,r}(y_t)$, $r = 1, \dots, l$, dropping the unnecessary dependence on l, which will be fixed for the remainder of the proof. Given f, we introduce the Jacobians[10] $\widetilde{J}f(t)$ with entries

$$\widetilde{J}f(t)_{rs} = \langle P_{T_t\partial_j M}\nabla f_r(t), P_{T_t\partial_j M}\nabla f_s(t)\rangle.$$

Note, for later use, that given y_t the Jacobians have conditional distributions (cf. footnote 3 of Chapter 13)

$$\widetilde{J}f(t)|y_t \sim \text{Wishart}(j, \langle\nabla F_r(y_t), \nabla F_s(y_t)\rangle) \stackrel{\Delta}{=} \text{Wishart}(j, \widetilde{J}F(y_t)). \qquad (15.3.2)$$

[9] It is not difficult to find functions F_l satisfying the conditions we require. For a concrete example, we argue as follows: Extend $\partial_{k-l}D$ geodesic distance δ along unit speed geodesics emanating from it. Denoting the extension by $\partial_{k-l}^{\delta}D$, note that since we are prepared to work locally, we can assume that $\partial_{k-l}D$ has positive critical radius, so that we can choose δ sufficiently small for the projection map $P_{\partial_{k-l}^{\delta}D}$ to be well defined on Tube$(\partial_{k-l}D, 2\delta)$. Choosing the standard orthonormal basis for $\mathbb{R}^k$, we can now define $F_l(t)$ coordinatewise by

$$(F_l(t))_j = ((t - P_{\partial_{k-l}^{\delta}D}(t))_j)^2, \quad 1 \le j \le l.$$

It is then easy to check that this F_l satisfies all the required conditions.

[10] While $\widetilde{J}f(t)$ is not strictly the Jacobian $Jf(t) : T_t\partial_j M \to T_{f_t}\mathbb{R}^l$, we call it the Jacobian because $(\det(\widetilde{J}f(t)))^{1/2} = |Jf(t)|$, where the norm $|Jf(t)|$ is the square root of the sum of all $l \times l$ principal minors of the matrix of the linear transformation $Jf(t)$ in terms of orthonormal bases of $T_t\partial_j M$ and $T_{f_t}\mathbb{R}^l$.

With notation now set, we can start our computation. As a first step, choosing $\widetilde{y}$ as a Morse function, note that $\mathcal{L}_0(M; M_{jl})$ is determined by the points[11]

$$\left\{t \in \partial_j M : f(t) = 0,\ y_t \in \partial_{k-l} D,\ P_{T_t M_{jl}} \nabla \widetilde{y} = 0\right\}, \qquad (15.3.3)$$

along with the indices of $\nabla^2 \widetilde{y}_t$ and the normal Morse indices of M_{jl} in M.

Applying an appropriately reworked version[12] of Corollary 12.4.5, an argument essentially identical to those appearing in the proofs of Theorems 13.3.1 and 13.3.2 shows that the expected value of $\mathcal{L}_0\left(M \cap y^{-1} D, M_{jl}\right)$ is given by

$$\frac{1}{(j-l)!} \int_{\partial_j M} \mathbb{E}\Big\{\alpha(\eta_t; M \cap y^{-1} D)\, \mathrm{Tr}^{T_t M_{jl}} \left((-\nabla^2 \widetilde{y}_{|M_{jl}})^{j-l}\right) |\widetilde{J} f(t)| \mathbb{1}_t \qquad (15.3.4)$$
$$\Big| f(t) = 0,\ P_{T_t M_{jl}} \nabla \widetilde{y} = 0\Big\} p_{f, P_{jl}}(0,0) \mathcal{H}_j(dt),$$

where in order to make the formulas a little more manageable, we write $\mathbb{1}_t$ for $\mathbb{1}_{\partial_{k-l} D}(y(t))$, $|\widetilde{J} f(t)|$ for $(\det(\widetilde{J} f(t)))^{1/2}$,

$$\eta_t \triangleq P^{\perp}_{T_t M_{jl}} \nabla \widetilde{y}_t, \qquad (15.3.5)$$

and $p_{f, P_{jl}}$ for the joint density[13] of $f(t)$ and $P_{T_t M_{jl}} \nabla \widetilde{y}$, which, it is easy to see, can be written as

$$(2\pi)^{-(j-l)/2} p_f(0), \qquad (15.3.6)$$

where p_f is the density of $f(y_t)$.

Our goal is to show that the expectation in (15.3.4) factors, in an appropriate fashion, into one factor related only to the Lipschitz–Killing curvature measures of M and another that is related only to $\partial_{k-l} D$. (Since the density term $p_{f, P_{jl}}(0,0)$ is actually independent of t, we can ignore it.) Thus, we focus on evaluating the conditional expectation in the integrand of (15.3.4).

To be more specific, we shall consider a finer conditional expectation, looking at

$$\mathbb{E}\Big\{\alpha(\eta_t; M \cap y^{-1} D)\, \mathrm{Tr}^{T_t M_{jl}} (-\nabla^2 \widetilde{y}_{|M_{jl}})^{j-l} |\widetilde{J} f(t)| \mathbb{1}_t \Big| y_t, y_{*,t}, \widetilde{y}_t, \widetilde{y}_{*,t}\Big\}, \qquad (15.3.7)$$

---

[11] Note that there is some redundancy in (15.3.3), since any $t \in \partial_{k-l} D$ also satisfies $f(t) = 0$. Nevertheless, it will be convenient for us to state things as we have done.

[12] As in the proof of Theorems 13.3.1 and 13.3.2, where we also allowed ourselves the latitude of quoting an unproven result, we again claim what while filling in all the details might take a few pages, it would involve nothing conceptually new and so is left to you.

[13] Actually, it is somewhat inappropriate to call $p_{f, P_{jl}}$ a joint "density" without being a little more careful. Clearly, $P_{T_t M_{jl}} \nabla \widetilde{y}_t$ lives in $T_t \partial_j M$, a j-dimensional space, and $f(y_t)$ lives in an l-dimensional space. However, conditional on $y_t, y_{*,t}$, we can find a subspace of $T_t \partial_j M$ that is perpendicular to $y_t^*(T_{y_t} \partial_{k-l} D)^{\perp}$, the pullback of the orthogonal complement of $T_{y_t} \partial_{k-l} D$ in $T_{y_t} \mathbb{R}^k$, as well as an orthonormal basis for this subspace $X_{1,t}, \ldots, X_{j-l,t}$. The vector $P_{T_t M_{jl}} \nabla \widetilde{y}_t$ can thus be written as a linear combination of the $X_{i,t}$'s and you should think of the density $p_{f, P_{jl}}$ as the joint density of f and the coefficients of $P_{T_t M_{jl}} \nabla \widetilde{y}_t$ in this basis. Working this way you can derive (15.3.6), which *is* well defined.

where $y_{*,t}$ and $\widetilde{y}_{*,t}$ are the push-forwards of y and $\widetilde{y}$. We shall see that this factors as required, which, after a little further manipulation of expectations, will achieve our goal.

This expectation is virtually identical to the one we evaluated in Theorem 13.3.2, except for two significant differences:

- The set $F_l^{-1}(0)$ may not be flat as in Theorem 13.3.2, where it was a subspace of $\mathbb{R}^k$.
- The normal Morse index α is not as simple as it was there, since now it can also depend on other strata of D.

We deal with the normal Morse index first. Theorem 9.2.6 implies that this Morse index factors into a product of two modified Morse indices, one for M and one for $y^{-1}D$. To set this up properly for the current scenario, note first that the support cone of $M \cap y^{-1}D$ at $t \in M_{jl}$ is the intersection of two support cones, that is,

$$\mathcal{S}_t M_{jl} = \mathcal{S}_t M \cap \mathcal{S}_t y^{-1} D \subset T_t \widetilde{M}.$$

Let

$$V_t \stackrel{\Delta}{=} \operatorname{span}\{\nabla f_i(t),\, 1 \le i \le l\}$$

be the normal to $T_t y^{-1}(\partial_{k-l} D)$ in all of $T_t \widetilde{M}$. By Theorem 9.2.6, for almost every $\nu_t \in T_t M_{jl}^{\perp}$,

$$\begin{aligned} \alpha(\nu_t; M \cap y^{-1}D) &= \alpha(\nu_t; \mathcal{S}_t(M \cap y^{-1}D)) \\ &= \alpha\left(\mathcal{T}_{V_t}\nu_t; \mathcal{S}_t M\right) \cdot \alpha\left(\nu_t; P_{T_t\partial_j M}\left(\mathcal{S}_t\left(y^{-1}D\right)\right)\right) \\ &\stackrel{\Delta}{=} \alpha_1(\nu_t) \cdot \alpha_2(\nu_t), \end{aligned} \tag{15.3.8}$$

where $\mathcal{T}_{V_t}$ is as defined in Theorem 9.2.6. Specifically,

$$\mathcal{T}_{V_t}\nu_t = P^{\perp}_{T_t\partial_j M}\nu_t - \sum_{r,s=0}^{l} \langle \nu_t, P_{T_t\partial_j M}\nabla f_r(t)\rangle \widetilde{J} f(t)^{rs} P^{\perp}_{T_t\partial_j M}\nabla f_s(t).$$

In particular, we have

$$\alpha(\eta_t; M \cap y^{-1}D) = \alpha_1(\eta_t) \cdot \alpha_2(\eta_t)$$

for the η_t of (15.3.5).

With regard to the second of these indices, note that the vectors in the cone $\mathcal{S}_t\left(y^{-1}D\right)$ are push-forwards under y_t^{-1} of vectors in a cone in $T_{y_t}\mathbb{R}^k$. Consequently,

$$\begin{aligned} \alpha_2(\eta_t) &= \alpha\left(\eta_t; P_{T_t\partial_j M}\left(\mathcal{S}_t\left(y^{-1}D\right)\right)\right) \\ &= \alpha\left(y_*(\eta_t); \mathcal{S}_{y_t}(D) \cap y_*(T_t\partial_j M)\right), \end{aligned}$$

where, with some abuse of notation, we have written $y_*(T_t\partial_j M)$ to denote the collection of push-forwards, by y, of vectors in $T_t\partial_j M$. It is clear that, as a random variable, $\alpha_2(\eta_t)$ is dependent only on the collection

$$\mathcal{F}_t \stackrel{\Delta}{=} \left\{y_t, y_{*,t}, \widetilde{y}_t, \widetilde{y}_{*,t}\right\}$$

of four random variables. Hence, the conditional expectation in (15.3.7) does not depend on t.

In view of the above observations, and appealing to Lemma 13.5.1, we have that the conditional expectation (15.3.7) can be written as

$$\begin{aligned}
&\mathbb{E}\left\{\alpha(\eta_t; M\cap y^{-1}D)\operatorname{Tr}^{T_tM_{jl}}(-\nabla^2\widetilde{y}_{|M_{jl}})^{j-l}|\widetilde{J}f(t)|\mathbb{1}_t\Big|\mathcal{F}_t\right\}\\
&\quad=\frac{l!(j-l)!}{j!}\mathbb{E}\left\{\alpha(\eta_t; M\cap y^{-1}D)\operatorname{Tr}^{T_t\partial_jM}\left((-\nabla^2\widetilde{y}_{|M_{jl}})\right)^{j-l}|\widetilde{J}f(t)|\mathbb{1}_t\Big|\mathcal{F}_t\right\}\\
&\quad=\frac{l!(j-l)!}{j!}\alpha_2(\eta_t)|\widetilde{J}f(t)|\mathbb{1}_t\mathbb{E}\left\{\alpha_1(\eta_t)\operatorname{Tr}^{T_t\partial_jM}\left((-\nabla^2\widetilde{y}_{|M_{jl}})\right)^{j-l}\Big|\mathcal{F}_t\right\}.
\end{aligned}$$

To evaluate the expectation here, we need to look a little more closely at the structure of the Hessian $\nabla^2\widetilde{y}_{|M_{jl}}$. As in the proof of Theorem 13.3.2, applying the Weingarten equation (7.5.12) gives us that

$$\nabla^2\widetilde{y}_{|M_{jl},t} = \nabla^2\widetilde{y}_t - \sum_{r,s=1}^{l}\langle\nabla\widetilde{y}_t, \nabla f_{r|\partial_jM}(t)\rangle\left(\widetilde{J}f(t)\right)^{rs}\nabla^2 f_{s|\partial_jM}(t).$$

This depends on the curvature of $F_l^{-1}(0)$. With some perversity (note how terms cancel) but with the future in mind, we acknowledge this dependence by writing

$$\nabla^2\widetilde{y}_{|M_{jl},t} = \Xi_t + \Theta_t,$$

where

$$\begin{aligned}
\Xi_t &\stackrel{\Delta}{=} \left(\nabla^2\widetilde{y}_t + \widetilde{y}_t\cdot I\right) - S_{\eta_t}\\
&\qquad - \sum_{r,s=1}^{l}\sum_{u=1}^{k}\langle\nabla\widetilde{y}_t, P_{T_t\partial_jM}\nabla f_r(t)\rangle\widetilde{J}f(t)^{rs}\frac{\partial F_s}{\partial y_u}\cdot\left(\nabla^2 y_u(t) + y_{u,t}\cdot I\right),\\
\Theta_t &\stackrel{\Delta}{=} \Theta_t^1 + \Theta_t^2,\\
\Theta_t^1 &\stackrel{\Delta}{=} \sum_{r,s=1}^{l}\sum_{u,v=1}^{k}\langle\nabla\widetilde{y}_t, P_{T_t\partial_jM}\nabla f_r(t)\rangle\widetilde{J}f(t)^{rs}\left.\frac{\partial^2 F_s}{\partial y_u\partial y_v}\right|_{y_t}\\
&\qquad\times\langle X_t, \nabla y_u(t)\rangle\cdot\langle Y_t, \nabla y_v(t)\rangle,\\
\Theta_t^2 &\stackrel{\Delta}{=} -\widetilde{y}_t\cdot I + \sum_{r,s=1}^{k-l}\sum_{u=1}^{k}\langle\nabla\widetilde{y}_t, P_{T_t\partial_jM}\nabla f_r(t)\rangle\widetilde{J}f(t)^{rs}\frac{\partial F_s}{\partial y_u}y_u, t\cdot I.
\end{aligned}$$

Written as above, the curvature of $F_l^{-1}(0)$ is contained in the double form Θ_t^1.

Continuing, we thus have that (15.3.7) is given by

$$\mathbb{E}\left\{\frac{l!}{j!}\alpha_1(\eta_t)\alpha_2(\eta_t)\,\mathrm{Tr}^{T_t\partial_j M}\left((-\nabla^2\widetilde{y}_{|M_{jl}})\right)^{j-l}|\widetilde{J}f(t)|\mathbb{1}_t\Big|\mathcal{F}_t\right\}$$

$$\begin{aligned}
&=\sum_{m=0}^{j-l}\frac{l!(j-l)!}{(j-l-m)!m!j!}\\
&\quad\times\mathrm{Tr}^{T_t\partial_j M}\left(\mathbb{E}\left\{\alpha_1(\eta_t)\alpha_2(\eta_t)\cdot\Xi_t^m\Theta_t^{j-l-m}|\widetilde{J}f(t)|\mathbb{1}_t\Big|\mathcal{F}_t\right\}\right)\\
&=\sum_{m=l}^{j}\frac{l!(j-l)!}{(m-l)!(j-m)!j!}\\
&\quad\times\mathrm{Tr}^{T_t\partial_j M}\left(\mathbb{E}\left\{\alpha_1(\eta_t)\alpha_2(\eta_t)\cdot\Xi_t^{j-m}\Theta_t^{m-l}|\widetilde{J}f(t)|\mathbb{1}_t\Big|\mathcal{F}_t\right\}\right),
\end{aligned}$$

where the first equality follows from Corollary 12.3.2 and the second is no more than a change of variables.

We shall apply Lemma 13.5.2, conditional on y_t, to each term in the sum above. To do so, we first identify variables and functions above with the variables and functions in Lemma 13.5.2. With this in mind, set

$$\begin{aligned}
Z_i&=\sum_{r=0}^{l}\langle\nabla\widetilde{y}_t,P_{T_t\partial_j M}\nabla f_r(t)\rangle\widetilde{J}f(t)_{ri}^{-1/2},\\
W&=\left(\widetilde{J}F(y_t)\right)^{-1/2}\widetilde{J}f(t)\left(\widetilde{J}F(y_t)'\right)^{-1/2},\\
U_j&=\left(\nabla^2 y_j+y_j\cdot I,\,S_{P^{\perp}_{T_t\partial_j M}\nabla y_j}\right),\quad 1\le j\le l,\\
U_{l+1}&=\left(\nabla^2\widetilde{y},\,S_{P^{\perp}_{T_t\partial_j M}\nabla\widetilde{y}}\right),\\
G_{j-m}&=\alpha_1\cdot\Xi_t^{j-m},\\
\widetilde{G}_{m-l}&=\alpha_2\cdot\Theta_t^{m-l}.
\end{aligned}$$

It is straightforward to verify that, conditional on y_t, the above random variables have the required conditional distributions. The result after application of the lemma with $\tilde{l}=j-m$, $\tilde{r}=m-l$, $\tilde{n}=j$, $\tilde{k}=l$ is

$$(2\pi)^{-m/2}\frac{j!}{(j-l)!}\frac{\omega_m}{\omega_0}\mathbb{E}\{G_{m-l}\}\mathbb{E}\{\widetilde{G}_{j-m}|y_t\}. \tag{15.3.9}$$

There is only one conditional expectation above because the U_j's are, in fact, independent of y_t.

Applying Lemma 13.5.2 again, along with the fairly obvious fact that $\mathbb{E}\left\{\Theta_t^m\right\}=C_m I^m$ for any m, we find constants, $\widehat{\rho}(m,\partial_{k-l}D)$, which may change from line to line, such that the expected value of the above (after multiplication by the factor of 2π in (15.3.6)) is equivalent to

$$\sum_{m=l}^{j} \frac{l!}{(m-l)!(j-m)!}(2\pi)^{-(j-m)/2}\widehat{\rho}(m, \partial_{k-l}D)$$
$$\times \operatorname{Tr}^{T_t\partial_j M}\left(\mathbb{E}\left\{\alpha(P^{\perp}_{T_t\partial_j M}\nabla\widetilde{y}_t; M)\cdot(\nabla^2\widetilde{y}_t + \widetilde{y}_t\cdot I)^{j-m}\right\} I^{m-l}\right)$$
$$= \sum_{m=l}^{j} \frac{l!}{(m-l)!(j-m)!}(2\pi)^{-(j-m)/2}\widehat{\rho}(m, \partial_{k-l}D)$$
$$\times \operatorname{Tr}^{T_t\partial_j M}\left(\mathbb{E}\left\{\alpha(P^{\perp}_{T_t\partial_j M}\nabla\widetilde{y}_t; M)\cdot(\nabla^2\widetilde{y}_t + \widetilde{y}_t)^{j-m}\right\}\right),$$

the last equality following from (7.2.11).

Above, $\widehat{\rho}(m, \partial_{k-l}D)$ is a complicated expression involving the expected value of products of the α_2 of (15.3.8), powers of Θ^1_t and Θ^2_t, and factors of 2π and ω_m's.

Finally, we see that as in Theorem 13.3.2, the term

$$\int_{\partial_j M} \operatorname{Tr}^{T_t\partial_j M}\left(\mathbb{E}\left\{\alpha(P^{\perp}_{T_t\partial_j M}\nabla\widetilde{y}_t; M)\cdot(\nabla^2\widetilde{y}_t + \widetilde{y}_t\cdot I)^{j-m}\right\}\right)$$

is proportional to $\mathcal{L}^1_m(M; \partial_j M)$, with a proportionality constant independent of $\partial_l D$. The reason that $\mathcal{L}^1$ appears instead of $\mathcal{L} = \mathcal{L}^0$ is because of the appearance of $\widetilde{y}_t \cdot I$ above. This completes the proof, since all that remains under the expectation are quantities whose distribution does not depend on t. □

15.4 Poincaré's Limit

There are two simple and well-known connections between independent standard Gaussian variables and the uniform distribution on spheres. The first we met already in Section 10.2 in discussing the relationship between tube formulas and excursion probabilities. Slightly rewriting what we had there gives the observation that if X is an n-vector the components of which are independent standard Gaussians, then the distribution of X given that $|X| = \lambda$ is uniform on $S_\lambda(\mathbb{R}^n)$, the sphere of radius λ in $\mathbb{R}^n$.

The second result is generally referred to as Poincaré's limit, although whether it really is *Poincaré's* limit is another question [46]. Its simplest version can be stated as follows:

If $\eta_n = (\eta_{n1}, \ldots, \eta_{nn})$ is uniformly distributed on $S_{\sqrt{n}}(\mathbb{R}^n)$ and $k \geq 1$ is fixed, then the joint distribution of $(\eta_{n1}, \ldots, \eta_{nk})$ converges weakly to that of k independent standard Gaussians as $n \to \infty$. One can prove this either by realizing η as conditioned normals as described above, or via the following elementary argument.

Set $k = 1$ and fix n. Choose $-\sqrt{n} \leq a < b \leq \sqrt{n}$ and note that $\mathbb{P}\{a \leq \eta_{n1} \leq b\}$ is given by the uniform measure of the spherical zone $a \leq \eta_{n1} \leq b$, which is

$$\frac{\int_a^b (n-u^2)^{n/2-1}\,du}{\int_{-\sqrt{n}}^{\sqrt{n}} (n-u^2)^{n/2-1}\,du} = \frac{\int_a^b (1-u^2/n)^{n/2-1}\,du}{\int_{-\sqrt{n}}^{\sqrt{n}} (1-u^2/n)^{n/2-1}\,du}, \tag{15.4.1}$$

as is easily seen from Figure 15.4.1. As $n \to \infty$, the right-hand ratio converges to

$$\frac{\int_a^b e^{-u^2/2}\,du}{\sqrt{2\pi}}$$

by elementary estimates, and so the one-dimensional case is done. The generalization to $k > 1$ is straightforward.

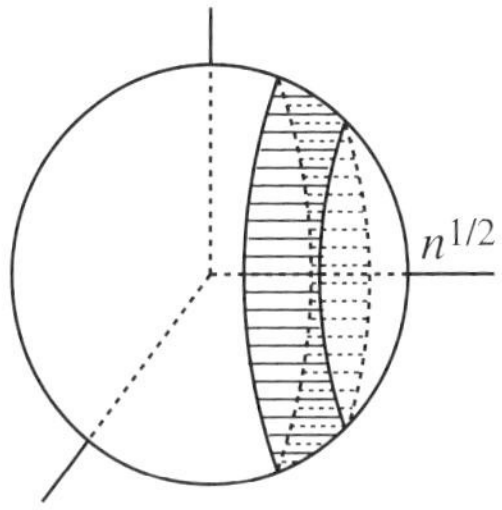

Fig. 15.4.1. The "laddered" region, the left and right edges of which project down to a and b on the right-pointing axis, is the spherical zone of (15.4.1) for $n = 3$.

This is the basic "Poincaré limit theorem," but we shall need more. For a start, we shall need to move from vectors to matrices, but this is really no more than a change of notation, which we shall handle in a moment. Furthermore, rather than talk about uniform random variables on spheres, it will be more convenient for us to choose Haar distributed random orthonormal matrices and use these to transform fixed points that will then have a spherically uniform distribution. This is also little more than a notational issue.

More importantly, however, we shall need to know about the convergence of expectations of functionals, and for this the proof itself becomes considerably more involved. Consequently, we present Theorem 15.4.1 below without a proof, referring you to [46, 47] for details.

Now for the notation. With η_n still uniformly distributed on $S_{\sqrt{n}}(\mathbb{R}^n)$, consider the random vector

$$X_{k,n} \stackrel{\Delta}{=} \pi_{\sqrt{n},n,k}(\eta_n),$$

where $\pi_{\lambda,n,k} : S_\lambda(\mathbb{R}^n) \to B_{\mathbb{R}^k}(0, \lambda)$, defined by

$$\pi_{\lambda,n,k}(x_1, \ldots, x_n) = (x_1, \ldots, x_k), \tag{15.4.2}$$

is projection from $S_\lambda(\mathbb{R}^n)$ onto the first $k \le n$ coordinates. Then the result we described above can be rephrased by writing that as $n \to \infty$, and for k fixed,

$$X_{k,n} \stackrel{\mathcal{L}}{\to} N(0, I_{k\times k}), \tag{15.4.3}$$

where $\stackrel{\mathcal{L}}{\to}$ denotes convergence in distribution (law).

Here is the generalization of Poincaré convergence that we shall need.

Theorem 15.4.1 (Poincaré's limit [47]). *Fix $l, k \geq 1$ and suppose that $g_n \in O(n)$ is a Haar distributed random orthonormal matrix. Consider the random $l \times k$ matrix $X_{l,k,n}$ with (i, j)th entry given by*

$$\left(\pi_{\sqrt{n},n,k}(\sqrt{n} g_n e_i)\right)_j,$$

where $\{e_1, \ldots, e_n\}$ is the usual orthonormal basis of $\mathbb{R}^n$. Then the matrix $X_{l,k,n}$ converges in total variation to $X_{l,k}$, an $l \times k$ matrix of i.i.d. $N(0,1)$ random variables. Furthermore, if F is a real-valued function of matrices for which

$$\mathbb{E}\{|F(X_{lk})|\} < \infty,$$

then

$$\lim_{n\to\infty} \mathbb{E}\left\{F(X_{l,k,n})\right\} = \mathbb{E}\left\{F(X_{l,k})\right\}. \tag{15.4.4}$$

In fact, the conclusion (15.4.4) is not what is usually referred to as "Poincaré's limit," but rather follows simply from the method of proof of a similar result in [47]. There it was shown that if $\mathbb{P}_{l,k}$ and $\mathbb{P}_{l,k,n}$ denote the distributions of $X_{l,k}$ and $X_{l,k,n}$, then the Radon–Nikodym derivative

$$\frac{d\mathbb{P}_{l,k,n}}{d\mathbb{P}_{l,k}}$$

is bounded and converges uniformly to 1 as $n \to \infty$. This is essentially all that is needed to lift the convergence in law of (15.4.3) to convergence in total variation, from which the convergence of the expectations in (15.4.4) follows.

15.5 Kinematic Fundamental Formulas

The kinematic fundamental formula (KFF) is undoubtedly one of the most general and fundamental results in integral geometry, of which many other well-known formulas are special cases or corollaries. For a full treatment of this result in a variety of scenarios you should turn to any of the classic references, including [24, 40, 65, 92, 139, 141].

We have already met a related result in the form of Crofton's formulas of Section 13.1. Crofton's formulas gave expressions for the Lipschitz–Killing curvatures of "typical" cross-sections of subsets M of $\mathbb{R}^n$, that is, of intersections $M \cap V$ where the V were subspaces of $\mathbb{R}^n$. The KFF, however, is essentially a formula giving the "average" Lipschitz–Killing curvatures of $M_1 \cap g_n M_2$, where g_n is a "typical" isometry (i.e., rigid motion) of $\mathbb{R}^n$ and M_1 and M_2 are both subsets of $\mathbb{R}^n$.

For us the KFF will be the key to identifying the functions $\widetilde{\rho}$ in results such as (15.1.1), which are themselves key to developing the final expression for the mean Lipschitz–Killing curvatures of excursion sets in Theorems 15.9.4 and 15.9.5 below.

15.5.1 The KFF on $\mathbb{R}^n$

For the first KFF we take two tame stratified subsets M_1 and M_2 of $\mathbb{R}^n$ with finite Lipschitz–Killing curvatures.[14] Although this is not actually the case we shall need, it is the simplest and most common version of the KFF and so a good place to start. We also take G_n, the isometry group of $\mathbb{R}^n$ with a Haar measure. Since G_n is isomorphic to $\mathbb{R}^n \times O(n)$, its Haar measures are not finite, making it somewhat misleading to talk of the "average" Lipschitz–Killing curvatures we described above or of "typical" isometries. Nevertheless, it is possible to normalize Haar measure on G_n in such a way as to obtain a measure ν_n for which, for any $x \in \mathbb{R}^n$ and any Borel set $A \subset \mathbb{R}^n$,

$$\nu_n\left(\{g_n \in G_n : g_n x \in A\}\right) = \mathcal{H}_n(A), \tag{15.5.1}$$

where $\mathcal{H}_n$ is the usual Lebesgue measure on $\mathbb{R}^n$.

With this normalization, the KFF reads[15]

$$\begin{aligned}\int_{G_n} \mathcal{L}_i\left(M_1 \cap g_n M_2\right) d\nu_n(g_n) &= \sum_{j=0}^{n-i} \begin{bmatrix} i+j \\ i \end{bmatrix} \begin{bmatrix} n \\ j \end{bmatrix}^{-1} \mathcal{L}_{i+j}(M_1)\mathcal{L}_{n-j}(M_2) \\ &= \sum_{j=0}^{n-i} \frac{s_{i+1}s_{n+1}}{s_{i+j+1}s_{n-j+1}} \mathcal{L}_{i+j}(M_1)\mathcal{L}_{n-j}(M_2),\end{aligned} \tag{15.5.2}$$

where, since we shall need them again soon, we recall that

$$\begin{bmatrix} n \\ k \end{bmatrix} = \frac{[n]!}{[k]![n-k]!}, \qquad [n]! = n!\omega_n, \qquad \omega_n = \frac{\pi^{n/2}}{\Gamma(\frac{n}{2}+1)}, \qquad s_n = \frac{2\pi^{n/2}}{\Gamma(\frac{n}{2})}.$$

---

[14] There is a fine point here about whether we should require that M_1 and M_2 also be piecewise C^2. Since our definition of the Lipschitz–Killing curvatures involves curvature, the KFFs that follow are poorly defined otherwise. On the other hand, there are alternative definitions of the Lipschitz–Killing curvatures (better called intrinsic volumes than curvatures in this case) without such strong smoothness assumptions along with corresponding KFFs. Thus, we shall leave out the reference to smoothness when it is not required, despite the fact that you will then have to look in the references given above to see what the KFF really means in these scenarios.

[15] In (15.5.2) the $\mathcal{L}_k$ are always *total* curvature measures, or intrinsic volumes. However, there are generalizations of (15.5.2) with curvature measures replacing intrinsic volumes. For further details, which we shall not require, see [80, 141].

Further to the previous footnote, note that there are also versions of (15.5.2) over more general classes of sets than tame stratified sets. Indeed, much of the recent work around the KFF has been devoted to seeing just how general the sets may be taken.

A proof of (15.5.2) at (and beyond) the level of generality that we have stated the KFF can be found, for example, in [33].

15.5.2 The KFF on $S_\lambda(\mathbb{R}^n)$

What will be far more important for us than the KFF on Euclidean space, for reasons that will soon become clear, is a version of the KFF for subsets of $S_{\sqrt{n}}(\mathbb{R}^n)$ in which G_n is replaced by $G_{n,\lambda}$, the group of isometries (i.e., rotations) on $S_\lambda(\mathbb{R}^n)$. Perhaps not surprisingly, with an appropriate definition of curvature measures, the KFF on the sphere ends up almost identical to the Euclidean version.

Noting that $G_{n,\lambda} \simeq O(n)$, we normalize Haar measure $\nu_{n,\lambda}$ on $G_{n,\lambda}$ much as we did for G_n in the Euclidean case. That is, for any $x \in S_\lambda(\mathbb{R}^n)$ and every Borel $A \subset S_\lambda(\mathbb{R}^n)$, we require that

$$\nu_{n,\lambda}\left(\left\{g_n \in G_{n,\lambda} : g_n x \in A\right\}\right) = \mathcal{H}_{n-1}(A). \tag{15.5.3}$$

The KFF on $S_\lambda(\mathbb{R}^n)$ then reads as follows, where M_1 and M_2 are tame stratified spaces in $S_\lambda(\mathbb{R}^n)$:

$$\begin{aligned}\int_{G_{n,\lambda}} &\mathcal{L}_i^{\lambda^{-2}}(M_1 \cap g_n M_2)\, d\nu_{n,\lambda}(g_n) \qquad (15.5.4)\\ &= \sum_{j=0}^{n-1-i} \begin{bmatrix} i+j \\ i \end{bmatrix} \begin{bmatrix} n-1 \\ j \end{bmatrix}^{-1} \mathcal{L}_{i+j}^{\lambda^{-2}}(M_1)\mathcal{L}_{n-1-j}^{\lambda^{-2}}(M_2) \\ &= \sum_{j=0}^{n-1-i} \frac{s_{i+1}s_n}{s_{i+j+1}s_{n-j}} \mathcal{L}_{i+j}^{\lambda^{-2}}(M_1)\mathcal{L}_{n-1-j}^{\lambda^{-2}}(M_2),\end{aligned}$$

where the functionals $\mathcal{L}_i^\kappa(\cdot)$ are from the one-parameter family defined in (10.5.8).

15.6 A Model Process on the l-Sphere

We now introduce a particularly simple process on the l-sphere $S(\mathbb{R}^l)$, for which we shall be able to compute all excursion set (and more) mean Lipschitz–Killing curvatures with calculations using no more than the KFF on the sphere. Ultimately, in Section 15.9, we shall use the results of this section to prove the main non-Gaussian results of the chapter. This will be preceded, in Section 15.7, by a development of the corresponding results for a special Gaussian process, the canonical isotropic Gaussian process on $S(\mathbb{R}^l)$. Although we do not really need it until then, we shall remind you of what it is and tell you what we need it for, so that the computations of the current section will be better motivated.

As we saw in Section 10.2, the canonical isotropic Gaussian process on the l-sphere $S(\mathbb{R}^l)$ is defined as the centered Gaussian process on $S(\mathbb{R}^l)$ with covariance function

$$\mathbb{E}\{f(s)f(t)\} = \langle s, t\rangle, \tag{15.6.1}$$

where $\langle\, ,\, \rangle$ is the usual Euclidean inner product.

We first met this process in Section 10.2, when we looked at the volume-of-tubes approach to computing Gaussian excursion probabilities. In particular, we saw there that any Gaussian field with a Karhunen–Loève expansion of order $l < \infty$ and constant unit variance can be mapped to a process on a subset of $S(\mathbb{R}^l)$, and that the suprema for the two processes are identical.

In this chapter, however, we have a different task for the isotropic process, for it will be the test process to which we referred earlier for computing the unknown functionals $\widetilde{\rho}$ in (15.1.1) and Theorem 15.2.1.

However, rather than dealing directly with the isotropic Gaussian process, we shall start with a series of approximations to it that are easier to handle and for which, as noted above, we can compute mean Lipschitz–Killing curvatures via the KFF.

15.6.1 The Process

We shall need a family, $\{y^{(n)})\}_{n\geq l}$, of smooth $\mathbb{R}^k$-valued processes on stratified subsets M of $S(\mathbb{R}^l)$. To define it, for each $n \geq l$ we first embed $S(\mathbb{R}^l)$ in $S(\mathbb{R}^n)$ in the natural way, by setting

$$S(\mathbb{R}^l) = \left\{t = (t_1, \ldots, t_n) \in S(\mathbb{R}^n) : t_{l+1} = \cdots = t_n = 0\right\}.$$

We next take $O(n)$, now thinking of it as the group of orthonormal $n \times n$ matrices, equipped with its normalized Haar measure μ_n, as our underlying probability space. Then the nth process $y^{(n)}$ is defined by

$$y^{(n)}(t, g_n) \stackrel{\Delta}{=} \pi_{\sqrt{n},n,k}\left(\sqrt{n}g_n t\right), \tag{15.6.2}$$

where $t \in S(\mathbb{R}^l)$, $g_n \in O(n)$, and $\pi_{\sqrt{n},n,k}$ is the projection from $S_{\sqrt{n}}(\mathbb{R}^n)$ to $\mathbb{R}^k$ given by (15.4.2).

From Theorem 15.4.1 above, it is clear that the processes $y^{(n)}$ converge in total variation norm[16] to an $\mathbb{R}^k$-valued Gaussian process

$$y(t) \stackrel{\Delta}{=} (y_1(t), \ldots, y_k(t)), \tag{15.6.3}$$

where the components have covariance

$$\mathrm{Cov}\left(y_i(t), y_j(s)\right) = \delta_{ij} \cdot \langle t, s\rangle_{\mathbb{R}^l}.$$

That is, each component of the limit is a version of the canonical isotropic Gaussian field on $\mathbb{R}^l$, and they are independent of one another.

As we are about to see, it is remarkably straightforward to compute the mean Lipschitz–Killing curvatures of many sets generated by the $y^{(n)}$, using only the KFF on $S_{\sqrt{n}}(\mathbb{R}^n)$. This is the content of the following subsection. Later, in Sections 15.7 and 15.9, we shall send $n \to \infty$ to see what happens for the Gaussian limit.

[16] Note that in Theorem 15.4.1 the convergence relates to a finite collection of random variables, whereas here we are talking about *processes* $y^{(n)}$ and y defined over an uncountable parameter space. However, since each of these processes depends on only a finite number of random variables (via its expansion), there is no real problem talking about "convergence in variation" of either the processes or, what is actually more important to us, of real-valued functionals defined on them.

15.6.2 Mean Curvatures for the Model Process

The following result is the key to all that follows in this section. Although it is based on the one-parameter family $\mathcal{L}_i^\lambda$ of Lipschitz–Killing curvatures rather than the ones we are more used to using, we have already seen in Section 15.5.2 that they are better suited to computations on spheres than are the usual ones. We shall make the change back to the usual curvatures before we leave this section.

Lemma 15.6.1. *Let $y^{(n)}$ be the model process (15.6.2) on a tame stratified space $M \subset S(\mathbb{R}^l)$, with $n \geq l$. Then, for any tame stratified space $D \subset \mathbb{R}^k$,*

$$\mathbb{E}\left\{\mathcal{L}_i^1(M \cap (y^{(n)})^{-1}D)\right\} \tag{15.6.4}$$

$$= \sum_{j=0}^{\dim M - i} n^{j/2} \begin{bmatrix} i+j \\ j \end{bmatrix} \begin{bmatrix} n-1 \\ j \end{bmatrix}^{-1} \mathcal{L}_{j+i}^1(M) \frac{\mathcal{L}_{n-1-j}^{n^{-1}}\left(\pi_{\sqrt{n},n,k}^{-1} D\right)}{s_n n^{(n-1)/2}}$$

$$= \sum_{j=0}^{\dim M - i} \frac{s_{i+1}}{s_{i+j+1}} \mathcal{L}_{j+i}^1(M) \frac{\mathcal{L}_{n-1-j}^{n^{-1}}\left(\pi_{\sqrt{n},n,k}^{-1} D\right)}{s_{n-j} n^{(n-1-j)/2}}.$$

Remark 15.6.2. *It is important to understand what the meaning of $\pi_{\sqrt{n},n,k}^{-1} D$ is in the above lemma, and in all that follows. The problem is that for all $t \in S_{\sqrt{n}}(\mathbb{R}^n)$, $\pi_{\sqrt{n},n,k}(t) \in B_{\sqrt{n}}(\mathbb{R}^k)$, which may, or may not, cover D. Thus, with the usual definition*

$$\pi_{\sqrt{n},n,k}^{-1} D = \left\{t \in S_{\sqrt{n}}(\mathbb{R}^n) : \pi_{\sqrt{n},n,k}(t) \in D\right\},$$

it follows that $\pi_{\sqrt{n},n,k}^{-1} D$ may be only the inverse image of a subset of D.

Proof. The first thing that we need to note is that $\pi_{\sqrt{n},n,k}^{-1} D$ is a tame stratified space in $S(\mathbb{R}^n)$, and so from the construction of $y^{(n)}$ we have

$$\begin{aligned}
&\mathbb{E}\{\mathcal{L}_i^1(M \cap (y^{(n)})^{-1}D)\} \\
&\quad = \int_{O(n)} \mathcal{L}_i^1(M \cap (y^{(n)})^{-1}D)(g_n)\, d\mu_n(g_n) \\
&\quad = \int_{O(n)} \mathcal{L}_i^1\left(M \cap n^{-1/2} g_n^{-1}\left(\pi_{\sqrt{n},n,k}^{-1} D\right)\right) d\mu_n(g_n) \\
&\quad = n^{-i/2} \int_{O(n)} \mathcal{L}_i^{n^{-1}}\left(\sqrt{n} M \cap g_n^{-1}\left(\pi_{\sqrt{n},n,k}^{-1} D\right)\right) d\mu_n(g_n) \\
&\quad = \frac{1}{s_n n^{(n-1+i)/2}} \int_{G_{n,n-1}} \mathcal{L}_i^{n^{-1}}\left(\sqrt{n} M \cap g_n\left(\pi_{\sqrt{n},n,k}^{-1} D\right)\right) d\nu_{n,n^{-1}}(g_n),
\end{aligned}$$

where the second-to-last line follows from the scaling properties of Lipschitz–Killing curvatures and the last is really no more than a notational change, using (15.5.3).

However, applying the KFF (15.5.4) to the last line above, we immediately have that it is equal to

$$\sum_{j=0}^{\dim M-i} n^{j/2} \begin{bmatrix} i+j \\ i \end{bmatrix} \begin{bmatrix} n-1 \\ j \end{bmatrix}^{-1} \frac{\mathcal{L}^{n-1}_{j+i}\left(\sqrt{n}M\right)}{n^{(i+j)/2}} \frac{\mathcal{L}^{n-1}_{n-1-j}\left(\pi^{-1}_{\sqrt{n},n,k} D\right)}{s_n n^{(n-1)/2}}$$

$$= \sum_{j=0}^{\dim M-i} n^{j/2} \begin{bmatrix} i+j \\ j \end{bmatrix} \begin{bmatrix} n-1 \\ j \end{bmatrix}^{-1} \mathcal{L}^{1}_{j+i}(M) \frac{\mathcal{L}^{n-1}_{n-1-j}\left(\pi^{-1}_{\sqrt{n},n,k} D\right)}{s_n n^{(n-1)/2}},$$

which proves the lemma. □

We cannot overemphasize how important and yet how simple is Lemma 15.6.1. The simplicity lies in the fact that despite the preponderance of subscripts and superscripts, the proof uses no more than the KFF and a few algebraic manipulations. If you compare this to the effort required to compute the mean Euler characteristic of excursion sets of univariate Gaussian processes in Chapters 11 and 12, not to mention the effort required in Chapter 13 to lift this to mean Lipschitz–Killing curvatures, you will have to agree that the above proof is indeed simple.

Of course, the price that we have paid for this simplicity is that we have a result only for a very special process, and not for the Gaussian and Gaussian-related processes of central interest to us. However, these are obtainable as appropriate limits of the $y^{(n)}$, as we shall now see.

Suppose we send $n \to \infty$ in (15.6.4), which by Poincaré's limit is effectively equivalent to replacing the model process $y^{(n)}$ with the Gaussian y of (15.6.3). Then, in order for the limiting process y to have finite expected values for the Lipschitz–Killing curvatures $\mathcal{L}_j(M \cap (y^{(n)})^{-1}D)$, we would like to have the following limits existing for each finite j:

$$\lim_{n\to\infty} n^{j/2} \begin{bmatrix} n-1 \\ j \end{bmatrix}^{-1} \frac{\mathcal{L}^{n-1}_{n-1-j}\left(\pi^{-1}_{\sqrt{n},n,k} D\right)}{s_n n^{(n-1)/2}}. \tag{15.6.5}$$

Assuming that these limits are indeed well defined and finite, we define

$$\begin{aligned} \widetilde{\rho}_j(D) &\triangleq \lim_{n\to\infty} \left(n^{j/2} \begin{bmatrix} n-1 \\ j \end{bmatrix}^{-1} \right) \frac{\mathcal{L}^{n-1}_{n-1-j}\left(\pi^{-1}_{\sqrt{n},n,k} D\right)}{s_n n^{(n-1)/2}} \\ &= (2\pi)^{-j/2}[j]! \lim_{n\to\infty} \frac{\mathcal{L}^{n-1}_{n-1-j}\left(\pi^{-1}_{\sqrt{n},n,k} D\right)}{s_n n^{(n-1)/2}}, \end{aligned} \tag{15.6.6}$$

where the second line comes from a Stirling's formula computation.

Sending $n \to \infty$ in Lemma 15.6.1 and applying Poincaré's limit of Theorem 15.4.1 (cf. footnote 16), we see that if

$$\mathbb{E}\left\{ |\mathcal{L}^1_i(M \cap y^{-1}D)| \right\} < \infty,$$

then

$$\mathbb{E}\left\{\mathcal{L}_i^1(M \cap y^{-1}D)\right\} = \lim_{n\to\infty} \mathbb{E}\left\{\mathcal{L}_i^1(M \cap (y^{(n)})^{-1}D)\right\} \tag{15.6.7}$$
$$= \sum_{j=0}^{\dim M - i} \begin{bmatrix} i+j \\ i \end{bmatrix} \mathcal{L}_{j+i}^1(M)\widetilde{\rho}_j(D).$$

This is starting to take the form of (15.0.5), the result that we are trying to prove. The combinatorial flag coefficients are in place, but both sides of the equation are based on the $\mathcal{L}_{j+i}^1$ curvatures rather than the $\mathcal{L}_{j+i}$, and we have yet to explicitly identify the functions $\widetilde{\rho}_j$. Note the important fact, however, that on the right-hand side of the equation we have already managed to split into product form factors that depend on the underlying manifold M and the set D.

As far as the fact that we are using the "wrong" Lipschitz–Killing curvatures is concerned, recall that we have already seen in Lemma 10.5.8 that the $\mathcal{L}_j^1$ can be expressed as linear combinations of the $\mathcal{L}_j$, and vice versa. Thus, in principle, it should not be hard to derive a version of (15.6.7) with the usual curvatures. In practice, the computation is not simple, but we shall carry it out in a moment. Once done, it will show that

$$\mathbb{E}\left\{\mathcal{L}_i(M \cap y^{-1}D)\right\} = \sum_{j=i}^{\infty}\sum_{l=0}^{\infty} \begin{bmatrix} i+j \\ j \end{bmatrix} \mathcal{L}_j(M) c_{i,j,l}\widetilde{\rho}_l(D), \tag{15.6.8}$$

for some universal constants $c_{i,j,l}$.

Understanding the geometric significance of the linear combinations

$$\sum_{l=0}^{\infty} c_{i,j,l}\widetilde{\rho}_l(\cdot) \tag{15.6.9}$$

is probably the most important part of this chapter, and is the closing step in proving the main result, Theorem 15.9.5. In particular, we shall see that the above sum depends only on $i+j$ and is intimately related to the volume-of-tubes results of Chapter 10.

Before doing this, however, we tackle the easier part of the problem, that of finding a version of (15.6.7) with the usual Lipschitz–Killing curvatures. Since the $\widetilde{\rho}_j$ of (15.6.7), assuming that they exist, are functionals of D and independent of M, we prove the following general result under the assumption that the limits (15.6.5) are indeed well defined.

Theorem 15.6.3. *Let $M \subset S(\mathbb{R}^l)$ be a tame stratified space, and assume that for $0 \le j \le \dim(M)$, the limits*

$$\widetilde{\rho}_j(D) = \lim_{n\to\infty}\left(n^{j/2}\begin{bmatrix} n-1 \\ j \end{bmatrix}^{-1}\right) \frac{\mathcal{L}_{n-1-j}^{n^{-1}}\left(\pi_{\sqrt{n},n,k}^{-1}D\right)}{s_n n^{(n-1)/2}} \tag{15.6.10}$$

are finite and

$$\mathbb{E}\left\{|\mathcal{L}_i(M \cap y^{-1}D)|\right\} < \infty, \tag{15.6.11}$$

where y is the Gaussian process on $S(\mathbb{R}^l)$ defined by (15.6.3). *Then*

$$\mathbb{E}\left\{\mathcal{L}_i(M \cap y^{-1}D)\right\} = \sum_{l=0}^{\dim M - i} \begin{bmatrix} i+l \\ l \end{bmatrix} \mathcal{L}_{i+l}(M)\rho_l(D), \tag{15.6.12}$$

where for $j \geq 1$,

$$\rho_j(D) = (j-1)! \sum_{l=0}^{\lfloor \frac{j-1}{2} \rfloor} \frac{(-1)^l}{(4\pi)^l l!(j-1-2l)!} \widetilde{\rho}_{j-2l}(D) \tag{15.6.13}$$

and

$$\rho_0(D) = \gamma_{\mathbb{R}^k}(D).$$

Proof. As usual, set $N = \dim(M)$, and recall the two basic formulas of Lemma 10.5.8 relating the $\{\mathcal{L}_j\}_{j \geq 0}$ and the $\{\mathcal{L}_j^1\}_{j \geq 0}$:

$$\mathcal{L}_i^1(\cdot) = \sum_{n=0}^{\infty} \frac{(-1)^n}{(4\pi)^n n!} \frac{(i+2n)!}{i!} \mathcal{L}_{i+2n}(\cdot) \tag{15.6.14}$$

and

$$\mathcal{L}_i(\cdot) = \sum_{n=0}^{\infty} \frac{1}{(4\pi)^n n!} \frac{(i+2n)!}{i!} \mathcal{L}_{i+2n}^1(\cdot). \tag{15.6.15}$$

Combining (15.6.14) with (15.6.7), we obtain

$$\begin{aligned}
&\mathbb{E}\left\{\mathcal{L}_i(M \cap y^{-1}D)\right\} \\
&\quad = \mathbb{E}\left\{\sum_{n=0}^{\lfloor \frac{N-i}{2} \rfloor} \frac{1}{(4\pi)^n n!} \frac{(i+2n)!}{i!} \mathcal{L}_{i+2n}^1(M \cap y^{-1}D)\right\} \\
&\quad = \sum_{n=0}^{\lfloor \frac{N-i}{2} \rfloor} \frac{1}{(4\pi)^n n!} \frac{(i+2n)!}{i!} \sum_{j=0}^{N-i-2n} \begin{bmatrix} i+2n+j \\ j \end{bmatrix} \mathcal{L}_{i+2n+j}^1(M)\widetilde{\rho}_j(D) \\
&\quad = \sum_{n=0}^{\lfloor \frac{N-i}{2} \rfloor} \frac{1}{(4\pi)^n n!} \frac{(i+2n)!}{i!} \sum_{j=0}^{N-i-2n} \begin{bmatrix} i+2n+j \\ j \end{bmatrix} \widetilde{\rho}_j(D) \\
&\qquad \times \sum_{l=0}^{\lfloor \frac{N-i-2n-j}{2} \rfloor} \frac{(-1)^l}{(4\pi)^l l!} \frac{(i+2n+j+2l)!}{(i+2n+j)!} \mathcal{L}_{i+2n+j+2l}(M).
\end{aligned}$$

Saving space, we suppress the dependence of $\widetilde{\rho}$ and $\mathcal{L}$ on D and M, respectively. Proceeding, $\mathbb{E}\{\mathcal{L}_i(M \cap y^{-1}D)\}$ can be written as

$$
\begin{aligned}
&\sum_{n=0}^{\lfloor \frac{N-i}{2} \rfloor} \frac{1}{(4\pi)^n n!} \frac{(i+2n)!}{i!} \sum_{j=0}^{N-i-2n} \begin{bmatrix} i+2n+j \\ j \end{bmatrix} \tilde{\rho}_j \\
&\qquad \times \sum_{l=0}^{\lfloor \frac{N-i-2n-j}{2} \rfloor} \frac{(-1)^l}{(4\pi)^l l!} \frac{(i+2n+j+2l)!}{(i+2n+j)!} \mathcal{L}_{i+2n+j+2l} \\
&= \sum_{j=0}^{N-i} \tilde{\rho}_j \sum_{n=0}^{\lfloor \frac{N-i-j}{2} \rfloor} \frac{1}{(4\pi)^n n!} \frac{(i+2n)!}{i!} \begin{bmatrix} i+2n+j \\ j \end{bmatrix} \\
&\qquad \times \sum_{l=0}^{\lfloor \frac{N-i-2n-j}{2} \rfloor} \frac{(-1)^l}{(4\pi)^l l!} \frac{(i+2n+j+2l)!}{(i+2n+j)!} \mathcal{L}_{i+2n+j+2l} \\
&= \sum_{j=0}^{N-i} \tilde{\rho}_j \sum_{\alpha=0}^{\lfloor \frac{N-i-j}{2} \rfloor} \sum_{\beta=0}^{\alpha} \frac{1}{(4\pi)^{\alpha-\beta}(\alpha-\beta)!} \frac{(i+2(\alpha-\beta))!}{i!} \\
&\qquad \times \begin{bmatrix} i+2(\alpha-\beta)+j \\ j \end{bmatrix} \frac{(-1)^\beta}{(4\pi)^\beta \beta!} \frac{(i+j+2\alpha)!}{(i+2(\alpha-\beta)+j)!} \mathcal{L}_{i+j+2\alpha},
\end{aligned}
$$

where the first and third equalities come from a change of order of summation and the second from the transformation $(\alpha, \beta) = (n+l, l)$. Again changing the order of summation we find that this is the same a

$$
\begin{aligned}
&\sum_{\alpha=0}^{\lfloor \frac{N-i}{2} \rfloor} \sum_{j=0}^{N-i-2\alpha} \sum_{\beta=0}^{\alpha} \tilde{\rho}_j \frac{1}{(4\pi)^\alpha (\alpha-\beta)!} \frac{(i+2(\alpha-\beta))!}{i!} \\
&\qquad \times \begin{bmatrix} i+2(\alpha-\beta)+j \\ j \end{bmatrix} \frac{(-1)^\beta}{\beta!} \frac{(i+j+2\alpha)!}{(i+2(\alpha-\beta)+j)!} \mathcal{L}_{i+j+2\alpha}.
\end{aligned}
$$

Making now the further change of variables $(m, k) = (\alpha, j+2\alpha)$, the above is equivalent to

$$
\begin{aligned}
&\sum_{k=0}^{N-i} \sum_{m=0}^{\lfloor \frac{k-1}{2} \rfloor} \sum_{\beta=0}^{m} \tilde{\rho}_{k-2m} \frac{1}{(4\pi)^m (m-\beta)!} \frac{(i+2(m-\beta))!}{i!} \\
&\qquad \times \begin{bmatrix} i+k-2\beta \\ k-2m \end{bmatrix} \frac{(-1)^\beta}{\beta!} \frac{(i+k)!}{(i+k-2\beta)!} \mathcal{L}_{i+k} \\
&= \sum_{k=0}^{N-i} \mathcal{L}_{i+k} \frac{(i+k)!}{i!} \sum_{m=0}^{\lfloor \frac{k-1}{2} \rfloor} \tilde{\rho}_{k-2m} \sum_{\beta=0}^{m} \frac{1}{(4\pi)^m (m-\beta)!} (i+2(m-\beta))!
\end{aligned}
$$

$$\times \begin{bmatrix} i+k-2\beta \\ k-2m \end{bmatrix} \frac{(-1)^\beta}{\beta!} \frac{1}{(i+k-2\beta)!},$$

the last line being just a minor reorganization of the preceding one.

We must therefore show that

$$\sum_{k=0}^{N-i} \mathcal{L}_{i+k} \frac{(i+k)!}{i!} \sum_{m=0}^{\lfloor \frac{k-1}{2} \rfloor} \tilde{\rho}_{k-2m} \sum_{\beta=0}^{m} \frac{1}{(4\pi)^m (m-\beta)!} (i+2(m-\beta))!$$
$$\times \begin{bmatrix} i+k-2\beta \\ k-2m \end{bmatrix} \frac{(-1)^\beta}{\beta!} \frac{1}{(i+k-2\beta)!}$$
$$= \sum_{k=0}^{N-i} \mathcal{L}_{i+k}(M) \begin{bmatrix} i+k \\ k \end{bmatrix} \rho_k,$$

where the functionals ρ_k are those of (15.6.13). Equivalently, we must show that

$$\rho_k = \begin{bmatrix} i+k \\ k \end{bmatrix}^{-1} \frac{(i+k)!}{i!} \sum_{m=0}^{\lfloor \frac{k-1}{2} \rfloor} \tilde{\rho}_{k-2m} \sum_{\beta=0}^{m} \frac{1}{(4\pi)^m (m-\beta)!} (i+2(m-\beta))!$$
$$\times \begin{bmatrix} i+k-2\beta \\ k-2m \end{bmatrix} \frac{(-1)^\beta}{\beta!} \frac{1}{(i+k-2\beta)!}$$
$$= \frac{k!\omega_k\omega_i}{\omega_{i+k}} \sum_{m=0}^{\lfloor \frac{k-1}{2} \rfloor} \tilde{\rho}_{k-2m} \sum_{\beta=0}^{m} \frac{(-1)^\beta}{(4\pi)^m (m-\beta)!\beta!(k-2m)!} \frac{\omega_{i+k-2\beta}}{\omega_{k-2m}\omega_{i+2m-2\beta}}.$$

This is equivalent, by (15.6.13), to proving that the following identity holds for all nonnegative integers $k \geq 1$, $i \geq 0$:

$$\frac{(-1)^m}{m!} \frac{\omega_{i+k}}{k\omega_i\omega_k/2} = \sum_{\beta=0}^{m} \frac{(-1)^\beta}{(m-\beta)!\beta!} \frac{\omega_{i+k-2\beta}}{\omega_{i+2m-2\beta} \cdot (k-2m)\omega_{k-2m}/2}.$$

Finally, after some further simple manipulations, this is equivalent to the identity

$$B\left(\frac{i}{2}+1, \frac{k}{2}\right) = \sum_{\beta=0}^{m} (-1)^{m-\beta} \binom{m}{\beta} B\left(\frac{i+2m-2\beta}{2}+1, \frac{k-2m}{2}\right),$$

where B is the beta function

$$B(\gamma, \delta) = \frac{\Gamma(\gamma)\Gamma(\delta)}{\Gamma(\gamma+\delta)} = \int_0^1 p^{\gamma-1}(1-p)^{\delta-1}\, dp.$$

That this identity holds is verified in the following lemma, and so the proof is complete. □

Lemma 15.6.4. *For every $\gamma, \delta > 0$ and every integer $0 \le m \le \gamma$,*

$$B(\gamma, \delta) = \sum_{\beta=0}^{m} (-1)^{m-\beta} \binom{m}{\beta} B(\gamma - m, \delta + m - \beta).$$

Proof. The identity is clearly equivalent to

$$1 = \frac{1}{B(\gamma, \delta)} \sum_{\beta=0}^{m} (-1)^{m-\beta} \binom{m}{\beta} \int_0^1 p^{\gamma-m-1}(1-p)^{\delta+m-\beta-1}\, dp.$$

Suppose, then, that P is a random variable with distribution $\text{Beta}(\gamma, \delta)$, so that it has density

$$f_P(p) = \frac{1}{B(\gamma, \delta)} p^{\gamma-1}(1-p)^{\delta-1}, \qquad 0 \le p \le 1.$$

Then,

$$1 = P^{-m}(1-(1-P))^m = \sum_{\beta=0}^{m} \binom{m}{\beta} (-1)^{m-\beta}(1-P)^{m-\beta} P^{-m}.$$

Since $m < \gamma$, we can take expectations of both sides above to obtain

$$\begin{aligned} 1 &= \mathbb{E}\{P^{-m}(1-(1-P))^m\} \\ &= \frac{1}{B(\gamma, \delta)} \sum_{\beta=0}^{m} \binom{m}{\beta} (-1)^{m-\beta} \int_0^1 p^{\gamma-1-m}(1-p)^{\delta+m-\beta-1}\, dp, \end{aligned}$$

and we are done. □

What remains now is to check that the limits (15.6.6) are indeed well defined, and to evaluate them. Before treating the general case, however, we shall work out one special one, for which the computations will be simpler and the result familiar.

15.7 The Canonical Gaussian Field on the l-Sphere

Our aim in this section is to carry out all the remaining computations needed to compute mean Lipschitz–Killing curvatures for the isotropic Gaussian process on the sphere when the set $D \subset \mathbb{R}^k$ is a half-space.

While this is merely a special case of what will follow in Section 15.9, the calculations in this case are considerably simpler and aid in understanding what is actually going on. Furthermore, as we shall see in a moment, the result that we shall obtain is also a special case of the main result of Chapter 12. Thus it can be thought of as an independent verification of earlier results, or as a corollary of them. We shall return to this point in more detail in Section 15.7.2 below. Either way, the current section should help in understanding the general case.

In terms of the notation of the previous section, we are dealing with the zero-mean Gaussian process from $S(\mathbb{R}^l)$ to $\mathbb{R}^k$ defined by (15.6.3), and the hitting set D taken to be the half-space

$$D = \left\{ y \in \mathbb{R}^k : \langle y, \eta \rangle \geq u \right\}, \tag{15.7.1}$$

for some unique unit vector $\eta \in S(\mathbb{R}^k)$ and $u \in \mathbb{R}_+$. The case $k = 1$ obviously yields precisely the excursion sets treated in Chapter 12. However, even for general k, the process f defined by

$$f(t) = \langle y(t), \eta \rangle$$

is a zero-mean, unit-variance Gaussian process, and so, writing

$$M \cap y^{-1} D = M \cap f^{-1}[u, +\infty),$$

we are still in the scenario of Chapter 12. Consequently, under appropriate regularity conditions on M, the results of Chapter 12 (especially Theorem 12.4.2) give us the mean Euler characteristic of $M \cap f^{-1}[u, +\infty)$. If we compare this with (15.6.8) with $i = 0$ (so as to obtain the Euler characteristic $\mathcal{L}_0$) we find that we must have

$$\sum_{l=0}^{\infty} c_{0,j,l} \widetilde{\rho}_l(D) = \rho_j(u) = \begin{cases} (2\pi)^{-(j+1)/2} H_j(u) e^{-u^2/2}, & j > 0, \\ 1 - \Phi(u), & j = 0, \end{cases}$$

where the H_j are, as usual, the Hermite polynomials (11.6.9), and so we have found our elusive constants for this case.

Nevertheless, our aim is not to use the results of Chapter 12 to help out in the current calculation, but rather to calculate everything afresh. However, at least now we know what kind of result we are looking for.

15.7.1 Mean Curvatures for Excursion Sets

Here is the theorem that we want to prove.

Theorem 15.7.1. *Let $M \subset S(\mathbb{R}^l)$ be a tame stratified space, y the canonical Gaussian process on $S(\mathbb{R}^l)$ defined by* (15.6.3), *and D the half-space in $\mathbb{R}^k$ defined by* (15.7.1) *for some unit vector $\eta \in \mathbb{R}^k$ and $u > 0$. Then*

$$\mathbb{E}\left\{ \mathcal{L}_i(M \cap y^{-1} D) \right\} = \sum_{l=0}^{\dim M - i} \begin{bmatrix} i+l \\ l \end{bmatrix} \mathcal{L}_{i+l}(M) \rho_l(D), \tag{15.7.2}$$

where for $j \geq 1$,

$$\rho_j(D) = (2\pi)^{-(j+1)/2} H_{j-1}(u) e^{-u^2/2}$$

and

$$\rho_0(D) = \gamma_{\mathbb{R}^k}(D).$$

The proof is based on the following lemma, which we state and prove first.

Lemma 15.7.2. *Under the assumptions of Theorem* 15.7.1, *the* $\widetilde{\rho}_l(D)$ *defined by* (15.6.6) *satisfy*

$$\widetilde{\rho}_j(D) = (2\pi)^{-(j+1)/2} u^{j-1} e^{-u^2/2}$$

for $j \geq 1$, *and*

$$\widetilde{\rho}_0(D) = 1 - \Phi(u).$$

Proof. We begin with the observation that as long as $n > u$, and adopting the notation of the preceding section, we have

$$\pi^{-1}_{\sqrt{n},n,k} D = B_{S_{\sqrt{n}}(\mathbb{R}^n)} \left(\sqrt{n}\eta, \cos^{-1}(u/\sqrt{n})\right), \tag{15.7.3}$$

a geodesic ball, or spherical cap, in $S_{\sqrt{n}}(\mathbb{R}^n)$ of radius $\cos^{-1}(u/\sqrt{n})$ centered at the point $\sqrt{n}\eta \in S_{\sqrt{n}}(\mathbb{R}^n)$. An example is given in the shaded region of Figure 15.7.1.

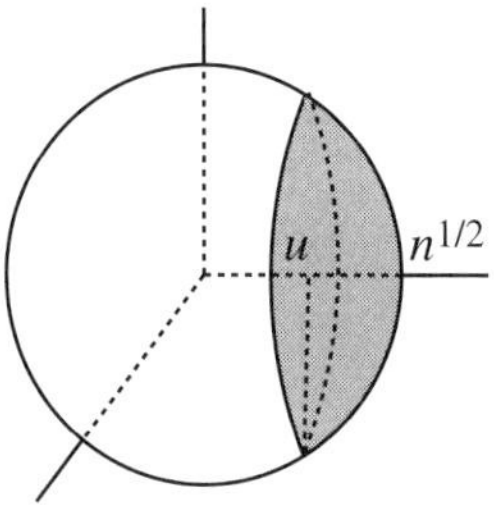

Fig. 15.7.1. The spherical cap (15.7.3) with $k = 1$, $n = 3$, and η taken along the right-pointing axis.

Consequently, it follows from the definition (15.6.6) of the $\widetilde{\rho}_l(D)$ that they are given by

$$\widetilde{\rho}_l(D) = (2\pi)^{-j/2}[j]! \lim_{n\to\infty} \frac{\mathcal{L}^{n-1}_{n-1-j}\left(\sqrt{n}B_{S(\mathbb{R}^n)}(\eta, \cos^{-1}(u/\sqrt{n}))\right)}{s_n n^{(n-1)/2}},$$

for all $j > 0$ and all $u > 0$.

In order to compute this limit, we move from spherical Lipschitz–Killing curvatures to Euclidean ones, with the claim that

$$\begin{aligned} &\mathcal{L}^{n-1}_{n-1-j}\left(\sqrt{n}B_{S(\mathbb{R}^n)}(\eta, \cos^{-1}(u/\sqrt{n}))\right) \\ &\quad = \left(\frac{u}{\sqrt{n}}\right)^{j-1} \mathcal{L}_{n-1-j}\left(B_{\mathbb{R}^{n-1}}(0, \sqrt{n-u^2})\right). \end{aligned} \tag{15.7.4}$$

To see this, first write

$$H \stackrel{\Delta}{=} \sqrt{n} B_{S(\mathbb{R}^n)} \left(\eta, \cos^{-1}(u/\sqrt{n})\right),$$

and compute $\mathcal{L}^{n-1}_{n-1-j}$ using the extrinsic formula (10.5.10). Note that for $j > 0$, the only contributions to $\mathcal{L}^{n-1}_{n-1-j}(H)$ come from ∂H,[17] which is a sphere of radius $\sqrt{n-u^2}$. (See Figure 15.7.1 again.)

Consequently, up to a constant, $\mathcal{L}^{n-1}_{n-1-j}(H)$ is the integral of

$$\mathrm{Tr}(S^{j-1}_{\eta})$$

over ∂H, where S is the second fundamental form of ∂H in $S_{\sqrt{n}}(\mathbb{R}^n)$.

However, although up to this point we have been considering ∂H as it sits in $S_{\sqrt{n}}(\mathbb{R}^n)$, it can also be treated as a subset of the hyperplane

$$L \stackrel{\Delta}{=} \left\{y \in \mathbb{R}^n : \langle y, \eta\rangle = u\right\}.$$

Consequently, we can also compute the Euclidean Lipschitz–Killing curvatures $\mathcal{L}_{n-i-j}(H)$, this time using the extrinsic representation (10.5.5). While these will not be the same as the $\mathcal{L}^{n-1}_{n-1-j}(H)$, they too will be given, up to constants, by the integral of the trace of a scalar second fundamental form over ∂H.

In this case, however, the scalar form comes from considering ∂H as a subset of L, and we write it as $\tilde{S}$. The two fundamental forms are related by

$$S = \frac{u}{\sqrt{n}} \tilde{S},$$

in the sense that if η is the unit outward-pointing normal of ∂H in $S_{\sqrt{n}}(\mathbb{R}^n)$ at $t \in \partial H$, and $\tilde{\eta}$ is the unit outward-pointing normal of ∂H in L at $x \in \partial H$, then for any $X_t, Y_t \in T_t\partial H$,

$$S_{\eta}(X_t, Y_t) = \frac{u}{\sqrt{n}} \tilde{S}_{\tilde{\eta}}(X_t, Y_t).$$

Using this equivalence and substituting into (10.5.10) and (10.5.5) to get the constants right, (15.7.4) now follows on noting that

$$\mathcal{L}_{n-1-j}(B_L(u\eta, \sqrt{n-u^2})) = \mathcal{L}_{n-1-j}(B_{\mathbb{R}^{n-1}}(0, \sqrt{n-u^2})).$$

Since we long ago computed the Lipschitz–Killing curvatures for $S(\mathbb{R}^n)$ (cf. (6.3.8)) and we know how they scale (cf. (6.3.1)), the right-hand side of (15.7.4) is now trivial to compute, and so for $j > 0$ and $n > u$, we now have

---

[17] There are only two terms that could possibly contribute to $\mathcal{L}^{n-1}_{n-1-j}(H)$, since H has only two strata, its interior and its boundary. The interior term makes no contribution, since in (10.5.10) the scalar second fundamental form for the interior term is identically zero. See the discussion following Lemma 10.5.8.

$$\begin{aligned}
&\mathcal{L}_{n-1-j}^{n-1}\left(\sqrt{n}B_{S(\mathbb{R}^n)}(\eta, \cos^{-1}(u/\sqrt{n}))\right) \qquad (15.7.5)\\
&\quad = \left(\frac{u}{\sqrt{n}}\right)^{j-1} \begin{bmatrix} n-1 \\ n-1-j \end{bmatrix} \omega_{n-1-j}(n-u^2)^{(n-1-j)/2}.
\end{aligned}$$

The rest of the proof for $j > 0$ is a straightforward application of Stirling's formula, specifically the fact that for any fixed α,

$$\lim_{n\to\infty} \frac{\Gamma\left(\frac{n}{2}\right)}{\Gamma\left(\frac{n-\alpha}{2}\right)} \sim \left(\frac{n}{2}\right)^{\alpha/2},$$

where by $a_n \sim b_n$ we mean that a_n/b_n converges to a finite constant as $n \to \infty$. Applying this asymptotic equivalence, we easily check that

$$\begin{bmatrix} n-1 \\ n-1-j \end{bmatrix} \sim \frac{(2\pi)^{j/2} n^{j/2}}{[j]!}$$

and

$$\frac{\omega_{n-1-j} n^{(n-1-j)/2}}{s_n n^{(n-1)/2}} \sim (2\pi)^{-(j+1)/2} n^{-1/2},$$

so that

$$\begin{aligned}
&\frac{[j]! u^{j-1} n^{-(j-1)/2}}{(2\pi)^{j/2}} \begin{bmatrix} n-1 \\ n-1-j \end{bmatrix} \frac{\omega_{n-1-j} n^{(n-1-j)/2}}{s_n n^{(n-1)/2}} \left(1 - \frac{u^2}{n}\right)^{(n-1-j)/2}\\
&\quad \sim (2\pi)^{-(j+1)/2} u^{j-1} e^{-u^2/2}.
\end{aligned}$$

Substituting this into (15.7.5) gives that for $j > 0$ and $u > 0$,

$$\widetilde{\rho}_l(D) = (2\pi)^{-(j+1)/2} u^{j-1} e^{-u^2/2},$$

as required.

It remains only to treat the case $j = 0$. But since it is an immediate implication of Poincaré's limit that for all $u > 0$,

$$\lim_{n\to\infty} \frac{\mathcal{L}_{n-1}^{n-1}\left(\sqrt{n}B_{S(\mathbb{R}^n)}(\eta, \cos^{-1}(u/\sqrt{n}))\right)}{s_n n^{(n-1)/2}} = 1 - \Phi(u),$$

this case is simple, and we are done. □

Proof of Theorem 15.7.1. The case $j = 0$ is an immediate consequence of Theorem 15.6.3 and Lemma 15.7.2. Relying on the same results, but now taking $j \geq 1$, we have

$$\rho_j(D) = (j-1)! \sum_{k=0}^{\lfloor \frac{j-1}{2} \rfloor} \frac{(-1)^k}{(4\pi)^k k!(j-1-2k)!} \widetilde{\rho}_{j-2k}(D)$$
$$= (j-1)! \sum_{k=0}^{\lfloor \frac{j-1}{2} \rfloor} \frac{(-1)^k (2\pi)^{-(j-1-2k)/2}}{(4\pi)^k k!(j-1-2k)!} u^{j-1-2k} e^{-u^2/2}$$
$$= (2\pi)^{-(j+1)/2} (j-1)! \sum_{k=0}^{\lfloor \frac{j-1}{2} \rfloor} \frac{(-1)^k}{2^k k!(j-1-2k)!} u^{j-1-2k} e^{-u^2/2}$$
$$= (2\pi)^{-(j+1)/2} H_{j-1}(u) e^{-u^2/2},$$

on noting the definition (11.6.9) of the Hermite polynomials. □

15.7.2 Implications for More General Fields

The previous subsection contains a new method for deriving the mean values of the Lipschitz–Killing curvatures of the excursion sets of the canonical isotropic process on $S(\mathbb{R}^N)$. Recall that our initial motivation for restricting attention to this one very special process over this rather special parameter space was a consequence of Theorem 15.2.1, which stated that we had only to identify the $\widetilde{\rho}_j$ in order to obtain a far more general result. The other terms, those involving the $\mathcal{L}_j(M)$, had already been identified by Theorem 15.2.1.

However, a quick revision of the proof will show that nowhere did we actually use Theorem 15.2.1 to prove Theorem 15.7.1. Rather, the proof of this theorem relied on no more than the KFF and some, essentially algebraic, manipulations.

A natural question, therefore, is to ask how far this approach can be extended. That is, can we actually manage without Theorem 15.2.1 at all, which was not an easy result to prove, and still handle more general scenarios?

In fact, we can manage with only the current techniques if we are prepared to limit ourselves to processes with finite orthogonal expansions. We describe the approach, without going into technical details.

Recall firstly the discussion of Section 10.2, where we saw that any unit-variance Gaussian process $\widetilde{f}$ with an orthonormal expansion of order $l < \infty$ can be realized as the canonical isotropic process f on an appropriately chosen subset of $S(\mathbb{R}^l)$. In fact, if $\widetilde{f}$ is defined over a tame stratified manifold M, then the corresponding subset of $S(\mathbb{R}^l)$, which we denote by $\psi(M)$ for consistency with the notation of Section 10.2, is also a tame stratified manifold.

Now suppose that on $S(\mathbb{R}^l)$ we take the usual Riemannian metric, which is also the metric induced on $S(\mathbb{R}^l)$ by f, and on M we take the metric induced by $\widetilde{f}$ (cf. Section 12.2). It then follows from the invariant nature of the Lipschitz–Killing curvatures that if $A \subset M$, then

$$\mathcal{L}_j^{\widetilde{f}}(A) \equiv \mathcal{L}_j^{f}(\psi(A)),$$

where superscripts have been added to emphasize that the Lipschitz–Killing curvatures on the left are computed with respect to the metric induced by $\widetilde{f}$ on M, while on the right they are computed with respect to the metric corresponding to f on $S(\mathbb{R}^l)$.

Putting all this together, we have that Theorem 15.7.2 holds not just for the isotropic process on $S(\mathbb{R}^l)$, but actually on all reasonable manifolds and for all smooth, unit-variance, Gaussian processes with a finite expansion. That is, we have reestablished all the main results of Chapters 12 and 13 for these processes.

The natural question to ask now is therefore whether we can extend this approach to processes without a finite expansion, thereby avoiding all the Morse-theoretic computations of the earlier chapters and of the crucial Theorem 15.2.1 in the current chapter.

Our answer is twofold: Firstly, we have not been able to do so, and it is not clear that the argument can be extended to processes without a finite expansion. Secondly, and, we think, more important, is the fact that the Morse-theoretic approach was crucial for establishing the accuracy of the expected Euler characteristic approximation to the excursion probability

$$\mathbb{P}\left\{\sup_{t\in M} f(t) \geq u\right\}$$

in Chapter 14. We see no way of adopting an approach based purely on geometric and tube-theoretic tools that will give the level of accuracy given by the Morse-theoretic approach.

Thus, despite the fact that in this chapter we are rederiving many of our earlier results, it seems that we did not waste our time previously. Apparently, there are some results that are still beyond the reach of the KFF.

15.8 Warped Products of Riemannian Manifolds

In view of the previous section, the light at the end of the tunnel, or chapter, should now be becoming clearer. In order to evaluate the mean Lipschitz–Killing curvatures that we are after, and in view of (15.6.6), what remains is to compute the limits

$$\lim_{n\to\infty} \frac{\mathcal{L}^{n-1}_{n-1-j}\left(\pi^{-1}_{\sqrt{n},n,k} D\right)}{s_n n^{(n-1)/2}}$$

for suitable sets D, a computation that is in part geometry and in part asymptotics. In this section we shall concentrate on the geometry.

As a first step we shall need to investigate the structure of the set $\pi^{-1}_{\sqrt{n},n,k} D \in S_{\sqrt{n}}(\mathbb{R}^n)$ a little more deeply than we have so far. In doing so, we shall also need to introduce the notion of the warped product of Riemannian manifolds, the last technical tool that we shall require from differential geometry in this book.

It is clear that the set $\pi^{-1}_{\sqrt{n},n,k} D$ is, topologically, a disjoint union

$$\pi^{-1}_{\sqrt{n},n,k} D \simeq \left(D \cap S_{\sqrt{n}}(\mathbb{R}^k) \right) \sqcup \left(D \cap \left(B_{\mathbb{R}^k}(0, \sqrt{n}) \right)^\circ \times S(\mathbb{R}^{n-k}) \right). \qquad (15.8.1)$$

For example, in Figure 15.7.1, the first part of this union is merely the point at the extreme right of the spherical cap, while the second part is the cap with this point removed.

Since we are assuming that D itself is a tame stratified space, the same is true of $\pi^{-1}_{\sqrt{n},n,k} D$ and of each of the two components above. Consequently, their intrinsic volumes are well defined. One is easy to compute. Since $D \cap S_{\sqrt{n}}(\mathbb{R}^k)$ is a tame stratified subset of $S_{\sqrt{n}}(\mathbb{R}^k)$, its intrinsic volumes can be computed using (10.5.8).

The second set in the union is, however, somewhat more complex, since we have written it as a product set and we have not yet developed tools for handling the Riemannian structure of products. Furthermore, what we have written in (15.8.1) is a topological equivalence, and ultimately, we shall need precise intrinsic volumes that are not topological invariants.

The way to handle these problems is twofold. First of all, we need to break the Riemannian structure of products into a product of structures, at least along each stratum of a stratified space. Secondly, we need to keep track of the fact that while (15.8.1) is topologically precise, at each point in $D \cap \left(B_{\mathbb{R}^k}(0, \sqrt{n}) \right)^\circ$ the corresponding $S(\mathbb{R}^{n-k})$ is likely to have a different radius.

In fact, the rightmost part of (15.8.1) is a subset of a *warped (Riemannian) product*, and each stratum of $D \cap (B_{\mathbb{R}^k}(0, \sqrt{n}))^\circ \times S(\mathbb{R}^{n-k})$ inherits this warped product structure. Once we choose the appropriate warp we can, and shall, relate to (15.8.1) no longer as a topologically correct relationship, but as if it were correct without qualifiers.

To understand what this means, we take a moment to investigate warped products and develop a general expression for the intrinsic volumes of their subsets. Once we have that, we shall be able to develop a concrete expression for the intrinsic volumes of $\pi^{-1}_{\sqrt{n},n,k} D$, which we do in Section 15.9.

15.8.1 Warped Products

We now develop some basic calculations needed to compute the intrinsic volumes of warped products. We were unable to find many of these calculations in the literature and so cannot adopt our usual policy of simply referring you to a good reference. Bear with us—they will not be too long.

A (Riemannian) warped product consists of two Riemannian manifolds (M_1, g_1) and (M_2, g_2) and a smooth function $\sigma^2 : M_1 \to [0, +\infty)$, known as the *warp*. The *warped product* is then defined to be the Riemannian manifold

$$(M_1, M_2, \sigma) \stackrel{\Delta}{=} (M_1 \times M_2, g_1 + \sigma^2 g_2). \qquad (15.8.2)$$

That is, the warped product of (M_1, g_1) and (M_2, g_2) with warp σ^2 is a Riemannian manifold $M_1 \times M_2$ such that for each $(t_1, t_2) \in M_1 \times M_2$ the Riemannian metric on

$M_1 \times M_2$ is given by[18]

$$g\left(X_{t_1}+Y_{t_2}, Z_{t_1}+W_{t_2}\right)=g_1\left(X_{t_1}, Z_{t_1}\right)+\sigma_{t_1}^2 g_2\left(Y_{t_2}, W_{t_2}\right).$$

Usually, as in our case, M_2 is a sphere.

As an example of a warped product consider

$$\widetilde{M}_\sigma=\left(B_{\mathbb{R}^k}(0, \sqrt{n})\right)^\circ \times S(\mathbb{R}^{n-k}),$$

where the Riemannian metric on the open ball is given by

$$g_\sigma = g_{\mathbb{R}^k} + \nabla\sigma \otimes \nabla\sigma \tag{15.8.3}$$

for

$$\sigma^2(t) = n - \|t\|^2_{\mathbb{R}^k}, \tag{15.8.4}$$

and the Riemannian metric $g_{\mathbb{R}^k}$ on $S(\mathbb{R}^{n-k})$ is the canonical one inherited from $\mathbb{R}^{n-k}$.

The reason we are interested in this example of a warped product is that each $(n-k-1+j)$-dimensional stratum $\widetilde{D}_{n-k-1+j}$ of

$$\left(D \cap \left(S(\mathbb{R}^k)\right)^c\right) \times S(\mathbb{R}^{n-k})$$

is isometrically embedded in this warped product and has the form

$$D_j \times S(\mathbb{R}^{n-k}) \tag{15.8.5}$$

for some j-dimensional submanifold D_j of the open ball $B_{\mathbb{R}^k}(0, \sqrt{n})^\circ$. Using this embedding we can compute

$$\mathcal{L}^{1/n}_{n-1-i}\left(\pi^{-1}_{\sqrt{n},n,k} D, \widetilde{D}_{n-k-1+j}\right),$$

the contribution of these strata to the intrinsic volumes of $\pi^{-1}_{\sqrt{n},n,k} D$.

The first step to computing these contributions is to compute the Levi-Civitá connection $\widetilde{\nabla}^\sigma$ of $\widetilde{M}_\sigma$, since this is needed in order to compute the second fundamental form of $D_j \times S(\mathbb{R}^{n-k})$ in $\widetilde{M}_\sigma$.

Consider, therefore, a general warped product (M_1, M_2, σ) and denote the Levi-Civitá connection on each M_j by ∇^j. Use E, or E_j, to denote vector fields on M_1, identified with their natural extensions on $M_1 \times M_2$. Similarly, F, or F_j, denotes fields on M_2 extended to $M_1 \times M_2$. Then the following relationships hold:

$$\widetilde{\nabla}^\sigma_{E_1} E_2 = \nabla^1_{E_1} E_2, \qquad \widetilde{\nabla}^\sigma_{F_1} F_2 = \nabla^2_{F_1} F_2, \tag{15.8.6}$$

and

---

[18] Since the tangent space $T_{(t_1,t_2)}(M_1 \times M_2)$ is equal to $T_{t_1}M_1 \oplus T_{t_2}M_2$, any vector $\tilde{X}_{(t_1,t_2)}$ in it can be written as $X_{t_1} + X_{t_2}$ for unique $X_{t_j} \in T_{t_j}M_j$.

$$\widetilde{\nabla}^{\sigma}_E F = \widetilde{\nabla}^{\sigma}_F E = E(\log \sigma) F. \tag{15.8.7}$$

The two relationships in (15.8.6) follow directly from the definition of the Levi-Civitá connection and the product structure of M. To prove (15.8.7) we apply Koszul's formula (7.3.12). If $F_1, \dots, F_k$ is an orthonormal frame on (M_2, g_2), then it is immediate that

$$\widehat{F}_i = \frac{1}{\sigma} F_i, \qquad 1 \le i \le k,$$

are orthonormal vector fields on (M, g).

Koszul's formula then gives us that for any vector field E on (M_1, g_1),

$$g(\nabla^{\sigma}_E F_i, F_j) = \frac{1}{2} E(g(F_i, F_j)),$$

since all the other terms in the formula are zero either by orthogonality or the product structure of (M, g). However,

$$E(g(F_i, F_j)) = E(\sigma^2 g_2(F_i, F_j)) = 2\sigma E(\sigma) g_2(F_i, F_j) = 2\sigma E(\sigma)\delta_{ij}.$$

Furthermore,

$$g(\nabla^{\sigma}_E F_i, E) = 0$$

for any other vector field E on (M_1, g_1). Therefore, for any $1 \le i \le k$,

$$\begin{aligned} \nabla^{\sigma}_E F_i &= \sum_{j=1}^{k} g(\nabla^{\sigma}_E F_i, \widehat{F}_j), \widehat{F}_j = \sum_{j=1}^{k} \frac{1}{\sigma} g(\nabla^{\sigma}_E F_i, F_j) \widehat{F}_j \\ &= \sum_{j=1}^{k} E(\sigma)\delta_{ij} \widehat{F_j} = \frac{E(\sigma)}{\sigma} F_i = E(\log \sigma) F_i, \end{aligned}$$

from which (15.8.7) now follows.

15.8.2 A Second Fundamental Form

With the Levi-Civitá connection $\widetilde{\nabla}^{\sigma}$ determined, the next step toward computing intrinsic volumes lies in understanding the second fundamental forms over the sets $D_j \times S(\mathbb{R}^k)$ as they sit in $\widetilde{M}_{\sigma}$, as well as their powers. For this we need to describe the normal spaces $T_{(t,\eta)} \widetilde{M}_{\sigma}^{\perp}$ for $t \in B_{\mathbb{R}^k}(0, \sqrt{n})$ and $\eta \in S(\mathbb{R}^{n-k})$. A simple argument shows that at a point $(t, \eta) \in D_j \times S(\mathbb{R}^{n-k})$,

$$(T_{(t,\eta)} \widetilde{M}_{\sigma})^{\perp} = \left(T_t D_j \oplus T_{\eta} S(\mathbb{R}^{n-k}) \right)^{\perp} \simeq T_t D_j^{\perp}, \tag{15.8.8}$$

where $T_t D_j^{\perp}$ is the orthogonal (with respect to g_{σ}) complement of $T_t D_j$ in $T_t B_{\mathbb{R}^k}(0, \sqrt{n})$. From this, it is now not hard to see that the normal Morse index $D_j \times S(\mathbb{R}^k)$ as it sits in $\widetilde{M}^{\sigma}$ is actually the same as the Morse index of D_j as it sits in $B_{\mathbb{R}^k}(0, \sqrt{n})$.

With the various tangent and normal spaces determined, we now let S denote the scalar second fundamental form of $D_j \times S(\mathbb{R}^{n-k})$ in $\widetilde{M}_\sigma$, and S_σ the scalar second fundamental form of D_j in $(B_{\mathbb{R}^k}(0, \sqrt{n}), g_\sigma)$. Then we have the following lemma.

Lemma 15.8.1. *Retaining the above notation, for $0 \le l \le n-1$, take*

$$(t,\eta) \in D_j \times S(\mathbb{R}^{n-k}), \qquad \nu_{k-j} \in (T_{(t,\eta)} D_j \times S(\mathbb{R}^{n-k}))^\perp.$$

Then

$$\frac{1}{l!}\operatorname{Tr}(S^l_{\nu_{k-j}}) = \sum_{r=0}^{l} \binom{n-k-1}{l-r} (-1)^{l-r} (\nu_{k-j}(\log\sigma_t))^{l-r} \operatorname{Tr}(S^r_{\sigma,\nu_{k-j}}).$$

Proof. Fix an orthonormal (under g_σ) basis $(E_1,\dots,E_k,F_1,\dots,F_{n-k-1})$ of $T_{(t,\eta)}\widetilde{M}_\sigma$ such that $(E_1,\dots,E_j)$ forms an orthonormal basis of $T_t D_j$. Equation (15.8.8) implies that any ν_{k-j} satisfying the conditions of the theorem can be expressed as

$$\nu_{k-j} = \sum_{r=1}^{j} a_r E_r$$

for some constants a_r. Applying (15.8.6) and (15.8.7) along with the Weingarten equation (7.5.12), we obtain

$$\begin{aligned}
S_{\nu_{k-j}}(E_r, E_s) &= -g(\nabla^\sigma_{E_r}\nu_{k-j}, E_s) = -g_\sigma(\nabla^\sigma_{E_r}\nu_{k-j}, E_s) = S_{\sigma,\nu_{k-j}}(E_r, E_s),\\
S_{\nu_{k-j}}(F_r, F_s) &= -g(\nabla^\sigma_{F_r}\nu_{k-j}, F_s) = -\nu_{k-j}(\log\sigma_t) g(F_r, F_s)\\
&= -\nu_{k-j}(\log\sigma_t)\delta_{rs},\\
S_{\nu_{k-j}}(E_r, F_s) &= 0.
\end{aligned}$$

Therefore, for each ν_{k-j}, the matrix of the shape operator $S_{\nu_{k-j}}$ in our chosen orthonormal basis is block diagonal with one block, of size j, being

$$\left\{S_{\sigma,\nu_{k-j}}(E_r, E_s)\right\}_{1\le r,s\le j}.$$

The other block, of size $n-k-1$, is

$$-\nu_{k-j}(\log\sigma_t) I_{(n-k-1)\times(n-k-1)}.$$

Therefore, applying (7.2.7), we have that for $l \le n-k-1$,

$$\frac{1}{l!}\operatorname{Tr}(S^l_{\nu_{k-j}}) = \sum_{r=0}^{l} \binom{n-k-1}{l-r} (-1)^{l-r} \left(\nu_{k-j}(\log\sigma_t)\right)^{l-r} \frac{1}{r!}\operatorname{Tr}(S^r_{\sigma,\nu_{k-j}}),$$

which completes the proof. □

Since Lipschitz–Killing curvatures are no more than integrals of powers of traces of second fundamental forms, we now have all we need to begin the final computation.

15.9 Non-Gaussian Mean Intrinsic Volumes

Our aim now is to compute the limits (15.6.6), that is,

$$\lim_{n\to\infty} \frac{\mathcal{L}^{n^{-1}}_{n-1-j}\left(\pi^{-1}_{\sqrt{n},n,k} D\right)}{s_n n^{(n-1)/2}}, \tag{15.9.1}$$

for which the following lemma is the most crucial step. It is in this step that the generalized Lipschitz–Killing curvature measures $\widetilde{\mathcal{L}}_j$ and integrals against them as in Definition 10.9.1 appear for the first time. The transition from this lemma to the final result in Theorem 15.9.5, in which the Gaussian Minkowski functionals appear, will then involve no more than some careful asymptotics.

Lemma 15.9.1. *As usual, D is a tame stratified subset of $\mathbb{R}^k$ with j-dimensional strata D_j. Suppose that $\widetilde{D}_{n-k-1+j} = D_j \times S(\mathbb{R}^{n-k})$ is an $(n-k-1+j)$-dimensional stratum of $\pi^{-1}_{\sqrt{n},n,k} D$ such that*

$$\widetilde{D}_{n-k-1+j} \cap S(\mathbb{R}^k) = \emptyset.$$

Then, for all $i \ge k - j \ge 0$,

$$\begin{aligned} &\mathcal{L}^{1/n}_{n-1-i}\left(\pi^{-1}_{\sqrt{n},n,k} D, \widetilde{D}_{n-k-1+j}\right) \\ &\quad = s_{n-k} \sum_{l=0}^{i+j-k} \frac{s_{k+l-j}}{s_i} \binom{n-k-1}{i+j-k-l} \\ &\qquad \times \widetilde{\mathcal{L}}_{j-l}\left(D^\sigma, \sigma^{n+k-2i-2j+2l-1}(2\pi)^{-k/2} h_{i+j-k-l} \mathbb{1}_{D_j}\right), \end{aligned} \tag{15.9.2}$$

where $D^\sigma = D \cap B_{\mathbb{R}^k}(0, \sqrt{n})$ is the tame stratified space obtained from the intersection of the embedding of D in $\mathbb{R}^k$ and the open ball of radius $\sqrt{n}$, endowed with the metric g_σ given by (15.8.3) *and*

$$h_l(t, \nu) \triangleq \langle \nu, t\rangle^l_{\mathbb{R}^k}.$$

(Note also that the final term in (15.9.2) *is an integral against the generalized curvature measure $\widetilde{\mathcal{L}}_{j-l}$.)*

Furthermore, suppose that for all $0 \le r, s \le i + k - j$,

$$\widetilde{\mathcal{L}}_r\left(D, |h_s|\varphi_k \mathbb{1}_{D_j}\right) < \infty, \tag{15.9.3}$$

writing φ_k for the k-dimensional Gaussian density $(2\pi)^{-k/2} e^{-|t|^2/2}$. Then

$$\begin{aligned} &\lim_{n\to\infty} \frac{1}{s_n n^{(n-1)/2}} \mathcal{L}^{1/n}_{n-1-i}\left(\pi^{-1}_{\sqrt{n},n,k} D, \widetilde{D}_{n-k-1+j}\right) \\ &\quad = \sum_{l=0}^{i+j-k} \frac{[k+l-j]!}{[i]!} \binom{i-1}{k+l-j-1} \widetilde{\mathcal{L}}_{j-l}\left(D, \varphi_k h_{i+j-k-l} \mathbb{1}_{D_j}\right). \end{aligned} \tag{15.9.4}$$

Proof. From (10.5.10),

$$\begin{aligned}
&\mathcal{L}^{1/n}_{n-1-i}\left(\pi^{-1}_{\sqrt{n},n,k}D, \widetilde{D}_{n-k-1+j}\right) \\
&\quad = \frac{C(k-j, i+j-k)}{(2\pi)^{(i+j-k)/2}(i+j-k)!}\int_{D_j\times S(\mathbb{R}^{n-k})}\int_{S(T_{(t,\eta)}D_j\times S(\mathbb{R}^{n-k})^{\perp})} \\
&\qquad \times \operatorname{Tr}(S^{i+j-k}_{\nu_{k-j}})\alpha(\nu_{k-j})\mathcal{H}_{k-j}(d\nu_{k-j})\mathcal{H}_{n-1-j+k}(dt, d\eta),
\end{aligned}$$

where

$$\mathcal{H}_{n-1-j+k}(dt, d\eta) = \sigma_t^{n-1-k}\mathcal{H}_{n-1-k}(d\eta)\mathcal{H}_j(dt)$$

is the Hausdorff measure that $D_j \times S(\mathbb{R}^{n-k})$ inherits from $\widetilde{D}^{\sigma}$, the warped product of (D^{σ}, g_{σ}) and $S(\mathbb{R}^k)$ with its usual metric and warp function σ^2 given by (15.8.4).

By Lemma 15.8.1, observation (15.8.8), and the subsequent remarks about the normal Morse index α,

$$\begin{aligned}
&\mathcal{L}^{1/n}_{n-1-i}\left(\pi^{-1}_{\sqrt{n},n,k}D, \widetilde{D}_{n-k-1+j}\right) \\
&\quad = C(k-j, i+j-k)(2\pi)^{-(i+j-k)/2}\int_{D_j}\int_{S(\mathbb{R}^{n-k})}\int_{S(T_tD_j^{\perp})} \\
&\qquad \times \sum_{l=0}^{i+j-k}\binom{n-1-k}{i+j-k-l}\sigma_t^{n-1-k}(-1)^{i+j-k-l}\left(\nu_{k-j}(\log\sigma_t)\right)^{i+j-k-l} \\
&\qquad \times \frac{1}{l!}\operatorname{Tr}(S^{l}_{\sigma,\nu_{k-j}})\alpha(\nu_{k-j})\mathcal{H}_{k-j}(d\nu_{k-j})\mathcal{H}_{n-1-k}(d\eta)\mathcal{H}_j(dt).
\end{aligned}$$

Equation (15.9.2) now follows from the fact that

$$\frac{C(k-j, i+j-k)(2\pi)^{-(i+j-k)/2}}{C(k-j, l)(2\pi)^{-l/2}} = \frac{s_{k+l-j}}{s_i},$$

followed by integrating over $S(\mathbb{R}^{n-k})$ and noting that

$$\nu_{k-j}(\log\sigma_t) = -\frac{\langle\nu_{k-j}, t\rangle_{\mathbb{R}^k}}{\sigma_t^2}.$$

As for the second conclusion of the lemma, (15.9.4), note that

$$\begin{aligned}
&\lim_{n\to\infty}\frac{s_{n-k}}{s_n n^{(n-1)/2}}\binom{n-k-1}{i+j-k-l}\sigma_t^{n+k-2i-2j+2l-1} \\
&\quad = \lim_{n\to\infty}\frac{s_{n-k}}{s_n n^{(n-1)/2}}\binom{n-k-1}{i+j-k-l}n^{(n+k-2i-2j-2l-1)/2} \\
&\qquad \times\left(1-\frac{\|t\|^2}{n}\right)^{(n+k-2i-2j-2l-1)/2} \\
&\quad = \frac{(2\pi)^{-k/2}}{(i+j-k-l)!}e^{-\|t^2\|/2}.
\end{aligned}$$

Also,

$$\frac{s_{k+l-j}}{s_i(i+j-k-l)!} = \frac{s_{k+l-j}(k+l-j-1)!}{s_i(i-1)!}\binom{i-1}{k+l-j}$$
$$= \frac{[k+l-j]!}{[i]!}\binom{i-1}{k+l-j}.$$

Finally, since it is not hard to see that there exists a finite K such that for all n large enough,

$$\left(1-\frac{|t|^2}{n}\right)^{n/2} \le Ke^{-|t|^2/2}$$

for all $t \in B_{\mathbb{R}^k}(0,\sqrt{n})^\circ$, dominated convergence yields (15.9.4) and we are done. □

In Lemma 15.9.1 we computed the contribution of the sets $\widetilde{D}_{n-k-1+j} = D_j \times S(\mathbb{R}^{n-k})$ to the curvatures $\mathcal{L}^{n^{-1}}_{n-1-i}(\pi^{-1}_{\sqrt{n},n,k}D)$ under the assumption that $\widetilde{D}_{n-k-1+j} \cap S(\mathbb{R}^k) = \emptyset$.

Our task now is to show that if in fact, $\widetilde{D}_{n-k-1+j} \cap S(\mathbb{R}^k) \neq \emptyset$, then there is actually no contribution to $\mathcal{L}^{n^{-1}}_{n-1-i}(\pi^{-1}_{\sqrt{n},n,k}D)$ for n large enough. We write this as shown in the following lemma.

Lemma 15.9.2. *Suppose that D satisfies the conditions of Lemma 15.9.1 and that $D_j \subset S_{\sqrt{n}}(\mathbb{R}^k)$ is a stratum of $D \cap B_{\mathbb{R}^k}(0,\sqrt{n})$ for some $n > 0$. Then, for $n > j+i+1$,*

$$\mathcal{L}^{n^{-1}}_{n-1-i}(\pi^{-1}_{\sqrt{n},n,k}D, \pi^{-1}_{\sqrt{n},n,k}D_j) = 0,$$

and so for all sufficiently large n,

$$\mathcal{L}^{n^{-1}}_{n-1-i}(\pi^{-1}_{\sqrt{n},n,k}D, h\mathbb{1}_{\pi^{-1}_{\sqrt{n},n,k}D_j}) \equiv 0$$

for all h.

Proof. Since $D_j \subset S_{\sqrt{n}}(\mathbb{R}^k)$,

$$\pi^{-1}_{\sqrt{n},n,k}D_j = D_j,$$

and so $\pi^{-1}_{\sqrt{n},n,k}D_j$ is a j-dimensional stratum of $\pi^{-1}_{\sqrt{n},n,k}D$. From Definition 10.9.1, we see that such strata contribute only to the intrinsic volumes of order 0 to j, as required. □

We now have all that we need to evaluate the illusive limits (15.6.6), namely,

$$\widetilde{\rho}_j(D) \stackrel{\Delta}{=} (2\pi)^{-j/2}[j]!\lim_{n\to\infty}\frac{\mathcal{L}^{n^{-1}}_{n-1-j}\left(\pi^{-1}_{\sqrt{n},n,k}D\right)}{s_n n^{(n-1)/2}},$$

and via them the all-important functions $\rho_j(D)$ of (15.6.13).

Theorem 15.9.3. *Suppose D satisfies the conditions of Lemma* 15.9.1. *Then*

$$\rho_i(D) = (2\pi)^{-i/2} \mathcal{M}_i^{\gamma}(D), \tag{15.9.5}$$

where the functionals $\mathcal{M}_i^{\gamma}(D)$ are the coefficients in the standard Gaussian tube formula derived in Section 10.9.3.

Proof. We start by computing the $\widetilde{\rho}_i$. By Lemmas 15.9.1 and 15.9.2,

$$\begin{aligned}
\widetilde{\rho}_i(D) &= (2\pi)^{-i/2}[i]! \sum_{j=k-i}^{k-1} \sum_{l=0}^{i+j-k} \frac{[k+l-j]!}{[i]!} \binom{i-1}{k+l-j-1} \\
&\quad \times \widetilde{\mathcal{L}}_{j-l}\left(D, \varphi_k h_{i+j-k-l} \mathbb{1}_{D_j}\right) \\
&= (2\pi)^{-i/2} \sum_{j=k-i}^{k-1} \sum_{l=0}^{i+j-k} \binom{i-1}{k+l-j-1} \\
&\quad \times \widetilde{\mathcal{M}}_{k+l-j}\left(D, \varphi_k h_{i+j-k-l} \mathbb{1}_{D_j}\right) \\
&= (2\pi)^{-i/2} \sum_{m=0}^{i-1} \binom{i-1}{m} \widetilde{\mathcal{M}}_{m+1}\left(D, \varphi_k h_{i-1-m}\right),
\end{aligned}$$

where the $\widetilde{M}$ are the generalized Minkowski curvature measures defined in (10.9.5).

With the $\widetilde{\rho}_i$ determined, we can now turn to the ρ_j. By (15.6.13) these are given by

$$\begin{aligned}
\rho_i(D) &= (i-1)! \sum_{l=0}^{\lfloor \frac{i-1}{2} \rfloor} \frac{(-1)^l}{(4\pi)^l l! (i-1-2l)!} \widetilde{\rho}_{i-2l}(D) \\
&= (i-1)! \sum_{l=0}^{\lfloor \frac{i-1}{2} \rfloor} \sum_{m=0}^{i-2l-1} \frac{(-1)^l}{(4\pi)^l l! (i-1-2l)!} (2\pi)^{-(i-2l)/2} \\
&\quad \times \binom{i-2l-1}{m} \widetilde{\mathcal{M}}_{m+1}\left(D, \varphi_k h_{i-2l-1-m}\right) \\
&= (2\pi)^{-(i+k)/2} \sum_{m=0}^{i-1} \binom{i-1}{m} \widetilde{\mathcal{M}}_{m+1}\left(D, H_{i-m-1}(\langle \eta, t \rangle) e^{-|t|^2/2}\right) \\
&= (2\pi)^{-i\text{Â½}} \mathcal{M}_i^{\gamma}(D),
\end{aligned}$$

the second-to-last line coming from the definition (11.6.9) of the Hermite polynomials, and the last from the definition (10.9.12) of the $\mathcal{M}_i^{\gamma}(D)$. □

The main results of this chapter, and indeed of the book, are now simple corollaries.

Theorem 15.9.4. *Let M be an N-dimensional, regular stratified space, D a regular stratified subset of $\mathbb{R}^k$. Let $y = (y_1, \ldots, y_k) : M \to \mathbb{R}^k$ be a vector-valued Gaussian* *field, with independent, identically distributed components satisfying the conditions of Corollary* 11.3.5. *Then*

$$\mathbb{E}\left\{\mathcal{L}_i\left(M \cap y^{-1}(D)\right)\right\} = \sum_{j=0}^{N-i} \begin{bmatrix} i+j \\ j \end{bmatrix} (2\pi)^{-j/2} \mathcal{L}_{i+j}(M) \mathcal{M}_j^{\gamma}(D), \qquad (15.9.6)$$

where the $\mathcal{L}_j$, $j = 0, \ldots, N$, *are the Lipschitz–Killing measures on* M *with respect to the metric induced by the* y_i, *and the* $\mathcal{M}_j^{\gamma}$ *are the generalized (Gaussian) Minkowski functionals on* $\mathbb{R}^k$.

Proof. Theorems 15.6.3 and 15.9.3, which actually require fewer conditions than we have assumed, immediately yield that (15.9.6) holds for the canonical isotropic process y of (15.6.3).

On the other hand, Theorem 15.2.1, which does require all the conditions of the theorem being proven, implies that in general, the mean Lipschitz–Killing curvatures $\mathbb{E}\left\{\mathcal{L}_i(M \cap y^{-1}(D)\right\}$ break up into a sum of Lipschitz–Killing curvatures of M and coefficients that are dependent only on D and some other constants. In particular, they are independent of the covariance structure of the underlying process and its parameter space. Consequently, these coefficients are the same as we computed for the canonical isotropic process on the sphere. That is, (15.9.6) holds in general. □

Finally, we have the main theorem of the book, which is a trivial consequence of the previous one. Indeed, we would have called it a corollary rather than a theorem, were it not for the fact that it is such an important result.

Theorem 15.9.5. *With the same notation and conditions as in Theorem* 15.9.4, *let* $F \in C^2(\mathbb{R}^k, \mathbb{R})$ *and define the non-Gaussian random field* $f = F(y(t))$. *Then*

$$\mathbb{E}\left\{\mathcal{L}_i\left(A_u(f, M)\right)\right\} = \sum_{j=0}^{N-i} \begin{bmatrix} i+j \\ j \end{bmatrix} (2\pi)^{-j/2} \mathcal{L}_{i+j}(M) \mathcal{M}_j^{\gamma}\left(F^{-1}[u, +\infty)\right).$$

Proof. This is a trivial consequence of Theorem 15.9.4, taking $D = F^{-1}([u, \infty))$.

15.10 Examples

To round off, we now want to look at three examples of Theorem 15.9.5, one familiar and two new. By "examples," what we mean are explicit computations of the coefficients $\mathcal{M}_j^{\gamma}\left(F^{-1}[u, +\infty)\right)$ for particular choices of the function F. There is no need to recompute the coefficents $\mathcal{L}_{i+j}$ since these are the same as in the Gaussian case and independent of the choice of F.

The familiar example will be the case of Gaussian f, in which case, in the notation of Theorem 15.9.5, $k = 1$ and $F(x) \equiv x$. Here we have seen the result many times, including Chapters 11 and 12 for the Euler characteristic and Chapter 13 for the remaining Lipschitz–Killing curvatures, not to mention in Section 15.7 of the present chapter. Thus our aim is simply to check that everything works as it should and to see how to derive the specific from the general.

For the remaining two examples, we take two of the examples of (15.0.2), specifically the so-called χ^2 and F random fields. In all three examples, we shall look for functions $\rho_{j,f}$ such that

$$\mathbb{E}\{\mathcal{L}_0(A_u(f,M))\} = \sum_{j=0}^{N} \rho_{j,f}(u)\mathcal{L}_j(M). \tag{15.10.1}$$

The $\rho_{j,f}$ are generally known as the *Euler characteristic* (*EC*) *densities for* f, and by Theorem 15.9.5 they are given by

$$\rho_{j,f}(u) = (2\pi)^{-j/2}\mathcal{M}_j^{\gamma}\left(F^{-1}[u,+\infty)\right). \tag{15.10.2}$$

Although they are defined via the mean Euler characteristic calculation, it follows from Theorem 15.9.5 that they determine all the other mean Lipschitz–Killing curvatures as well, since

$$\mathbb{E}\{\mathcal{L}_i(A_u(f,M))\} = \sum_{j=0}^{N-i} \begin{bmatrix} i+j \\ j \end{bmatrix} \rho_{j,f}(u)\mathcal{L}_{i+j}(M). \tag{15.10.3}$$

Throughout, we shall assume, as we have done so far and without further comment, that the assumptions of Theorem 15.9.5 hold regarding both the manifold M and the process y.

15.10.1 The Gaussian Case

The simplest of all examples is given by y being a real-valued, centered, unit-variance Gaussian random field and F the identity function, so that $f \equiv y$. In this case we already know by (10.9.13) that

$$\mathcal{M}_j^{\gamma}\left(F^{-1}[u,\infty)\right) = \mathcal{M}_j^{\gamma}([u,\infty)) = \frac{1}{\sqrt{2\pi}}H_{j-1}(u)e^{-u^2/2}, \tag{15.10.4}$$

and so

$$\rho_{j,f}(u) = (2\pi)^{-(j+1)/2}H_{j-1}(u)e^{-u^2/2}. \tag{15.10.5}$$

Despite this easy path, we shall rederive the result via the Taylor series mentioned in Chapter 10, as a preliminary to the more complicated χ^2 and F cases.

Recall that the $\mathcal{M}^{\gamma}$ arise as coefficents in the formal tube expansion (10.9.11). That is,

$$\gamma_{\mathbb{R}^l}(\text{Tube}(D,\rho)) = \gamma_{\mathbb{R}^l}(D) + \sum_{j=1}^{\infty} \frac{\rho^j}{j!}\mathcal{M}_j^{\gamma_{\mathbb{R}^l}}(D). \tag{15.10.6}$$

Taking $D_u = [u, \infty)$ here, and recalling the fact that

$$\frac{d^j}{dx^j} e^{-x^2/2} = (-1)^j H_j(x) e^{-x^2/2}$$

(cf. (11.6.11)), we have, via a standard Taylor series expansion of the exponential function, that

$$\begin{aligned}\gamma_{\mathbb{R}^k}\big(T(D_u, \rho)\big) &= 1 - \Phi(u-\rho)\\ &= 1 - \left(\Phi(u) + \sum_{j=1}^{\infty} \frac{(-\rho)^j}{j!} \frac{(-1)^{j-1}}{\sqrt{2\pi}} H_{j-1}(u) e^{-u^2/2}\right)\\ &= 1 - \Phi(u) + \sum_{j=1}^{\infty} \frac{\rho^j}{j!} \frac{1}{\sqrt{2\pi}} H_{j-1}(u) e^{-u^2/2},\end{aligned}$$

so that on comparison with (15.10.6), we find that

$$\mathcal{M}_j^{\gamma_{\mathbb{R}^k}}(D_u) = \frac{1}{\sqrt{2\pi}} H_{j-1}(u) e^{-u^2/2},$$

which is, of course, the same as (15.10.4) and so also implies (15.10.5).

Substituting (15.10.5) into (15.10.1) gives the mean Euler characteristic results of Chapters 11 and 12 (cf. (11.7.15) and Theorems 12.4.1 and 12.4.2), while substituting into (15.10.3) recaptures the main results of Chapter 13 for the remaining Lipschitz–Killing curvatures (cf. Theorems 13.2.1 and 13.4.1).

15.10.2 The χ^2 Case

The χ^2 random fields arise in a number of applications (cf. [172]) and are easily defined by taking F, in our standard notation, to be

$$G(y) = |y|^2.$$

(The desire for a change of notation will become clear soon.) We shall write the corresponding random processes as

$$g(t) = G(y(t)).$$

The EC densities for these processes[19] were originally derived in [2] for $N = 2$ and in [172] in general, in both cases from a first-principles approach and with considerable and complicated computation. We shall work via Theorem 15.9.5, an approach first adopted in [157, 158]. As you will see in a moment, the general approach is basically the same as in the simple Gaussian case, although the details of the computation are a little more complicated.

[19] The EC densities for noncentral χ^2 processes are also known, and available in [172] and [157, 158].

We start the actual computation by noting that if we take $F(x) = |x|$, then $f = F \circ y$ is the square root of the χ^2_k random field g on M. Since the EC densities of f are related to those of the χ^2_k field g by

$$\rho_{j,g}(u) = \rho_{j,f}(\sqrt{u}),$$

it suffices to calculate the EC densities of f, which turns out to lead to slightly tidier computations.[20]

Following the same path we took for the Gaussian example, we note that for this f we have

$$D_u = F^{-1}\left([u, +\infty)\right) = \overline{\mathbb{R}^k \backslash B_{\mathbb{R}^k}(0, u)},$$

and $T(D_u, \rho) = D_{u-\rho}$, so that

$$\gamma_{\mathbb{R}^k}\left(T(D_u, \rho)\right) = \gamma_{\mathbb{R}^k}(D_{u-\rho}).$$

It remains to express the right-hand side here as a Taylor series in ρ. Elementary computations give the probability density p_k of f as

$$p_k(x) = \frac{1}{\Gamma(k/2)2^{(k-2)/2}} x^{k-1} e^{-x^2/2},$$

and so

$$\gamma_{\mathbb{R}^k}\left(D_{u-\rho}\right) = \gamma_{\mathbb{R}^k}\left(D_u\right) - \sum_{j=1}^{\infty} \frac{(-\rho)^j}{j!} \frac{d^{j-1} p_k}{dx^{j-1}}\bigg|_{x=u}.$$

Direct calculations, again exploiting (11.6.11) as in the Gaussian case, show that

$$\begin{aligned}
\frac{d^{j-1} p_k}{dx^{j-1}} &= \frac{1}{\Gamma(k/2)2^{(k-2)/2}} \sum_{i=0}^{j-1} \binom{j-1}{i} (-1)^i \frac{d^{j-1-i} x^{k-1}}{dx^{j-1-i}} H_i(x) e^{-x^2/2} \\
&= \frac{e^{-x^2/2}}{\Gamma(k/2)2^{(k-2)/2}} \sum_{i=0}^{j-1} \binom{j-1}{i} (-1)^i \frac{d^{j-1-i} x^{k-1}}{dx^{j-1-i}} H_i(x).
\end{aligned}$$

The summation can be rewritten as

[20] Of course, we have introduced a small problem here, since $F(x) = |x|$ is not C^2, as required by Theorem 15.9.5. Nevertheless, we shall ignore the problem on the basis that a quick check of the proof of Theorem 15.9.5 should convince you that for each u, F need be C^2 only in a neighborhood of u. Thus, if we restrict ourselves to $u > 0$, all is fine. In any case, the problem is an artifact of choosing to work with $|x|$ instead of the more natural $|x|^2$, which should also help convince you that it is not too serious.

$$\sum_{i=0}^{j-1} \mathbb{1}_{\{k\geq j-i\}}\binom{j-1}{i}(-1)^i \frac{(k-1)!}{(k+i-j)!} x^{k+i-j} H_i(x)$$

$$= x^{k-j}\sum_{i=0}^{j-1}\sum_{l=0}^{\lfloor i/2\rfloor} \mathbb{1}_{\{k\geq j-i\}}\binom{j-1}{i}(-1)^{i+l}\frac{(k-1)!}{(k+i-j)!}\frac{i!}{(i-2l)!l!2^l}x^{2i-2l}$$

$$= x^{k-j}\sum_{l=0}^{\lfloor\frac{j-1}{2}\rfloor}\sum_{i=2l}^{j-1} \mathbb{1}_{\{k\geq j-i\}}\binom{j-1}{i}(-1)^{i+l}\frac{(k-1)!}{(k+i-j)!}\frac{i!}{(i-2l)!l!2^l}x^{2i-2l}$$

$$= x^{k-j}\sum_{l=0}^{\lfloor\frac{j-1}{2}\rfloor}\sum_{m=0}^{j-1-2l} \mathbb{1}_{\{k\geq j-m-2l\}}\binom{k-1}{j-1-m-2l}\frac{(-1)^{m+l}(j-1)!}{m!l!2^l}x^{2m+2l}.$$

Combining the last three displays gives

$$\begin{aligned}&\gamma_{\mathbb{R}^k}\left(D_{u-\rho}\right)\\ &\quad= \gamma_{\mathbb{R}^k}\left(D_u\right) + \sum_{j=1}^{\infty}\frac{\rho^j}{j!}\frac{u^{k-j}e^{-u^2/2}}{\Gamma(k/2)2^{(k-2)/2}}\sum_{l=0}^{\lfloor\frac{j-1}{2}\rfloor}\sum_{m=0}^{j-1-2l}\\ &\qquad\times \mathbb{1}_{\{k\geq j-m-2l\}}\binom{k-1}{j-1-m-2l}\frac{(-1)^{j-1+m+l}(j-1)!}{m!l!2^l}u^{2m+2l}.\end{aligned}$$

It now follows from the definition (15.10.2) of the EC densities that for $j\geq 1$,

$$\begin{aligned}\rho_{j,f}(u) &= \frac{u^{k-j}e^{-x^2/2}}{(2\pi)^{j/2}\Gamma(k/2)2^{(k-2)/2}}\sum_{l=0}^{\lfloor\frac{j-1}{2}\rfloor}\sum_{m=0}^{j-1-2l}\\ &\quad\times \mathbb{1}_{\{k\geq j-m-2l\}}\binom{k-1}{j-1-m-2l}\frac{(-1)^{j-1+m+l}(j-1)!}{m!l!2^l}u^{2m+2l},\end{aligned}$$

and so we have the following result.

Theorem 15.10.1. *The EC densities of the χ^2_k random field g are given, for $j\geq 1$ and $u>0$, by*

$$\begin{aligned}\rho_{j,g}(u) &= \frac{u^{(k-j)/2}e^{-u/2}}{(2\pi)^{j/2}\Gamma(k/2)2^{(k-2)/2}}\sum_{l=0}^{\lfloor\frac{j-1}{2}\rfloor}\sum_{m=0}^{j-1-2l}\\ &\quad\times \mathbb{1}_{\{k\geq j-m-2l\}}\binom{k-1}{j-1-m-2l}\frac{(-1)^{j-1+m+l}(j-1)!}{m!l!2^l}u^{m+l}.\end{aligned}$$

When $j=0$,

$$\rho_{0,g}(u) = \mathbb{P}\left\{\chi^2_k\geq u\right\}.$$

Perhaps the most important thing to note from this example is that although many of the formulas above may be long, there is nothing at all difficult in them. All that was required, once the structure of the sets D_u was understood, was basic calculus.

Indeed, this is often the case. However, there are cases, as we shall see from the following example, for which, as the function F becomes more complex, it becomes more efficient to backtrack a little and to compute the Minkowski curvature functionals from first principles.

15.10.3 The F Case

To define the $F_{m,n}$ random field, we take $y = (y^1, \dots, y^{m+n})$ as in Theorem 15.9.5, and set $f = F \circ y$, where

$$F(y) = \frac{n}{m} \frac{\sum_{i=1}^{m} (y^i)^2}{\sum_{i=1}^{n} (y^{m+i})^2}.$$

There is problem with the $F_{m,n}$ field, however, that does not arise with the $F_{m,n}$ random variable (i.e., the field at a specific point). This is that for the field it may well happen that there are t for which $y(t) = 0$, in which case $f(t)$ is undefined. To avoid this, we shall assume that $n + m > N$, so that by Theorem 11.2.10 we are assured that, with probability one, there are no such points. Then, restricting to $u > 0$ and using the same logic as in footnote 20, we have no problem.

In computing the EC densities of an $F_{m,n}$ field we return to first principles, and the definition (10.9.10) of $\mathcal{M}^{\gamma_{\mathbb{R}^k}}$ as

$$\mathcal{M}_j^{\gamma_{\mathbb{R}^{m+n}}}(M) \triangleq \sum_{k=0}^{j-1} \binom{j-1}{k} (-1)^{j-1-k} \widetilde{\mathcal{M}}_{k+1}\left(M, \frac{d^{j-1-k}}{d\eta^{j-1-k}} \varphi_{m+n}\right).$$

To differentiate φ_{m+n} we again apply (11.6.11). With $D_u = F^{-1}[u, +\infty)$ and $\partial D_u = F^{-1}(\{u\})$, noting the facts that $\langle \nabla F, \nabla \|y\|^2/2 \rangle = 0$ and

$$H_n(0) = \begin{cases} 0, & n \text{ odd}, \\ (-1)^l \frac{(2l)!}{l!2^l}, & n = 2l, \end{cases}$$

it is now not hard (although it involves some patience playing with constants) to see from (10.9.2) and (10.9.5) that for $j \geq 1$,

$$\mathcal{M}_j^{\gamma_{\mathbb{R}^{m+n}}}(D_u) \tag{15.10.7}$$

$$= (j-1)! \sum_{l=0}^{\lfloor \frac{(j-1)}{2} \rfloor} \frac{(-1)^l}{l!2^l} \frac{1}{(2\pi)^{(m+n)/2}} \int_{\partial D_u} e^{-|x|^2/2} \mathcal{M}_{j-2l}(D_u, dx)$$

$$= (j-1)! \sum_{l=0}^{\lfloor \frac{(j-1)}{2} \rfloor} \frac{(-1)^l}{l!2^l} \frac{1}{(2\pi)^{(m+n)/2}}$$

$$\times \int_{\partial D_u} e^{-|x|^2/2} \frac{1}{(j-2l-1)!} \operatorname{Tr}^{\partial D_u}\left(S_{\partial D_u}^{j-2l-1}\right) d\mathcal{H}_{m+n-1}(x).$$

The computation of this last integral is not simple, so we separate it out as a lemma.

Lemma 15.10.2. *In the notation above,*

$$\frac{1}{(2\pi)^{(m+n)/2}} \int_{\partial D_u} \frac{1}{k!} \operatorname{Tr}^{\partial D_u} \left(S^k_{\partial D_u} \right) e^{-|x|^2} \, d\mathcal{H}_{m+n-1}(x)$$

$$= \frac{\Gamma\left(\frac{m+n-k-1}{2}\right)}{2^{(k-1)/2}\Gamma\left(\frac{m}{2}\right)\Gamma\left(\frac{n}{2}\right)} \left(\frac{mu}{n}\right)^{(m-1-k)/2} \left(1+\frac{mu}{n}\right)^{-(m+n-2)/2}$$

$$\times \sum_{i=0}^{k} (-1)^{k-i} \left(\frac{mu}{n}\right)^i \binom{m-1}{k-i}\binom{n-1}{i}.$$

Proof. We start by writing

$$\operatorname{Tr}^{\partial D_F}\left(S^k_{\partial D_F}\right)(y) = \operatorname{Tr}^{\partial D_{F(y)}}\left(\left(-|\nabla F(y)|^{-1}\nabla^2 F(y)_{|\partial D_{F(y)}}\right)^k\right).$$

Then it is not hard to see that

$$\frac{1}{k!}\operatorname{Tr}^{\partial D_F}\left(S^k_{\partial D_F}\right) = \frac{1}{k!}\operatorname{Tr}^{\partial D_F}\left(\left(-|\nabla F|^{-1}\nabla^2 F_{|\partial D_F}\right)^k\right). \tag{15.10.8}$$

In order to compute this trace, define

$$U(y) = \sum_{j=1}^{m}\left(y^j\right)^2, \qquad V(y) = \sum_{j=1}^{n}\left(y^{j+m}\right)^2, \quad G(y) = \frac{U(y)}{V(y)}.$$

Since $F = nU/mV$, straightforward calculations show that

$$|\nabla F| = \frac{2n}{m}\sqrt{G(1+G)}\cdot\frac{1}{\sqrt{V}},$$

$$|\nabla F|^{-1}\nabla F = \sqrt{\frac{1}{1+G}}\frac{1}{2\sqrt{U}}\nabla U - \sqrt{\frac{G}{1+G}}\frac{1}{2\sqrt{V}}\nabla V,$$

$$|\nabla F|^{-1}\nabla^2 F = \frac{1}{2\sqrt{G(1+G)}}\left(\frac{1}{\sqrt{V}}\nabla^2 U - \frac{1}{V^{3/2}}(dV\otimes dU + dU\otimes dV)\right.$$
$$\left. + \frac{2G}{V^{3/2}}dV\otimes dV - \frac{G}{\sqrt{V}}\nabla^2 V\right).$$

We now evaluate the matrix of $|\nabla F|^{-1}\nabla^2 F$ in a specific set of orthonormal frames of $\nabla F^{\perp}$. Specifically, viewing $\mathbb{R}^{m+n}$ as the product $\mathbb{R}^m \times \mathbb{R}^n$ with the product metric, we fix subspaces L_1, the kernel of ∇U in $\mathbb{R}^m$, and similarly L_2, the kernel of ∇V in $\mathbb{R}^n$, for which we choose orthonormal frames B_1, B_2. The desired set of frames for the kernel of ∇F is then $B = \{B_1, B_2, X\}$, where

$$X = \sqrt{\frac{G}{1+G}}\frac{1}{2\sqrt{U}}\nabla U - \sqrt{\frac{1}{1+G}}\frac{1}{2\sqrt{V}}\nabla V.$$

The matrix of $-|\nabla F|^{-1}\nabla^2 F$ in this set of frames is diagonal, with entries

$$\frac{1}{\sqrt{VG(1+G)}}\left(\overbrace{-1,\ldots,-1}^{m-1 \text{ times}},\overbrace{G,\ldots,G}^{n-1\text{ times}},0\right).$$

By expanding the trace of the kth power of such a diagonal matrix we now obtain, from (15.10.8), that

$$\frac{1}{k!}\operatorname{Tr}^{\partial D_F}\left(S^k_{\partial D_F}\right)=\left(\frac{1}{\sqrt{VG(1+G)}}\right)^k\sum_{i=0}^{k}(-1)^{k-i}G^i\binom{m-1}{k-i}\binom{n-1}{i}.$$

Federer's coarea formula (7.4.13) now implies that

$$\begin{aligned}
&\frac{1}{(2\pi)^{(m+n)/2}}\int_{\partial D_u} e^{-|x|^2/2}\frac{1}{k!}\operatorname{Tr}^{\partial D_u}\left(S^k_{\partial D_u}\right)d\mathcal{H}_{m+n-1}(x)\\
&=\lim_{\varepsilon\to 0}\frac{1}{2\varepsilon}\int_{\mathbb{R}^{m+n}}|\nabla F|\mathbb{1}_{\{|F-u|<\varepsilon\}}\frac{1}{k!}\operatorname{Tr}^{\partial D_F}\left(S^k_{\partial D_F}\right)d\gamma_{\mathbb{R}^{m+n}}(x)\\
&=\mathbb{E}\left\{|\nabla F|\frac{1}{k!}\operatorname{Tr}^{\partial D_F}\left(S^k_{\partial D_F}\right)\Big|F=u\right\}\frac{dF_*(\gamma_{\mathbb{R}^{m+n}})}{d\lambda_{\mathbb{R}}}(u).
\end{aligned}$$

The lemma now follows after substituting for the density $dF_*(\gamma_{\mathbb{R}^{m+n}})/d\lambda_{\mathbb{R}}$ and noting that $V(1+G)=U+V\sim\chi^2_{m+n}$ is independent of G and

$$\mathbb{E}\left\{(U+V)^p\right\}=2^p\frac{\Gamma\left(\frac{m+n}{2}+p\right)}{\Gamma\left(\frac{m+n}{2}\right)}.$$

□

Combining (15.10.7) and Lemma 15.10.2, we have proven the following.

Theorem 15.10.3. *For $m+n\geq N$ and $u>0$ the EC densities for the $F_{m,n}$ random field are given by*

$$\begin{aligned}
&(2\pi)^{j/2}\rho_{j,f}(u)\\
&=\frac{\Gamma\left(\frac{m+n-j}{2}\right)}{2^{(j-2)/2}\Gamma\left(\frac{m}{2}\right)\Gamma\left(\frac{n}{2}\right)}\left(\frac{mu}{n}\right)^{(m-j)/2}\left(1+\frac{mu}{n}\right)^{-(m+n-2)/2}\\
&\quad\times(-1)^{j-1}(j-1)!\sum_{l=0}^{\lfloor\frac{(j-1)}{2}\rfloor}\frac{\Gamma\left(\frac{m+n-j}{2}+l\right)}{\Gamma\left(\frac{m+n-j}{2}\right)l!}\\
&\quad\times\sum_{i=0}^{j-2l-1}(-1)^{i+l}\left(\frac{mu}{n}\right)^{i+l}\binom{m-1}{j-1-2l-i}\binom{n-1}{i},
\end{aligned}$$

for $j\geq 1$, and for $j=0$ by

$$\rho_{j,f}(u)=\mathbb{P}\{F_{m,n}\geq u\}.$$

There are many more formulas like these, for a wide variety of non-Gaussian processes, some simpler, many even more complex.

Complexity, however, is a relative concept. The underlying formula behind all of the special cases of this last section was the simple result of Theorem 15.9.5, namely,

$$\mathbb{E}\{\mathcal{L}_i(A_u(f,M))\} = \sum_{j=0}^{N-i} \begin{bmatrix} i+j \\ j \end{bmatrix} (2\pi)^{-j/2} \mathcal{L}_{i+j}(M) \mathcal{M}_j^{\gamma}\left(F^{-1}[u,+\infty)\right).$$

To a geometer, this is simple. To a statistician or probabilist, this last formula is complex, since it involves unfamilar geometric quantities. Our hope is that after reading this book, the statistician and the probabilist will now feel more comfortable with the geometry underlying applications of random fields. Perhaps we have even motivated the geometers and others among you to take a look at [8], which will show what can be done with these results on a practical level. Now that this book is finally done, together with Keith Worsley we can finally begin work on [8], as we promised back in the preface. Look for it in your favorite bookstore, but be prepared to wait a little. In the meantime, look for a work in progress on our websites, where we shall also keep updates on this book.

References

[1] R. J. Adler, Excursions above a fixed level by n-dimensional random fields, *J. Appl. Probab.*, **13** (1976), 276–289.

[2] R. J. Adler, *The Geometry of Random Fields*, Wiley Series in Probability and Mathematical Statistics, Wiley, Chichester, UK, 1981.

[3] R. J. Adler, *An Introduction to Continuity, Extrema, and Related Topics for General Gaussian Processes*, IMS Lecture Notes–Monograph Series, Vol. 12, Institute of Mathematical Statistics, Hayward, CA, 1990.

[4] R. J. Adler, On excursion sets, tube formulae, and maxima of random fields, *Ann. Appl. Probab.*, **10** (2000), 1–74.

[5] R. J. Adler and L. D. Brown, Tail behaviour for suprema of empirical processes, *Ann. Probab.*, **14**-1 (1986), 1–30.

[6] R. J. Adler and R. Epstein, Some central limit theorems for Markov paths and some properties of Gaussian random fields, *Stochastic Proc. Appl.*, **24** (1987), 157–202.

[7] R. J. Adler and A. M. Hasofer, Level crossings for random fields, *Ann. Probab.*, **4** (1976), 1–12.

[8] R. J. Adler, J. E. Taylor, and K. J. Worsley, *Random Fields and Geometry: Applications*, in preparation.

[9] J. M. P. Albin, On extremal theory for stationary processes, *Ann. Probab.*, **18**-1 (1990), 92–128.

[10] D. Aldous, *Probability Approximations via the Poisson Clumping Heuristic*, Applied Mathematical Sciences, Vol. 77, Springer-Verlag, New York, 1989.

[11] C. B. Allendoerfer and A. Weil, The Gauss–Bonnet theorem for Riemannian polyhedra, *Trans. Amer. Math. Soc.*, **53** (1943), 101–129.

[12] T. W. Anderson, *An Introduction to Multivariate Statistical Analysis*, 3rd ed., Wiley Series in Probability and Statistics, Wiley–Interscience, Hoboken, NJ, 2003.

[13] T. M. Apostol, *Mathematical Analysis*, Addison–Wesley, Reading, MA, 1957.

[14] J.-M. Azaïs, J.-M. Bardet, and M. Wschebor, On the tails of the distribution of the maximum of a smooth stationary Gaussian process, *ESAIM Probab. Statist.*, **6** (2002), 177–184.

[15] J.-M. Azaïs and M. Wschebor, On the distribution of the maximum of a Gaussian field with d parameters, *Ann. Appl. Probab.*, **15**-1A (2005), 254–278.

[16] Yu. K. Belyaev, Point processes and first passage problems, in *Proceedings of the 6th Berkeley Symposium on Mathematics, Statistics, and Probability*, Vol. 2, University of California Press, Berkeley, CA, 1972, 1–17.

[17] S. M. Berman, Asymptotic independence of the numbers of high and low level crossings of stationary Gaussian processes, *Ann. Math. Statist.*, **42** (1971), 927–945.

[18] S. M. Berman, Excursions above high levels for stationary Gaussian processes, *Pacific J. Math.*, **36** (1971), 63–79.

[19] S. M. Berman, Maxima and high level excursions of stationary Gaussian processes, *Trans. Amer. Math. Soc.*, **160** (1971), 65–85.

[20] S. M. Berman, An asymptotic bound for the tail of the distribution of the maximum of a Gaussian process, *Ann. Inst. H. Poincaré Probab. Statist.*, **21**-1 (1985), 47–57.

[21] S. M. Berman, An asymptotic formula for the distribution of the maximum of a Gaussian process with stationary increments, *J. Appl. Probab.*, **22**-2 (1985), 454–460.

[22] S. M. Berman, *Sojourns and Extremes of Stochastic Processes*, Wadsworth–Brooks/Cole Statistics/Probability Series, Wadsworth–Brooks/Cole, Pacific Grove, CA, 1992.

[23] S. M. Berman and N. Kôno, The maximum of a Gaussian process with nonconstant variance: A sharp bound for the distribution tail, *Ann. Probab.*, **17**-2 (1989), 632–650.

[24] A. Bernig and L. Bröcker, Lipschitz–Killing invariants, *Math. Nachr.*, **245** (2002), 5–25.

[25] N. H. Bingham, C. M. Goldie, and J. L. Teugels, *Regular Variation*, Cambridge University Press, Cambridge, UK, 1987.

[26] I. F. Blake and W. C. Lindsey, Level-crossing problems for random processes, *IEEE Trans. Inform. Theory*, **IT-19** (1973), 295–315.

[27] S. Bochner, Monotone Funktionen Stieltjessche Integrale and harmonische Analyse, *Math. Ann.*, **108** (1933), 378.

[28] V. I. Bogachev, *Gaussian Measures*, Mathematical Surveys and Monographs, Vol. 62, American Mathematical Society, Providence, RI, 1998.

[29] W. M. Boothby, *An Introduction to Differentiable Manifolds and Riemannian Geometry*, Academic Press, San Diego, 1986.

[30] C. Borell, The Brunn–Minkowski inequality in Gauss space, *Invent. Math.*, **30** (1975), 205–216.

[31] C. Borell, The Ehrhard inequality, *C. R. Math. Acad. Sci. Paris*, **337**-10 (2003), 663–666.

[32] C. Borell, Minkowski sums and Brownian exit times, *Ann. Fac. Sci. Toulouse*, to appear.

[33] L. Bröcker and M. Kuppe, Integral geometry of tame sets, *Geom. Dedicata*, **82**-1–3 (2000), 285–323.

[34] P. J. Brockwell and R. A. Davis, *Time Series: Theory and Methods*, Springer-Verlag, New York, 1991.

[35] E. V. Bulinskaya, On the mean number of crossings of a level by a stationary Gaussian process, *Theory Probab. Appl.*, **6** (1961), 435–438.

[36] R. Cairoli and J. B. Walsh, Stochastic integrals in the plane, *Acta Math.*, **134** (1975), 111–183.

[37] S. Chatterjee, An error bound in the Sudakov–Fernique inequality, preprint, math.PR/0510424, 2005.

[38] L. Chaumont and M. Yor, *Exercises in Probability: A Guided Tour from Measure Theory to Random Processes, via Conditioning*, Cambridge Series in Statistical and Probabilistic Mathematics, Cambridge University Press, Cambridge, UK, 2003.

[39] J. Cheeger, W. Müller, and R. Schrader, On the curvature of piecewise flat spaces, *Comm. Math. Phys.*, **92** (1984), 405–454.

[40] S. S. Chern, On the kinematic formula in integral geometry, *J. Math. Mech.*, **16** (1966), 101–118.

[41] H. Cramér, A limit theorem for the maximum values of certain stochastic processes, *Teor. Verojatnost. i Primenen.*, **10** (1965), 137–139.

[42] H. Cramér and M. R. Leadbetter, *Stationary and Related Stochastic Processes*, Wiley, New York, 1967.

[43] N. Cressie and R. W. Davis, The supremum distribution of another Gaussian process, *J. Appl. Probab.*, **18**-1 (1981), 131–138.

[44] R. C. Dalang and T. Mountford, Jordan curves in the level sets of additive Brownian motion, *Trans. Amer. Math. Soc.*, **353** (2001), 3531–3545.

[45] R. C. Dalang and J. B. Walsh, The structure of a Brownian bubble. *Probab. Theory Related Fields*, **96** (1993), 475–501.

[46] P. Diaconis and D. Freedman, A dozen de Finetti-style results in search of a theory, *Ann. Inst. H. Poincaré Probab. Statist.*, **23**-2 (suppl.) (1987), 397–423.

[47] P. W. Diaconis, M. L. Eaton, and S. L. Lauritzen, Finite de Finetti theorems in linear models and multivariate analysis, *Scand. J. Statist.*, **19**-4 (1992), 289–315.

[48] V. Dobrić, M. B. Marcus, and M. Weber, The distribution of large values of the supremum of a Gaussian process, in *Colloque Paul Lévy sur les Processus Stochastiques* (*Palaiseau*, 1987), Astérisque, Vols. 157–158, Société Mathématique de France, Paris, 1988, 95–127.

[49] R. L. Dobrushin, Gaussian and their subordinated self-similar random generalized fields, *Ann. Probab.*, **7** (1979), 1–28.

[50] J. L. Doob, *Stochastic Processes*, Wiley, New York, 1953.

[51] R. M. Dudley, Sample functions of the Gaussian process, *Ann. Probab.*, **1** (1973), 66–103.

[52] R. M. Dudley, Metric entropy of some classes of sets with differentiable boundaries, *J. Approx. Theory*, **10** (1974), 227–236.

[53] R. M. Dudley, Central limit theorems for empirical measures, *Ann. Probab.*, **6** (1978), 899–929.

[54] R. M. Dudley, Lower layers in $\Re^2$ and convex sets in $\Re^3$ are not GB classes, in A. Beck, ed., *The Second International Conference on Probability in Banach Spaces*, Lecture Notes in Mathematics, Vol. 709, Springer-Verlag, Berlin, New York, Heidelberg, 1978, 97–102.

[55] R. M. Dudley, *Real Analysis and Probability*, Wadsworth, Belmont, CA, 1989.

[56] R. M. Dudley, *Uniform Central Limit Theorems*, Cambridge University Press, Cambridge, UK, 1999.

[57] E. B. Dynkin, Markov processes and random fields, *Bull. Amer. Math. Soc.*, **3** (1980), 957–1000.

[58] E. B. Dynkin, Markov processes as a tool in field theory, *J. Functional Anal.*, **50** (1983) 167-187.

[59] E. B. Dynkin, Gaussian and non-Gaussian random fields associated with a Markov process, *J. Functional Anal.* **55** (1984) 344-376.

[60] E. B. Dynkin, Polynomials of the occupation field and related random fields, *J. Functional Anal.*, **58** (1984), 20–52.

[61] A. Erdélyi, *Higher Transcendential Functions*, Vol. II, Bateman Manuscript Project, McGraw–Hill, New York, 1953.

[62] V. R. Fatalov, Exact asymptotics of large deviations of Gaussian measures in a Hilbert space, *Izv. Nats. Akad. Nauk Armenii Mat.*, **27**-5 (1992), 43–61 (in Russian).

[63] V. R. Fatalov, Asymptotics of the probabilities of large deviations of Gaussian fields: Applications. *Izv. Nats. Akad. Nauk Armenii Mat.*, **28**-5 (1993), 25–51 (Russian).

[64] H. Federer, Curvature measures, *Trans. Amer. Math. Soc.*, **93** (1959), 418–491.

[65] H. Federer, *Geometric Measure Theory*, Springer-Verlag, New York, 1969.

[66] X. Fernique, Regularité des trajectoires des fonctions aléatoires gaussiennes, in *École d'Été de Probabilités de Saint-Flour* IV: 1974, Lecture Notes in Mathematics, Vol. 480, Springer-Verlag, Berlin, 1975, 1–96.

[67] X. Fernique, *Fonctions Aléatoires Gaussiennes Vecteurs Aléatoires Gaussiens*, Centre de Recherches Mathématiques, Université de Montréal, Montreal, 1997.

[68] A. M. Garsia, E. Rodemich, and H. Rumsey, A real variable lemma and the continuity of paths of some Gaussian processes, *Indiana Univ. Math. J.*, **20** (1970), 565–578.

[69] Y. Gordon, Some inequalities for Gaussian processes and applications, *Israel J. Math.*, **50** (1985), 265–289.

[70] Y. Gordon, Elliptically contoured distributions, *Probab. Theory Related Fields*, **76** (1987), 429–438.

[71] M. Goresky, *Geometric Cohomology and Homology of Stratified Objects*, Ph.D. thesis, Brown University, Providence, RI, 1976.

[72] M. Goresky and R. MacPherson, *Stratified Morse Theory*, Ergebnisse der Mathematik und ihrer Grenzgebiete (3) (Results in Mathematics and Related Areas (3)), Vol. 14, Springer-Verlag, Berlin, 1988.

[73] A. Gray, *Tubes*, Addison–Wesley, Redwood City, CA, 1990.

[74] H. Hadwiger, *Vorlesüngen über Inhalt, Oberfläche und Isoperimetrie*, Springer-Verlag, Berlin, 1957.

[75] H. Hadwiger, Normale Körper im euklidischen Raum und ihre topologischen und metrischen Eigenschaften, *Math. Z.*, **71** (1959), 124–140.

[76] E. J. Hannan, *Multiple Time Series*, Wiley, New York, 1970.

[77] T. Hida and M. Hitsuda, *Gaussian Processes*, Translations of Mathematical Monographs, Vol. 120, Americal Mathematical Society, Providence, RI, 1993.

[78] M. L. Hogan and D. Siegmund, Large deviations for the maxima of some random fields, *Adv. Appl. Math.*, **7**-1 (1986), 2–22.

[79] H. Hotelling, Tubes and spheres in n-spaces and a class of statistical problems, *Amer. J. Math.*, **61** (1939), 440–460.

[80] D. Hug and R. Schneider, Kinematic and Crofton formulae of integral geometry: Recent variants and extensions, in C. Barceló i Vidal, ed., *Homenatge al Professor Luís Santaló i Sors*, Universitat de Girona, Girona, Spain, 2002, 51–80.

[81] K. Itô, The expected number of zeros of continuous stationary Gaussian processes, *J. Math. Kyoto Univ.*, **3** (1964), 206–216.

[82] K. Itô, Distribution valued processes arising from independent Brownian motions, *Math. Z.*, **182** (1983), 17–33.

[83] K. Itô and M. Nisio, On the convergence of sums of independent Banach space valued random variables, *Osaka J. Math.*, **5** (1968), 35–48.

[84] G. Ivanoff and E. Merzbach, *Set-Indexed Martingales*, Monographs on Statistics and Applied Probability, Vol. 85, Chapman and Hall/CRC, Boca Raton, FL, 2000.

[85] N. C. Jain and M. B. Marcus, Continuity of sub-Gaussian processes, in *Probability on Banach Spaces*, Advances in Probability and Related Topics, Vol. 4, Dekker, New York, 1978, 81–196.

[86] S. Janson, *Gaussian Hilbert Spaces*, Cambridge University Press, Cambridge, UK, 1997.

[87] S. Johansen and I. M. Johnstone, Hotelling's theorem on the volume of tubes: Some illustrations in simultaneous inference and data analysis, *Ann. Statist.*, **18**-2 (1990), 652–684.

[88] J. Jost, *Riemannian Geometry and Geomtric Analysis*, 2nd ed., Springer-Verlag, Berlin, 1998.

[89] M. Kac, On the average number of real roots of a random algebraic equation, *Bull. Amer. Math. Soc.*, **43** (1943), 314–320.

[90] O. Kallenberg, *Random Measures*, 4th ed., Akademie-Verlag, Berlin, 1986.

[91] D. Khoshnevisan, *Multi-Parameter Processes: An Introduction to Random Fields*, Springer-Verlag, New York, 2002.

[92] D. A. Klain and G.-C. Rota, *Introduction to Geometric Probability*, Cambridge University Press, Cambridge, UK, 1997.

[93] M. Knowles and D Siegmund, On Hotelling's approach to testing for a nonlinear parameter in a regression, *Internat. Statist. Rev.*, **57** (1989), 205–220.

[94] A. N. Kolmogorov and V. M. Tihomirov, ϵ-entropy and ϵ-capacity of sets in functional spaces, *Uspekhi Mat. Nauk*, **14** (1959), 1–86 (in Russian); *Amer. Math. Soc. Transl.*, **17** (1961), 277–364 (in English).

[95] M. F. Kratz and H. Rootzén, On the rate of convergence for extremes of mean square differentiable stationary normal processes, *J. Appl. Probab.*, **34**-4 (1997), 908–923.

[96] H. Landau and L. A. Shepp, On the supremum of a Gaussian process, *Sankya*, **32** (1970), 369–378.

[97] M. R. Leadbetter, G. Lindgren, and H. Rootzén, *Extremes and Related Properties of Random Sequences and Processes*, Springer-Verlag, New York, 1983.

[98] M. Ledoux, *The Concentration of Measure Phenomenon*, Mathematical Surveys and Monographs, Vol. 89, American Mathematical Society, Providence, RI, 2001.

[99] M. Ledoux and M. Talagrand, *Probability in Banach Spaces: Isoperimetry and Processes*, Springer-Verlag, Berlin, 1991.

[100] J. M. Lee, *Introduction to Topological Manifolds*, Graduate Texts in Mathematics, Vol. 202, Springer-Verlag, New York, 2000.

[101] J. M. Lee, *Introduction to Smooth Manifolds*, Graduate Texts in Mathematics, Vol. 218, Springer-Verlag, New York, 2003.

[102] N. Leonenko, *Limit Theorems for Random Fields with Singular Spectrum*, Mathematics and Its Applications, Vol. 465, Kluwer Academic Publishers, Dordrecht, The Netherlands, 1999.

[103] G. Letac, Problèmes classiques de probabilité sur un couple de Gelfand, in E. Dugue, E. Lukacs, and V. K. Rohatgi, eds., *Analytical Methods in Probability Theory: Proceedings, Oberwolfach*, 1980, Lecture Notes in Mathematics, Vol. 861, Springer-Verlag, Berlin, 1981, 93–120.

[104] M. A. Lifshits, Tail probabilities of Gaussian suprema and Laplace transform, *Ann. Inst. H. Poincaré Probab. Statist.*, **30**-2 (1994), 163–179.

[105] M. A. Lifshits, *Gaussian Random Functions*, Kluwer, Dordrecht, The Netherlands, 1995.

[106] M. S. Longuet-Higgins, On the statistical distribution of the heights of sea waves, *J. Marine Res.*, **11** (1952), 245–266.

[107] M. S. Longuet-Higgins, The statistical analysis of a random moving surface, *Philos. Trans. Roy. Soc.*, **A249** (1957), 321–387.

[108] T. L. Malevich, A formula for the mean number of intersections of a surface by a random field, *Izv. Akad. Nauk UzSSR*, **6** (1973), 15–17 (in Russian).

[109] M. B. Marcus, Level crossings of a stochastic process with absolutely continuious sample paths, *Ann. Probab.*, **5** (1977), 52–71.

[110] M. B. Marcus and L. A. Shepp, Sample behaviour of Gaussian processes, in *Proceedings of the 6th Berkeley Symposium on Mathematics, Statistics, and Probability*, Vol. 2, University of California Press, Berkeley, CA, 1972, 423–442.

[111] B. Matérn, *Spatial Variation: Stochastic Models and Their Application to Some Problems in Forest Surveys and Other Sampling Investigations*, Meddelanden Fran Statens Skogsforskningsinstitut, Band 49, Number 5, Swedish Forestry Research Institute, Stockholm, 1960.

[112] Y. Matsumoto, *An Introduction to Morse Theory*, Translations of Mathematical Monographs, Vol. 208, American Mathematical Society, Providence, RI, 2002.

[113] K. S. Miller, Complex random fields, *Inform. Sci.*, **9** (1975), 185–225.

[114] R. S. Millman and G. D. Parker, *Elements of Differential Geometry*, Prentice–Hall, Englewood Cliffs, NJ, 1977.

[115] F. Morgan, *Geometric Measure Theory: A Beginner's Guide*, 3rd ed., Academic Press, San Diego, 2000.

[116] S. Morita, *Geometry of Differential Forms*, Translations of Mathematical Monographs, Vol. 201, American Mathematical Society, Providence, RI, 2001.

[117] M. Morse and S. Cairns, *Critical Point Theory in Global Analysis and Differential Topology: An Introduction*, Academic Press, New York, 1969.

[118] J. Nash, The imbedding problem for Riemannian manifolds, *Ann. Math.* (2), **63** (1956), 20–63.

[119] H. Natsume, The realization of abstract stratified sets, *Kodai Math. J.*, **3**-1 (1980), 1–7.

[120] L. Noirel and D. Trotman, Subanalytic and semialgebraic realisations of abstract stratified sets, in *Real Algebraic Geometry and Ordered Structures* (*Baton Rouge, LA*, 1996), Contemporary Mathematics, Vol. 253, American Mathematical Society, Providence, RI, 2000, 203–207.

[121] B. O'Neill, *Elementary Differential Geometry*, 2nd ed., Academic Press, San Diego, 1997.

[122] M. J. Pflaum, *Analytic and Geometric Study of Stratified Spaces*, Lecture Notes in Mathematics, Vol. 1768, Springer-Verlag, Berlin, 2001.

[123] J. Pickands III, Maxima of stationary Gaussian processes, *Z. Wahrscheinlichkeitstheorie Verw. Gebiete*, **7** (1967), 190–223.

[124] J. Pickands III, Asymptotic properties of the maximum in a stationary Gaussian process, *Trans. Amer. Math. Soc.*, **145** (1969), 75–86.

[125] J. Pickands III, Upcrossing probabilities for stationary Gaussian processes, *Trans. Amer. Math. Soc.*, **145** (1969), 51–73.

[126] V. I. Piterbarg, *Asymptotic Methods in the Theory of Gaussian Processes and Fields*, Translations of Mathematical Monographs, Vol. 148, American Mathematical Society, Providence, RI, 1996.

[127] V. I. Piterbarg, On the distribution of the maximum of a Gaussian field with constant variance on a smooth manifold, *Theory Probab. Appl.*, **41** (1996), 367–379.

[128] R. Pyke, Partial sums of matrix arrays and Brownian sheets, in E. F. Harding and D. G. Kendall, eds., *Stochastic Analysis: A Tribute to the Memory of Rollo Davidson*, Wiley, London, 1973, 331–348.

[129] R. Pyke, The Haar-function construction of Brownian motion indexed by sets, *Z. Wahrscheinlichkeitstheorie Verw. Gebiete*, **64**-4 (1983), 523–539.

[130] D. Rabinowitz and D. Siegmund, The approximate distribution of the maximum of a smoothed Poisson random field, *Statist. Sinica*, **7** (1997), 167–180.

[131] S. O. Rice, Mathematical analysis of random noise, *Bell System Tech. J.*, **24** (1945), 46–156; also in N. Wax, ed., *Selected Papers on Noise and Stochastic Processes*, Dover, New York, 1954.

[132] R. Riesz and B. Sz-Nagy, *Functional Analysis*, Ungar, New York, 1955.

[133] W. Rudin, *Fourier Analysis on Groups*, Wiley, New York, 1962; reprint, 1990.

[134] W. Rudin, *Real and Complex Analysis*, 2nd ed., McGraw–Hill, New York, 1974.

[135] G. Samorodnitsky, *Bounds on the Supremum Distribution of Gaussian Processes: Exponential Entropy Case*, preprint, 1987.

[136] G. Samorodnitsky, *Bounds on the Supremum Distribution of Gaussian Processes: Polynomial Entropy Case*, preprint, 1987.

[137] G. Samorodnitsky, Probability tails of Gaussian extrema, *Stochastic Process. Appl.*, **38**-1 (1991), 55–84.
[138] G. Samorodnitsky and M. S. Taqqu, *Stable Non-Gaussian Random Processes*, Chapman and Hall, New York, 1994.
[139] L. A. Santaló, *Integral Geometry and Geometric Probability*, Encyclopedia of Mathematics and Its Applications, Vol. 1, Addison–Wesley, Reading, MA, 1976.
[140] I. J. Schoenberg, Metric spaces and completely monotone functions, *Ann. Math.*, **39** (1938), 811–841.
[141] R. Schneider, *Convex Bodies: The Brunn–Minkowski Theory*, Encyclopedia of Mathematics and Its Applications, Vol. 44, Cambridge University Press, Cambridge, UK, 1993.
[142] K. Shafie, D. Siegmund, B. Sigal, and K. J. Worsley, Rotation space random fields with an application to FMRI data, *Ann. Statist.*, **31**-6 (2003), 1732–1771.
[143] D. O. Siegmund and K. J. Worsley, Testing for a signal with unknown location and scale in a stationary Gaussian random field, *Ann. Statist.*, **23** (1995), 608–639.
[144] D. Slepian, On the zeroes of Gaussian noise, in M. Rosenblatt, ed., *Time Series Analysis*, Wiley, New York, 1963, 104–115.
[145] D. Slepian and L. Shepp, First passage time for a particular stationary periodic Gaussian process, *J. Appl. Probab.*, **13** (1976), 27–38.
[146] D. R. Smart, *Fixed Point Theorems*, Cambridge University Press, Cambridge, UK, 1974.
[147] J. Sun, Tail probabilities of the maxima of Gaussian random fields, *Ann. Probab.*, **21**-1 (1993), 34–71.
[148] A. Takemura and S. Kuriki, On the equivalence of the tube and Euler characteristic methods for the distribution of the maximum of Gaussian fields over piecewise smooth domains, *Ann. Appl. Probab.*, **12**-2 (2002), 768–796.
[149] A. Takemura and S. Kuriki, Tail probability via the tube formula when the critical radius is zero, *Bernoulli*, **9**-3 (2003), 535–558.
[150] M. Talagrand, Regularity of Gaussian processes, *Acta Math.*, **159** (1987), 99–149.
[151] M. Talagrand, Small tails for the supremum of a Gaussian process, *Ann. Inst. H. Poincaré Probab. Statist.*, **24**-2 (1988), 307–315.
[152] M. Talagrand, A simple proof of the majorizing measure theorem, *Geom. Functional Anal.*, **2** (1992), 118–125.
[153] M. Talagrand, Sharper bounds for Gaussian and empirical processes, *Ann. Probab.*, **22** (1994), 28–76.
[154] M. Talagrand, Majorizing measures: The generic chaining, *Ann. Probab.*, **24** (1996), 1049–1103.
[155] M. Talagrand, Majorizing measures without measures, *Ann. Probab.*, **29** (2001), 411–417.
[156] M. Talagrand, *The Generic Chaining: Upper and Lower Bounds of Stochastic Processes*, Springer Monographs in Mathematics, Springer-Verlag, Berlin, 2005.
[157] J. E. Taylor, *Euler Characteristics for Gaussian Fields on Manifolds*, Ph.D. thesis, Department of Mathematics and Statistics, McGill University, Montreal, 2001.
[158] J. E. Taylor, A Gaussian kinematic formula, *Ann. Probab.*, **34**-1 (2006), 122–158.
[159] J. E. Taylor, A. Takemura, and R. J. Adler, Validity of the expected Euler characteristic heuristic, *Ann. Probab.*, **33**-4 (2005), 1362–1396.
[160] B. S. Tsirelson, I. A. Ibragimov, and V. N. Sudakov, Norms of Gaussian sample functions, in *Proceedings of the 3rd Japan–USSR Symposium on Probability Theory* (*Tashkent*, 1975), Lecture Notes in Mathematics, Vol. 550, Springer-Verlag, Berlin, 1976, 20–41.
[161] M. Turmon, *Assessing Generalization of Feedforward Neural Networks*, Ph.D. thesis, Cornell University, Ithaca, NY, 1995.

[162] V. N. Vapnik, *The Nature of Statistical Learning Theory*, 2nd ed., Springer-Verlag, New York, 2000.

[163] V. N. Vapnik, and A. Ya. Červonenkis, On the uniform convergence of relative frequencies of events to their probabilities, *Theory Probab. Appl.*, **16** (1971), 264–280.

[164] V. N. Vapnik and A. Ya. Červonenkis, *Theory of Pattern Recognition: Statistical Problems in Learning*, 1974, Nauka, Moscow (in Russian).

[165] J. B. Walsh, An Introduction to Stochastic Partial Differential Equations, *École d'été de probabilités de Saint-Flour* XIV—1984, Lecture Notes in Mathematics 1180, Springer-Verlag, New York, 1986, 265–439.

[166] S. Watanabe, *Stochastic Differential Equations and Malliavan Calculus*, Springer-Verlag, Berlin, 1984.

[167] M. Weber, The supremum of Gaussian processes with a constant variance, *Probab. Theory Related Fields*, **81**-4 (1989), 585–591.

[168] H. Weyl, On the volume of tubes, *Amer. J. Math.*, **61** (1939), 461–472.

[169] E. Wong, *Stochastic Processes in Information and Dynamical Systems*, McGraw–Hill, New York, 1971.

[170] E. Wong and M. Zakai, Martingales and stochastic integrals for processes with a multi-dimensional parameter, *Z. Wahrscheinlichkeitstheorie Verw. Gebiete*, **29** (1974), 109–122.

[171] E. Wong and M. Zakai, Weak martingales and stochastic integrals in the plane, *Ann. Probab.*, **4** (1976), 570–587.

[172] K. J. Worsley, Local maxima and the expected Euler characteristic of excursion sets of χ^2, F, and t fields. *Adv. Appl. Probab.*, **26** (1994), 13–42.

[173] K. J. Worsley, Boundary corrections for the expected Euler characteristic of excursion sets of random fields, with an application to astrophysics, *Adv. Appl. Probab.*, **27** (1995), 943–959.

[174] K. J. Worsley, Estimating the number of peaks in a random field using the Hadwiger characteristic of excursion sets, with applications to medical images, *Ann. Statist.*, **23** (1995), 640–669.

[175] K. J. Worsley, The geometry of random images, *Chance*, **9**-1 (1997), 27–40.

[176] K. J. Worsley and K. J. Friston, A test for a conjunction, *Statist. Probab. Lett.*, **47**-2 (2000), 135–140.

[177] A. M. Yaglom, Some classes of random fields in n-dimensional space, related to stationary random processes, *Theory Probab. Appl.*, **2** (1957), 273–320.

[178] A. M. Yaglom, Second-order homogeneous random fields, in *Proceedings of the 4th Berkeley Symposium on Mathematics, Statistics, and Probability*, Vol. 2, University of California Press, Berkeley, CA, 1961, 593–622.

[179] K. Ylinen, Random fields on noncommutative locally compact groups, in H. Heyer, ed., *Probability Measures on Groups* VIII (*Oberwolfach*), Lecture Notes in Mathematics, Vol. 1210, Springer-Verlag, Berlin, New York, Heidelberg, 1986, 365–386.

[180] N. D. Ylvisaker, The expected number of zeros of a stationary Gaussian process, *Ann. Math. Statist.*, **36** (1965), 1043–1046.

[181] A. C. Zaanen, *Linear Analysis*, North-Holland, Amsterdam, 1956.

[182] M. Zahle, Approximation and characterization of generalized Lipshitz–Killing curvatures, *Ann. Global Anal. Geom.*, **8** (1990), 249–260.

Notation Index

Subject Index

Springer Monographs in Mathematics

This series publishes advanced monographs giving well-written presentations of the "state-of-the-art" in fields of mathematical research that have acquired the maturity needed for such a treatment. They are sufficiently self-contained to be accessible to more than just the intimate specialists of the subject, and sufficiently comprehensive to remain valuable references for many years. Besides the current state of knowledge in its field, an SMM volume should also describe its relevance to and interaction with neighbouring fields of mathematics, and give pointers to future directions of research.

Abhyankar, S.S. **Resolution of Singularities of Embedded Algebraic Surfaces** 2nd enlarged ed. 1998
Adler, R.J.; Taylor, J. **Random Fields and Geometry** 2007
Alexandrov, A.D. **Convex Polyhedra** 2005
Andrievskii, V.V.; Blatt, H.-P. **Discrepancy of Signed Measures and Polynomial Approximation** 2002
Angell, T.S.; Kirsch, A. **Optimization Methods in Electromagnetic Radiation** 2004
Ara, P.; Mathieu, M. **Local Multipliers of C^*-Algebras** 2003
Armitage, D.H.; Gardiner, S.J. **Classical Potential Theory** 2001
Arnold, L. **Random Dynamical Systems** corr. 2nd printing 2003 (1st ed. 1998)
Arveson, W. **Noncommutative Dynamics and E-Semigroups** 2003
Aubin, T. **Some Nonlinear Problems in Riemannian Geometry** 1998
Auslender, A.; Teboulle, M. **Asymptotic Cones and Functions in Optimization and Variational Inequalities** 2003
Banagl, M. **Topological Invariants of Stratified Spaces** 2006
Banasiak, J.; Arlotti, L. **Perturbations of Positive Semigroups with Applications** 2006
Bang-Jensen, J.; Gutin, G. **Digraphs** 2001
Baues, H.-J. **Combinatorial Foundation of Homology and Homotopy** 1999
Böttcher, A.; Silbermann, B. **Analysis of Toeplitz Operators** 2006
Brown, K.S. **Buildings** 3rd printing 2000 (1st ed. 1998)
Carl, S.; Le, V.K.; Motreanu, D., **Nonsmooth Variational Problems and Their Inequalities**
Chang, K.-C. **Methods in Nonlinear Analysis** 2005
Cherry, W.; Ye, Z. **Nevanlinna's Theory of Value Distribution** 2001
Ching, W.K. **Iterative Methods for Queuing and Manufacturing Systems** 2001
Chudinovich, I. **Variational and Potential Methods for a Class of Linear Hyperbolic Evolutionary Processes** 2005
Coates, J.; Sujatha, R. **Cyclotomic Fields and Zeta Values** 2006
Crabb, M.C.; James, I.M. **Fibrewise Homotopy Theory** 1998
Dineen, S. **Complex Analysis on Infinite Dimensional Spaces** 1999
Dugundji, J.; Granas, A. **Fixed Point Theory** 2003
Ebbinghaus, H.-D.; Flum, J. **Finite Model Theory** 2006
Edmunds, D.E.; Evans, W.D. **Hardy Operators, Function Spaces and Embeddings** 2004
Elstrodt, J.; Grunewald, F.; Mennicke, J. **Groups Acting on Hyperbolic Space** 1998
Engler, A.J.; Prestel, A. **Valued Fields** 2005
Fadell, E.R.; Husseini, S.Y. **Geometry and Topology of Configuration Spaces** 2001
Fedorov, Yu. N.; Kozlov, V.V. **A Memoir on Integrable Systems** 2001
Flenner, H.; O'Carroll, L.; Vogel, W. **Joins and Intersections** 1999
Gelfand, S.I.; Manin, Y.I. **Methods of Homological Algebra** 2nd ed. 2003 (1st ed. 1996)
Griess, R.L. Jr. **Twelve Sporadic Groups** 1998
Gras, G. **Class Field Theory** corr. 2nd printing 2005
Greuel, G.-M.; Lossen, C.; Shustin, E. **Introduction to Singularities and Deformations** 2007
Hida, H. ***p*-Adic Automorphic Forms on Shimura Varieties** 2004
Ischebeck, E; Rao, R.A. **Ideals and Reality** 2005

Ivrii, V. **Microlocal Analysis and Precise Spectral Asymptotics** 1998
Jakimovski, A.; Sharma, A.; Szabados, J. **Walsh Equiconvergence of Complex Interpolating Polynomials** 2006
Jech, T. **Set Theory** (3rd revised edition 2002)
Jorgenson, J.; Lang, S. **Spherical Inversion on SLn (R)** 2001
Kanamori, A. **The Higher Infinite** corr. 2nd printing 2005 (2nd ed. 2003)
Kanovei, V. **Nonstandard Analysis, Axiomatically** 2005
Khoshnevisan, D. **Multiparameter Processes** 2002
Koch, H. **Galois Theory of *p*-Extensions** 2002
Komornik, V. **Fourier Series in Control Theory** 2005
Kozlov, V.; Maz'ya, V. **Differential Equations with Operator Coefficients** 1999
Lam, T.Y. **Serre's Problem on Projective Modules** 2006
Landsman, N.P. **Mathematical Topics between Classical & Quantum Mechanics** 1998
Lapidus, M.L.; van Frankenhuijsen, M., **Fractal Geometry, Complex Dimensions and Zeta Functions**
Leach, J.A.; Needham, D.J. **Matched Asymptotic Expansions in Reaction-Diffusion Theory** 2004
Lebedev, L.P.; Vorovich, I.I. **Functional Analysis in Mechanics** 2002
Lemmermeyer, F. **Reciprocity Laws: From Euler to Eisenstein** 2000
Malle, G.; Matzat, B.H. **Inverse Galois Theory** 1999
Mardesic, S. **Strong Shape and Homology** 2000
Margulis, G.A. **On Some Aspects of the Theory of Anosov Systems** 2004
Miyake, T. **Modular Forms** 2006
Murdock, J. **Normal Forms and Unfoldings for Local Dynamical Systems** 2002
Narkiewicz, W. **Elementary and Analytic Theory of Algebraic Numbers** 3rd ed. 2004
Narkiewicz, W. **The Development of Prime Number Theory** 2000
Neittaanmaki, P.; Sprekels, J.; Tiba, D. **Optimization of Elliptic Systems. Theory and Applications** 2006
Onishchik, A.L. **Projective and Cayley-Klein Geometries** 2006
Parker, C.; Rowley, P. **Symplectic Amalgams** 2002
Peller, V. (Ed.) **Hankel Operators and Their Applications** 2003
Prestel, A.; Delzell, C.N. **Positive Polynomials** 2001
Puig, L. **Blocks of Finite Groups** 2002
Ranicki, A. **High-dimensional Knot Theory** 1998
Ribenboim, P. **The Theory of Classical Valuations** 1999
Rowe, E.G.P. **Geometrical Physics in Minkowski Spacetime** 2001
Rudyak, Y.B. **On Thorn Spectra, Orientability and Cobordism** 1998
Ryan, R.A. **Introduction to Tensor Products of Banach Spaces** 2002
Saranen, J.; Vainikko, G. **Periodic Integral and Pseudodifferential Equations with Numerical Approximation** 2002
Schneider, P. **Nonarchimedean Functional Analysis** 2002
Serre, J-P. **Complex Semisimple Lie Algebras** 2001 (reprint of first ed. 1987)
Serre, J-P. **Galois Cohomology** corr. 2nd printing 2002 (1st ed. 1997)
Serre, J-P. **Local Algebra** 2000
Serre, J-P. **Trees** corr. 2nd printing 2003 (1st ed. 1980)
Smirnov, E. **Hausdorff Spectra in Functional Analysis** 2002
Springer, T.A.; Veldkamp, F.D. **Octonions, Jordan Algebras, and Exceptional Groups** 2000
Székelyhidi, L. **Discrete Spectral Synthesis and Its Applications** 2006
Sznitman, A.-S. **Brownian Motion, Obstacles and Random Media** 1998
Taira, K. **Semigroups, Boundary Value Problems and Markov Processes** 2003
Talagrand, M. **The Generic Chaining** 2005
Tauvel, P.; Yu, R.W.T. **Lie Algebras and Algebraic Groups** 2005
Tits, J.; Weiss, R.M. **Moufang Polygons** 2002
Uchiyama, A. **Hardy Spaces on the Euclidean Space** 2001
Üstünel, A.-S.; Zakai, M. **Transformation of Measure on Wiener Space** 2000
Vasconcelos, W. **Integral Closure. Rees Algebras, Multiplicities, Algorithms** 2005
Yang, Y. **Solitons in Field Theory and Nonlinear Analysis** 2001
Zieschang, P.-H. **Theory of Association Schemes** 2005

Printed in the United States of America